Rice: Cultivation and Processing

Rice: Cultivation and Processing

Editor: Dwayne Wilson

MURPHY & MOORE
www.murphy-moorepublishing.com

Murphy & Moore Publishing,
1 Rockefeller Plaza,
New York City, NY 10020, USA

Visit us on the World Wide Web at:
www.murphy-moorepublishing.com

ISBN: 978-1-63987-492-7 (Hardback)

Cataloging-in-Publication Data

Rice : cultivation and processing / edited by Dwayne Wilson.
p. cm.
Includes bibliographical references and index.
ISBN 978-1-63987-492-7
1. Rice. 2. Rice--Processing. I. Wilson, Dwayne.
SB191.R5 R53 2022
641.331 8--dc23

Table of Contents

Preface

Rice is a seed of Oryza sativa, which is a grass species. Domesticated rice is a main staple food in Asia and Africa. Traditionally, rice is cultivated by flooding the fields during and after setting the seedlings. This method reduces the growth of those weed and pest plants which have no submerged growth state and also controls vermin. Rice cultivation is favorable in those regions where rainfall is high and labor costs are low. There are some diseases that affect rice cultivation such as rice blast, sheath blight, rice ragged stunt and tungro. Rice blast is caused by the fungus Magnaporthe grisea. This book traces the progress of rice cultivation and processing, and highlights some of its key concepts. It will also provide interesting topics for research which interested readers can take up. The extensive content of this book provides the readers with a thorough understanding of the subject.

Significant researches are present in this book. Intensive efforts have been employed by authors to make this book an outstanding discourse. This book contains the enlightening chapters which have been written on the basis of significant researches done by the experts.

Finally, I would also like to thank all the members involved in this book for being a team and meeting all the deadlines for the submission of their respective works. I would also like to thank my friends and family for being supportive in my efforts.

Editor

1

Physiological Responses and Expression Profile of *NADPH Oxidase* in Rice (*Oryza Sativa*) Seedlings under Different Levels of Submergence

Yu-Sian Wu and Chin-Ying Yang*

Abstract

Background: Flooding due to global climate change is a serious problem that frequently decreases crop yields. Rice fields in flood-prone areas often experience full or partial submergence. Submergence has an adverse effect on internal oxygen availability, sugar status and survival. Complete submergence imposes severe pressure on plants, principally because the excess water in their surroundings deprives them of certain basic resources such as oxygen, carbon dioxide and light for photosynthesis. To better understand the mechanisms involved under different levels of flooding, it is necessary to further observe physiological responses and to identify the Rboh genes involved and determine how they are regulated during submergence.

Results: In this study, significant physiological changes were observed in plant height, leaf sheath elongation and chlorophyll a, b and total content under partial and full submergence treatments. Senescence-regulating genes were severely affected under full submergence. Additionally, intracellular oxidative homeostasis was disrupted by overproduction of H_2O_2 and O_2^-, which affected cell viability and antioxidant enzyme activity, under different levels of submergence. Quantitative RT-PCR analyses revealed that complex regulation of Rboh genes is involved under different levels of submergence.

Conclusion: Our results demonstrated that the effect of physiological and the transcript levels of OsRboh genes were presented different responses to different levels of submergence in rice seedlings. There have different mechanism in intracellular to response different levels of submergence. Finally we discuss effects of the regulation of OsRboh expression and ROS production which was important to maintain homeostasis to help rice seedlings face different levels of submergence.

Keywords: Rice, Partial submergence, Full submergence, ROS, Antioxidant enzyme activity, Rboh, Hypoxia

Background

The effect of global climate change on crop yields is an urgent issue. Waterlogging is a serious problem that affects crop gas exchange in low-lying rain-fed areas. Approximately 30 % of the world's rice (*Oryza sativa*) farmlands are at a low elevation and irrigated by rain (Bailey-Serres et al. 2010). Under limited oxygen availability, photosynthesis and respiration are restricted, leading to an energy crisis, toxic products from anaerobic respiration and the accumulation of reactive oxygen species (ROS) in plant cells (Licausi and Perata 2009; Pucciariello et al. 2012a; Yang and Hong 2015).

Several studies have focused on the important role of ROS during hypoxia signaling under full submergence conditions (Baxter et al. 2014; Fukao et al. 2011; Liu et al. 2015; Yang and Hong 2015). Plasma membrane NADPH oxidases in plants were discovered through their sequence similarity to mammalian respiratory burst NADPH oxidase subunit $gp91^{phox}$ and have been named respiratory burst oxidase

* Correspondence: emiyang@dragon.nchu.edu.tw
Department of Agronomy, National Chung Hsing University, Taichung 40227, Taiwan

homologs (Rbohs) (Sagi and Fluhr 2006). Rbohs participate in ROS production and are involved in the oxidative burst in response to pathogens (Lherminier et al. 2009; Sagi and Fluhr 2006). Some reports have indicated that Rboh proteins possess a resistance function, such as AtRboh D and AtRboh F, which may be involved in triggering cell death in inoculated plants (Pogany et al. 2009; Vellosillo et al. 2010). AtRboh D and AtRboh F are involved not only in the hypersensitive response but also the ABA- and ethylene-induced stomatal response and are associated with a very complicated interaction between the ABA-, ethylene-, JA- and SA-signaling pathways (Desikan et al. 2006; Maruta et al. 2011). Both ethylene and H_2O_2 were demonstrated promote aerenchyma formation in rice stems in a dose-dependent manner (Steffens et al. 2011). In a recent study, the NADPH oxidase Rboh D was found to be involved in primary hypoxia signaling and modulate down-stream gene expression (Yang and Hong 2015).

Floods are major constraints to crop production worldwide that affect oxygen availability in the environment and lead to serious physiological damage. Transient flash floods leading to partial or complete submergence can affect crop growth and yield (Kato et al. 2014). In recent years, many reports have indicated that complete submergence has an adverse effect on internal oxygen availability, sugar status and survival (Bailey-Serres and Voesenek 2010; Winkel et al. 2013). Complete submergence imposes severe pressure on plants, principally because excess water in their surroundings deprives them of certain basic resources such as oxygen, carbon dioxide and light for photosynthesis (Sakagami et al. 2013). Paddy field rice is generally tolerant to waterlogging and partial submergence compared with other economic crops such as maize and soybean. Lowland rice (Flood Resistant 13A; FR13A) can restrict its elongation growth, economizing its carbohydrate reserves to enable the development of new leaves upon desubmergence (Fukao et al. 2011).

So far, these results have mainly been obtained from complete submergence studies. To better understand the mechanisms involved under different levels of flooding, it is necessary to further identify the Rboh genes involved and to determine how they are regulated during submergence. In this study, we characterized the physiological responses and the transcript profiles of nine *Rbohs* under different levels of flooding treatments in rice (*Oryza sativa* L. *japonica*, Tai keng 9). Moreover, we examined the accumulation of H_2O_2 and O_2^- using DAB and NBT staining and the activities of antioxidant enzymes. This study provides evidence of differential regulation of OsRboh expression

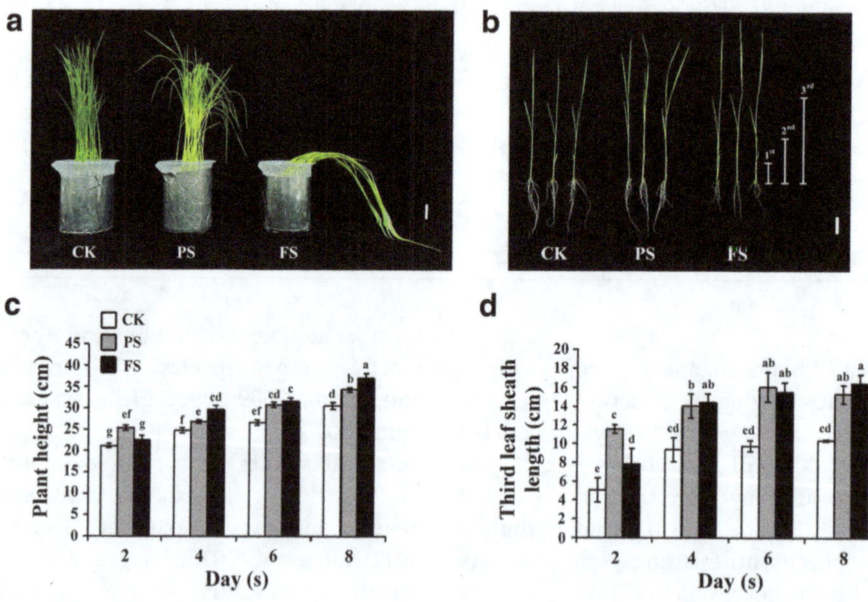

Fig. 1 Characterization of rice (*Oryza sativa*) seedlings growth under different levels of submergence. **a** Photographs of 14 days-old rice seedlings after exposed different levels of submergence for 8 days. **b** The phenotypes of plant height after exposed different levels of submergence treatment for 8 days. (1st, 2nd and 3rd; length of the first, second and third leaf sheath). **c** Plant height of 14 days-old rice seedlings after treated submergence for 2, 4, 6 and 8 days. **d** The third leaves sheath length of 14 days-old rice seedlings after treated submergence for 2, 4, 6 and 8 days. Control check (CK), partial submergence (PS) and full submergence (FS). Bar = 3 cm. The data represent average values ± SD from 30 seedlings of each treatment obtained from six biologically independent experiments. Values with the different letters are significantly different at $P < 0.05$, according to post-hoc LSD test

and ROS accumulation under partial or full submergence in rice.

Results

Partial Submergence Leads to Faster Plant Height Increases and Leaf Sheath Elongation at a Relatively Early Stage

To assess several physiological mechanisms associated with partial and full submergence in rice, plant height and leaf sheath length were determined in 14 days-old rice seedlings under partial submergence (PS) or full submergence (FS) for 2, 4, 6 and 8 days. The leaves of the rice seedlings were yellowish, thin and long, and could not remain upright under FS for 8 days compared with regular growth (control check; CK) and PS conditions (Fig. 1a and b). After PS treatment, the yellowish color was only present in tissue that had been submerged (Fig. 1a and b). During the partial submergence treatment period, plant height and the length of the 3rd sheath were significantly increased after 2 days treatment. Subsequently, plant height and the length of the 3rd leaf sheath increased in both PS and FS conditions compared with the control (Fig. 1c and d). There were no significant differences in the lengths of the 1st and 2nd leaf sheaths after PS or FS treatment compared with the control (Additional file 1: Figure S1). These results indicated that the plant height increase of rice seedlings during submergence was mainly affected by elongation of the 3rd leaf sheath. Notably, partial submergence led to faster elongation at a relatively early stage compared with full submergence.

The Chlorophyll Content and Senescence-regulating Genes are Severely Affected under Full Submergence

The photosynthetic apparatus and particularly PS II in rice seedlings is affected by submergence, which causes ROS damage to the photosystem and a decreased photosynthetic rate and chlorophyll content (Panda et al. 2008). The submergence tolerance regulator, SUBMERGENCE1A (SUB1A), is involved in acclimation responses during leaf senescence caused by prolonged darkness in rice (Fukao et al. 2012).

To dissect the influence of chlorophyll content and the expression of senescence-regulating genes under different levels of submergence, we determined the chlorophyll a, b and total content after PS or FS treatment for 2, 4, 6 and 8 d in 14 days-old rice seedlings. The results showed that the chlorophyll a, b and total contents were significantly decreased with submergence treatment, particularly in the FS treatment (Fig. 2). To evaluate whether senescence-associated gene (SAG) expression was affected by different levels of submergence, we used quantitative RT-PCR to determine the mRNA expression of *Stay-Green* (*SGR*), which is necessary for chlorophyll degradation in light-harvesting complex II (Hortensteiner 2009), *Red Chlorophyll Catabolite Reductase* (*RCCR*), which functions in chlorophyll degradation in chloroplasts (Pruzinska et al. 2007), and *Osl85*, which encodes

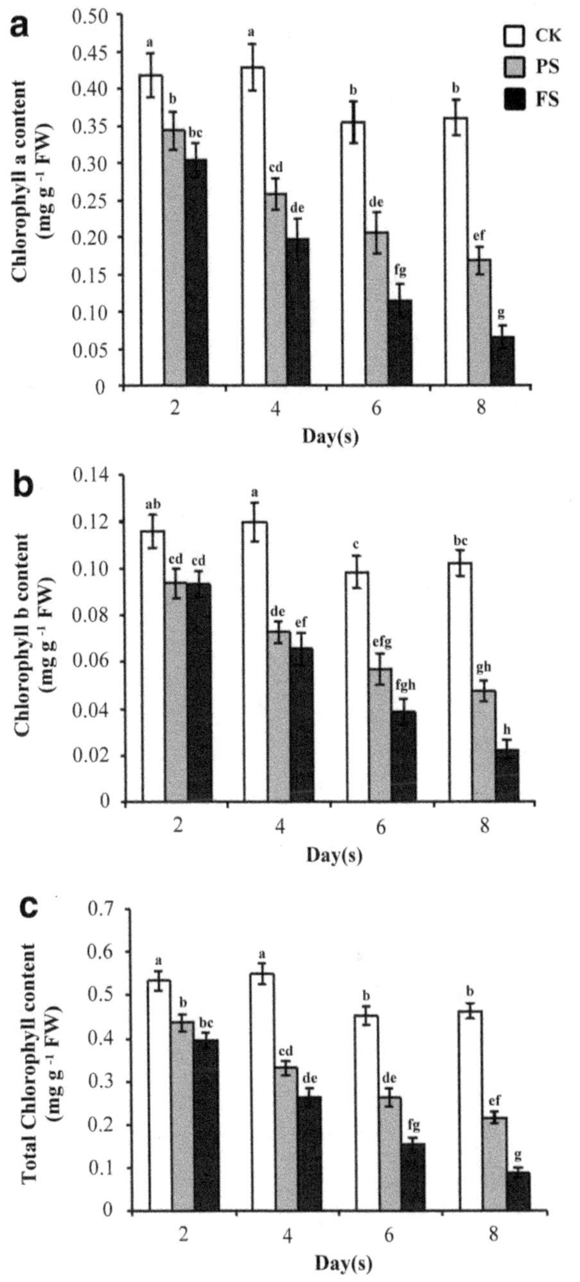

Fig. 2 Different levels of submergence influence the content of chlorophyll in rice seedlings. The chlorophyll content of 14-days old seedlings after treated control (CK), partial submergence (PS) and full submergence (FS) conditions for 2, 4, 6 and 8 days. The content of chlorophyll a (**a**), chlorophyll b (**b**) and total chlorophyll (**c**) were determined. Values represent means ± standard deviation from six biologically independent experiments ($n = 3$). Values with the different letters are significantly different at $P < 0.05$, according to post-hoc LSD test

isocitrate lyase, a marker gene that is upregulated during dark-induced and natural leaf senescence (Yamada et al. 2014). The quantitative RT-PCR results revealed that the transcripts of these senescence-regulating genes were highly induced by FS treatment compared with PS treatment (Fig. 3). These results demonstrated that FS conditions severely affect the chlorophyll content, which is accompanied by senescence induction, in rice seedlings more than PS conditions.

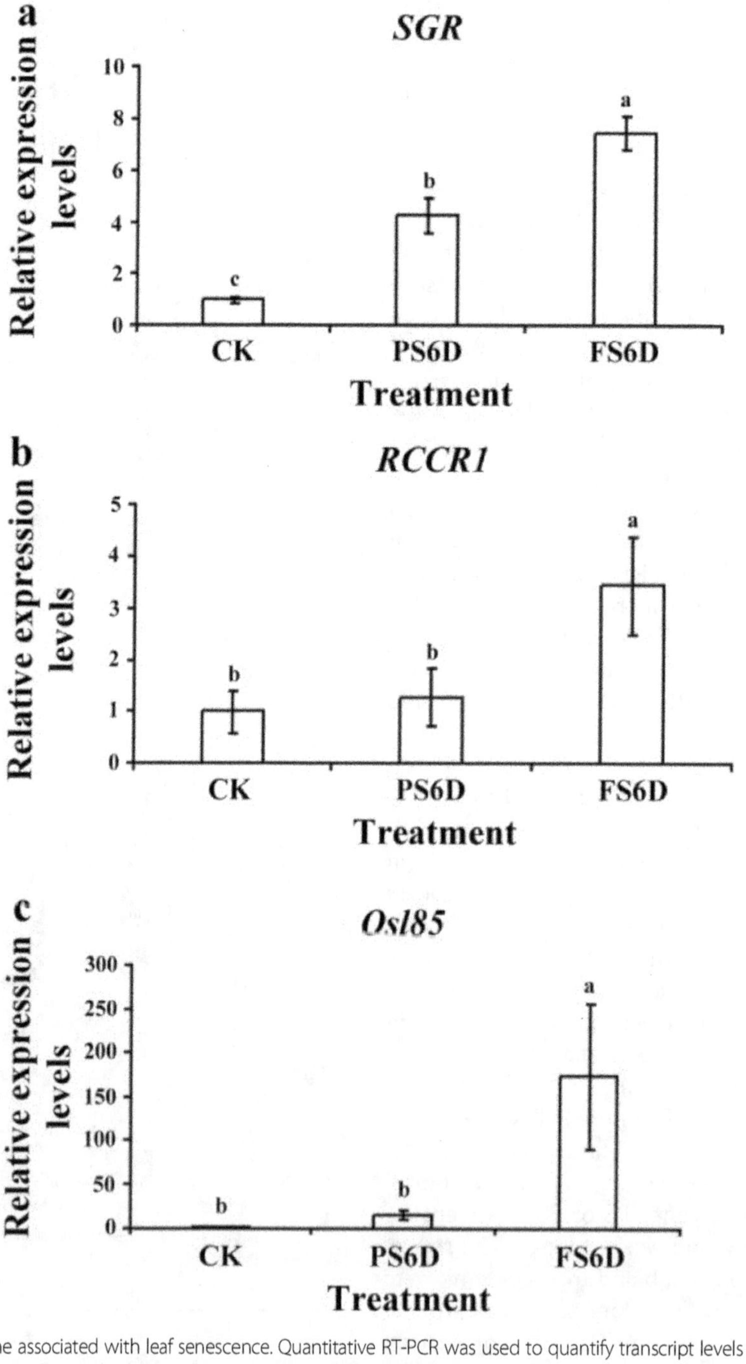

Fig. 3 Transcript levels of gene associated with leaf senescence. Quantitative RT-PCR was used to quantify transcript levels of *Stay-Green* (*SGR*), *Red Chlorophyll Catabolite Reductase* (*RCCR*) and *Isocitrate Lyase* (*Osl85*) genes in 14 days-old rice seedlings after treated control (CK), partial submergence (PS) and full submergence (FS) conditions for 6 days. Relative amounts of transcripts were calculated and normalized to that of Ubiquitin mRNA. Values represent means ± standard deviation from five biologically independent experiment. Values with the different letters are significantly different at $P < 0.05$, according to post-hoc LSD test

Full Submergence Leads to H_2O_2 and O_2^- Overproduction that Disrupts Intracellular Oxidative Homeostasis

Several studies have shown that ROS are common components of biochemical changes in chloroplasts, mitochondria and peroxisomes when plants are subjected to oxygen deprivation conditions (Blokhina et al. 2003; Damanik et al. 2010). To investigate ROS accumulation, H_2O_2 and O_2^- accumulation under different levels of submergence was evaluated by DAB and NBT staining. The 3rd leaves in the submerged area appeared yellow after PS or FS treatment for 6 days (Fig. 4). As shown in Fig. 4b and c, H_2O_2 and O_2^- were accumulated to higher levels in FS than in PS conditions, especially in the tips of leaves. We also determined the cell viability by Evans blue staining, and the results showed significant cell death in the tips and blades of leaves in FS conditions (Fig. 4d and e).

To investigate the cellular scavenging capacity of the antioxidative enzymes in rice seedlings under different levels of submergence, the activities of antioxidant enzymes such as catalase (CAT), ascorbate peroxidase (APX), superoxide dismutase (SOD) and total peroxidase (POX) were evaluated. The CAT, SOD and APX activities were decreased under PS and FS treatment for 6 days and recovery for 1 day compared with the control plants. The APX activity was decreased more under FS treatment for 6 days than PS treatment. The POX activity was significantly increased under PS and FS treatment for 6 days and recovery for 1 day (Fig. 5). Thus, our results suggest that the overproduction of H_2O_2 and O_2^- disrupts intracellular oxidative homeostasis, which affects cell viability and antioxidant enzyme activity under different levels of submergence.

Transcript Profiles of *OsRboh* Genes under Different Levels of Submergence

H_2O_2participates in hypoxia signal transduction to modulate the expression of a set of genes encoding heat-shock proteins and other groups of ROS-mediated proteins (Pucciariello et al. 2012b). NADPH oxidase enzymes are key players in H_2O_2 production. Previously, we showed that *AtRboh* genes are involved in hypoxia signaling, and revealed the different transcript profiles of *Rboh* genes under hypoxic stress in *Arabidopsis*. AtRboh D plays a major role at an early stage to modulate the expression of down-stream hypoxia-inducible genes under hypoxic stress (Yang and Hong 2015). Phylogenetic analysis of the corresponding corrected protein sequences of

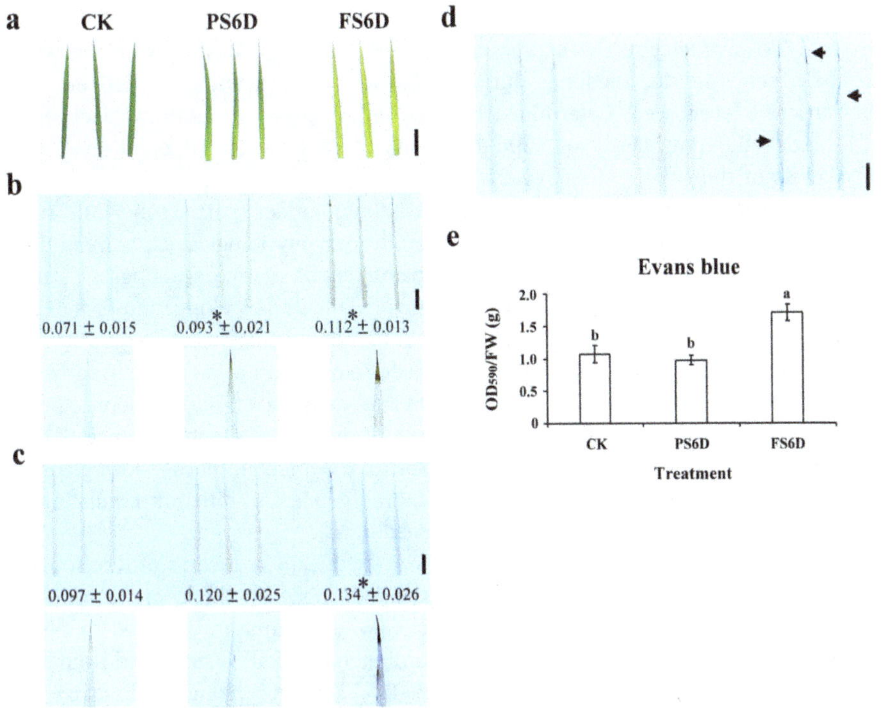

Fig. 4 DAB, NBT and Evans blue staining for the detection of H_2O_2, O_2^- and cell death in rice seedlings under different levels of submergence. **a** Phenotypes of 14 days-old rice seedlings after treated control (CK), partial submergence (PS) and full submergence (FS) for 6 days. **b** and **c**, DAB and NBT staining of H_2O_2 and O_2^- in detached leaves of 14 days-old rice seedlings after treated control (CK), partial submergence (PS) and full submergence (FS) for 6 days. Same results were obtained in three independent experiments. Photograph shows results of three independent leaves. **d** and **e**, Evans blue staining and cell death analysis (see materials and methods for details). Bar = 1 cm. The data represent average values ± SD from five biologically independent experiments. Values with the different letters are significantly different at $P < 0.05$, according to post-hoc LSD test

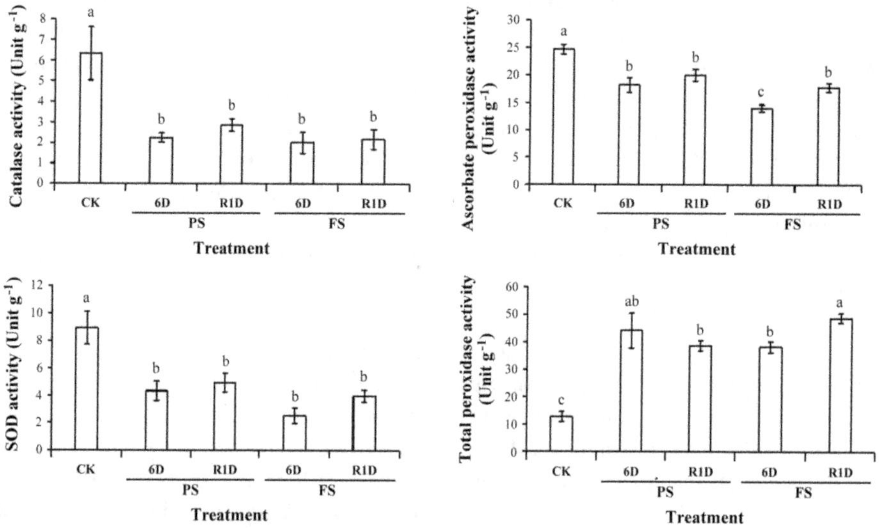

Fig. 5 The activities of CAT, APX, SOD and POX in rice seedlings under different levels of submergence. Enzyme activity was detected from detached shoots of 14 days-old rice seedlings after treated control (CK), partial submergence (PS) and full submergence (FS) for 6 days then recovery 1 day (R1D). The catalase (CAT), ascorbate peroxidase (APX), superoxide dismutase (SOD) and total peroxidase (POX) activity were determined, respectively. The data represent average values ± SD from six biologically independent experiments. Values with the different letters are significantly different at $P < 0.05$, according to post-hoc LSD test

ten *Arabidopsis* and nine rice Rboh proteins were obtained from the TAIR (http://www.arabidopsis.org/) and Phytozome (https://phytozome.jgi.doe.gov/) databases (Additional file 2: Figure S2). To characterize the expression profiles of the nine Rboh genes in rice seedlings during submergence, their transcript levels were determined by quantitative RT-PCR under different levels of submergence. The transcript levels of the *Rboh C* (*Os05g45210*), *D* (*Os05g38980*), *E* (*Os01g61880*) and *F* (*Os08g35210*) genes increased under PS conditions, the transcript levels of *Rboh H* (*Os12g35610*) and *I* (*Os11g33120*) genes increased under FS conditions, the transcript levels of *Rboh A* (*Os01g53294*) decreased under PS treatment for 24 h and FS treatment for 24 or 48 h, the transcript levels of the *Rboh B* (*Os01g25820*) and *G* (*Os09g26660*) genes decreased under FS conditions (Fig. 6a and b). These results not only demonstrate that Rboh genes are involved in different levels of submergence, but also reveal the complex regulation of Rboh genes in hypoxia signaling in rice seedlings.

Discussion

Rice fields in flood-prone areas often experience full or partial submergence. The many visible symptoms of injury caused by full or partial submergence include an initial phase of fast elongation by one or more leaves accompanied by yellowing of old leaves, and slow growth in the dry mass of roots and shoots (Jackson and Ram 2003; Kato et al. 2014; Sarkar and Bhattacharjee 2012). The survival strategies of flood-tolerant plants are characterized as low-O_2 escape syndrome (LOES) and low-O_2 quiescence syndrome (LOQS) (Voesenek and Bailey-Serres 2015). LOES facilitates escape from submergence stress and includes upward growth of leaves (hyponastic growth) and petiole/stem elongation initiated by ethylene accumulation (Jackson 2008). To further dissect the effects of different levels of flooding, we investigated the physiological and molecular responses of rice seedlings under partial and full submergence. In this study, our physiological experiment demonstrated that the plant height of rice seedlings was increased, which was mainly affected by elongation of the 3rd leaf sheath, during submergence. Furthermore, partial submergence led to faster elongation at a relatively early stage compared with full submergence (Fig. 1). Therefore, an early physiological response to partial submergence is the elongation of leaf sheaths, which implies sensing of different levels of submergence and different regulation mechanisms in rice seedlings.

A significant reduction of Rubisco activity, inhibition of the CO_2 photosynthetic rate and ROS accumulation occurs when plants are exposed to flooding conditions for prolonged periods (Kamal and Komatsu 2015; Panda et al. 2008; Yang and Hong 2015; Yang 2014). The catabolism of chlorophyll is regulated by natural leaf senescence and environmental stress (Lim et al. 2007). Our results showed that the chlorophyll a, b and total contents were significant decreased by submergence treatment, particularly FS treatment (Fig. 2). The transcript accumulation of genes associated with senescence was highly induced by

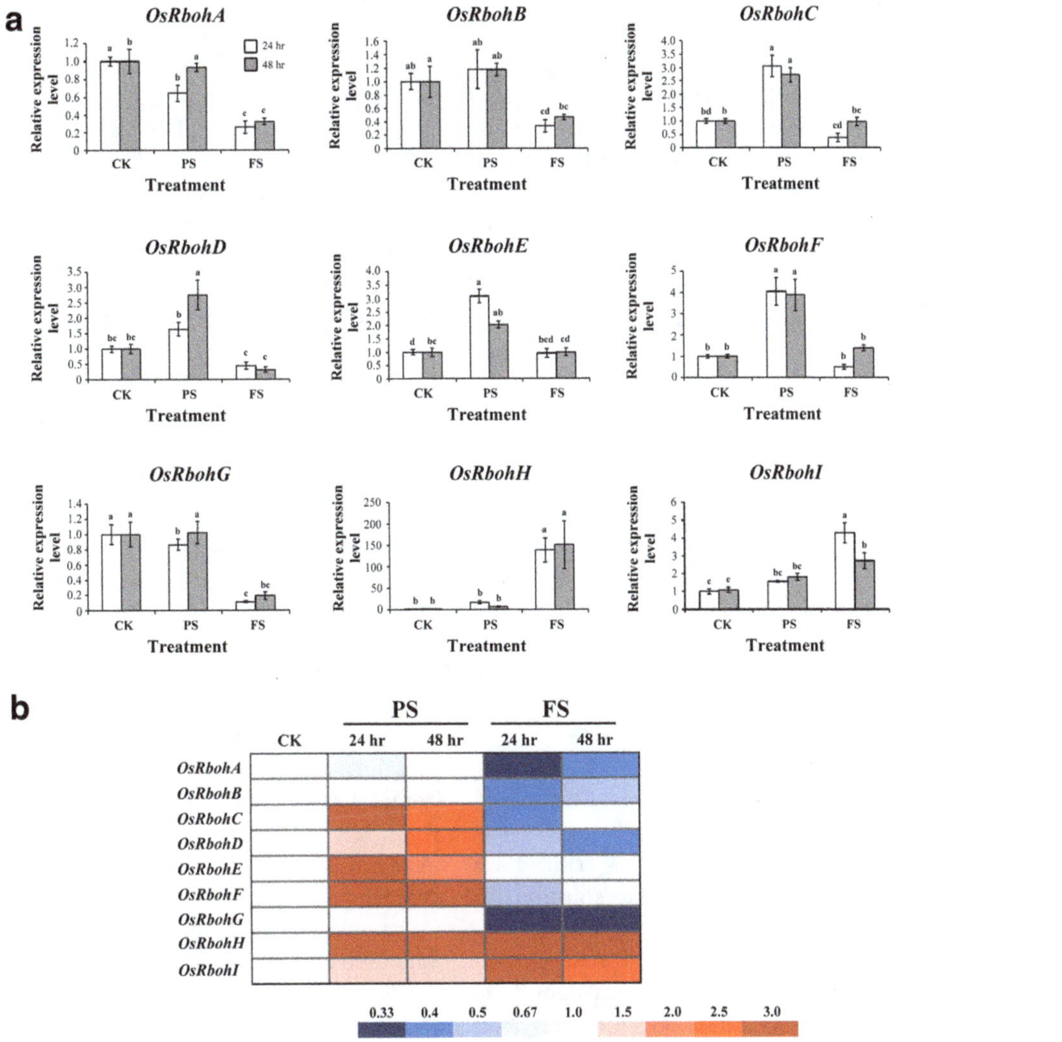

Fig. 6 Transcript profiles of *Rboh* genes in rice seedlings under different levels of submergence. **a** quantitative RT-PCR analyses of transcript levels of *Rbohs* genes in 14 days-old rice seedlings after treated control (CK), partial submergence (PS) and full submergence (FS) conditions. **b** fold change were presented by different color. Total RNAs were isolated from shoots of 14-days-old seedlings after 24 and 48 h submergence treatment and levels of *Rboh A–I* mRNA were determined. Relative amounts of transcripts were calculated and normalized to that of *ubiquitin* mRNA. Values represent means standard deviation from five biologically independent experiments. Values with the different letters are significantly different at $P < 0.05$, according to post-hoc LSD test

FS treatment compared with PS treatment (Fig. 3). These results reveal that different levels of submergence not only affect SAG expression, but also lead to a reduction of chlorophyll content. Notably, a characteristic enzyme gene of the glyoxylate cycle, *Osl85* (*isocitrate lyase*), was significantly induced after FS treatment. This seems to indicate that the glyoxylate cycle is involved in the response of rice seedlings to different levels of submergence.

Under submergence, the O_2 concentration of water is lower than that of air, leading to suppressed respiration in roots and enhanced fermentation (Nishiuchi et al. 2012; Yin et al. 2013). This metabolic shift can cause toxicity from ethanol accumulation. The most common group of toxic intermediates produced is ROS. We demonstrated that H_2O_2 and O_2^- accumulation was highest in the tips of leaves by histochemical staining and observed significant cell death in FS conditions (Fig. 4). Antioxidative enzyme activity detection showed that POX activity was significantly increased under PS and FS treatment for 6 days and recovery for 1 day (Fig. 5). Thus, the overproduction of H_2O_2 and O_2^- disrupts intracellular oxidative homeostasis, which affects cell viability and antioxidant enzyme activity under different levels of submergence.

Previously, we characterized the transcription profiles of ten *Rbohs* in *Arabidopsis* under hypoxic stress and found Rboh D is involved in primary hypoxia signaling to modulate the expression of hypoxia-inducible genes under hypoxic stress (Yang and Hong 2015). Here, we showed that the expression profiles of nine Rboh genes in rice seedlings differed under different levels of submergence (Fig. 6), indicating Rboh genes are involved in responses to different levels of submergence. As in *Arabidopsis*, our data reveal the complex regulation of Rboh genes in hypoxia signaling in rice seedlings.

Conclusions

Although flooding leads to chlorophyll content decreased and accompanied senescence induction, the transcript levels of OsRboh genes was presented different responses to different levels of submergence in rice seedlings. Our results demonstrate that have different mechanism in intracellular to response different levels of submergence. The regulation of OsRboh expression and ROS production was important to help rice seedlings face different levels of submergence.

Methods

Plant Materials, Growth Conditions and Stress Treatment

Rice (*Oryza sativa, japonica.*) Tai-keng 9 (TK9) was used in this study. Seeds were surface sterilized via immersion in a 3 % sodium hypochlorite solution for 30 min, and were then thoroughly rinsed with distilled water. Seeds were subsequently placed on wet filter paper for 3 day at 28 °C in a 16-h-light (236 $\mu molm^{-2}s^{-1}$)/8-h-dark cycle in a growth chamber. The germinated seeds were transplanted onto iron grid in the beaker with Kimura B solution. For submergence treatment, 14-day old seedlings (2–3 leaf stage) were transferred into water box (W:L:H, 40 cm × 40 cm × 60 cm) filled with 23 cm high of water for partial submergence (PS) which the water level reached half of plants height or filled with 50 cm high of water for full submergence (FS). After each treatment, sampled tissue was immediately frozen in liquid nitrogen and stored in −80 °C until use.

Measurement of Plant Height, Length of Leaf Sheath and Chlorophyll Content

After 14-day old seedlings were treatments by PS or FS for 2, 4, 6 and 8 days, the length of each shoot and leaf sheath (at least 30 plants each) were recorded for indicated times. Experiments were repeated five times independently. For chlorophyll content assays, chlorophyll a, b and total contents were extracted from 50 mg of shoot tissue in 2 mL of sodium phosphate buffer (50 mM pH 6.8). Add 40 μL extracted solution in 960 μL 99 % ethanol then incubated for 30 min at RT in the dark with gentle shaking. After centrifugation at 4 °C for 15 min at 1000 *g*, the absorbance of the supernatant was measured at 665 and 649 nm with a spectrophotometer (Metertec SP8001).

Assay of O_2^- and H_2O_2 Accumulation and Cell Viability by Histochemical Staining Method

Accumulation of O_2^- in cell was visualized by the nitroblue tetrazolium (NBT) staining method. 14-day old seedlings after submergence treatment for 6 days, the 3 rd leaves were immersed into 1 mM NBT solution prepared in 10 mM phosphate buffer (pH 7.8) at RT under dark for 8 h. When blue spots appeared, the pigments

Table 1 Primers used for quantitative RT-PCR experiments

Gene name	NCBI accession	Forward primer sequence	Reverse primer sequence	Product size (bp)
ubiquitin	Os03g13170	5′-aaccagctgaggcccaaga-3′	5′-acgattgatttaaccagtccatga-3′	77
SGR	Os09g36200	5′-cgcatgcaatgtcgccaaatg-3′	5′-gctcaccacactcattccctaaag-3′	138
RCCR1	Os10g25030	5′-gcaccttctcactgacagcaatac-3′	5′-accacgcactatctcttccaagg-3′	83
Osl85	Os07g34520	5′-catgggcaaaggagttactgaagag-3′	5′-ggatttggcaagaacatggctg-3′	88
OsRbohA	Os01g53294	5′-gtcttatgcagtcatgaatgtaca-3′	5′-gaataatatacagttaattagcct-3′	129
OsRbohB	Os01g25820	5′-cctagtggaagaagctgtgct-3′	5′-cactatgaaagggaacatcacaa-3′	119
OsRbohC	Os05g45210	5′-tgttttagggatggttttacac-3′	5′-tgtacagacagaaggttaacgt-3′	94
OsRbohD	Os05g38980	5′-gaccagaccaggaaaaaaacaccaa-3′	5′-acacagaaagagttgctaaccgatg-3′	85
OsRbohE	Os01g61880	5′-catcgtgcatagattctgga-3′	5′-catgcattcccactgttcca-3′	62
OsRbohF	Os08g35210	5′-tcgtctatcatagatatacatg-3′	5′-cgtgtactttggtgacctcag-3′	85
OsRbohG	Os09g26660	5′-aagcgttgctaattttcgctat-3′	5′-gagaggatgttttttgaacgg-3′	117
OsRbohH	Os12g35610	5′-gtacaattatatacagattaatg -3′	5′-cgaacaaccaatcactcactaa-3′	67
OsRbohI	Os11g33120	5′-tggccagataatttcatcggtt-3′	5′-gctactctaagtattacaaagta-3′	101

were removed from leaves by boiling in 75 % ethanol for 5 min. Accumulation of H_2O_2 in cells was visualized by the 3, 3′ -diaminobenzidine (DAB) staining method as described in detail previously (Yang and Hong 2015). In the stained leaves, H_2O_2 is visualized as reddish-brown coloration. The leaves were photographed with a digital camera and the intensity of the colour in the NBT- or DAB-stained area was measured using Photoshop image software. Experiments were repeated three times independently. The stained intensity was measured at least 10 leaves for each biological repeat. Cell viability assay was visualized by the Evans blue staining. 14-day old seedlings after submergence treatment for 6 days, the 3 rd leaves were collected, weighed and immersed into 0.2 % Evans blue solution for 3 h with gentle shaking. The de-stained leaves (0.05 g/each) were homogenized with liquid nitrogen in 500 μL of 10 % SDS followed by the addition of 500 μL of de-ionized water. The solutions were centrifuged at 15,800 *g* for 30 min at 4 °C, the absorbance of the supernatant was measured at 590 nm with a spectrophotometer (Metertec SP8001). Experiments were repeated six times independently.

Determination of Antioxidative Enzyme Activity

The shoot samples (50 mg) were excised and immediately used for enzyme extraction. The shoot tissues were homogenized with 50 mM sodium phosphate buffer (pH 6.8) with liquid nitrogen. The homogenate was centrifuged at 12,000 *g* for 20 min and the resulted supernatant was used for determination of enzyme activity. For CAT activity was performed as described previously (Chao et al. 2012; Kato and Shimizu 1985). The decrease in H_2O_2 was determined as the decrease in absorbance at 240 nm with a spectrophotometer (Metertec SP8001). The activity was calculated by the extinction coefficient (40 mM^{-1} cm^{-1} at 240 nm) for H_2O_2. One unit of CAT was defined as the amount of enzyme that degraded 1 μmol H_2O_2 per min. For SOD activity was determined according to Chao et al. (2012). The reaction mixture contained 100 mM triethanolamine- diethanolamine buffer (pH 7.4), EDTA/$MnCl_2$ (100 mM/50 mM, pH 7.4), 7.5 mM β-NADH, 10 mM 2-mercaptoethanol, and enzyme extract. The reaction was started by the addition of β-NADH and absorbance was measured at 340 nm for 1 min. One unit of SOD was defined as the amount of enzyme that inhibited by 50 % the rate of β-NADH oxidation. For APX activity, the decrease in ascorbic acid (AsA) concentration was determined as the decline in absorbance at 290 nm and activity was calculated by the extinction coefficient (2.8 mM^{-1} cm^{-1} at 290 nm) for AsA (Chao et al. 2012; Nakano and Asada 1981). For total peroxidase activity was performed as described previously (Lin and Kao 1999; MacAdam et al. 1992). The shoot tissues were homogenized with 50 mM potassium phosphate buffer (pH 5.8) containing 0.8 M KCl with liquid nitrogen to extract both soluble and ionically bound ('total') peroxidase. The reaction mixture contained 50 mM potassium phosphate buffer (pH 5.8), 21.6 mM guaiacol, 39 mM H_2O_2 and enzyme extract. The reaction was started by the addition of H_2O_2 and absorbance was measured at 470 nm for 1 min. The activity was calculated by the extinction coefficient (26.6 mM^{-1} cm^{-1} at 470 nm) for tetraguaiacol. One unit of peroxidase was defined as the amount of enzyme that caused the formation 1 μmol tetraguaiacol per min.

Quantitative RT-PCR Analyses

Shoot samples were collected from seedlings and frozen until analysis. RNA was isolated from frozen tissues by TRIzol reagent (Invitrogen, Carlsbad, CA, USA). Total RNA samples from shoots were first treated with DNase I and then reverse transcribed into cDNA by Moloney murine leukemia virus reverse transcriptase (Invitrogen). Quantitative RT-PCR was performed using an Illumina Eco Real-Time PCR system (Biogenesis Technology, San Diego, USA) with Power SYBR Green PCR Master Mix (Applied Biosystems), according to the manufacturer's instructions. The relative expression level of each gene was quantified using the comparative threshold cycle method, as described in the manufacturer's instructions for the Eco™ Real-Time PCR System User Guide. The *ubiquitin* (*Os03g13170*) gene was used as an internal control to normalize the cDNA levels. Data were analyzed using the EcoStudy software, version 4.0 (Biogenesis Technology). Amplification conditions were as follows: 94 °C for 5 min, and then 45 cycles at 94 °C for 15 s, 55 °C for 15 s and 72 °C for 30 s. The primers used for quantitative RT-PCR analyses were listed in Table 1. Quantitative RT-PCR experiments were repeated least six times independently in duplicate, and the data were averaged.

Additional files

Additional file 1: Figure S1. Characterization of rice (*Oryza sativa*) seedlings growth under different levels of submergence. The first (a) and second (b) leaves sheath length of 14 days-old rice seedlings after treated submergence for 2, 4, 6 and 8 days. Control check (CK), partial submergence (PS) and full submergence (FS). The data represent average values ± SD from 30 seedlings of each treatment obtained from six biologically independent experiments. Values with the different letters are significantly different at $P < 0.05$, according to post-hoc LSD test.

Additional file 2: Figure S2. Phylogenetic tree of members of Rbohs protein family. Phylogenetic tree of 10 *Arabidopsis* Rboh proteins and 9 rice (Oryza sativa) related Rboh proteins. The phylogenetic tree was constructed by the Neighbor Joining algorithm (Saitou and Nei 1987) implemented in the MEGA 6 software package (Tamura et al. 2013).

Abbreviations

APX: ascorbate peroxidase; CAT: catalase; FS: full submergence; H_2O_2: hydrogen peroxide; POX: total peroxidase; PS: partial submergence; Rbohs: respiratory burst oxidase homologs; ROS: reactive oxygen species; RT-

PCR: reverse transcription-polymerase chain reaction; SOD: superoxide dismutase.

Competing interests
The authors declare that they have no competing interests.

Authors' Contributions
YSW carried out the experiments and analyzed the data. Dr. CYY defined the research theme, designed experiments and wrote the manuscript. Both authors read and approved the final manuscript.

Acknowledgment
The author thanks Dr. Jun-Yi Yang for instrument assistance of quantitative RT-PCR.

References

Bailey-Serres J, Voesenek LA (2010) Life in the balance: a signaling network controlling survival of flooding. Curr Opin Plant Biol 13(5):489–494

Bailey-Serres J, Fukao T, Ronald P, Ismail A, Heuer S, Mackill D (2010) Submergence tolerant rice: SUB1's journey from Landrace to Modern Cultivar. Rice 3(2–3):138–147

Baxter A, Mittler R, Suzuki N (2014) ROS as key players in plant stress signalling. J Exp Bot 65(5):1229–1240

Blokhina O, Virolainen E, Fagerstedt KV (2003) Antioxidants, oxidative damage and oxygen deprivation stress: a review. Ann Bot 91(2):179–194

Chao YY, Chou TS, Kao CH (2012) Involvement of abscisic acid and hydrogen peroxide in regulating the activities of antioxidant enzymes in leaves of rice seedlings under magnesium deficiency. Plant Growth Regul 66(1):1–8

Damanik RI, Maziah M, Ismail MR, Ahmad S, Zain A (2010) Responses of the antioxidative enzymes in Malaysian rice (*Oryza sativa* L.) cultivars under submergence condition. Acta Physiol Plant 32(4):739–747

Desikan R, Last K, Harrett-Williams R, Tagliavia C, Harter K, Hooley R, Hancock JT, Neill SJ (2006) Ethylene-induced stomatal closure in Arabidopsis occurs via AtrbohF-mediated hydrogen peroxide synthesis. Plant J 47(6):907–916

Fukao T, Yeung E, Bailey-Serres J (2011) The submergence tolerance regulator SUB1A mediates crosstalk between submergence and drought tolerance in rice. Plant Cell 23(1):412–427

Fukao T, Yeung E, Bailey-Serres J (2012) The submergence tolerance gene *SUB1A* delays leaf senescence under prolonged darkness through hormonal regulation in rice. Plant Physiol 160(4):1795–1807

Hortensteiner S (2009) Stay-green regulates chlorophyll and chlorophyll-binding protein degradation during senescence. Trends Plant Sci 14(3):155–162

Jackson MB (2008) Ethylene-promoted elongation: an adaptation to submergence stress. Ann Bot 101(2):229–248

Jackson MB, Ram PC (2003) Physiological and molecular basis of susceptibility and tolerance of rice plants to complete submergence. Ann Bot 91(2):227–241

Kamal AH, Komatsu S (2015) Involvement of reactive oxygen species and mitochondrial proteins in biophoton emission in roots of soybean plants under flooding stress. J Proteome Res 14(5):2219–2236

Kato M, Shimizu S (1985) Chlorophyll metabolism in higher plants VI. Involvement of peroxidase in chlorophyll degradation. Plant Cell Physiol 26(7):1291–1301

Kato Y, Collard BC, Septiningsih EM, Ismail AM (2014) Physiological analyses of traits associated with tolerance of long-term partial submergence in rice. AoB Plants 6:plus058. doi: 10.1093/aobpla/plu058

Lherminier J, Elmayan T, Fromentin J, Elaraqui KT, Vesa S, Morel J, Verrier JL, Cailleteau B, Blein JP, Simon-Plas F (2009) NADPH oxidase-mediated reactive oxygen species production: subcellular localization and reassessment of its role in plant defense. Mol Plant Microbe Interact 22(7):868–881

Licausi F, Perata P (2009) Low oxygen signaling and tolerance in plants. Adv Bot Res 50:139–198

Lim PO, Kim HJ, Nam HG (2007) Leaf senescence. Annu Rev Plant Biol 58:115–136

Lin CC, Kao CH (1999) NaCl induced changes in ionically bound peroxidase activity in roots of rice seedlings. Plant and Soil 216:147–153

Liu B, Rennenberg H, Kreuzwieser J (2015) Hypoxia induces stem and leaf nitric oxide (NO) emission from poplar seedlings. Planta 241(3):579–589

MacAdam JW, Nelson CJ, Sharp RE (1992) Peroxidase activity in the leaf elongation zone of tall fescue. Plant Physiol 99(3):872–878

Maruta T, Inoue T, Tamoi M, Yabuta Y, Yoshimura K, Ishikawa T, Shigeoka S (2011) *Arabidopsis* NADPH oxidases, AtrbohD and AtrbohF, are essential for jasmonic acid-induced expression of genes regulated by MYC2 transcription factor. Plant Sci 180(4):655–660

Nakano Y, Asada K (1981) Hydrogen peroxide is scavenged by ascorbate-specific peroxidase in spinach chloroplasts. Plant Cell Physiol 22(5):867–880

Nishiuchi S, Yamauchi T, Takahashi H, Kotula L, Nakazono M (2012) Mechanisms for coping with submergence and waterlogging in rice. Rice 5:2. doi:10.1186/1939-8433-5-2

Panda D, Sharma SG, Sarkar RK (2008) Chlorophyll fluorescence parameters, CO_2 photosynthetic rate and regeneration capacity as a result of complete submergence and subsequent re-emergence in rice (*Oryza sativa* L.). Aquat Bot 88(2):127–133

Pogany M, von Rad U, Grun S, Dongo A, Pintye A, Simoneau P, Bahnweg G, Kiss L, Barna B, Durner J (2009) Dual roles of reactive oxygen species and NADPH oxidase RBOHD in an *Arabidopsis-Alternaria* pathosystem. Plant Physiol 151(3):1459–1475

Pruzinska A, Anders I, Aubry S, Schenk N, Tapernoux-Luthi E, Muller T, Krautler B, Hortensteiner S (2007) In vivo participation of red chlorophyll catabolite reductase in chlorophyll breakdown. Plant Cell 19(1):369–387

Pucciariello C, Banti V, Perata P (2012a) ROS signaling as common element in low oxygen and heat stresses. Plant Physiol Bioch 59:3–10

Pucciariello C, Parlanti S, Banti V, Novi G, Perata P (2012b) Reactive oxygen species-driven transcription in *Arabidopsis* under oxygen deprivation. Plant Physiol 159(1):184–196

Sagi M, Fluhr R (2006) Production of reactive oxygen species by plant NADPH oxidases. Plant Physiol 141(2):336–340

Saitou N, Nei M (1987) The neighbor-joining method: a new method for reconstructing phylogenetic trees. Mol Biol Evol 4(4):406–425

Sakagami JI, John Y, Sone C (2013) Complete submergence escape with shoot elongation ability by underwater photosynthesis in African rice, *Oryza glaberrima* Steud. Field Crop Res 152:17–26

Sarkar RK, Bhattacharjee B (2012) Rice genotypes with *SUB1* QTL differ in submergence tolerance, elongation ability during submergence and re-generation growth at re-emergence. Rice 5:7

Steffens B, Geske T, Sauter M (2011) Aerenchyma formation in the rice stem and its promotion by H_2O_2. New Phytol 190(2):369–378

Tamura K, Stecher G, Peterson D, Filipski A, Kumar S (2013) MEGA6: molecular evolutionary genetics analysis version 6.0. Mol Biol Evol 30(12):2725–2729

Vellosillo T, Vicente J, Kulasekaran S, Hamberg M, Castresana C (2010) Emerging complexity in reactive oxygen species production and signaling during the response of plants to pathogens. Plant Physiol 154(2):444–448

Voesenek LA, Bailey-Serres J (2015) Flood adaptive traits and processes: an overview. New Phytol 206(1):57–73

Winkel A, Colmer TD, Ismail AM, Pedersen O (2013) Internal aeration of paddy field rice (*Oryza sativa*) during complete submergence—importance of light and floodwater O_2. New Phytol 197(4):1193–1203

Yamada Y, Furusawa S, Nagasaka S, Shimomura K, Yamaguchi S, Umehara M (2014) Strigolactone signaling regulates rice leaf senescence in response to a phosphate deficiency. Planta 240(2):399–408

Yang CY (2014) Hydrogen peroxide controls transcriptional responses of *ERF73/HRE1* and *ADH1* via modulation of ethylene signaling during hypoxic stress. Planta 239(4):877–885

Yang CY, Hong CP (2015) The NADPH oxidase Rboh D is involved in primary hypoxia signalling and modulates expression of hypoxia-inducible genes under hypoxic stress. Environ Exp Bot 115:63–72

Yin D, Chen S, Chen F, Jiang J (2013) Ethylene promotes induction of aerenchyma formation and ethanolic fermentation in waterlogged roots of *Dendranthema* spp. Mol Biol Rep 40(7):4581–4590

2

The Rice *TCM5* Gene Encoding a Novel Deg Protease Protein is Essential for Chloroplast Development under High Temperatures

Kailun Zheng[1†], Jian Zhao[1,2†], Dongzhi Lin[1†], Jiaying Chen[1,2], Jianlong Xu[3], Hua Zhou[1,4], Sheng Teng[2] and Yanjun Dong[1*]

Abstract

Background: High temperature affects a broad spectrum of cellular components and metabolism in plants. The Deg/HtrA family of ATP-independent serine endopeptidases is present in nearly all organisms. Deg proteases are required for the survival of *Escherichia coli* at high temperatures. However, it is still unclear whether rice Deg proteases are required for chloroplast development under high temperatures.

Results: In this study, we reported the first rice *deg* mutant *tcm5* (*thermo-sensitive chlorophyll-deficient mutant 5*) that has an albino phenotype, defective chloroplasts and could not survive after the 4–5 leaf seedling stage when grown at high temperature (32 °C). However, when grown at low temperatures (20 °C), *tcm5* has a normal phenotype. Map-based cloning showed that *TCM5* encoding a chloroplast-targeted Deg protease protein. The *TCM5* transcripts were highly expressed in all green tissues and undetectable in other tissues, showing the tissue-specific expression. In *tcm5* mutants grown at high temperatures, the transcript levels of certain genes associated with chloroplast development especially PSII-associated genes were severely affected, but recovered to normal levels at low temperatures. These results showed important role of *TCM5* for chloroplast development under high temperatures.

Conclusions: The *TCM5* encodes chloroplast-targeted Deg protease protein which is important for chloroplast development and the maintenance of PSII function and its disruption would lead to a defective chloroplast and affected expression levels of genes associated with chloroplast development and photosynthesis at early rice seedling stage under high temperatures.

Keywords: Chloroplast development, Deg protease, Photosystem, Rice

Background

Chloroplast is a semi-autonomous organelle and carries out photosynthesis: the capture of light energy and its conversion into chemical energy. It is known that chloroplast development consists of a series of complex events related to chloroplast differentiation and can be divided into three steps which are coordinately regulated by plastid and nucleus genes (Jarvis and López-Juez 2013; Kusumi et al. 2011; Kusumi and Iba, 2014). The first step involves the activation of plastid replication and plastid DNA synthesis. The second step is the chloroplast 'build-up', characterized by the establishment of chloroplast genetic system. At this step nuclear-encoded plastid RNA polymerase (NEP) preferentially transcribes plastid genes encoding plastid gene expression machineries (Hajdukiewicz et al. 1997), and the transcription/translation activities in the chloroplast are dramatically increased. At third step, the plastid and nuclear genes encoding photosynthetic apparatus are expressed at very high levels. Plastid genes for photosynthetic apparatus are mainly transcribed by plastid-encoded RNA polymerase (PEP) (Santis-Maciossek et al. 1999). All expressions of these genes lead to the synthesis and assembly of photosynthetic machineries. In spite of these, the

* Correspondence: dong@shnu.edu.cn
†Equal contributors
[1]Development Center of Plant Germplasm Resources, College of Life and Environmental Sciences, Shanghai Normal University, Shanghai 200234, China
Full list of author information is available at the end of the article

mechanisms of the major genes in higher plants remain largely unknown (Pfalz and Pfannschmidt 2013).

The family of Deg proteases (for degradation of periplasmic proteins) (Strauch and Beckwith 1988), also named as HtrA proteases (for high temperature requirement A) (Lipinska et al. 1988), is one important group among proteolytic enzymes. The Deg proteases, consisting of three representative proteins (DegP, DegQ and DegS), were initially found in *Escherichia coli* and well-studied in *Escherichia coli* and mammals. Most Deg members contain one to four PDZ protein-protein interaction domains (Clausen et al. 2002). A first *E.coli* DegP was identified based on the fact that it is required for the survival at high temperatures (Lipinska et al. 1988, 1990; Strauch and Beckwith 1988). A second *E.coli* protease, DegS, acts in a stress signaling cascade sensing misfolded proteins in the periplasm and transducing the signal to the cytoplasm (Walsh et al. 2003). A third *E.coli* protease, DegQ, is a periplasmic protease but also shares some functional features with DegP (Kolmar et al. 1996). However, relatively little is known about members of this family in plants (Kato and Sakamoto, 2010). To date, although it was predicated to contain 16 genes encoding putative Deg proteases in *Arabidopsis thaliana* (Huesgen et al. 2005), 15 in *Oryza sativa* (Tripathi and Sowdhamini 2006) and 20 in *Populus trichocarpa* (Garcia-Lorenzo et al. 2006), but barely a few Deg proteases from Arabidopsis have been well-studied. Six AtDeg proteases are located in chloroplasts (Itzhaki et al. 1998), one in peroxisomes (Schuhmann et al. 2008), one in mitochondria (for review, see Schuhmann et al. 2012), and one in the nucleus (Pendle et al. 2005). AtDeg3 and AtDeg10 were reported to locate in both chloroplast and mitochondrion, while AtDeg7 is located in both nucleus and mitochondrion (Tanz et al. 2014). Additionally, the Deg proteases were reported to chloroplast-located be involved in the degradation of damaged photosynthetic proteins, especially the photosystem II (PSII) reaction center D1 protein (Kato et al. 2012). For example, the chloroplast-targeted AtDeg1 acts as a chaperone, assisting in the assembly of PSII dimers and supercomplexes (Sun et al. 2010b). The AtDeg15 is a processing enzyme, cleaving the N-terminal peroxisomal targeting signal 2 (Helm et al. 2007; Schuhmann et al. 2008). Recently, it was precisely predicated to contain 10 Deg proteases and 6 DegP-like proteases (Schuhmann et al. 2012) in *Oryza sativa*. However, to our knowledge, any *deg* rice mutants have not been reported yet, still less the function.

Here we report a detailed phenotypic analysis of the rice *deg* mutant, *tcm5*, which exhibited the albino phenotype before the 3-leaf stage at 32 °C, whereas normal green at 20 °C, and the cloning of *TCM5*, encoding a Deg protease protein. Additionally, the transcripts of certain genes for chlorophyll biosynthesis, photosynthesis and chloroplast development were severely affected in *tcm5* mutants at high temperatures. Our work implicates that rice *TCM5* plays an important role in chloroplast development and maintenance of PSII function under high temperatures.

Results

Characterization of the *tcm5* Mutant

The leaves of *tcm5* mutants displayed an albino phenotype (Fig. 1a) before the 3-leaf-stage, and could not survive after the 4–5 leaf stage at 32 °C (Fig. 1b), but they turned yellowish green at 28 °C. Interestingly, the *tcm5* mutants showed no obvious phenotypic differences from the wild type (WT) plants at 20 °C. Similarly, at 32 °C, the Chl a, Chl b and carotenoid (Car) contents in *tcm5* mutants at the 3-leaf stage were drastically lower than those in WT plants (Fig. 1e), however, the pigment levels were similar to WT plants at 20 °C (Fig. 1f). In addition, during the whole growth periods, leaf Chl contents (SPAD values) in *tcm5* plants always showed lower levels than those in WT plants, moreover, the difference ranges become smaller when climate temperatures gradually decreased from summer to autumn in 2010 (Shanghai, China) (Additional file 1: Figure S1). These results indicated that the mutant trait was high-temperature sensitive

PSII complexes are located in the chloroplast. PSII activity of the 3-leaf-stage *tcm5* and WT seedlings at 20 and 32 °C was surveyed by measuring the ratio of variable fluorescence to maximum fluorescence as follows: $Fv/Fm = (Fm\text{-}Fo)/Fm$, where Fo and Fm are minimum and maximum chlorophyll a fluorescence of dark-adapted leaves, respectively. Fv/Fm reflects the maximum potential capacity of the PSII photochemical reactions (Krause and Weis 1991). At 32 °C, the Fv/Fm was 0.787 ± 0.03 in WT plants and 0.365 ± 0.02 in *tcm5* mutants, indicating that the photochemical efficiency of PSII was significantly suppressed in the *tcm5* mutants. In addition, $\Phi PSII$ represents the actual photosynthesis efficiency. At 32 °C, the ΦPSII value was 0.624 ± 0.04 in WT plants and 0.361 ± 0.04 in *tcm5* mutants, showing that the photosynthesis efficiency was significantly reduced in *tcm5* mutants. By contrast, at 20 °C, all respective values in the *tcm5* mutants were similar to WT levels (Additional file 2: Table S1). Thus, the *tcm5* mutation severely inhibited the PSII activity under high temperatures.

We next examined whether the lack of photosynthetic pigments in *tcm5* mutants was accompanied by ultrastructural changes in the chloroplasts. At 32 and 20 °C, we observed well-developed chloroplasts, with dense and well-structured grana lamella stacks, in WT plants, (Fig. 2a,c). By contrast, at 32 °C we did not observe any well-developed chloroplasts and well-structured grana lamella stacks (Fig. 2b) in *tcm5* mutants. However, there

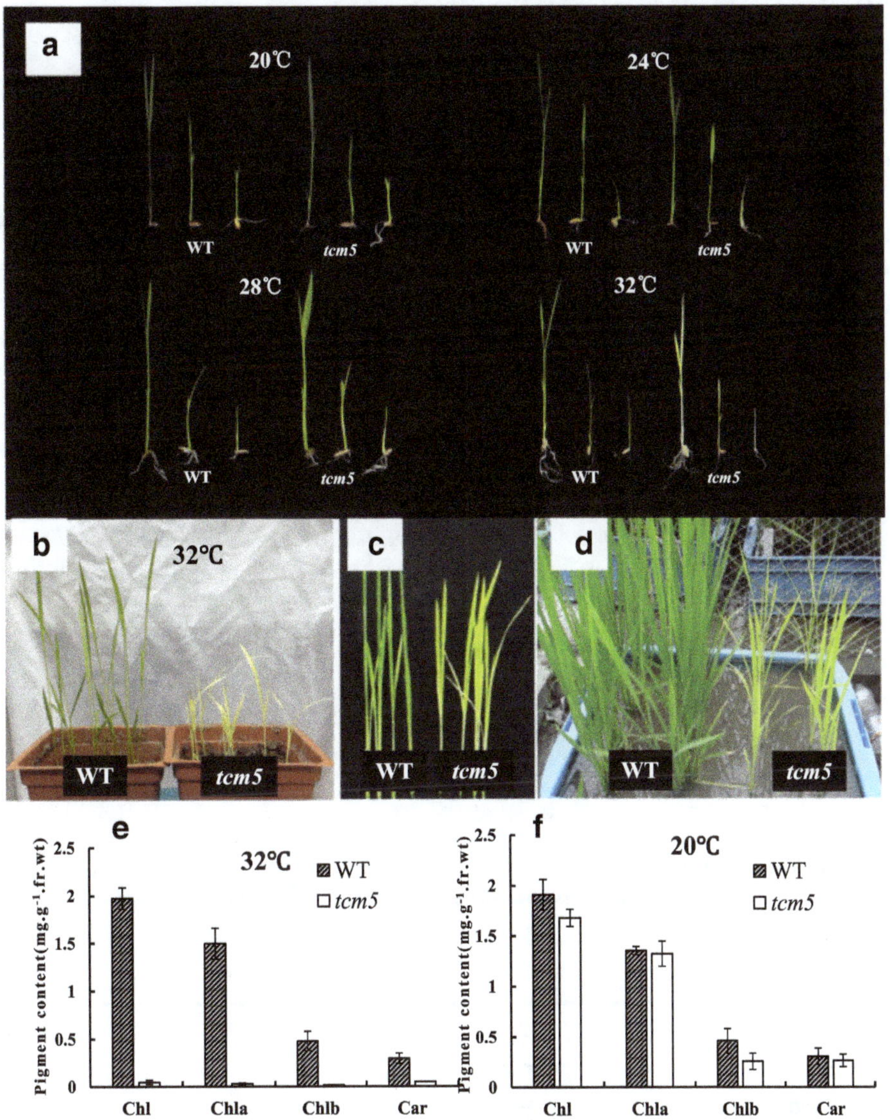

Fig. 1 Characterization of the *tcm5* mutants; **a**, 1-, 2-, 3-leaf stage seedlings of wild type (WT) (*left*) and *tcm5* mutants (*right*) grown at 20, 24, 28 and 32 °C, respectively; **b**, 4-leaf-stage plants of wild type (*left*) and *tcm5* mutants (*right*) grown at 32 °C; **c**,**d**, indicate 25-day-old (**c**) and 50-day-old (**d**) plants grown at rice paddy field, respectively (2010, Shanghai, China); **e**, **f**, indicate the pigment contents in the 3rd leaves of WT and mutants grown at 32 and 20 °C, respectively. *Chl* total chlorophyll, *Chla* chlorophyll a, *Chlb* chlorophyll b, *Car* carotenoid, *WT* wide type

was no obvious difference between *tcm5* and WT plants at 20 °C (Fig. 2d). These results indicated that the *tcm5* mutation causes abnormal chloroplast development at high temperatures.

We also surveyed some agronomic characters of the *tcm5* and WT plants and found that, under field condition, the grain and panicle number per plant (Additional file 1: Figure S2) was significantly decreased, showing that the mutation inhibited the plant growth in *tcm5* mutants.

Map-based Cloning of *TCM5*

We used map-based cloning to identify the *TCM5* locus. To this end, we isolated 96 individuals with the albino phenotype from the segregating 412 F_2 plants derived from the cross of Pei'ai 64S and *tcm5* mutant. The segregation ratio (3:1) of green to albino plants showed that the mutant phenotype is controlled by a single recessive gene (*tcm5*) (Additional file 2: Table S2). First, by genotyping the 96 mutant individuals, we localized *tcm5* between RM2474 and RM18692 on chromosome 5 (Fig. 3a). To fine-map *TCM5* gene, we developed new InDel and CAPS markers between RM2474 and RM18692 (Additional file 2: Table S3). The *TCM5* locus was finally narrowed to an 18.9-kb between P1 and P3 markers using 6400 F_2 mutant individuals (Fig. 3b). The target region was predicated to contain two

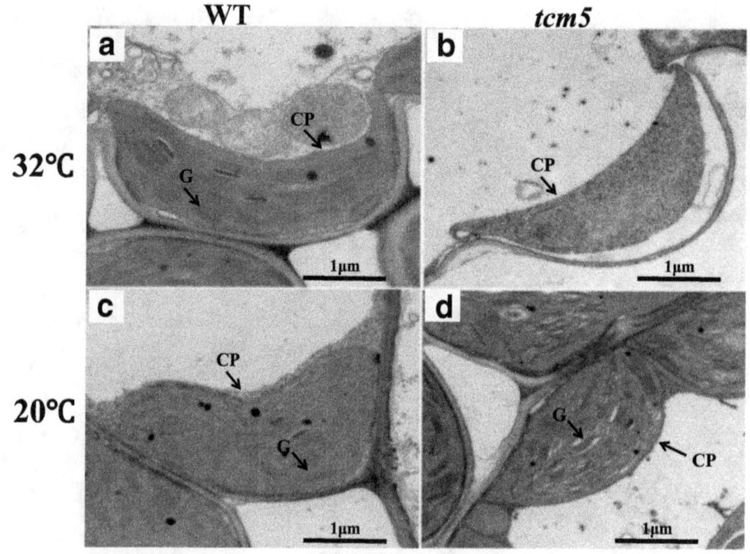

Fig. 2 Transmission electron microscopic images of chloroplasts in WT and *tcm5* mutant; **a**, **b**, Chloroplast structure in WT(**a**) and *tcm5* (**b**) cell at 32 °C; **c**, **d**, Chloroplast structure in WT(**c**) and *tcm5*(**d**) cell at 20 °C. *CP, chloroplast; G, grana lamella stacks*

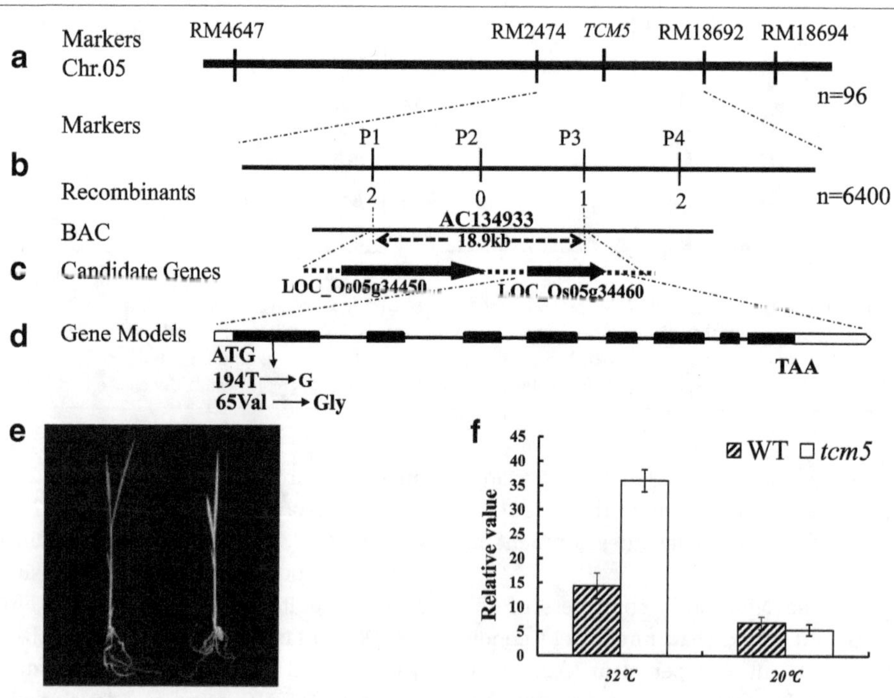

Fig. 3 Genetic analysis and cloning of the *TCM5* gene; **a**, The *TCM5* gene was initially located between the RM2474 and RM18692 SSR markers on chromosome 5 using 96 F_2 mutant individuals; **b**, *TCM5* was narrowed to a 18.9 kb between InDel marker P1 and CAPS marker P3 and co-segregated with the CAPS marker P2 using 6400 F_2 mutant individuals; **c**, The target region contains two predicted candidate genes (*LOC_Os05g34450, LOC_Os05g34460*); **d**, A single nucleotide mutation at codon 194(T→G) in the first exon in *LOC_Os05g34460*, consisting of eight exons; **e**, The complemented plants with pCAMBIA1301:CaMV35S:TCM5 (*left*) and transgenic T_0 plants with empty vector pCAMBIA1301(*right*); **f**, Transcript levels of *TCM5* in WT and *tcm5* at the 3-leaf stage grown at 32 and 20 °C, *OsActin* was used as a control for qPCR

candidate genes (*LOC_Os05g34450, LOC_Os05g34 460*) (Fig. 3c). We sequenced both candidate genes and only found a single nucleotide mutation (T→G) at position 194 bp from the ATG start codon in *LOC_Os05g34460*, which resulted in an amino acid change from Val to Gly (Fig. 3d).

To further prove that the lack of *TCM5* was responsible for the mutant phenotype in the *tcm5* mutants, genetic complementation was performed. Resultantly, all of the 11 transgenic plants with pCAMBIA1301:CaMV35S:TCM5 were obtained and displayed wild-type phenotype under high temperatures (Fig. 3e). Meanwhile, 13 independent lines were transformed with empty vector, pCAMBIA1301, all failed to rescue the *tcm5* mutant. These results confirmed that *LOC_Os05g34460* is *TCM5*.

Characterization of TCM5

The *TCM5* encoded a putative Deg protease homologue, also predicated and named as OsDeg10 (Schuhmann et al. 2012). Searching the rice genome database revealed that *TCM5* is a single-copy gene. The *TCM5* gene is comprised of eight exons and encodes a 614-amino acid protein with the molecular mass of approximately 67kD. In addition, TCM5 was predicated to contain not only a chloroplast transit peptide of 94 amino acids at the N terminus, but also two PDZ domains of trypsin-like serine proteases. We found that the *tcm5* mutation site occurred at chloroplast transit peptides.

BLAST searches found that there are 10 Deg protease and 6 Deg-like protease proteins in the whole rice genome as described by Schuhmann et al.(2012) and TCM5 shares the highest sequence similarity to LOC_O s02g50880 (OsDeg9.1) (52.6 % amino acid identity) (Additional file 1: Figure S3). Additionally, close homology of TCM5 were also indented in *Arabidopsis thaliana, Brachypodium distachyon, Glycine max, Sorghum bicolor*, and *Zea mays*, showing the high conservative property within higher plants. And, Bradi2g25260 from *Brachypodium distachyon* shares the highest homology to TCM5 (88.1 % amino acid identity) (Fig. 4a). However, the functions of these homologs remain unclear. As shown in Fig. 4b, TCM5 homologs can be clearly divided into two groups: monocots and dicotyledons, consistent with the biological taxonomy (Fig. 4b).

TCM5 Expression

To examine the expression pattern of *TCM5*, we used RT-PCR to investigate the transcript levels of *TCM5* in

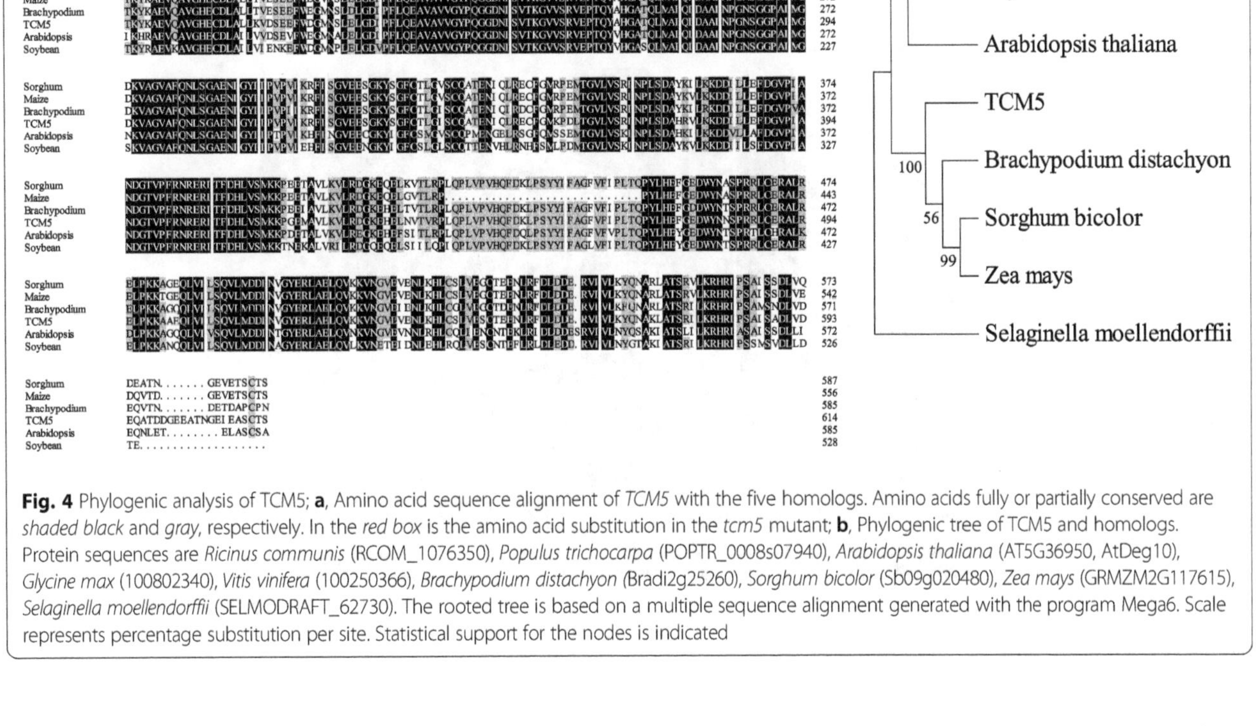

Fig. 4 Phylogenic analysis of TCM5; **a**, Amino acid sequence alignment of *TCM5* with the five homologs. Amino acids fully or partially conserved are *shaded black* and *gray*, respectively. In the *red box* is the amino acid substitution in the *tcm5* mutant; **b**, Phylogenic tree of TCM5 and homologs. Protein sequences are *Ricinus communis* (RCOM_1076350), *Populus trichocarpa* (POPTR_0008s07940), *Arabidopsis thaliana* (AT5G36950, AtDeg10), *Glycine max* (100802340), *Vitis vinifera* (100250366), *Brachypodium distachyon* (Bradi2g25260), *Sorghum bicolor* (Sb09g020480), *Zea mays* (GRMZM2G117615), *Selaginella moellendorffii* (SELMODRAFT_62730). The rooted tree is based on a multiple sequence alignment generated with the program Mega6. Scale represents percentage substitution per site. Statistical support for the nodes is indicated

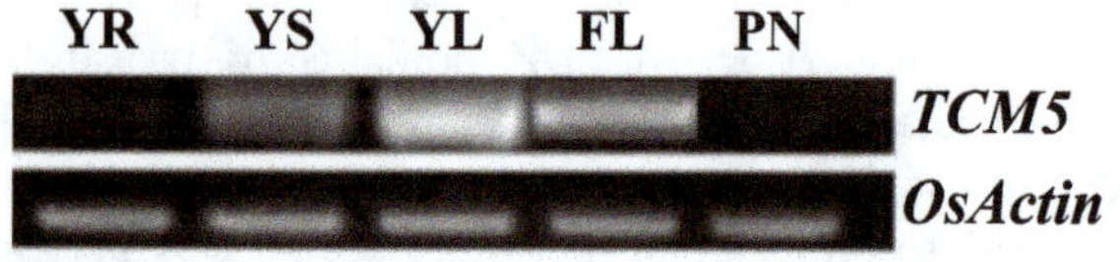

Fig. 5 Expression analysis of *TCM5* by RT-PCR analysis; YR, young-seedling roots; YS, young-seedling stem; YL, young-seedling leaf; FL, flag leaf at heading; PN, young panicles. *OsActin* was used as a control (cycle number for *OsActin* was 28, cycle number for *TCM5* was 35)

various tissues (Fig. 5). Resultantly, *TCM5* was highly expressed in all green tissues (leaf and stem) and undetectable in roots and panicles, showing its tissue-specific expression.

To further explore whether the *tcm5* mutation and temperature influence *TCM5* expression, we examined *TCM5* transcript levels in 3-leaf *tcm5* and WT seedlings grown at 20 and 32 °C (Fig. 3f). Interestingly, regardless of WT and *tcm5* plants, *TCM5* transcript levels were more strongly expressed at high temperatures than those at low temperatures. Specifically, at 32 °C, *TCM5* transcript levels in the *tcm5* mutants were over 2-fold higher than WT levels. However, at 20 °C, *TCM5* expression levels were relatively low and comparable to WT plants (Fig. 3f). These results suggested that *TCM5* function is more needed for chloroplast development under high temperatures.

Subcellular Localization of TCM5 Protein

The TCM5 protein was predicted to localize to chloroplasts by TargetP (Emanuelsson et al. 2000, http://www.cbs.dtu.dk/services/TargetP/). To investigate the actual subcellular localization of TCM5, the cDNA fragment encoding the N-terminal region (amino acids 1–135) of TCM5 was introduced into the N-terminal of the GFP gene in the expression vector pMON530-GFP. Then we transformed the pMON530:TCM5-GFP plasmids into tobacco cells by Agrobacterium-mediated infection, and used empty pMON530-GFP vector without a specific targeting sequence as the positive control. Confocal laser scanning microscopy was used to observe the fluorescent signals after 48 h transformation. The green fluorescent signal of the TCM5-GFP fusion protein perfectly co-localized with the chlorophyll autofluorescence (Fig. 6a) in tobacco cells. By contrast, the cells transformed with the empty GFP vector had green fluorescent signals in both the cytoplasm and the nucleus (Fig. 6b). These findings confirmed that TCM5 was localized to the chloroplast.

The Transcript Expressions of Functionally Related Genes in the *tcm5* Mutants

We examined the transcript levels of genes associated with Chl biosynthesis, photosynthesis and chloroplast development (*PORA, HEMA, CAO1, YGL1, cab1R, rbcL, rbcS, psaA, psbA, LhcpII,V1, V2, V3, OsRpoTp, OsPoLp1, rpoB, 23SrRNA, rps7, rps21, FtsZ*) in the *tcm5* mutants. Under high temperatures, the transcript levels of all tested genes for Chl biosynthesis (Fig. 7a), such as glutamyl tRNA reductase (*HEMA*) and chlorophyllide a oxygenase (*CAO1*), NADPH:protochlorophyllide oxidoreductase (*PORA*) and Chl synthase(*YGL1*), were obviously reduced in *tcm5* mutants, compared with WT plants, consistent with the decreased of Chl content (Fig. 1e) and the albino phenotype (Fig. 1a). We also investigated the photosynthesis-related genes (Fig. 7b) *cab1R*, encoding the Chl a/b binding protein, *rbcL*, encoding the large subunit of ribulose-1,5-bisphosphate carboxylase (Rubisco), *rbcS*,

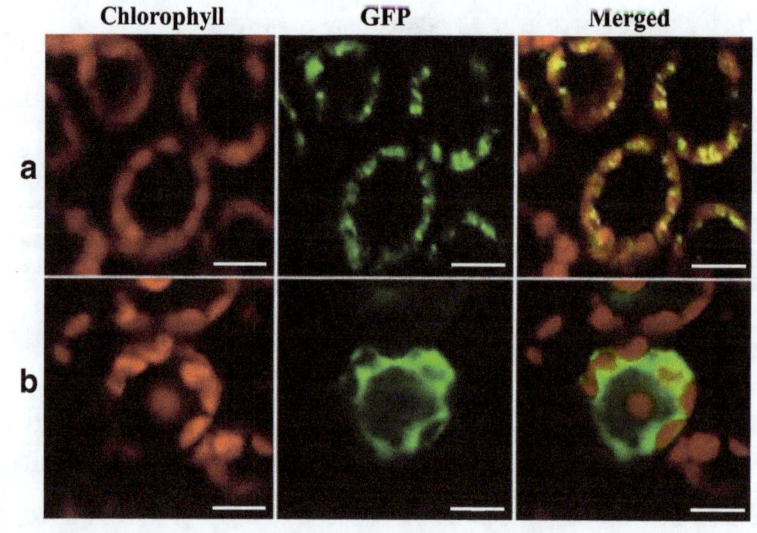

Fig. 6 Subcellular localization of TCM5 protein; **a**, TCM5-GFP fusion; **b**, Empty GFP vector without a specific targeting sequence. The scale bar represents 20 μm

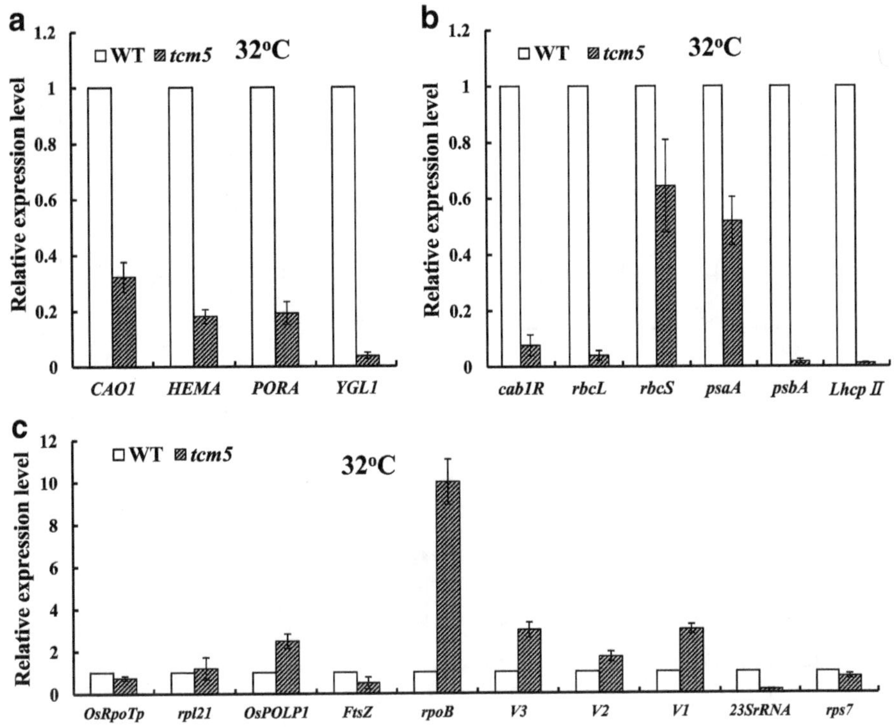

Fig. 7 qPCR analysis of those genes associated with Chl biosynthesis, photosynthesis and chloroplast development at 32 °C; **a**, **b**, **c** Expression levels of genes related to Chl biosynthesis (**a**), photosynthesis (**b**) and chloroplast development (**c**) in WT and *tcm5* mutant in the 3rd leaves, respectively. The relative expression level of each gene in WT and mutant was analyzed by qPCR and normalized using the *OsActin* as an internal control. Data are means ± SD ($n = 3$)

encoding the small subunit of Rubisco, *psaA* and *psbA* for the reaction center polypeptides in photosystems and *LhcpII* encoding light harvesting complex protein in PS II, and found that all these genes were obviously suppressed in *tcm5* mutants especially PSII-associated *psbA* and *LhcpII*, consistent with the low photochemical efficiency and photosynthesis efficiency in *tcm5* mutants under high temperatures (Additional file 2: Table S1).

For chloroplast development-associated transcripts, we investigated the expressions of *V3 (RNRL)* encoding the large subunits of ribonucleotide reductase (Yoo et al. 2009), *V2*, encoding plastid/mitochondrial guanylate kinase (pt/mt GK) (Sugimoto et al. 2007), *V1(NUS1)*, encoding protein NUS1 (Kusumi et al. 2011), *OsRpoTp*, encoding NEP core subunits (Hiratsuka et al. 1989), and *rpoB* encoding one PEP core subunit (Inada et al. 1996; Kusumi et al. 2011) and *OsPoLP1*, encoding one plastidal DNA polymerase (Takeuchi et al. 2007), and *FtsZ*, encoding a component of the plastid division machinery (Vitha et al. 2001). Also we examined the transcript levels of ribosome-associated genes, *23S rRNA* (23S ribosomal RNA), encoding one component of the plastid translation machinery, and *rpl21* and *rps7*, encoding the large and small subunits ribosomal protein, respectively. Resultantly, with the exception of the not significant effects on *FtsZ*, *OsRpoTp*, *rps7* and *rpl2* in the *tcm5* mutants grown at 32 °C, other genes (*rpoB*, *V3*, *V2*, *V1*, *OsPoLP1*, and *23S rRNA*) were significantly affected (Fig. 7c), resulting in abnormal chloroplast (Fig. 2b). Overall, the *tcm5* mutation affected the transcriptional levels of some genes associated with not only Chl biosynthesis, photosynthesis but also the chloroplast development under high temperatures. By contrast, in the *tcm5* mutants grown at 20 °C, all transcripts of the affected genes at 32 °C (Fig. 8) were, at least, partly recovered to WT levels. Accordingly, the differentially expressed levels of so many key genes mentioned above under low or high temperatures might lead to the thermo-sensitivity for leaf-color in the *tcm5* mutant.

Discussion

In this study, we reported a rice nuclear-encoded Deg proteases protein *TCM5*, which is required for chloroplast development under high temperatures. Its disruption led to abnormal chloroplast development and albino phenotype of rice at high temperatures, resulting from abnormally expressed transcripts of those genes associated with chlorophyll biosynthesis, photosynthesis and chloroplast development. Our work demonstrated

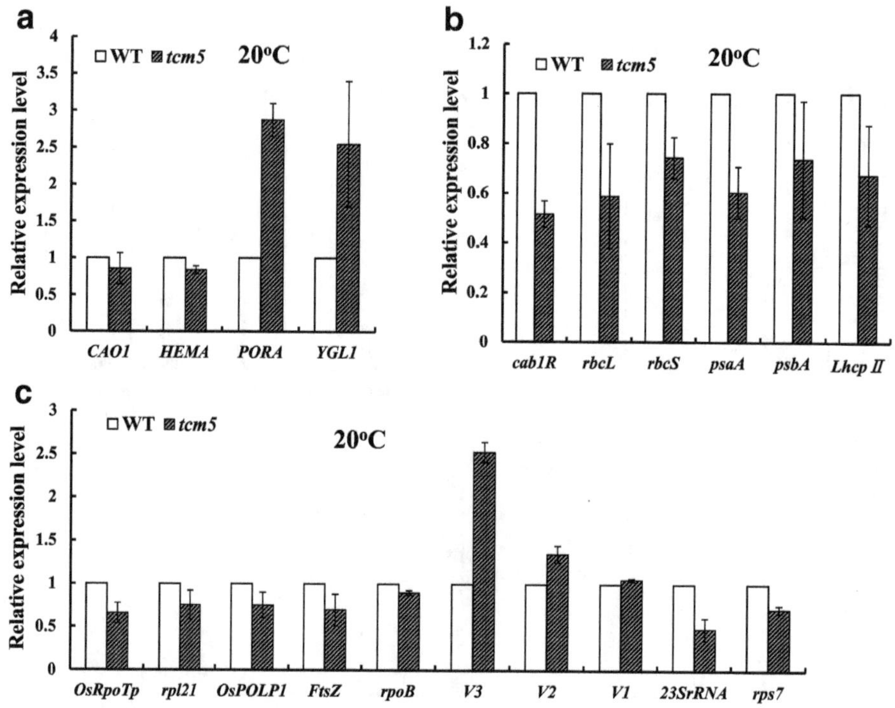

Fig. 8 qPCR analysis of those genes associated with Chl biosynthesis, photosynthesis and chloroplast development at 20 °C. **a**, **b**, **c** Expression levels of genes related to Chl biosynthesis (**a**), photosynthesis (**b**) and chloroplast development (**c**) in WT and *tcm5* mutants in the 3rd leaves, respectively. The relative expression level of each gene in WT and mutant was analyzed by qPCR and normalized using the *OsActin* as an internal control. Data are means ± SD (*n* = 3)

that *TCM5* plays an important role in chloroplast development under high temperatures.

TCM5 is Essential for Chloroplast Development under High Temperatures

The chloroplast is a semi-autonomous organelle, which contains about 100 genes, more than 3000 proteins function within it (Leister 2003). Thus, tightly coordinated gene expression between the plastid and nuclear genomes is essential for chloroplast development. The mutation of those genes could result in chlorophyll deficient/chloroplast defects in plants. At present, over 70 chlorophyll/chloroplast deficient mutants in rice have been identified (for review, see Kurata et al. 2005). Among them *v1, v2, v3, v5, v7, chs1, chs2, chs3, chs4, and chs5* mutants have been identified to be low temperature sensitive (Kusumi and Iba 2014). Recently, two genes, *TCD9* encoding α subunit of chaperonin protein 60 (Jiang et al. 2014) and *OsV4* encoding a PPR protein (Gong et al. 2014), essential for chloroplast development at low temperature were identified in rice. In contrast, six high-temperature sensitive rice mutants were reported, of which five mutants (Yatou and Cheng 1989) exhibited the chlorotic type over 35 °C, and *cde1(t)* mutant (Liu et al. 2007) displayed the chlorotic type over 26 °C, most likely resulted from the mutation of *OsGluRS*. In this study, the *tcm5* mutants displayed the abnormal chloroplast and even could not survive after the 4–5 leaf stage at 32 °C, but are normal under 20 °C. Judging from these results, there exists different shift sensitive-temperatures and molecular bases for chloroplast development under high temperatures in rice. In addition, the impaired chloroplasts (Fig. 2b) and the abnormal transcripts of genes for chloroplast development (Fig. 7) were observed in *tcm5* mutants at high temperatures, and those affected genes under high temperatures, at least, in part could recover to the comparable levels to WT plants at low temperatures (Fig. 8). These results can explain the changes of chloroplast development and leaf color between 20 and 32 °C (Figs. 1 and 2). Additionally, the highly induced mRNA accumulation (Fig. 3f) of *TCM5* at high temperatures showed that *TCM5* was temperature-regulated gene, consistent with the previous results that Deg proteases are required for the survival of *Escherichia coli* at high temperatures (Lipinska et al. 1988, 1990; Strauch and Beckwith 1988). Our study showed that the *TCM5* was constitutively expressed with high levels of transcripts in all green tissues (Fig. 5) and TCM5 protein distributes throughout the entire chloroplast (Fig. 6a), showing the functions of TCM5 in chloroplast. Also, the lethality of *tcm5* mutants may be due to the block of chloroplast development under high temperatures (Fig. 1b). These results strongly confirm the notion that rice *TCM5* is needed for chloroplast development under high temperatures. However, the reasons why the abnormal chloroplast/albino phenotypes of rice, such as *tcm5* mutant in this study and *cde1(t)* mutant, occur under solely higher temperatures have not been well documented (Yatou and Cheng 1989; Liu et al. 2007). A possible explanation in this study is that the *TCM5* function possibly is not prerequisite under lower temperatures, but is essential under higher temperatures for chloroplast development. This explanation was strongly supported by the results from transcriptional analysis (Fig. 3f) that (a): all low expressions and the existence of no discriminated difference at 20 °C regardless of WT and *tcm5* mutant; (b): high expressions at 32 °C regardless of WT and *tcm5* mutant despite of the higher level in the mutant caused by feedback effects.

As stated above, the process of the chloroplast development from proplastids to mature chloroplasts can be divided into three steps (Jarvis and López-Juez 2013; Kusumi et al. 1997; Kusumi and Iba, 2014). The functions of *OsPOLP1*, *FtsZ* and *V3(RNRL)* encoding plastidal DNA polymerase, a component of plastid division machinery and the large subunits of ribonucleotide reductase respectively (Vitha et al. 2001; Takeuchi et al. 2007; Yoo et al. 2009), are known to involve in the first step. In addition, *OsRpoTp*, *V2*, and *rpoB*, encoding NEP, plastidal guanylate kinase, and PEP β core subunit, respectively (Hiratsuka et al. 1989; Kusumi et al. 2004; Sugimoto et al. 2007) are involved in the second step. In view of not significant decreases of transcripts of key genes mentioned above (Fig. 7), *TCM5* could not function in the first and second steps. In addition, the transcripts of all PEP-transcribed plastid genes (*cab1R*, *rbcL*, *psaA*, *psbA*, *LhcpII*) and nuclear gene *rbcS* involved in the third step were severely suppressed under high temperatures in *tcm5* mutants (Fig. 7), showing that *TCM5* functions the third step. Also, the significantly decreased accumulation of *23S rRNA* (Fig. 7c) in *tcm5* mutants under high temperatures indicated that *TCM5* affects the plastid translation machinery in chloroplasts. These results revealed that *TCM5* is necessary in the process of chloroplast development under high temperatures.

TCM5 may be Involved in PSII Function under High Temperatures

The Deg/HtrA family is present in nearly all organisms from bacteria to human and plants. It was known that Deg proteases are essential for the survival of *Escherichia coli* at high temperatures (Lipinska et al. 1988, 1990; Strauch and Beckwith 1988). The chloroplast-targeted Deg proteases were reported to be involved in the degradation of damaged photosynthetic proteins, especially the PSII reaction center D1 protein (for review, see Schuhmann et al. 2012, Kato et al. 2012). In Arabidopsis, AtDeg1 acts as a chaperone, assisting in the assembly of PSII dimers and supercomplexes (Sun et al. 2010b). More interestingly, Arabidopsis *deg* mutants *deg5*, *deg7*, and *deg8* all showed decreased growth including chloroplast development, compared with the WT plants under high light, however, have no obvious differences under normal conditions, all AtDeg5, AtDeg7 and AtDeg8 participate in the repair of PSII after photo inhibition, and *AtDeg5*, *AtDeg7*, and *AtDeg8* were all high-light-dependent genes (Sun et al. 2010). Regrettably, any Arabidopsis *deg10* mutant whose AtDeg10 shares the highest homology to TCM5 (74.7 % amino acid identity, Additional file 1: Figure S4) has not reported yet, still less the function. Judging from the recovered normal chloroplast at low temperatures (<24 °C) in *tcm5* mutants, it could be concluded that *TCM5* is functionally required under high temperatures. The significant decreases of *Fv/Fm*, reflecting the maximum potential capacity of the PSII photochemical reactions, and *ΦPSII*, representing the actual photosynthesis efficiency, in *tcm5* mutants at 32 °C and the similar degrees as WT plants at 20 °C showed that the malfunction of TCM5 barely affected the photosynthesis efficiency under high temperatures. In addition to the similar levels as WT plants at low temperatures (Fig. 8), the transcripts of *psbA* (>50-fold) and *LhcpII* (>80-fold), both encoding the PSII core proteins (D1, Lhcp II), were more significantly suppressed compared to those of *psaA* (<2-fold) encoding the PSI core protein (PsaA) in *tcm5* mutants at high temperatures (Fig. 7b). Thus, it could be reasonably speculated that the TCM5 disruption maybe hinder the PSII function, ultimately leading to the significantly decreased PSII activities (*Fv/Fm* and *ΦPSII*) under high temperatures (32 °C), which was consistent with the previous results of high correlation between thermotolerance and photosystem II activity (Chen et al. 2014). In addition, Arabidopsis Deg7 (AtDeg7, At3g03380), with the low homology to TCM5 (16 % amino acid identity) (data not shown), involving in the primary cleavage of photodamaged D1, D2, CP47, and CP43 in PSII complexes, was reported to repair the PSII function under high-light conditions (Sun et al. 2010a). Considering that *TCM5* is high-temperature-dependent gene, different from high-light-dependent *AtDeg7* (Sun et al. 2010a), we might speculate that, at least, the mechanisms of temperature response are different among them. In addition, TCM5/OsDeg10 was solely located in thylakoid membrane (Fig. 6), not only different from AtDeg1, 5, 8 (on the stromal side) and, AtDeg2, 7(on the lumenal side), but also AtDeg10 (on both chloroplasts and mitochondrion) (Tanz et al. 2014), which clearly showed the different functions even between OsDeg10/TCM5 and AtDeg10.

Taken together, these results show that the *tcm5* mutation affected the third stage of chloroplast development under high temperatures, and the activation of PSII-associated genes is severely weakened, resulting in the suppression of chloroplast development, Chl-biosynthesis and photosynthesis, finally leading to the death under high temperatures. To our knowledge, this is the first experimental evidence that there exists a unique rice Deg/HtrA protease protein essential for chloroplast development under high temperatures. As for *TCM5*, how to participate in chloroplast development and maintain PSII function under high temperatures is needed in the further study.

Conclusions

The TCM5 encodes chloroplast-targeted Deg protease protein which is important for chloroplast development and the maintenance of PSII function and its disruption would lead to a defective chloroplast and affected expression levels of genes associated with chloroplast development and photosynthesis of rice plants under high temperatures.

Methods

Plant Materials and Growth Conditions

The thermo-sensitive chlorophyll-deficient mutant, *tcm5*, used in this study, was obtained from *japonica* rice variety, Jiahua 1(WT), and induced by ^{60}Co gamma-radiation in 2006. The F_2 genetic mapping population was generated from a cross between Pei'ai 64S (*indica*) and the *tcm5* mutant. Rice plants were cultivated in rice paddy fields in Shanghai (summer-autumn season, temperate climate), China, following conventional methods. For phenotypic characterization, chlorophyll (Chl) and carotenoid content measurement and RNA extraction, the wild-type (WT) and *tcm5* seedlings were grown in growth chambers. In the F_2 population, thermo-sensitive Chl deficient albino plants were selected for genetic mapping.

Phenotype Characterization, Chlorophyll (Chl) and Carotenoid Measurements

The rice seeds were grown in the growth chambers with 12 h-dark/12 h-light at four temperatures (20, 24, 28, 32 °C) and the light intensity of 120 μmol of photons m^{-2} s^{-1}. The extracts were obtained from 100 mg of the third fresh leaves at the 3-leaf stage seedlings and homogenised in 10 ml of 100 % acetone. Spectrophotometric quantification was carried out in BECKMANCOULTER-DU720. Total Chl, Chla, Chlb and carotenoid contents were determined according to the methods of Arnon (1949) and Alan (1994). In 2010, WT and *tcm5* plants were grown in experimental rice fields of Shanghai Normal University by conventional methods. Leaf chlorophyll SPAD values (Additional file 1: Figure S1) by means of CHLOROPHYLL METER (SPAD-502, Minolta Co., Ltd, Japan), which can provide a simple, quick portable and non-destructive method for estimating leaf Chl content (Peng et al. 1993; Turner and Jund 1991; Dwyer et al. 1991) were measured every week from transplanting (summer) to heading (autumn). Finally, agronomic traits of rice plants (Additional file 1: Figure S2) were measured at maturity.

Chlorophyll Fluorescence Analysis and Transmission Electron Microscopy

Rice seedlings were grown at 20 °C or 32 °C in growth chambers. Chlorophyll fluorescence analyses for the third leaves at the 3-leaf stage were performed with a PAM-2000 portable chlorophyll fluorometer (MINI-PAM; Walz; http://walz.com). The variables *Fo, Fm, Fv,* the *Fv/Fm* ratio and *ΦPSII* were measured and calculated according to Meurer et al. (1996) after seedlings were dark-adapted for 20 min. Meanwhile, transverse sections of top leaves were sampled from the 3-leaf-stage seedlings grown at 20 °C or 32 °C. Samples were fixed in 4 % glutaraldehyde buffer, 2.5 % glutaraldehyde and 1 % osmic acid phosphate buffer at 4 °C for 5 h after vacuum. After staining with uranyl acetate, tissues were further dehydrated in an ethanol series and finally embedded in Spurr's medium prior to ultrathin sectioning. Samples were stained again and examined with a Hitachi-7650 transmission electron microscope.

Positional Cloning of *TCM5*

For genetic analysis, F_2 population from the cross between *tcm5* and Pei'ai 64S was used for fine mapping of the *TCM5* locus. We adopted 53 SSR primers based on data in Gramene (http://www.gramene.org). New SSR, InDel and CAPS markers (Additional file 1: Table S3) were developed based on the entire genomic sequences of the *japonica* Nipponbare variety (Goff et al. 2002) and the *indica* variety 9311 (Yu et al. 2002). The candidate genes' function and full length cDNA were acquired using TIGR (http://rice.plantbiology.msu.edu/cgi-bin/gbrowse/rice/) and KOME (http://cdna01.dna.affrc.go.jp/cDNA/index.html) respectively.

RT-PCR and Quantitative Real-time PCR

To understand the tissue expression pattern of *TCM5*, we extracted wild type RNA from young roots, young stems, young leaves, flag-leaves and young panicles with TRIzol Reagent (Invitrogen; http://www.invitrogen.com) and DNase I treated RNA using an RNeasy kit (Qiagen; http://www.qiagen.com) following the manufacturer's instructions. The first-strand cDNA was synthesized with the Revert-Aid first-strand cDNA synthesis kit (Toyobo; http://www.toyobo.co.jp) following the manufacturer's instructions. *OsActin* were internal controls. The specific primers are listed in Additional file 1: Table S4. RT-PCR analysis was carried out to assess the transcriptional levels.

For transcriptional expression analysis of *TCM5* and 20 genes (*PORA, CAO1, YGL1, cab1R, rbcL, rbcS, psaA, psbA, Lhcp II, V3 (RNRL), V1, V2, OsRpoTp, OsPoLp1, rpoB, 23SrRNA, rps7, rps21, FtsZ*) associated with Chl-biosynthesis, chloroplast development and photosynthesis, total RNA was obtained from the third fresh leaves of WT and *tcm5* seedlings grown at 20 °C or 32 °C. Quantitative real-time PCR amplifications were carried out with ABI7500 Real-Time PCR System (Applied Biosystems; http://www.appliedbiosystems.com), and the relative quantification of gene expression data was analyzed as described by Livak and Schmittgen (2001). The data set was normalized using *OsActin* as a control. The specific primers were listed in Additional file 1: Table S4.

Complementation Analysis

For complementation analysis, cDNA clone (AK073199) plasmid vector (J0133023K08) (http://cdna01.dna.affrc.go.jp) carrying the full-length *TCM5* (*LOC_Os05g34460*) cDNA sequence, was purchased from National Institute of Agrobiological Sciences of Japan in 2011. Then, the 2426 bp DNA fragment, carrying the full-length CDS (1845 bp) of *TCM5*, was amplified using the primers 5′- CCGGAATTCGTCGACCTCTAT -3′ (*Eco*RI) and 5′-TGGCCGGATCCGGACTGTTT-3′ (*BamH*I). The underlined sequences represent cleavage sites of the enzymes. Then, the resulted DNA fragment was cloned into the pMD18-T vector (TaKaRa). Following sequence verification, the fragments were cloned into the pCAMBIA1301 binary vector (CAMBIA; www.cambia.org.au). Finally, the pCAMBIA1301:CaMV35S:TCM5-cDNA plasmids were transferred into *Agrobacterium tumefaciens* EHA105 and introduced into the *tcm5* mutants by Agrobacterium-mediated transformation (Hiei et al. 1994). In this study, transgenic plants were generated from recovered resistant calli by the procedure of Lu et al. (2001).

Subcellular Localization of GFP Proteins

To investigate the subcellular localization of TCM5 protein, the cDNA fragment encoding the N-terminal region (amino acids 1–135) of TCM5 was amplified from cDNA clone (AK073199) using the primer pairs (pF 5′-GGGGTACCATGCTCGCCTCCGTCC-3′ (*Kpn*I) and pR 5′- CGGCCATGGCGAGTCCAGCGCCAGCT-3′ (*Nco*I) the underlined sequence represents cleavage sites of enzymes) and ligated into the pMON530-GFP vector, in frame with GFP. The resultant pMON530-TCM5-GFP plasmids were introduced into tobacco (*Nicotiana tabacum*) leaves and co-cultured at 25 °C for 2 days. At the same time, the pMON530-GFP empty carrier was used as a control. Then the GFP fluorescences in tobacco cells were observed using a Zeiss confocal laser scanning microscope (LSM 5 PASCAL; http://www.zeiss.com).

Sequence and Phylogenetic Analysis

Gene prediction and structure analysis were performed using TIGR (http://rice.plantbiology.msu.edu/cgi-bin/gbrowse/rice/).The chloroplast transit peptides (CTP) was predicted by the ChloroP program (http://www.cbs.dtu.dk/services/TargetP/). The full-length amino acid sequences of the TCM5 and homologs identified via BLAST search were aligned with the MUSCLE tool (Edgar 2004) using the default parameters. The phylogenetic tree was constructed and tested by MEGA6 (http://www.megasoftware.net) based on the neighbor-joining method.

Additional files

Additional file 1: Figure S1. The change of chlorophyll contents (SPAD values) in *tcm5* and WT plants from the 1st (summer) to 12th weeks (heading date, autumn) after transplanting (2010, Shanghai, China). **Figure S2.** Comparative agronomic characteristics of *tcm5* and wild type of rice plants grown under field condition (2010, Shanghai, China). *PH, plant height; PN, panicle number per plant; GW: 1000-grain weight (g); GN, grains per panicle.* **Figure S3.** Phylogenic analysis of rice Deg proteases inculding 10 Deg proteases and 6 Deg-like proteases proteins (Schuhmann et al., 2012) in the rice genome. **Figure S4.** Phylogenic analysis of TCM5 with all Arabidopsis Deg proteases.

Additional file 2: Table S1. The fluorescence induction parameters. **Table S2.** Genetic segregation analysis of tcm5 mutants in the F2 population. **Table S3.** The PCR-based molecular markers designed for fine mapping. **Table S4.** Markers designed for RT-PCR.

Competing interests

The authors declare that they have no competing interests.

Authors' contributions

DL and YD provided the mutant rice seeds. DL, YD, KZ, JC and JZ generated F_2 and F_3 seeds for genotyping and phenotyping. KZ, JZ, DL, HZ and JC performed the experiments of phenotype assays and molecular analysis. KZ, DL, YD, ST and JX designed and discussed the research. DL, YD, KZ and JX wrote the manuscript. All authors approved the manuscript.

Acknowledgments

We sincerely thank Dr. Youbin Xiang (Stowers Institute for Medical Research, USA) for her critical reading of the manuscript. We thank DNA Bank of National Institute of Agrobiological Sciences, Japan, which kindly offered the rice cDNA clone (AK073199). We are grateful to Dr. Zhongnan Yang from Shanghai Normal University for her skillful technical assistances. The project was financially supported by Natural Science Foundation of China (No. 30971552), Shanghai Municipal Education Commission (No. 09YZ167, No.J50401), Shanghai Municipal Science and Technology Commission (No. 9391912300, 10DZ2271800,12ZR1422000), and the Scientific Program of Shanghai Normal University (SK20159).

Author details

[1]Development Center of Plant Germplasm Resources, College of Life and Environmental Sciences, Shanghai Normal University, Shanghai 200234, China. [2]Shanghai Institute for Biological Sciences, Chinese Academy of Sciences, Shanghai 200032, China. [3]Institute of Crop Sciences, Chinese Academy of Agricultural Sciences, 12 South Zhong-Guan Cun Street, Beijing 100081, China. [4]Present address: Agricultural Faculty, Hokkaido University, Sappro 060-0817, Japan.

References

Alan RW (1994) The spectral determination of chlorophylls a and b, as well as total carotenoids, using various solvents with spectrophotometers of different resolution. J Plant Physiol 144:307–311

Arnon DI (1949) Copper enzymes in isolated chloroplasts. polyphenoloxidase in beta vulgaris. Plant Physiol 24:1–15

Chen K, Sun XY, Amombo E, Zhu Q, Zhao ZG, Liang Chen L, Xu QG, Fu JM (2014) High correlation between thermotolerance and photosystem II activity in tall fescue. Photosynth Res 122:305–314

Clausen T, Southan C, Ehrmann M (2002) The HtrA family of proteases: implications for protein composition and cell fate. Mol Cell 10:443–455

Dwyer LM, Tollenaar M, Houwing L (1991) A nondestructive method to monitor leaf greenness in corn. Can J Plant Sci 71:505–509

Edgar RC (2004) MUSCLE: multiple sequence alignment with high accuracy and high throughput. Nucleic Acids Res 32:1792–1797

Emanuelsson O, Nielsen H, Brunak S, Heijn GV (2000) Predicting subcellular localization of proteins based on their N-terminal amino acid sequence. J Mol Biol 300:1005–1016

Garcia-Lorenzo M, Sjödin A, Jansson S, Funk C (2006) Protease gene families in Populus and Arabidopsis. BMC Plant Biol 6:30

Goff SA, Ricke D, Lan TH, Presting G, Wang R, Dunn M, Glazebrook J, Sessions A, Oeller P, Varma H, Hadley D, Hutchison D, Martin C, Katagiri F, Lange BM, Moughamer T, Xia Y, Budworth P, Zhong J, Miguel T, Paszkowski U, Zhang S, Colbert M, Sun WL, Chen L, Cooper B, Park S, Wood TC, Mao L, Quail P, Wing R, Dean R, Yu Y, Zharkikh A, Shen R, Sahasrabudhe S, Thomas A, Cannings R, Gutin A, Pruss D, Reid J, Tavtigian S, Mitchell J, Eldredge G, Scholl T, Miller RM, Bhatnagar S, Adey N, Rubano T, Tusneem N, Robinson R, Feldhaus J, Macalma T, Oliphant A, Briggs S (2002) A draft sequence of the rice genome (*Oryza sativa* L. ssp. *japonica*). Science 296:92–100

Gong X, Su Q, Lin D, Jiang Q, Xu J, Zhang J, Teng S, Dong Y (2014) The rice *OsV4* encoding a novel pentatricopeptide repeat protein is required for chloroplast development during the early leaf stage under cold stress. J Integr Plant Biol 56:400–410

Hajdukiewicz PT, Allison LA, Maliga P (1997) The two RNA polymerases encoded by the nuclear and the plastid compartments transcribe distinct groups of genes in tobacco plastids. EMBO J 16:4041–4048

Helm M, Luck C, Prestele J, Hierl G, Huesgen PF, Fröhlich T, Arnold GJ, Adamska I, Görg A, Lottspeich F (2007) Dual specificities of the glyoxysomal/peroxisomal processing protease Deg15 in higher plants. Proc Natl Acad Sci U S A 104:11501–11506

Hiei Y, Ohta S, Komari KT (1994) Efficient transformation of rice (*Oryza sativa* L.) mediated by Agrobacterium and sequence analysis of the boundaries of the T-DNA. Plant J 6:271–282

Hiratsuka J, Shimada H, Whittier R, Ishibashi T, Sakamoto M, Mori M, Kondo C, Honji Y, Sun CR, Meng BY (1989) The complete sequence of the rice (*Oryza sativa*) chloroplast genome: intermolecular recombination between distinct tRNA genes accounts for a major plastid DNA inversion during the evolution of the cereals. Mol Gen Gene 217:185–194

Huesgen PF, Schuhmann H, Adamska I (2005) The family of Deg proteases in cyanobacteria and chloroplasts of higher plants. Physiol Plant 123:413–420

Inada H, Kusumi K, Nishimura M, Iba K (1996) Specific expression of the chloroplast gene for RNA polymerase (*rpoB*) at an early stage of leaf development in rice. Plant Cell Physiol 37:229–232

Itzhaki H, Naveh L, Lindahl M, Cook M, Adam Z (1998) Identification and characterization of DegP, a serine protease associated with the luminal side of the thylakoid membrane. J Biol Chem 273:7094–7098

Jarvis P, López-Juez E (2013) Biogenesis and homeostasis of chloroplasts and other plastids. Nat Rev Mol Cell Biol 14:787–802

Jiang Q, Mei J, Gong XD, Xu JL, Zhang JH, Teng S, Lin D, Dong Y (2014) Importance of the rice *TCD9* encoding alpha subunit of chaperonin protein60 (Cpn60α) for the chloroplast development during the early leaf stage. Plant Sci 215–216:172–179

Kato Y, Sakamoto W (2010) New insights into the types and function of proteases in plastids. Int Rev Cell Mol Biol 280:185–218

Kato Y, Sun X, Zhang LX, Sakamoto W (2012) Cooperative D1 degradation in the photosystem II repair mediated by chloroplastic proteases in Arabidopsis. Plant Physiol 159:1428–1439

Kolmar H, Waller P, Sauer R (1996) The DegP and DegQ periplasmic endoproteases of *Escherichia coli*: specificity for cleavage sites and substrate conformation. J Bacteriol 178:5925–5929

Krause GH, Weis E (1991) Chlorophyll fluorescence and photosynthesis: the basics. Annu Rev Plant Physiol Plant Mol Biol 42:313–349

Kurata N, Miyoshi K, Nonomura KI, Yamazaki Y, Ito Y (2005) Rice mutants and genes related to organ development, morphogenesis and physiological traits. Plant Cell Physiol 46:48–62

Kusumi K, Iba K (2014) Establishment of the chloroplast genetic system in rice during early leaf development and at low temperatures. Front Plant Sci 5:386–391

Kusumi K, Mizutani A, Nishimura M, Iba K (1997) A virescent gene *V1* determines the expression timing of plastid genes for transcription/translation apparatus during early leaf development in rice. Plant J 12:1241–1250

Kusumi K, Yara A, Mitsui N, Tozawa Y, Iba K (2004) Characterization of a rice nuclear-encoded plastid RNA polymerase gene *OsRpoTp*. Plant Cell Physiol 45:1194–1201

Kusumi K, Sakata C, Nakamura T, Kawasaki S, Yoshimura A, Iba K (2011) A plastid protein NUS1 is essential for build-up of the genetic system for early chloroplast development under cold stress conditions. Plant J 68:1039–1050

Leister D (2003) Chloroplast research in the genomic age. Trends Genet 19:47–56

Lipinska B, Sharma S, Georgopoulos C (1988) Sequence analysis and regulation of the HtrA gene of *Escherichia coli*: a sigma 32-independent mechanism of heat-inducible transcription. Nucleic Acids Res 16:10053–10067

Lipinska B, Zylicz M, Georgopoulos C (1990) The HtrA (DegP) protein, essential for *Escherichia coli* survival at high temperatures, is an endopeptidase. J Bacteriol 172:1791–1797

Liu W, Fu Y, Hu G, Si H, Zhu L, Wu C, Sun Z (2007) Identification and fine mapping of a thermo-sensitivechlorophyll deficient mutant in rice (*Oryza sativa* L.). Planta 226:785–795

Livak KJ, Schmittgen TD (2001) Analysis of relative gene expression data using real time quantitative PCR and the method. Methods 25:402–408

Lu HJ, Zhou XR, Gong ZX, Upadhyaya NM (2001) Generation of selectable marker-free transgenic rice using double right-border (DRB) binary vectors. Aust J Plant Physiol 28:241–248

Meurer J, Meierhoff K, Westhoff P (1996) Isolation of high-chlorophyll fluorescence mutants of *Arabidopsis thaliana* and their characterisation by spectroscopy, immunoblotting and northern hybridisation. Planta 198:385–396

Pendle AF, Clark GP, Boon R, Lewandowska D, Lam YW, Andersen J, Mann M, Lamond AI, Brown JW, Shaw PJ (2005) Proteomic analysis of the Arabidopsis nucleolus suggests novel nucleoli functions. Moll Boil Cell 16:260–269

Peng S, Garcia FV, Lasa RC, Classman KG (1993) Adjustment for specific leaf weight improves chlorophyll meter's estimate of rice leaf nitrogen concentration. Agron J 85:987–990

Pfalz J, Pfannschmidt T (2013) Essential nucleoid proteins in early chloroplast development. Trends Plant Sci 18:186–194

Santis-Maciossek GD, Kofer W, Bock A, Schoch S, Maier RM, Wanner G, Rüdiger W, Koop HU (1999) Targeted disruption of the plastid RNA polymerase genes *rpoA*, B and C1: molecular biology, biochemistry and ultrastructure. Plant J 18:477–489

Schuhmann H, Huesgen PF, Gietl C, Adamska I (2008) The DEG15 serine protease cleaves peroxisomal targeting signal 2-containing proteins in *Arabidopsis thaliana*. Plant Physiol 148:1847–1856

Schuhmann H, Huesgen PF, Adamska I (2012) The family of Deg/HtrA proteases in plants. BMC Plant Biol 12:52

Strauch KL, Beckwith J (1988) An *Escherichia coli* mutation preventing degradation of abnormal periplasmic proteins. Proc Natl Acad Sci U S A 85:1576–1580

Sugimoto H, Kusumi K, Noguchi K, Yano M, Yoshimura A, Iba K (2007) The rice nuclear gene, *VIRESCENT2*, is essential for chloroplast development and encodes a novel type of guanylate kinase targeted to plastids and mitochondria. Plant J 52:512–527

Sun X, Fu T, Chen N, Guo J, Ma J, Zou M, Lu C, Zhang L (2010a) The stromal chloroplast Deg7 protease participates in the repair of photosystem II after photoinhibition in Arabidopsis. Plant Physiol 152:1263–1273

Sun X, Ouyang M, Guo J, Ma J, Lu C, Adam Z, Zhang L (2010b) The thylakoid protease Deg1 is involved in photosystem-II assembly in *Arabidopsis thaliana*. Plant J 62:240–249

Takeuchi R, Kimura S, Saotome A, Sakaguchi K (2007) Biochemical properties of a plastidial DNA polymerase of rice. Plant Mol Biol 64:601–611

Tanz SK, Castleden I, Hooper CM, Small I, Millar AH (2014) Using the SUBcellular database for Arabidopsis proteins to localize the Deg protease family. Front Plant Sci 5:396

Tripathi LP, Sowdhamini R (2006) Cross genome comparisons of serine proteases in Arabidopsis and rice. BMC Genomics 7:200

Turner FT, Jund MF (1991) Chlorophyll meter to predict nitrogen topdress requirement for semidwarf rice. Agron J 83:926–928

Vitha S, Mcandrew RS, Osteryoung KW (2001) *FtsZ* ring formation at the chloroplast division site in plants. J Cell Biol 153:111–120

Walsh NP, Alba BM, Bose B, Gross CA, Sauer RT (2003) OMP peptide signals initiate the envelope-stress response by activating DegS protease via relief of inhibition mediated by its PDZ domain. Cell 113:61–71

Yatou O, Cheng XY (1989) Temperature sensitive chlorophyll mutations. Rice Genet Newsletter 6:131

Yoo Y, Cho SH, Sugimoto H, Li J, Kusumi K, Koh H, Iba K, Paek N (2009) Rice *Virescent3* and *Stripe1* encoding the largeand small subunits of ribonucleotide reductase are required for chloroplast biogenesis during early leaf development. Plant Physiol 150:388–401

Yu J, Hu S, Wang J, Wong GK, Li S, Liu B, Deng Y, Dai L, Zhou Y, Zhang X, Cao M, Liu J, Sun J, Tang J, Chen Y, Huang X, Lin W, Ye C, Tong W, Cong L, Geng J, Han Y, Li L, Li W, Hu G, Huang X, Li W, Li J, Liu Z, Li L et al (2002) A draft sequence of the rice genome (*Oryza sativa* L. ssp. *indica*). Science 296:79–92

3

Identification of novel major and minor QTLs associated with *Xanthomonas oryzae* pv. *oryzae* (African strains) resistance in rice (*Oryza sativa* L.)

Gustave Djedatin[1*], Marie-Noelle Ndjiondjop[2], Ambaliou Sanni[5], Mathias Lorieux[4], Valérie Verdier[3†] and Alain Ghesquiere[4†]

Abstract

Background: *Xanthomonas oryzae* pv. *oryzae* (*Xoo*) is the causal agent of Bacterial Leaf Blight (BB), an emerging disease in rice in West-Africa which can induce up to 50 % of yield losses. So far, no specific resistance gene or QTL to African *Xoo* were mapped. The objectives of this study were to identify and map novels and specific resistance QTLs to African *Xoo* strains.

Results: The reference recombinant inbred lines (RIL) mapping population derived from the cross between IR64 and Azucena was used to investigate *Xoo* resistance. Resistance to African and Philippine *Xoo* strains representing different races was assessed on the RIL population under greenhouse conditions. Five major quantitative trait loci (QTL) for resistance against African *Xoo* were located on different chromosomes. Loci on chromosomes 1, 7, 9, 10 and 11 explained as much as 13 %, 37 %, 13 %, 11 % and 15 % of resistance variation, respectively. A major novel QTL located on chromosome 7 explained 37 % of the phenotypic variance to the African *Xoo* corresponding to race A3 whereas that on chromosome 11 is effective to all African races tested. Together with genes and QTLs for resistance to bacterial blight previously described, the QTLs described here were mapped onto the reference *O. sativa* subs japonica (var. Nipponbare) physical map.

Conclusion: We characterized new resistance QTLs. While some co-localize with known resistance genes/QTLs to Asian strains, others are specific to African strains. We result with new information on genes and QTLs for resistance to bacterial blight that will be useful for controlling the disease.

Keywords: Molecular mapping, QTL, disease resistance, *Xanthomonas oryzae* pv. *oryzae*, *Oryza sativa*

Background

Rice is a staple food for much of the world's population, including that of sub-Saharan Africa. Population growth and migration have recently lead to a rapid growth in rice consumption in dozens of African countries. With the recent expansion and intensification of rice cultivation in Africa, rice diseases have concomitantly increased in most African rice growing areas (Sere et al. 2005; Traoré et al. 2009). Bacterial Blight (BB), caused by *Xanthomonas oryzae* pv. *oryzae* (hereafter, *Xoo*), is one of the most serious rice bacterial disease in Africa. The disease is prevalent in irrigated and rain-fed lowland rice growing areas. BB was first reported in Mali in 1979 and later in Senegal, Niger, Nigeria, Gabon, Mauritania, Benin, Burkina and Cameroon. Over the last decade a significant recurrence of this disease was observed in several regions in Africa (Reckhaus et al. 1983; Gonzalez et al. 2007; Basso et al. 2011; Verdier et al. 2011).

Resistance *(R)* genes are largely used in rice breeding programs in Asia to control BB disease. More than 30 *R* genes, which are given the prefix Xa for Xanthomonas, have been identified so far (for a review see Verdier et al.

* Correspondence: djedatingustave@yahoo.fr; gdjedatin@gmail.com
†Equal contributors
[1]Université Polytechnique d'Abomey, BP 2282 Abomey, Benin
Full list of author information is available at the end of the article

2011) among which a few are deployed in breeding programs. Important prerequisites to the deployment of *R* genes are as follows: 1) to have an intensive knowledge of *Xoo* population structure, race distribution and frequency 2) to determine the durability of resistance of *R* genes to be deployed.

Most of the BB *R* genes provide complete race-specific resistance to *Xoo* strains. Different combinations of *Xa4, xa5, Xa7, xa13* and *Xa21* have been incorporated in popular rice commercial varieties in different countries in Asia (Century et al. 1999; Singh et al. 2006; Swamy et al. 2006; Perez et al. 2008; Sundaram et al. 2009, Shanti 2010; Suh et al. 2013; Ruengphayak et al. 2015). Few examples indicate that some R genes used for controlling BB disease are overcome by virulent strains as shown in Korea with the resistant gene *Xa21* (Lee et al. 1999; Zhang et al. 2006). *Xa4* is a gene used for more than 30 years and has introgressed in high yielding varieties in Asia but has lost efficacy in many cultivated areas (Mew et al. 1992). Although durability of BB *R* genes is, in part, because mutation of *Xoo* to overcome R genes (Vera Cruz et al. 2000), recent field and laboratory studies have also shown the influence of temperature on the interactions of rice *R* gene with *Xoo*. High temperatures are conducive to BB disease, and most BB *R* genes, including *Xa4*, are less effective at controlling BB disease at high temperatures (Vera Cruz et al. 2000; Webb et al. 2010).

Xanthomonas oryzae (*Xo*) is a diverse species, with distinct phylogenetic lineages comprising US *Xo*, Asian *Xoo*, African *Xoo*, and *Xanthomonas oryzae* pv. *oryzicola* (*Xoc*) (Triplett et al. 2011; Hajri et al. 2012). Another lineage improperly named *Xanthomonas campestris* pv. *leersiae* (Xcl) comprises strains isolated on weeds (Wonni et al. 2014). Previous work highlighted differences in the race structure between Asian and African *Xoo* strains (Gonzalez et al. 2007). Virulence assays revealed three races (A1, A2 and A3) present in Mali, Burkina-Faso, Niger and Cameroon that do not represent any of the known *Xoo* races characterized in Asia so far (Gonzalez et al. 2007; Triplett et al. 2011). According to experiments conducted on BB isogenic lines (IRBB), BB resistance genes *Xa4, xa5* and *Xa7* provide resistance to some African *Xoo* strains (Gonzalez et al. 2007). Although in absence of a complete overview of *Xoo* race prevalence in Africa, we anticipated that *Xa4, xa5* and *Xa7* could provide resistance against strains of *Xoo* in Burkina-Faso, Cameroun and Niger. Despite the increasing importance of BB in Africa, little is known on the genetic determinism of resistance. *O. glaberrima* and *O. sativa* accessions were screened for resistance to African *Xoo* strains. The tropical japonica landrace Azucena is susceptible to all African *Xoo* strains. Few accessions, among them the indica cultivar IR64, are highly resistant to African *Xoo* strains. None of these accessions had the *xa5* or *Xa21* resistance alleles (Djedatin et al. 2011) suggesting that these accessions carry new resistance genes that could be good targets for R gene discovery and further deployment.

With the completion of genome sequences for *japonica* and *indica* rice (Kawahara et al. 2013) and for *O. glaberrima* (Wang et al. 2014a, b), it is essential to have a better picture of the different *Xa* resistance genes and QTLs characterized so far and their positions in the rice genome.

The objectives of this study are to:

1. Identify and analyse the genetic basis of rice resistance to African *Xanthomonas oryzae* pv. *oryzae* strains by developing a QTL approach using the reference mapping population made of recombinant inbred lines (RIL) derived from the cross between IR64 and Azucena.
2. Map novel and known bacterial blight resistance genes and QTLs to *Xoo* strains and analyze their colocalization on the reference Nipponbare physical map.

For the first time in history, we report on specific resistance QTLs to African *Xoo* strains. These QTLs will be used in breeding program to enhance rice genetic resistance to BB in Africa.

Results

Study of inheritance of BB resistance

The average lesion length induced by *Xoo* African strains on IR64 and Azucena are respectively: 0.25 ± 0.1 and 16.1 ± 2 cm with *Xoo* MAI1; 0.76 ± 0.2 and 22.36 ± 2.7 cm with *Xoo* BAI4; 4.07 ± 1 and 22.29 ± 3 cm with *Xoo* BAI3; 2.07 ± 0.5 and 26.29 ± 3.3 cm with *Xoo* NAI8. IR64 is observed to be highly resistant and Azucena highly susceptible to all virulent strains of African *Xoo* tested so far. The disease scores of the RILs range widely, from as low as 0.12 cm to as high as 30.05 cm, 0.14 to 32.2 cm, 0.2 to 32.6 cm and 0.44 to 37.9 cm with *Xoo* strains MAI1, BAI4, BAI3 and NAI8, respectively (Table 1).

Table 1 Lesion length induced by African and Asian *Xoo* on recombinant inbred lines and their parents

Xoo strains	Lesion length (cm) induced on		
	IR64	Azucena	RIL
MAI1	0.25 ± 0.1	16.1 ± 2	0.12 to 30.05
BAI4	0.76 ± 0.2	22.36 ± 2.7	0.14 to 32.2
BAI3	4.07 ± 1	22.29 ± 3	0.2 to 32.6
NAI8	2.07 ± 0.5	26.29 ± 3.3	0.44 to 37.9
PXO86	16.46 ± 1.5	26.0 ± 3	6.95 to 34.5
PXO61	1.92 ± 0.4	28.32 ± 3.1	0.5 to 30.8

Xoo Xanthomonas oryzae pv. *oryzae*, *RIL* Recombinant Inbred Lines

This continuous variation of lesion lengths indicates the existence of QTLs underlying the segregation of resistance. Both parents, IR64 and Azucena, are susceptible to Asian *Xoo* strain PXO86 with an average lesion length of 16.46 ± 1.5 and 26 ± 3 cm, respectively. Conversely, IR64 is resistant to PXO61; the Philippines race 1, with an average lesion length of 1.92 ± 0.4 cm, whereas Azucena is susceptible with an average lesion length of 28.32 ± 3.1 cm. The lesion length of the 178 RILs lines shows a continuous variation with an average lesion length of 6.95 to 34.5 cm and 0.5 to 30.8 cm with PXO86 and PXO6, respectively (Table 1), indicating the resistance to Asian strains is controlled by QTLs.

Mapping QTLs using SSR markers

The IR64 x Azucena genetic map used in this study were comprised of 226 SSR markers and covered 1652.06 cM of the genome with an average intermarker interval of 7.31 cM. QTL mapping, based on ANOVA, evidences twelve putative QTLs induced by African *Xoo* strains (MAI1, BAI4, BAI3 and NAI8). Two of them (*qABB-7* and *qABB-11*) have a large effect on chromosomes 7 and 11, respectively. The others induce small effect on chromosomes 1, 3, 4, 8, 9, 10. The composite interval mapping using WinQTL-Cartographer 2.5 reveals five specific QTLs underlying resistance to African *Xoo* strains including those detected by ANOVA on chromosomes 1, 7, 9, 10 and 11. The estimated additive effect indicates that these loci derive from the resistant parent IR64. *qABB-1*, specific to MAI1, is linked to SSR markers RM129 and RM493, with a LOD score of 4.72 and a percentage of variance explained (R^2) of 8 %. *qABB-7* is linked to RM125 and RM214 with a LOD score of 16.20 and R^2 of 30 % for MAI1, and a LOD score of 13.98 and R^2 of 30 % for BAI4. *qABB-11* is close to RM224 and RM144 with a LOD score of 5.32, 5.87 and 4.18 with *Xoo* MAI1, BAI3 and BAI4, respectively. *qABB-11* controls 7, 9, and 14 % of the phenotypic variation explained by *Xoo* BAI4, MAI1, and BAI3, respectively. The Inclusive Composite Interval Mapping using Qgene-4.3.0, the more accurate QTL analysis method, confirms five specific QTLs mapped on chromosomes 1, 7, 9, 10 and 11 underlying resistance to African *Xoo* strains. The estimated additive effects confirm that these loci derive from the resistant parent IR64. Additive effect, linked markers, LOD score value and PVE are summarized in Table 2. *qABB-11* on chromosome 11 was involved in the resistance to all African *Xoo* strains tested so far. QTLs on chromosomes 9 (*qABB-9*) and 10 (*qABB-10*) are specific to *Xoo* strain NAI8 (race A1). Asian *Xoo* strains induce five resistance QTLs different from those induced by African's strains except that on chromosome 11. This is also induced by *Xoo* strain PXO61 (Philippines race 1). Three resistance QTLs to *Xoo* strain PXO86 (Phil race 2) are mapped on chromosomes 5 (*qBB-5*), 8 (*qBB-8*) and 12 (*qBB-12*) (Table 2). As indicated by the estimated additive effects, the QTLs on chromosomes 5 and 8 are controlled by Azucena allele while that on chromosome 12 is underlined by IR64 allele. A resistance QTL (*qBB-4*) is detected on chromosome 4 for the Asian *Xoo* strain PXO61.

Table 2 Novel QTLs mapped in IR64 x Azucena population using African and Asian *Xanthomonas oryzae pv.oryzae*

Xoo[a] strains	Country of origin	QTL localization: chromosome	QTL name	LOD[b] score	Closely Linked marker	Marker position (cM)	Additive effect	Donor Allele	PVE[c] (%)
MAI1	Mali	1	*qABB-1*	5.068	RM129	58.45	1.7	IR64	13.4
		7	*qABB-7*	16.006	RM125	5.75	3.2	IR64	36.6
		11	*qABB-11*	4.666	RM144	100.21	1.5	IR64	12.4
BAI4	Burkina Faso	7	*qABB-7*	13.943	RM125	5.75	3.5	IR64	33.4
		11	*qABB-11*	4.68	RM144	100.21	1.8	IR64	12.8
BAI3	Burkina Faso	11	*qABB11*	5.728	RM144	100.21	2.5	IR64	15.3
NAI8	Niger	9	*qABB-9*	4.359	RM242	69.28	2.3	IR64	12.9
		10	*qABB-10*	3.606	RM294A	74.36	-2.23	Azucena	10.8
		11	*qABB-11*	6.03	RM144	100.21	2.7	IR64	17.4
PXO86	Philippines	5	*qBB-5*	4.209	RM440	70.61	-1.68	Azucena	11.5
		8	*qBB-8-2*	4.957	RM281	129.96	-2.16	Azucena	13.4
		12	*qBB-12*	4.869	RM512	38.32	1.9	IR64	13.2
PXO61	Philippines	4	*qBB-4*	3.403	RM252	84.70	-1.7	Azucena	10
		11	*qABB-11*	36.78	RM144	100.21	7.3	IR64	67.9

Xoo Xanthomonas oryzae pv. *oryzae*, *LOD* logarithm of odds, *PVE* Percentage of variance explained

Heredity studies

The screening of the recombinant inbred lines using Asian *Xoo* strain PXO61 evidenced the same QTL previously induced on chromosome 11 (*qABB-11*) by all African *Xoo* strains (with LOD score = 36.78 and R2 = 67) (Table 2).

The segregation ratio obtained by screening F_2: IR24 x IRBB4 population with the Asian *Xoo* strain PXO61 is 3 resistant for 1susceptible. This is the segregation ratio of a dominant gene in a F_2 population. It was the *Xa4* dominant gene which was specific to PXO61. On the contrary, the segregation studies in the same population using the African *Xoo* strain BAI3 revealed a ratio of 3 susceptible for 1 resistant, that is the segregation ratio of a recessive gene in a F_2 population. Then, the African *Xoo* strain BAI3 induces a recessive gene at the locus of *qABB-11*.

The study of bacterial growth and of the xylem colonization speed by the bacteria shows that both strains PXO61 and BAI3 grown in the same way in IR24 rice variety which is susceptible to bacterial blight. This fact is expressed by a more or less equal number of PXO61 and BAI3 colonies detected at each time. It's the case on the twelfth day after inoculation when the number of PXO61 and BAI3 colonies is about Log10 (8) in the fraction A and Log10 (9) in the fraction B. This number increases to Log10 (10) and Log10 (10.5), respectively, in the fractions C and D (Fig. 1). On the other hand, in the resistant IRBB4 isogenic line, the colonies number of these two strains is far inferior to those got in IR24 variety (Fig. 2). In this case, both strains grown unequally. On the 12th day after inoculation, we count Log10 (8) of BAI3 and Log10 (6) of PXO61 in the fraction A, Log10 (8) of BAI3 against Log10 (4) of PXO61 in the fraction B. The Asian *Xoo* strain PXO61 is then stopped at the level of fraction B, which is expressed by the absence of colonies in the fractions C and D. On the other hand, Log10 (6) colonies of the African *Xoo* strains BAI3 have been shown in the fraction C and D (Fig. 2).

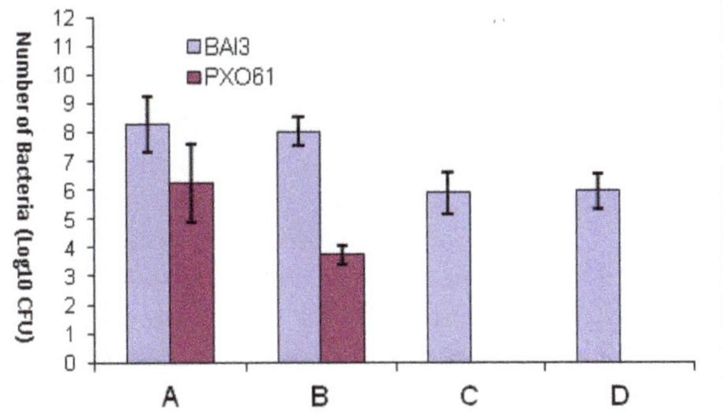

Fig. 2 Bacteria growth during twelve days after inoculation of rice isogenic line IRBB4. A = 1st fragment of 5 cm (from the inoculation point) of infected leaves. B = 2nd fragment of 5 cm (from 5 to 10 cm to the inoculation point) of infected leaves. C = 3rd fragment of 5 cm (from 10 to 15 cm to the inoculation point) of infected leaves. D = 4th fragment of 5 cm (from 15 to 20 cm to the inoculation point) of infected leaves

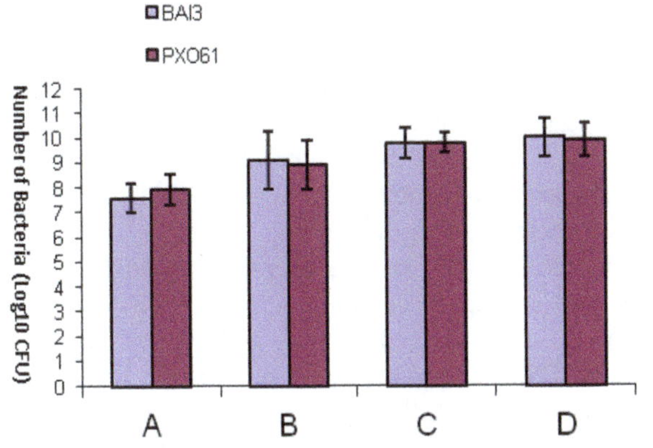

Fig. 1 Bacteria growth during twelve days after inoculation of IR24 rice variety. A = 1st fragment of 5 cm (from the inoculation point) of infected leaves. B = 2nd fragment of 5 cm (from 5 to 10 cm to the inoculation point) of infected leaves. C = 3rd fragment of 5 cm (from 10 to 15 cm to the inoculation point) of infected leaves. D = 4th fragment of 5 cm (from 15 to 20 cm to the inoculation point) of infected leaves

The results of the quantification and of bacterial kinetic show clearly that the African *Xoo* strain BAI3 is quite different from the one of the Asian *Xoo* strain PXO61 that lights a dominant gene which was the *Xa4* gene. In fact, in the IRBB4 isogenic line, resistant to both strains, the lesion caused by BAI3 was three times bigger than the one caused by PXO61. *Xa4* prevent the multiplication and the progression of PXO61 in the vascular tissues which is expressed on the one hand by the reduced number of PXO61 in these tissues and, on the other hand by the total absence of this bacteria 10 cm to the inoculation point. This is in accordance with the functioning of dominant resistance gene. On the other hand, BAI3 colonies have been observed even at 20 cm to the inoculation point. This strain managed to multiply and to colonize the vascular tissues thus leading to a less resistant phenotype (Fig. 2). The African *Xoo* strain BAI3 would thus induced a recessive gene.

Discussion

In this study, novel resistance QTLs to BB are identified and mapped. Moreover, the information on known bacterial blight genes/QTLs characterized so far was updated (Additional file 1) and their genetic and physical localization positioned on the reference rice physical map (var Nipponbare).

IR64 has been confirmed as highly resistant to African *Xoo* strains. The IR64 x Azucena derived mapping populations segregated for BB resistance. Novel QTLs were discovered, mapped to chromosomes 1, 4, 5, 7, 8, 9, 10, 11, 12. Most were of relatively small effect except on chromosomes 7 and 11. *qBB-4, qBB-5, qABB-9* and *qBB-12* mapped to genomic regions where BB resistance QTL and or *Xa* genes were previously characterized in other crosses, indicating that BB-resistance genes to Asian and African *Xoo* strains may be shared by several rice accessions. Out of five, four QTLs induced by African *Xoo* strains are different from those induced by the Philippines *Xoo* strain PXO61 indicating these strains induced different resistant genes. QTLs induced by African *Xoo* strains are underlined by the allele of the indica rice IR64 while Azucena *(japonica)* is the donor parent of resistance against Asian *Xoo* strains except with strain PXO61.

The genetic characterization of *Xoo* strains together with the recent advances in *X. oryzae* genomic studies indicate that African *X. oryzae* pv. *oryzae* strains form a separate group within the *X. oryzae* (Gonzalez et al. 2007; Triplett et al. 2011; Hajri et al. 2012; Wonni et al. 2014). Also their repertoire of transcription activator-like (TAL) effectors is reduced compared to the Asian *Xoo* one (Gonzalez et al. 2007).

Two resistance QTLs with main effects were identified. The first, *qABB-7,* induced by the African *Xoo* strains MAI1 and BAI4 on chromosome 7 and the second, *qABB-11,* induced on chromosome 11 by all African strains tested so far are particularly interesting. The first one, *qABB-7,* controls 37 % of the phenotypic variance with a high LOD score of 16 (Table 2) compared to 2.5 to 3 fixed LOD score in other QTL studies (Wang et al. 2006a, b; Sakraborty and Zeng 2011). Our preliminary data indicate that the single resistance genes *Xa4, xa5* and *Xa7* provide strong levels of resistance to African *Xoo* strains collected in the 1980's and in 2003 (Gonzalez et al. 2007). *Xa4* is located in the *qABB-11* region. The fine mapping of *qABB-7* and *qABB-11* is in progress. Our study also reveals the difference between African and Asian *Xoo* strains in terms of virulence. The resistant and susceptible phenotypes of the parents (IR64 and Azucena) induced by African *Xoo* strains greatly contrast with those induced by Asian *Xoo* strains (Table 1). According to the previous results, resistance to BB depends on the rice genetic background. *O. sativa* subspecies *indica* appeared to be the best source of resistance to bacterial blight in rice conversely to the African cultivated rice *O. glaberrima* that showed a relative narrow resistance basis to BB (Djedatin et al. 2011).

Preliminary segregation and bacteria growth studies revealed that the two most important QTLs induced by African *Xoo* on rice chromosomes 11 are recessive while most of the resistance genes characterized so far are dominant with the exception of *xa5, xa8, xa13, xa19, xa20, xa24, xa33(t), xa34(t)* and *xa35(t)* (Verdier et al. 2011). These results are compatible with the hypothesis that African and Asian strains have different effector genes that induced different resistance genes (Yu et al. 2011a, b; Hajri et al. 2011).

So far, 40 R genes and 17 QTLs conferring host resistance against various strains of *Xoo* have been identified (Li et al. 1999; Chen at al. 2002; Gu et al. 2004; Blair et al. 2003; Ramalingam et al. 2003; Wu et al. 2008a, b; Cheema et al. 2008; Ruan et al. 2008; Korinsak et al. 2009; Wang et al. 2009; Sundaram et al. 2009; Chen et al. 2011; Bhasin et al. 2012; Han et al. 2014; Kim et al. 2015). Approximately one third of naturally occurring R genes against *Xoo* (*xa5, xa8, xa13, xa15, xa19, xa20, xa24, xa26, xa28, xa31(t), xa33(t)* and *xa34*) are recessive (Sanchez et al. 1999; Wu et al. 2008a, b; Ruan et al. 2008; Korinsak et al. 2009). More than 20 R genes were mapped onto rice chromosomes, and some of them have been well characterized. Nine resistance genes have been molecularly cloned including six dominant genes, *Xa21* (Song et al., b), *Xa1* (Yoshimura et al. 1998), *Xa3/Xa26* (Sun et al. 2004; Xiang et al. 2006), *Xa27* (Gu et al. 2004; Bimolata et al. 2013), *Xa10* (Tian et al. 2014), *Xa23* (Wang et al. 2014a, b) and three recessive *xa5* (Iyer and McCouch 2004) and *xa13* (Chu et al. 2006a, b), *xa25* (Liu et al. 2011) have been cloned. All known resistance genes/QTLs and their flanking markers mapped on the reference Nipponbare physical map show that they are unequally distributed on rice chromosomes. Chromosomes 4 and 11 appeared to carry most of the known BB resistance genes. Indeed, these chromosomes are known to carry clusters of resistance genes analogs (RGA) (Mago et al. 1999; Ghazi et al. 2009). Resistance QTLs to rice sheath blight resistance (Zou et al. 2000) and resistance genes to blast (Wang et al. 1994) also clustered on chromosome 11. Many QTLs mapped closely to single dominant or recessive *Xa* genes. It is the case of *Xa1* and *Xa2* which bracketed *AQBT008* on chromosome 4. Some of them are considered as a single gene like *AQBT023* designated as *Xa4* on the Lemont x Teqing map (Li et al. 1999). After interpolation on the Nipponbare physical map, *AQBT023* was shown to be distinct to *Xa4* located on chromosome 11.

This interpolation on the reference Nipponbare physical map highlights common, specific and novel QTL/genes for resistance to African and Asian *Xoo* strains. Indeed, *xa34(t)* resistance gene to Chinese *Xoo* races V co-localize with qABB-1, the resistance QTL induced by the African *Xoo* strain MAI1 on rice chromosome 1. In the same way, *AQBT021* effective on Asian *Xoo* strains, co-localize with *qABB-10,* a resistance QTL induce by the African *Xoo* strain NAI8 on chromosome 10. *xa8* (Ogawa and Yamamoto 1987), *qBB7* (QTL identified on chromosome 7, Ramalingam et al. 2003) and *qABB-7* overlap on chromosome 7 as well as *xa24* (Wu et al. 2008a, b) and *AQBT001* on chromosome 2. Lemont and Azucena, the donors' parents of *AQBT021* and *qABB-10* respectively, belong to the japonica subspecies. These resistance QTLs may be the same. On chromosome 8, *xa13*, *qBB8* and *qBB-8-2* co-localize perfectly. *qBB8* and *qBB-8-2* may be underlined by the recessive *xa13* gene, but the heredity of *qBB8* and *qBB-8-2* has not been studied yet. On chromosome 11, the QTL induced by African *Xoo* strains co-localized with *Xa4, Xa3, Xa32* and *xa35(t)*, the known single resistance genes to Asian *Xoo* strains. These *Xa* genes indicated that some major R genes also contribute to quantitative resistance as reported in common bean in which RGAs co-localized with anthracnose-specific QTL (Geffroy et al. 2000). This was also the case of QTL associated with resistance to stripe rust and Barley yellow dwarf virus in barley (Toojinda et al. 2001), and partial resistance to Cucumber mosaic virus in pepper (Pflieger et al. 1999). Specifics and novels QTLs inducing resistance to African and/or Asian *Xoo* strains are identified as *qABB-9* induced by the African *Xoo* strain NAI8 on chromosome 9, *qBB-4* and *qBB-5* induced on chromosomes 4 and 5 by Asian *Xoo* strains PXO61 and PXO86 respectively, (Fig. 3).

In addition to specific resistance QTLs to African *Xoo* trains, the known resistance genes/QTLs to Asian *Xoo* strains which co-localized with QTL induced by African *Xoo* will be used in rice breeding programs to develop bacterial blight resistant cultivars for Africa.

Conclusion

We mapped several resistance QTLs to bacterial blight in rice using a reference recombinant inbred lines derived from the cross between Azucena and IR64 rice varieties. Some of them are specific and novel to African or Asian strain of *Xanthomonas oryzae* pv. *oryzae* such as *qABB-9* induced by the African *Xoo* strain NAI8 on chromosome 9, *qBB-4* and *qBB-5* induced on chromosomes 4 and 5 by Asian *Xoo* strains PXO61 and PXO86, respectively. The others co-localize with known *Xoo* resistance genes/QTLs. It is the case of *qABB-1*, the resistance QTL induced by the African *Xoo* strain MAI1 on rice chromosome 1 which co-localize with *xa34(t)* resistance gene to Chinese *Xoo* races V. In the same way, *AQBT021* effective on Asian *Xoo* strains, co-localize with *qABB-10*, a resistance QTL induced by the African *Xoo* strain NAI8 on chromosome 10.

So far, and for the first time, specific resistance QTLs to African *Xoo* are identified and mapped. The fine mapping of the QTL induced by African *Xoo* MAI1 and BAI4 on chromosome 7 which controlled 37 % of phenotypic variance as well as the one induced on chromosome 11 by all African *Xoo* tested is ongoing and will provide news markers for breeding program.

Methods

Plant materials

The reference mapping population consisted in 172 recombinant inbred lines (RIL) obtained by single seed descent (SSD) from the cross between the tropical japonica landrace Azucena (susceptible parent) and the indica cultivar IR64 (resistant parent) was used to identify and map the bacterial blight (BB) resistance QTLs. IR64 is an elite improved indica rice grown in tropical Asia. It carries the BB resistance gene *Xa4* and has a large spectrum of resistance to *Xoo*. The Azucena landrace is susceptible to BB disease, and tolerant to drought and other abiotic stresses. The IR64 x Azucena mapping population was used to develop genetic map for characterizing gene/QTLs associated to drought, iron toxicity tolerance and resistance to the Rice yellow Mottle Virus (Boisnard A et al. 2007).

Bacterial strains and plant inoculations

Six *Xanthomonas oryzae* pv. *oryzae* (*Xoo*) strains were used to evaluate resistance to BB in the mapping population. These strains included four African *Xoo* strains: BAI3 and NAI8 (race A1) from Burkina-Faso and Niger respectively, BAI4 (race A2) from Burkina-Faso, MAI1 (race A3) from Mali and two Asian strains PXO61 (Philippines race 1) and PXO86 (Philippines race 2). The bacterial strains were cultured on PSA medium (per liter of H_2O, 10 g of peptone, 10 g of sucrose, 1 g of glutamic acid, and 16 g of bacto agar at pH 7.0) overnight. Inoculum was prepared by re-suspending bacterial culture in sterile, distilled water at an optical density of 0.2 (DO_{600}), bearing approximately 10^8 cfu per ml. The progenies and their parents were grown under controlled conditions (28 °C; 80 % humidity and 12 h day length) in greenhouse at IRD Montpellier/France. They were inoculated at the booting stage (6 weeks after germination) using the leaf-clipping method (Kauffman et al. 1973) in which the fully-opened leaves were cut with scissors dipped in the bacterial suspension. Ten plants per genotype and two uppermost fully expanded and randomly chosen leaves were inoculated with each strain. Phenotypic evaluations, consisting of the disease scoring, were done 3 weeks after inoculation by measuring the leaf lesion length. Disease reactions were classified according to the mean lesion length (LL) as following: resistant (R) if the lesion length was < 5 cm, moderately resistant (MR) with LL of 5 to 10 cm; moderately susceptible (MS) with LL of 10 to 15 cm, susceptible (S) with LL > 15 cm.

Molecular mapping analysis using SSR markers

Leaves of rice plants were harvested and ground in liquid nitrogen. Genomic DNA was extracted as previously described (Edwards et al. 1991). PCR were performed in

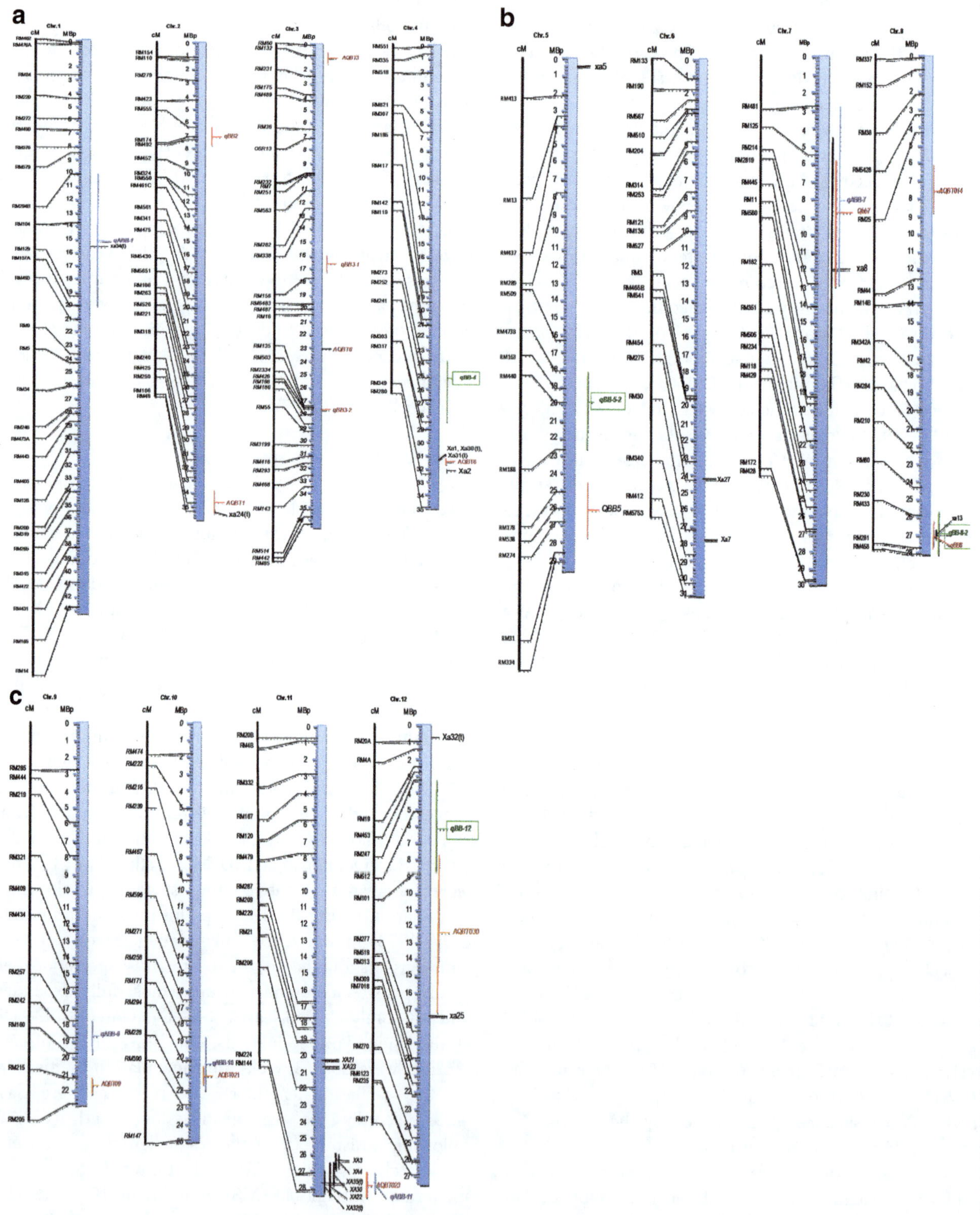

Fig. 3 a, b and **c**: Integrative map showing all known resistance QTLs/genes to African and Asian *Xoo*. The vertical graduated blue thick lines represent the physical map of each chromosome and is linked to the genetic map on the left showing the microsatellites markers (RM). On the right size of each physical map the genes and QTLs are indicated in red, black and green. In black are the R genes that were previously identified with Asian *Xoo* strains. R genes that were cloned are indicated by a horizontal line to a specific location on the physical map. The genes that are not cloned yet are represented by a vertical line indicating the confidence interval. In red we indicated QTLs that were detected in others rice breeding populations. Novel QTLs identified using the African *Xoo* strains are represented in blue those induced by the Philippines strains PXO61 and PXO86 are shown in green

15 μL reactions in an automated thermal cycler and the program consisted of the following cycles: initial denaturation at 94 °C for 5 min; 30 cycles of denaturation at 94 °C for 30 s, annealing temperature for 30 s and extension at 72 °C for 45 s; and a final extension step at 72 °C for 5 min. Amplified products were analyzed by electrophoresis on 6.5 % polyacrylamide gels, using electrophoresis system LICOR; or by electrophoresis in a 2 % agarose gel.

Statistical analysis and *Xoo* resistance QTLs mapping

A linkage map comprising 226 SSR markers and constructed from the RIL population was used for mapping resistance QTL to *Xoo*. An analysis of variance, using marker genotypes as the groups, was carried out using MapDisto (Lorieux 2007). Data files were prepared using the Export map and data function of MapDisto. Analyses of distribution of the phenotypic traits as well as QTL detection were mainly performed using the Qgene v. 4.3.0 program (Nelson 2005, http://www.qgene.org) and Windows QTL cartographer 2.5 (Wang et al. 2006b). Different methods were compared such as Single-marker regression (SMR), Simple interval mapping (SIM), and Composite interval mapping (CIM). The Forward cofactor selection option was used in CIM. The LOD score statistic was used for all methods in order to make the results comparable. Empirical thresholds to declare presence of a QTL were obtained using the resampling by permutation method, performing 1,000 iterations for each trait/chromosome combination (log-likelihood of odds (LOD) score of 3).

Heredity studies

QTL mapping using Asian Xoo strain PXO61

At the locus of *qABB-11*, the QTL on chromosome 11 that was involved in the resistance on all African *Xoo* tested, were localized a cluster of *Xa* genes including *Xa3, Xa4* and *Xa21. Xa3* was not effective against *Xoo* race 1 (Gonzalez et al. 2007). *Xa21* was identified in *Oryza longistaminata*, a wild rice race. Therefore, *Xa4* would be the only one *Xa* candidate gene at the above locus. In order to validate the presence of *Xa4* gene at this locus, the Asian *Xoo* strain PXO61 belonging to Philippines race 2 was used to screen the RIL population according to Kauffman et al. (1973). The resistance of rice to PXO61 strain was specifically under *Xa4* control.

Development and screening of F_2: IR24 x IRBB4 population

The Asian rice variety IR24 belonging to indica subspecies was crossed with isogenic line IRBB4 carrying *Xa4* gene. IRBB4 was used as donor while IR24 represent a recurrent parent. The Hybrid F_1 obtained was used to generate a F_2 population. Two sets of 100 F_2 individuals were screened with the African *Xoo* strain BAI3 and the Asian one PXO61, respectively.

In planta growth experiments

Rice variety IR24 with its isogenic line IRBB4 were screened using African *Xoo* strain BAI3 and Asian *Xoo* strain PXO61. Two, three and four pieces of 5 cm from the apex to the base of infected leaf were harvested 4, 8 and 12 days after inoculation, respectively. On each day, infected leaves fragments were harvested on three different plants. Infected leaves collected were briefly rinsed in 70 % of ethanol for 10 s followed by submersion in sterilized water. Leaves were put into 2 ml eppendorf tubes containing 2 metallic beads ($\phi = 3$ mm), frozen by submersion into liquid nitrogen and ground into fine powder using the Qiagen Tissue Lyser system (30 rounds/s for 2 min). Ground material was resuspended in 1 ml of sterilized water and 10 μl drops of a dilution series were spotted onto PSA medium plate in triplicates. The plates were incubated at 28 °C until colonies could be counted. This experiment was performed three times.

Mapping of known resistance gene/QTLs on the reference Nipponbare physical map

In a first step, information on all known BB resistance genes and QTLs was reviewed. This review included gramene accessions, number of genes/QTLs, their names, synonyms and symbols, the genetic populations in which they were mapped. Their donor's parents as well as their genetic position and their co-localized markers in various mapping populations were also reported here. In the same way, physical positions were recorded if available (Supplementary data). The different genetic maps used were SSR Cornell 2000, RIL IR64 x Azucena, DH IR4 x Azucena, RIL Lemont x Tequin, RIL Zhenshan 97 x Minghui 63, JRGP RFLP 2000 Nipponbare x Kasalath, Cornell RFLP 2001 *O. sativa* x *O. longistaminata*, and the reference MSU7. Physical positions of the cloned genes such as *Xa1, Xa7, Xa10, Xa23, Xa26, Xa27, xa5, xa13* and *xa25* were directly reported on the integrative map. For the non cloned genes and QTLs, we chose the closest ones with known genetic positions among the co-localized maker and interpolated them on the reference Nipponbare MSU7 physical map. Other genes and QTLs with conflicting positions, as well as those that have been not mapped such as *Xa11, Xa12, Xa15, xa32(t), AQBT026* and *AQBT030*, were not positioned on the integrative map.

Abbreviations

ANOVA: analysis of variance; BB: bacterial blight; CFU: colony-forming units; CIM: composite interval mapping; DO: optical density; LL: lesion length; LOD: log-likelihood of odds; MR: moderately resistant; MS: moderately susceptible; PSA: peptone sucrose agar; QTL: quantitative trait loci; R: resistant; RIL: recombinant inbred lines; S: susceptible; SIM: simple interval mapping; SMR: single marker regression; SSD: single seed descent; SSR: simple sequence repeats; TAL: transcription activator-like; *Xoo*: Xanthomonas oryzae pv. *oryzae*.

Competing interests

The authors declare that they have no competing interests.

Authors' contributions

GD carried out the phenotyping assays, participated in QTL analysis and physical maps drawing and drafted the manuscript. MN, AS, W and AG participated in the design of the study and its coordination. ML participated in QTL analysis. AG drowns the physical maps. W conceived and participated in phenotyping assays. All authors contributed to the manuscript draft, read and approved the final manuscript.

Authors' information

GD: Assistant Professor at Université Polytechnique d'Abomey/Benin. MN: head of biotechnology lab of AfricaRice. AS: Professor and head of Applied biochemistry and molecular biology lab of University of Abomey-Calavi/ Benin ML: Currently works overseas at CIAT, Cali, Colombia, where he is leading the Rice Genetics and Genomics group. W: Director of research at IRD Montpellier, head of IPME unit. AG: Director of research at IRD Montpellier, head of DIADE unit.

Acknowledgments

We thank USAID which partly supported this work through the project "Marker-Assisted Selection for improvement of rice varieties resistance to RYMV for West Africa" conducted by the second author (AfricaRice) and the Generation Challenge Program (subprogram SP5), which partly funded this study. We are grateful to Africa Rice and IRD for their support and Thierry Mathieu (IRD) for his help during greenhouse experiments and screening. We thank also Daniel Fonséca and Nour Ahmadi for providing information about the SSR markers used in this study.

Author details

[1]Université Polytechnique d'Abomey, BP 2282 Abomey, Benin. [2]Africa Rice Center (AfricaRice), 01 BP2031 Cotonou, Benin. [3]IPME Interactions Plantes Microorganismes, Environnement, IRD - Cirad - University Montpellier, 34394 Montpellier, France. [4]UMR Diversité, Adaptation et Développement des plantes (DIADE), Institut de Recherche pour le Développement, 911 Avenue Agropolis BP 64501, 34394 Montpellier Cedex 5, France. [5]Université d'Abomey Calavi, 01 BP526 Cotonou, Benin.

References

Basso A, Onasany A, Issaka S, Sido AY, Haougui A, Adam T, Séré Y, Saadou M (2011) Le flétrissement bactérien du riz au Niger: diversité pathologique d'isolats collectés sur les périmètres irrigués. J Appl Biosci 38:2551–2563

Bhasin H, Bhatia D, Raghuvanshi S, Lore J, Sahi G, Kaur B, Vikal Y, Singh K (2012) New PCR-based sequence-tagged site marker for bacterial blight resistance gene *Xa38* of rice. Mol Breed 30:607 611

Bimolata W, Kumar A, Sundaram RM, Laha GS, Qureshi IA, Reddy GA, Ghazi IA (2013) Analysis of nucleotide diversity among alleles of the major bacterial blight resistance gene *Xa27* in cultivars of rice (Oryza sativa) and its wild relatives. Planta 238(2):293–305. doi:10.1007/s00425-013-1891-3

Blair MW, Garris AJ, Iyer AS, Chapman B, Kresovich S, McCouch SR (2003) High resolution genetic mapping and candidate gene identification at the *xa5* locus for bacterial blight resistance in rice (Oryza sativa L.). Theor Appl Genet 107:62–73

Boisnard A, Albar L, Thiéméle D, Tondeau M, Ghesquiere A (2007) Evaluation of genes from eIF4E and eIF4G multigenic families as potential candidates for partial resistance QTLs to Rice yellow mottle virus in rice. Theoretical and Applied Genetics 116:53–62.

Century KS, Lagman RA, Adkisson M, Morlan J, Tobias R, Schwartz K, Smith A, Love J, Ronald PC, Whalen MC (1999) Developmental control of *Xa21*-mediated disease resistance in rice. Plant J 20:231–236

Cheema KK, Grewal NK, Vikal Y, Sharma R, Lore JS, Das A, Bhatia D, Mahajan R, Gupta V, Bharaj T, Singh K (2008) A novel bacterial blight resistance gene from *Oryza nivara* mapped to 38 kb region on chromosome 4 L and transferred to *Oryza sativa* L. Genet Res 90:397–407

Chen H, Wang S, Zhang Q (2002) A new gene for bacterial blight resistance in rice located on chromosome 12 identified from Minghui 63, and elite restorer line. Phytopathology 92:750–754

Chen S, Liu X, Zeng L, Ouyang D, Yang J, Zhu X (2011) Genetic analysis and molecular mapping of a novel recessive gene *xa34(t)* for resistance against Xanthomonas oryzae pv. oryzae. Theor Appl Genet. DOI 10.1007/ s00122-011-1534-7

Chu ZH, Fu BY, Yang H, Xu CG, Li ZK, Sanchez A, Park YJ, Bennetzen JL, Zhang QF, Wang SP (2006a) Targeting *xa13*, a recessive gene for bacterial blight resistance in rice. Theor Appl Genet 112:455–461

Chu ZH, Yuan M, Yao LL, Ge XJ, Yuan B, Xu CG, Li XH, Fu BY, Li ZK, Bennetzen JL, Zhang QF, Wang SP (2006b) Promoter mutations of an essential gene for pollen development result in disease resistance in rice. Genes Dev 20:1250–1255

Djedatin G, Ndjiondjop MN, Mathieu T, Vera Cruz CM, Sanni A, Ghesquière A, Verdier V (2011) Evaluation of African cultivated rice *Oryza glaberrima* for resistance to bacterial blight. Plant Dis 95:441–447

Edwards K, Johnstone C, Thompson C (1991) A simple and rapidmethod for the preparation of plant genomic DNA for PCR analysis. Nucleic Acids Res 19:1349

Geffroy V, Sevignac M, d'Oliveira J, Fouilloux G, Skroch P, Thoquet P, Gepts P, Langin T, Dron M (2000) Inheritance of partial resistance against *Colletotrichum lindemuthianum* in *Phaseolus vulgaris* and co-localization of quantitative trait loci with genes involved in specific resistance. Mol Plant Microbe Interact 3:287–296

Ghazi IA, Srivastava PS, Dalal V, Gaikwad K, Singh AK, Sharma TR, Singh NK, Mohapatra T (2009) Physical mapping, expression analysis and polymorphism survey of resistance gene analogues on chromosome 11 of rice. J Biosci 34:251–261

Gonzalez C, Szurek B, Manceau C, Mathieu T, Sere Y, Verdier V (2007) Molecular and pathotypic characterization of new Xanthomonas oryzae strains from West Africa. Mol Plant Microbe Interact 20:534–546

Gu K, Tian D, Yang F, Wu L, Sreekala C, Wang D, Wang GL, Yin Z (2004) High-resolution genetic mapping of *Xa27(t)*, a new bacterial blight resistance gene in rice, Oryza sativa L. Theor Appl Genet 108:800–807

Hajri A, Brin C, Zhao S, David P, Feng J, Koebnik R, Szurek B, Verdier V, Boureau T, Poussier S (2012) Multilocus sequence analysis and type III effector repertoire mining provide new insights into the evolutionary history and virulence of Xanthomonas oryzae. Molecular Plant Pathology 13:288–302

Han X, Yang Y, Wang X, Zhou J, Zhang W, Yu C, Cheng C, Cheng Y, Yan C, Chen J (2014) Quantitative Trait Loci Mapping for Bacterial Blight Resistance in Rice Using Bulked Segregant Analysis. Int J Mol Sci 15(7):11847–11861

Iyer AS, McCouch SR (2004) The rice bacterial blight resistance gene *xa5* encodes a novel form of disease resistance. Mol Plant Microbe Interact 17:1348–1354

Kauffman HE, Reddy APK, Hsieh SPY, Merca SD (1973) An improved technique for evaluating resistance to rice varieties of Xanthomonas oryzae. Plant Dis Rep 57:537–541

Kawahara Y et al (2013) Improvement of the Oryza sativa Nipponbare reference genome using next generation sequence and optical map data. Rice 6:4

Kim S, Suh J, Qin Y, Noh T, Reinke RF, Jena KK (2015) Identification and fine-mapping of a new resistance gene, Xa40, conferring resistance to bacterial blight races in rice (Oryza sativa L.) Theoretical and Applied Genetics 128:1933–1943.

Korinsak S, Sriprakhon S, Sirithanya P, Jairin J, Korinsak S, Vanavichit A, Toojinda T (2009) Maejo Int J Sci Technol 3:235–247

Lee SW, Choi SH, Han SS, Lee DG, Lee BY (1999) Distribution of Xanthomonas oryzae pv. oryzae strains virulent to *Xa21* in Korea. Phytopathology 89:928–933

Li ZK, Luo LJ, Mei HW, Paterson AH, Zhao XZ, Zhong DB, Wang YP, Yu XQ, Zhu L, Tabien R, Stansel JW, Ying CS (1999) A "defeated" rice resistance gene acts as a QTL against a virulent strain of *Xanthomonas oryzae* pv. *oryzae*. Mol Gen Genet 261:58–63

Liu Q, Yuan M, Zhou Y, Li X, Xiao J, Wang S (2011) A paralog of the MtN3/saliva family recessively confers race-specific resistance to *Xanthomonas oryzae* in rice. Plant Cell Environ 34:1958–1969. doi:10.1111/j.1365-3040.2011.02391.x

Lorieux M (2007) MapDisto, A Free User-Friendly Program For Computing Genetic Maps. Computer demonstration (P958) given at the Plant and Animal Genome XV conference, Jan 13-17 2007, San Diego, CA., URL: http://mapdisto.free.fr/

Mago R, Nair S, Mohan M (1999) Resistance gene analogues from rice: cloning, sequencing and mapping. Theor Apply Genet 99:50–57

Mew TW, Vera Cruz CM, Medalla ES (1992) Changes in race frequency of Xanthomonas oryzae pv. oryzae in response to the planting of rice cultivars in the Philippines. Plant Dis 76:1029–1032.

Nelson J (2005) Methods and Software for Genetic Mapping, pp. 53–74 in The Handbook of Plant Genome Mapping; Genetic and Physical Mapping, edited by Meksem K., Kahl G., editors. Wiley-VCH Verlag GmbH and Co, Weinheim, Germany

Ogawa T, Yamamoto T (1987) Selection of Recurrent Parents to Develop near-Isogenic Lines Resistant to Bacterial Leaf-Blight of Rice. Jarq Japan Agric Res Q 21:65–69

Perez LM, Redoña ED, Mendioro MS, Vera Cruz CM and Leung H (2008) Introgression of Xa4, Xa7 and Xa21 for resistance to bacterial blight in thermosensitive genetic male sterile rice (Oryza sativa L.) for the development of two-line hybrids. Euphytica 164:627–636.

Pflieger S, Lefebvre V, Caranta C, Blattes A, Goffinet B, Palloix A (1999) Disease resistance gene analogs as candidates for QTLs involved in pepper–pathogen interactions. Genome 42:1100–1110

Ramalingam J, Cruz CMV, Kukreja K, Chittoor JM, Wu JL, Lee SW, Baraoidan M, George ML, Cohen MB, Hulbert SH, Leach JE, Leung H (2003) Candidate Defense genes from rice, barley, and maize and their association with qualitative and quantitative resistance in rice. Mol Plant Microbe Interact 16:14–24

Reckhaus PM (1983) Occurrence of bacterial blight of rice in Niger, West Africa. Plant Dis 67:1039

Ruan H,Yan C, An D, Liu R, Chen J (2008) Identifying and Mapping New Gene *xa32(t)* for Resistance to Bacterial Blight(*Xanthomonas oryzae* pv. *oryzae, Xoo*) from Oryza meyeriana L. Acta Agriculturae Boreali-Occidentalis Sinica. DOI CNKI:SUN:XBNX.0.2008-06-035

Ruengphayak S, Chaichumpoo E, Phromphan S, Kamolsukyunyong W, Sukhaket W, Phuvanartnarubal E, Korinsak S, Korinsak S, Vanavichit A (2015) Pseudo-backcrossing design for rapidly pyramiding multiple traits into a preferential rice variety. Rice 8:7. doi:10.1186/s12284-014-0035-0

Sakraborty S, Zeng Z (2011) QTL Mapping for Days to Flowering under Drought Condition in Rice (Oryza sativa L.) Genome. Notulae Botanicae Horti Agrobotanici Cluj-Napoca 39:58–63

Sanchez AC, Ilag LL, Yang D, Brar DS, Ausubel F, Khush GS, Yano M, Saskai T, Li Z, Huang N (1999) Genetic and physical mapping of *xa13*, a recessive bacterial blight resistance gene in rice. Theor Appl Genet 98:1022–1028

Sere Y, Onasanya A, Verdier V, Akator K, Ouedraogo LS, Segda Z, Mbare MM, Sido AY, Baso A (2005) Rice Bacterial Leaf Blight in West Africa: Preliminary studies on disease in farmer's field and screening. Asian J Plant Sci 4:577–579

Shanti ML, Shenoy W, Lalitha Devi G, Mohan Kumar V, Premalatha P, Naveen Kumar G, Shashidhar HE, Zehr UB, Freeman WH (2010) Marker-assisted breeding for resistance to bacterial leaf blight in popular cultivar and parental lines of hybrid rice. J Plant Pathol 92:495–501

Singh S, Sidhu JS, Huang N, Vikal Y, Li Z, Brar DS, Dhaliwal HS and Khush GS (2006) Pyramiding three bacterial blight resistance genes (xa5, xa13 and Xa21) using marker-assisted selection into indica rice cultivar PR106. Theoretical and Applied Genetics. 102:1011–1015

Suh JP, Jeung JU, Noh TH, Cho YC, Park SH, Park HS, Shin MS, Kim CK, Jena KK (2013) Development of breeding lines with three pyramided resistance genes that confer broad-spectrum bacterial blight resistance and their molecular analysis in rice. Rice 6:5. doi:10.1186/1939-8433-6-5

Sun XL, Cao YL, Yang ZF, Xu CG, Li XH, Wang SP, Zhang QF (2004) *Xa26*, a gene conferring resistance to *Xanthomonas oryzae* pv. *oryzae* in rice, encodes an LRR receptor kinase-like protein. Plant J 37:517–527

Sundaram RM, Vishnupriya MR, Laha GS, Rani NS, Rao PS, Balachandran SM, Reddy GA, Sarma NP, Sonti RV (2009) Introduction of bacterial blight resistance into Triguna, a high yielding, mid-early duration rice variety. Biotechnol J 4:400–407

Swamy P, Panchbhai AN, Dodiya P, Naik V, Panchbhai SD, Zehr UB, Azhakanandam K, Char BR (2006) Evaluation of bacterial blight resistance in rice lines carrying multiple resistance genes and *Xa21* transgenic lines. Curr Sci 90:818–824

Tian D, Wang J, Zeng X, Gu K, Qiu C, Yang X, Zhou Z, Goh M, Luo Y, Murata-Hori M et al (2014) The rice TAL effector-dependent resistance protein *XA10* triggers cell death and calcium depletion in the endoplasmic reticulum. Plant Cell 26:497–515

Toojinda T, Broers LH, Chen XM, Hayes PM, Kleinhofs A, Korte J, Kudrna D, Leung H, Line R, Powell W, Ramsey L, Vivar H, Waugh R (2001) Mapping quantitative and qualitative disease resistance genes in a doubled haploid population of barley (*Hordeum vulgare*). Theor Appl Genet 101:580–589

Traoré O, Pinel-Galzi A, Sorho F, Sarra S, Rakotomalala M, Sangu E, Kanyeka Z, Séré Y, Konaté G, Fargette D (2009) A reassessment of the epidemiology of Rice yellow mottle virus following recent advances in field and molecular studies. Virus Res 141:258–67

Triplett LR, Hamilton JP, Buell CR, Tisserat NA, Verdier V, Zink F, Leach JE (2011) Genomic Analysis of *Xanthomonas oryzae* from US Rice Reveals Substantial Divergence from Known X. oryzae Pathovars. Appl Environ Microbiol 77:3930–7

Vera Cruz C M, Bai J, Oña I, Leung H, Nelson RJ, Mew T and Leach JE. 2000. Predicting durability of a disease resistance gene based on an assessment of the fitness loss and epidemiological consequences of avirulence gene mutation. PNAS 97:13500–13505

Verdier V, Vera Cruz C, Leach JE (2012) Controlling rice bacterial blight in Africa: Needs and prospects. J Biotechnol 159(4):320-328

Wang G, Mackill DJ, Bonman JM, McCouch SR, Nelson RJ (1994) RFLP mapping of genes conferring complete and partial resistance to blast resistance in a durably resistant rice cultivar. Genetics 136:1421–1434

Wang G, Wan X, Crossa J, Crouch J, Weng J, Zhai H, Wan J (2006a) QTL mapping of grain length in rice (Oryza sativa L.) using chromosome segment substitution lines. Genet Res 88:93–104

Wang S, Basten CJ, Zeng ZB (2006b) Windows QTL Cartographer2.5. Department of Statistics, North Carolina Stat University, Raleigh, http://www.statgen.ncsu.edu/~Shchwang/WQTLCart.htm

Wang CL, Xu AB, Gao Y, Fan YL, Liang YT, Zheng CK, Sun LQ, Wang WQ, Zhao KJ (2009) Generation and characterisation of Tn5-*tagged Xanthomonas oryzae* pv. *oryzae* mutants that overcome *Xa23*-mediated resistance to bacterial blight of rice. Eur J Plant Pathol 123:343–351

Wang C, Fan Y, Zheng C, Qin T, Zhang X, Zhao K (2014a) High-resolution genetic mapping of rice bacterial blight resistance gene Xa23. Mol Genet Genomics 289:745–753. doi:10.1007/s00438-014-0848-y

Wang M, Yu Y et al (2014b) The genome sequence of African rice (Oryza glaberrima) and evidence for independent domestication. Nat Genet 46(9):982–911

Webb KM et al (2010) A benefit of high temperature: Increased effectiveness of a rice bacterial blight disease resistance gene. New Phytol 185:568–576

Wonni I, Cottyn B, Detemmerman L, Dao S, Ouedraogo L, Sarra S, Tekete C, Poussier S, Corral R, Triplett L, Koita O, Koebnik R, Leach J, Szurek B, Maes M, Verdier V (2014) Analysis of *Xanthomonas oryzae* pv. *oryzicola* population in Mali and Burkina Faso reveals a high level of genetic and pathogenic diversity. Phytopathology 104:520–531

Wu L, Goh ML, Sreekala C, Yin Z (2008a) *XA27* Depends on an Amino-Terminal Signal-Anchor-Like Sequence to Localize to the Apoplast for Resistance to *Xanthomonas oryzae* pv *oryzae*. Plant Physiol 148:1497–1509

Wu X, Li X, Xu C, Wang S (2008b) Fine genetic mapping of *xa24*, a recessive gene for resistance against *Xanthomonas oryzae* pv. *oryzae* in rice. Theor Appl Genet 118:185–191

Xiang Y, Cao YL, Xu CG, Li XH, Wang SP (2006) *Xa3*, conferring resistance for rice bacterial blight and encoding a receptor kinase-like protein, is the same as *Xa26*. Theor Appl Genet 113:1347–1355

Yoshimura S, Yamanouchi U, Katayose Y, Toki S, Wang ZX, Kono I, Kurata N, Yano M, Iwata N, Sasaki T (1998) Expression of *Xa1*, a bacterial blight-resistance gene in rice, is induced by bacterial inoculation. Proc Natl Acad Sci U S A 95:1663–1668

Yu Y, Streubel J, Balzergue S, Champion A, Boch J, Koebnik R, Feng J, Verdier V, Szurek B (2011a) Colonization of rice leaf blades by an African strain of *Xanthomonas oryzae* pv. *oryzae* depends on a new TAL effector which induces the rice nodulin-3 *Os11N3* gene. Mol Plant Microbe Interact 24(9): 1102–1113

Yu Y, Streubel J, Balzergue S, Champion A, Boch J, Koebnik R, Feng J, Verdier V, Szurek B (2011) Colonization of rice leaf blades by an African strain of *Xanthomonas oryzae* pv. *oryzae* depends on a new TAL effector which induces the rice nodulin-3 *Os11N3* gene. Mol Plant Microbe Interact. Epub ahead of print

Zhang J, Li X, Jiang G, Xu Y, He Y (2006) Pyramiding of *Xa7* and *Xa21* for the improvement of disease resistance to bacterial blight in hybrid rice. Plant Breed 125:600–605

Zou JH, Pan XB, Chen Z, Xu JY, Lu JF, Zhai WX, Zhu LH (2000) Mapping quantitativre trait loci controlling sheath blight resistance in two rice cultivars (Oryza sativa L.). Theor Appl Genet 101:569–573

4

Rice panicle plasticity in Near Isogenic Lines carrying a QTL for larger panicle is genotype and environment dependent

Dewi Erika Adriani[1,2], Michael Dingkuhn[1,3], Audrey Dardou[1], Hélène Adam[4], Delphine Luquet[1] and Tanguy Lafarge[1*]

Abstract

Background: Panicle architectural traits in rice (branching, rachis length, spikelet number) are established between panicle initiation and heading stages. They vary among genotypes and are prone to Genotype x Environment interactions. Together with panicle number, panicle architecture determines sink-based yield potential. Numerous studies analyzed genetic and environmental variation of plant morphology, but the plasticity of panicle structure is less well understood. This study addressed the response of rice panicle size and structure to limited light availability at plant level for near-isogenic lines (NILs) with IR64 or IRRI146 backgrounds, carrying the QTL *qTSN4* (syn. SPIKE) for large panicles. Full light and shading in the greenhouse and two population densities in the field were implemented. The image analysis tool P-TRAP was used to analyze the architecture of detached panicles.

Results: The *qTSN4* increased total branch length, branching frequency and spikelet number per panicle in IRRI146 background in the field and greenhouse, and in IR64 background in the greenhouse, but not for IR64 in the field. In the field, however, *qTSN4* reduced panicle number, neutralizing any potential yield gains from panicle size. Shading during panicle development reduced spikelet and branch number but *qTSN4* mitigated partly this effect. Spikelet number over total branch length (spikelet density) was a stable allometry across genotypes and treatments with variation in spikelet number mainly due to the frequency of secondary branches. Spikelet number on the main tiller was correlated with stem growth rate during panicle development, indicating that effects on panicle size seemed related to resources available per tiller.

Conclusions: The *qTSN4* effects on panicle spikelet number appear as indirect and induced by upstream effects on pre-floral assimilate resources at tiller level, as they were (1) prone to G x E interactions, (2) non-specific with respect to panicle architectural traits, and (3) associated with pre-floral stem growth rate.

Keywords: Panicle plasticity, Panicle architecture, Branch number, Branch length, Spikelet number per panicle, Pre-floral stem vigor, *qTSN4*

Background

Growth of the rice plant is divided into three phases: the vegetative (from germination to panicle initiation), reproductive (from panicle initiation to heading) and grain filling or ripening phase (from heading to maturity) (Li, 1979). Yield components are progressively determined along the rice cycle, first during the vegetative phase (tiller number), second during the reproductive phase (number of fertile tillers and size of panicle sinks) and finally during the grain filling phase (spikelet filling rate). Sink-based yield potential is thus fixed by the flowering stage (Xiong et al. 2011) and depends mostly on the dimensioning of crucial organs that occurs during the reproductive phase. This is the time when the morphological structure of the panicle bearing the sinks, but also the size of internodes that store C reserves, become effective.

Panicle development consists of branching, branch elongation and spikelet deployment. These elemental processes are governed by traits that are crucial for crop

* Correspondence: tanguy.lafarge@cirad.fr
[1]CIRAD, UMR AGAP, F-34398 Montpellier, France
Full list of author information is available at the end of the article

breeding in getting to higher yield and improving food security (Yamagishi et al. 2004, Miura et al. 2010, Endo-Higashi and Izawa 2011, Ohsumi et al. 2011, Peng et al. 2014). At heading stage onwards, lower carbon availability to the plant like shading reduces shoot dry matter and grain filling percentage of Nipponbare rice (Kobata et al. 2000) because sink capacity is already fixed. However, during panicle development, Choi et al. (2013) reported that temporary lack of sunshine (10 – 14 days) adversely affects panicle morphogenesis (differentiated spikelet number, branch number, panicle length). According to the authors, the fact that this could be mitigated with elevated ambient [CO_2] indicates that panicle development is driven by carbon resources although the process itself consumes little assimilate. Indeed, trait expression has to be plastic to ensure balanced source-sink relationships under variable resources. Also, Kamiji et al. (2011) reported that N top-dressing during the first stage of panicle development has a large effect on spikelet production, and genotypic differences in spikelet production could be explained by the crop growth rate during the 14-day period before heading. The sensitivity to carbon supply during this period was confirmed by Lafarge et al. (2010) on rice by looking at the detrimental effect on yield components of a 10-day shading period imposed in the field at early reproductive stage. Shiratsuchi et al. (2007) revealed that yield is correlated with the ratio of spikelet number to tiller dry weight after spikelet differentiation, and Endo-Higashi and Izawa (2011) suggested that pre-floral photosynthate accumulation determines reproductive sink capacity. This was recently confirmed by Adriani et al. (2016) analyzing near isogenic lines (NILs) with larger panicles developed by Qi et al. (2008) and Fujita et al. (2009, 2012) carrying *qTSN4* on the long arm of chromosome 4, a QTL identified for high total spikelet number (TSN). Adriani et al. (2016) reported that panicle size was strongly affected by (i) shading imposed during the reproductive phase in the greenhouse or (ii) change in plant density in the field. These authors also revealed that the effect of qTSN was associated with an earlier tillering cessation and the development of subsequent larger internodes and leaf blades and so was already visible before panicle initiation. Consequently, the panicle sink is not only generated during the reproductive phase, but appears also adjusted to the plant's (or tiller's) internal resources, which in turn depends on the environment. It also highlights that there is a direct response of panicle development to tiller and plant vigor that can be related, either directly to a reduction in light access and C acquisition through cloudiness or planting density, as also reported on sorghum (Lafarge et al. 2002), or indirectly to another abiotic constraint affecting C source-sink balance in the plant, for example drought (Luquet et al. 2008; Pallas et al. 2013).

Plant phenotypic plasticity is commonly observed as both plant morphology and phenology adjust to resource availability and environmental signals (Sultan 2000, Dingkuhn et al. 2005, Luquet et al. 2005), causing GxE interactions. Whatever the nature of a constraint, the plant response through adaptive plasticity ultimately modifies assimilate allocation. A relevant way to analyze processes involved in such plasticity is to directly affect plant C assimilates availability, as experienced by the plant growing under limited resource availability. In this sense, previous studies addressed the phenotypic plasticity of rice tillering, e.g., under phosphorus deficiency (Luquet et al. 2005), variable crop fertilization and population (Shimizu et al. 2010), or shading at vegetative and early reproductive stage (Lafarge et al. 2010) or grain filling (Kobata et al. 2000, Okawa et al. 2003). However, less is known on the adaptive plasticity of panicle architecture and its genetic control. A major objective of rice breeders was to develop large panicles, in terms of many fertile spikelets, in order to raise the yield ceiling, starting with the conception of new plant types with varieties having limited tillering ability (New Plant Type, NPT: Dingkuhn et al. 1991; Peng et al. 1994; Super Hybrid Rice: Peng et al. 2008; Li et al. 2014). These large panicles raised, however, questions on the adequate panicle structure and size for efficient grain filling (Peng et al. 2008) and the trade-off between panicle size and number (Dingkuhn et al. 2015).

The aim of the present study was to better quantify the variation of panicle size and architecture caused by *qTSN4* and to assess to which extent its genetic background and environmental factors affect phenotypic plasticity. For this purpose, the NILs developed by Fujita et al. (2009, 2012, 2013) and their recurrent parents, differing in panicle size, were grown into two experiments with two treatments providing differential plant access to light: a greenhouse experiment (Montpellier, France) with a full light and shading treatment during panicle development, and a field trial (IRRI, Philippines) characterized with two population densities. Panicle structure was phenotyped using detached panicles characterized by the software P-TRAP (**P**anicle **TRA**it **P**henotyping) developed by Al-Tam et al. (2013). The specific objectives were to (1) determine the plasticity of *qTSN4* phenotypes in the different environments, (2) analyze which structural traits of the panicle vary as panicle size varies, and (3) relate variation of panicle size to pre-floral stem growth rates.

Results

The qTSN4 and treatment effect on yield components

Grain production per unit area (filled grain dry weight per square meter) in GH significantly increased in the presence of *qTSN4* for both genetic backgrounds (Table 2). For IR64 background, this was associated with a significant increase in filled grain number per panicle and a significant decrease in 1000 grain dry weight. For IRRI146 background, none of these yield components alone made a significant contribution to the increase in grain production. Panicle number per plant and per unit ground area was not affected by the QTL in GH.

In the field, however, *qTSN4* did not affect grain production per unit area in IR64 background and even significantly reduced it in IRRI146 background (Table 3). In IR64 background, the absence of *qTSN4* effect on grain production was associated with a reduction in panicle number (significant per unit of ground area) and in 1000-grain dry weight, which were compensated by a significant increase in the filling rate and a non-significant but numerically large increase in the number of filled grains per panicle. In IRRI146 background, the reduction in grain production in the presence of *qTSN4* was the consequence of a significant reduction in panicle number (per ground area and per plant) and filling rate, which were not compensated by the positive effects on the other yield components. The contrasting effect of *qTSN4* on grain production between the GH (positive) and field (negative or neutral) environments mainly resulted from (i) an increase in grain number per panicle that occurred in GH but not in the field, and (ii) a reduction in panicle number that occurred in the field but not in the GH.

Shading during panicle development in GH caused a significant ($P < 0.001$) reduction in grain production for both genetic backgrounds (Table 2). The reduction, averaging 47 % across genotypes, was explained by the associated reduction in 1000-grain weight (−7 %) and filled grain number per panicle (−46 %), the effects being similar among genetic backgrounds. Panicle number was not affected as shading was only imposed after panicle initiation (Table 2). No *qTSN4* x treatment interactions were observed (Table 2), indicating that QTL and shading effects acted independently. In fact, shading had been implemented after panicle initiation to avoid affecting panicle number, so that yield effects of shading would be a result of competition among panicles. Both shading and QTL effects on yield were thus mainly due to spikelet number per panicle and 1000-grain weight.

Increased plant population in the field (Table 3), which reduced access to light resources per plant but not per ground area, had no effect on grain production but strongly affected yield components. It significantly ($P < 0.001$) increased panicle number per ground area, but by compensation reduced panicle number per plant and filled grain number per panicle for all genotypes ($P < 0.001$). No *qTSN4* x treatment interactions was observed when plant density changed from 25 to 100 plants m^{-2}.

The interaction between genetic backgrounds, the presence or absence of the QTL and the treatments was further analyzed with a three-stages nested design ANOVA conducted in each trial as described with P values in Additional file 1: Table S1 (electronic annex). In GH, grain production was significantly affected by shading within QTL ($P < 0.0001$) as were affected filled grain number per panicle, 1000 filled grain dry weight and filling rate in the main tiller, whereas panicle number per plant was not affected. In the field, however, grain production was not affected by plant density within QTL whereas panicle number per ground area was significantly affected ($P < 0.0001$) (Additional file 1: Table S1). This difference in performance between shading in the greenhouse and density in the field highlights an interaction between QTL and environments. A QTL effect within genetic backgrounds was observed with 1000 filled grain dry weight and filling rate for both GH and the field, but with filled grain number per panicle only in GH (Additional file 1: Table S1).

Traits observed on individual tillers

The increase in grain production at the whole plant level caused by *qTSN4* in GH (Table 2) was also observed at the single tiller level (main culm and tiller 4) in both genetic backgrounds (Table 4). This was associated with an increase in the dry weight fraction of the panicle (compared to the stem), although not significant for T4 in IRRI146 background, and in final panicle dry matter and filling ratio (this one was not measured for tiller 4) (Table 4). In the field, the absence of *qTSN4* impact on grain production in IR64 background (Table 3) was supported by the absence of any effect at tiller level on panicle dry matter (Table 5) although the dry weight fraction (panicle/stem) was even significantly reduced in the presence of the *qTSN4*. By contrast, in IRRI146 background, the reduction in grain production (Table 3) was surprisingly associated with an increase in panicle dry matter of the main tiller but a decrease of that of tiller 4 (Table 5). This suggests a strong dominance of the main tiller over primary tillers in IRRI146 background. At the same time, dry weight fraction (panicle/stem) was not affected by *qTSN4* for both tillers.

The reduction in grain production observed at plant level due to shading in GH in both genotype backgrounds (Table 2) was associated at single tiller level with a reduced panicle dry matter and dry weight fraction (panicle/stem) in both tiller types (Table 4). In the field, the lower grain production per plant (Table 3) at high population was also confirmed at tiller level but the reduced panicle dry matter was significant only for IR64 (Table 5). No *qTSN4* x treatment effect interactions were observed at single tiller level (Tables 4 and 5).

Traits constituting panicle architecture

The image processing of main-tiller panicles by the P-TRAP software resulted in the extraction of parameters related to panicle structure (panicle length, branch number) and spikelet number (as an example with IR64 background, Fig. 1). Visually the QTL effect was not discernible, and differences were only detected by measurements (as described thereafter). Regarding treatment effect, the panicle architecture of IR64 was notably modified by shading in the greenhouse and by high density in the field, while the panicle of the NIL appeared not as affected as that of the parent line. Panicle architectural traits extracted from the image collection through P-TRAP are summarized in Additional file 1: Table S2 (electronic annex) and Fig. 2. Shading in GH significantly reduced spikelet number (Fig. 2A) and total branch length (Fig. 2B) for all genotypes. The *qTSN4* numerically increased spikelet number and total length in both treatments and both genetic backgrounds but the effect was significant only for IR64 background and for the shaded treatment. Interestingly, patterns of *qTSN4* and treatment effects were similar for spikelet number and total length, both in GH and field, indicating a conserved proportionality between the number of spikelets

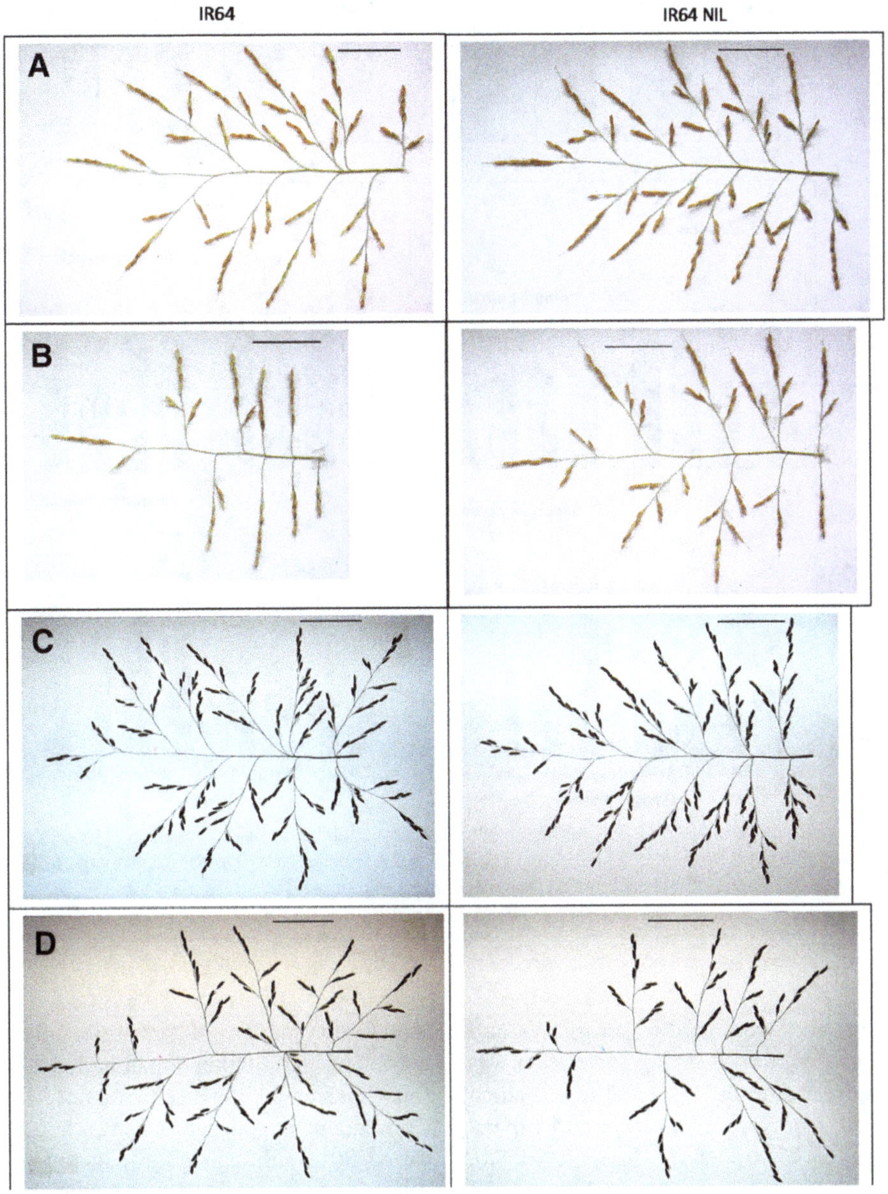

Fig. 1 Panicle architecture analysis by P-TRAP of IR64 parent line and its NIL. **a** Greenhouse – control. **b** Greenhouse – shading. **c** Field – low density. **d** Field – high density. Scale bar 5 cm

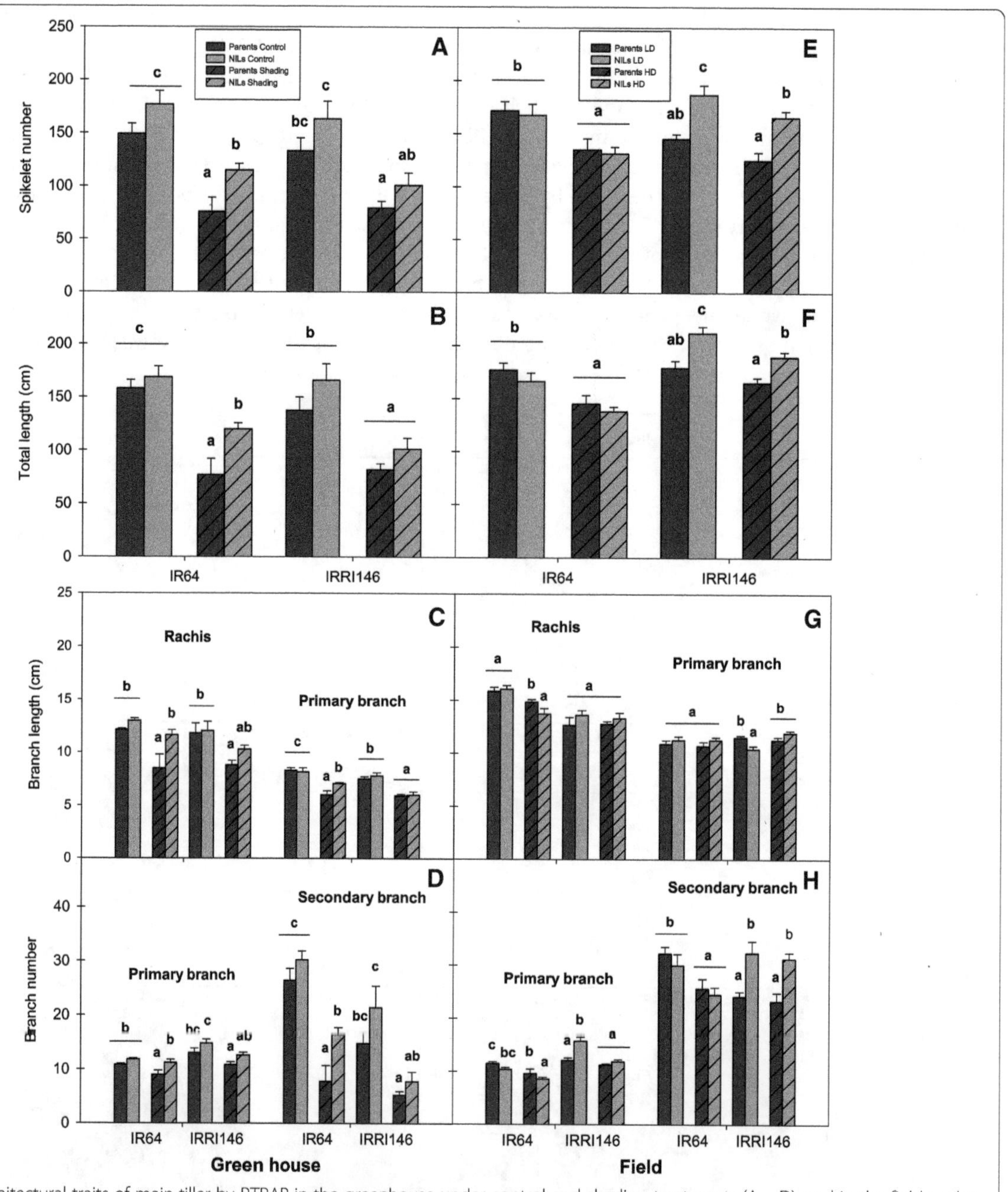

Fig. 2 Panicle architectural traits of main tiller by PTRAP in the greenhouse under control and shading treatments (A – D), and in the field under low density (LD) and high density (HD) treatments (E – H). A and E Spikelet number per panicle. B and F Total length of branches (cm). C and G Rachis and primary branch length (cm). D and H Primary and secondary branches number. The values are mean ± SE. Different letters indicate significant differences at $P < 0.05$ according to Duncan test for multiple comparisons of each genotype ($n = 5$ for GH-CNRS, $n = 8$ for field-IRRI)

and their supporting structure regardless of *qTSN4* and treatment effects. Consequently, a strong correlation was observed between spikelet number (SN) and total branch length (TL) across all genotypes, treatments and experiments (Fig. 3) with SN = 15.00 + 0.83TL [cm] ($R^2 = 0.88$). The relative effects of *qTSN4* vs. parent on SN and TL (Additional file 1: Table S2) were also correlated ($R^2 = 0.96$) and varied proportionally (Y = 0.02 + 1.07X) across the 16 combinations of background, treatment and experiment. Spikelet number and cumulative branch length were thus linked and the QTL did not affect this allometry. Consequently, spikelet density (number/branch length) varied little, although *qTSN4* increased it marginally but significantly in two cases (Additional file 1: Table S2).

The length of the rachis in GH was reduced by shading with the parents but not with the NILs (Fig. 2C),

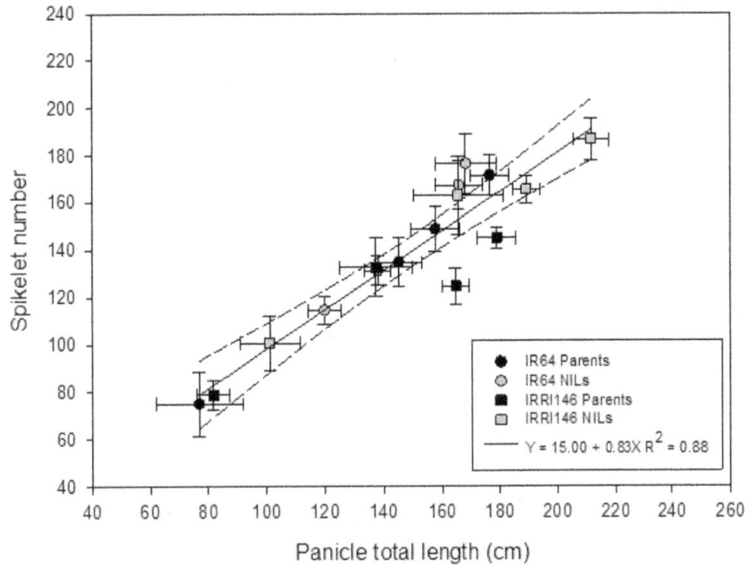

Fig. 3 Relationships between panicle total length (cm) and spikelet number of main tiller across treatments and experiments of the parents (black symbols) and the NILs (grey symbols) of two genetic backgrounds. Regression curves are associated with confidence interval at P = 0.05 (n = 5 for GH-CNRS, n = 8 for field-IRRI)

and, in the field, was not consistently modified by population treatment (Fig. 2G). Primary branch length was slightly reduced by shading in GH (Fig. 2C) but not affected by increased population in the field (Fig. 2G). No consistent positive or negative effect of *qTSN4* was observed on rachis and primary branch length. Secondary branch length was not affected by *qTSN4* in either experiment, except for a positive *qTSN4* effect observed in IRRI146 under high population in the field (data not presented). While primary branches (6–12 cm) were nearly as long as the rachis (8–16 cm), secondary branches were much shorter (2–3 cm). All branch types were longer in the field than in GH.

Patterns of *qTSN4* and treatment effects on primary branch number (Fig. 2D and G) were small and resembled those described for rachis length (Fig. 2C and G, Additional file 1: Table S2). The number of secondary branches was much more variable than that of primary branches, particularly in GH. Shading reduced it strongly (Fig. 2D), whereas increased population did so only for IR64 (both NILs). The *qTSN4* generally stimulated secondary branch number, particularly in IRRI146 (both in GH and in the field), which appeared to be the main trait of panicle plasticity. Overall, secondary branching was the main source of panicle structural plasticity, and it was significantly affected by both factors shading and QTL. The spikelet number on the main tiller was positively correlated with the main stem dry weight-based growth rate during panicle development for both genetic backgrounds (Fig. 4a and b) and both GH and field (Fig. 4c and d). In each representation, the average value of the *qTSN4* -NILs was higher than that of the parent lines, for both spikelet number and growth rate. The correlation was stronger in GH than that in the field (Figs. 4c vs. d) and for IRRI146 than for IR64 background (Figs. 4b vs. a). On average for all experiments and treatments, the QTL increased the main stem growth rate by 38 % and SN by 15 %. Results were similar for correlations between secondary branch number and main stem growth rate (data not presented).

Discussion

Panicle size and structure, in addition to panicle number, determine the sink components of yield potential in cereals like rice. It is known that panicle number and size are plastic traits prone to mutual compensation under competition for resources (Dingkuhn et al. 2015). The present study aimed at exploring interactions among (1) *qTSN4*, known to increase spikelet number, (2) genetic background and (3) environments involving different levels of competition for light as they affect panicle architectural traits. The choice of NILs thereby allowed us to study a single-locus effect on the behavior of panicle traits and their interactions with other traits participating into phenotype plasticity.

The study was conducted on two recipient lines (IR64, IRRI146) and their isolines (NIL) carrying *qTSN4* (syn. SPIKE), a QTL enhancing spikelet number per panicle (Fujita et al. 2009, 2012, 2013). This QTL was reported to improve, in addition to panicle size and sink capacity,

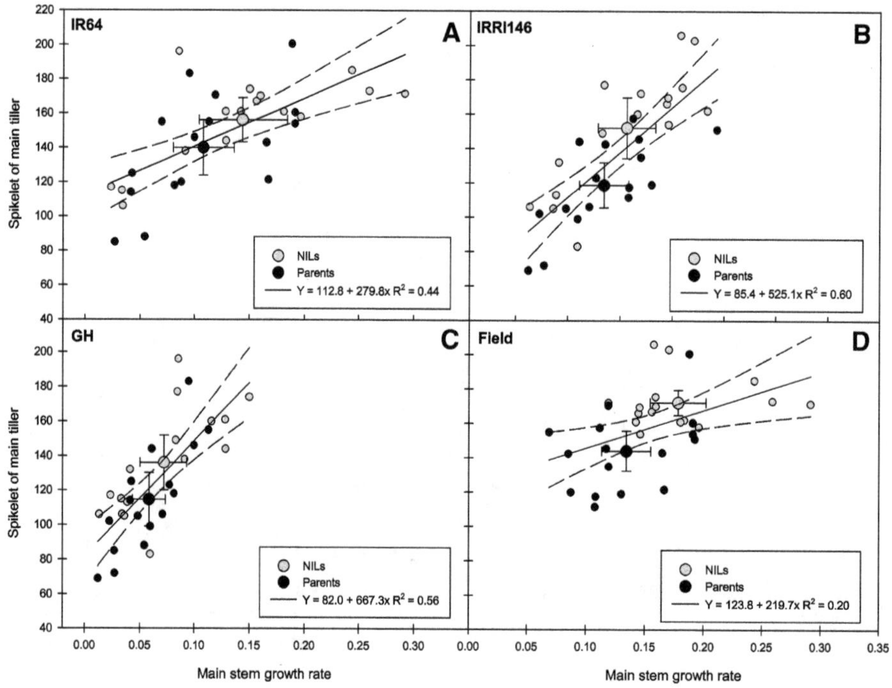

Fig. 4 Relationships between main stem growth rate and spikelet number (SN) of main tiller of the parents (black symbols) and the NILs (grey symbols) of two genetic backgrounds. **a** IR64 background in both experiments. **b** IRRI146 background in both experiments. **c** Green house experiment in both backgrounds. **d** Field experiment in both backgrounds. Regression curves are associated with confidence interval at P = 0.05 (*n* = 16)

assimilate source traits such as flag leaf size (Adriani et al. 2016) and carboxylation capacity (Fabre et al. 2016), owing to the association of *Nal1* gene within this QTL (Fujita et al. 2013). This might suggest that *qTSN4* effects on panicle size are not directly controlled by *qTSN4* but indirectly *via* whole-plant assimilate source-sink relations. If its effect on panicle size is indirect, we can expect it to be (1) highly prone to GxE interactions (Zhuang et al. 1997; Kobayashi et al. 2004; Chen et al. 2008), (2) non-specific with respect to panicle architectural traits (conservation of structural allometric relationships as panicle size varies), and (3) associated with pre-floral growth rate which is known to physiologically control panicle size (Horie et al. 2003; Kamiji et al. 2011). To test these hypotheses, we challenged NILs for *qTSN4* with variable light resources, focused on structural panicle traits and determined their relationship with pre-floral growth rates. The fine characterization of panicle architectural plasticity was made possible here by using an original, dedicated software P-TRAP (Al-Tam et al. 2013) and comparing i) control and shaded plants during panicle formation in a greenhouse experiment in Montpellier and ii) low and high-density crops in the field at IRRI in the Philippines.

Overall, the results confirmed a positive effect of *qTSN4* on panicle spikelet number, which in turn was correlated with the main stem growth rate during panicle development. However, in the present study, this effect suggests a dominance of the main tiller over the other tillers as it is mainly observed at main tiller level, and led to plant- and population-level yield increases only in GH but not in the field which questions the yield advantage of qTSN4 over that of parent lines IR64 and IRRI146 reported across four seasons in IRRI fields (Fujita et al. 2013).

Hypothesis 1 – *qTSN4* phenotype depends on GxE interactions

Within each of the two experiments, no significant GxE effect was observed for any of the observed variables. In IRRI146 background, the positive *qTSN4* effect on panicle size and spikelet number per panicle was stable across environments. Fujita et al. (2013) also reported, in one growing season in the Philippines, that spikelet number per panicle was increased in IRRI146-SPIKE, the NIL backcrossed to IRRI146 background. However in this study, in IR64 background, the *qTSN4* effect was observed only in GH but it was absent in the field. This contrasts with the result reported by Fujita et al. (2013) over four seasons in the field (two trials in each the dry and the wet season) showing systematically a positive effect of *qTSN4* on panicle size and spikelet number in

IR64 background. Despite some differences between studies in total N application (90 kg N in Fujita's case *vs.* 160 kg N here) and plant density (20 plants m^{-2} with 3 seedlings per hill in Fujita's case *vs.* 25 or 100 plants m^{-2} with 1 seedling per hill in the present case), panicle number was consistent between the two studies, being intermediate in Fujita's case (400–525 m^{-2}) compared to the values of the present study (344–362 m^{-2} under low and 471–598 m^{-2} under high population). In both field studies, *qTSN4* reduced panicle number, which was over-compensated by the increased panicle size in Fujita's case (leading to increased yield) but under-compensated in our study. By contrast, in GH in the present study in both light treatments, *qTSN4* did not affect panicle number but increased panicle size. The interaction between *qTSN4* and the environment on plant performance was highlighted when comparing the contrasted behavior between the greenhouse and the field.

Effects of *qTSN4* on other plant morphological traits than panicle size were also observed in a variable fashion, and panicle size may not be the primary phenotypic expression of this QTL. Both panicle number and size effects may even be the results of trade-offs induced by other, phenologically earlier *qTSN4* effects. In this sense, the positive effect of *qTSN4* was also reported on blade area and internode length of the last phytomers of the tiller, associated with higher photosynthesis and starch storage in internodes (Adriani et al. 2016; Fabre et al. 2016). These results highlight the role of the tiller internal trophic status on panicle size and support the G x E interaction observed here on panicle size.

Hypothesis 2 – *qTSN4* affects panicle size while panicle structural allometries are conserved

The panicle architectural traits that were increased in GH (together with SN) in the presence of *qTSN4* were more frequently branch number than branch length, whatever the light treatment. In the field, this was confirmed only for IRRI146 background. This is in line with Fujita et al. (2012) who observed a greater spikelet number in *qTSN4* lines, associated with an exclusive increase in panicle branch number, and with Zhang et al. (2014) who reported the increase of secondary branches, panicle length and grain number per panicle in lines carrying *qLSCHL4*, a QTL allelic to Nal1 that regulates leaf morphology and chlorophyll content. Similar observations, indicating an important role of secondary branch number in supporting panicle spikelet number, were also made for other QTLs. Mei et al. (2006), Hashida et al. (2013), Terao et al. (2010) and Ohsumi et al. (2011) reported higher spikelet number per panicle in the presence of QTL caused by more primary (*qPBN6*) or secondary branches (*qSBN1*). Interestingly, Peng et al. (2014) reported a QTL specifically for high primary branch number (*qPPB3*).

The *qTSN4* increased rachis and primary branch length in IR64 background under shade, but this effect only mitigated the length reduction induced by shading. Panicle morphogenesis starts with the formation of primary branches on successive nodes of the rachis, followed by higher-order branches (5 – 14 days after PI), and finally by floret differentiation and branch elongation (14 or 16 days after PI until heading) (Yamagishi et al. 2004; Itoh et al. 2005; Wang and Li 2005). Indeed, branch number is the main trait driving panicle sink size, and we hypothesize that light (or assimilate) resources predominantly affect numbers (of branches or spikelets) rather than their size. An analogy can be raised with the effect of shading on tillering between PI and flowering in the field, where tiller and panicle number in IR64 were mostly affected whereas stem elongation and spikelet number per panicle were comparatively conserved (Lafarge et al. 2010).

In the present study, the ratio (allometry) of spikelet number over total branch length per panicle was conserved across large variations of both variables, were they caused by *qTSN4* or experiment or treatment. The frequency of secondary branches, which provides for supplementary spikelet sites if resources allow them, was the most variable trait of panicle architecture. Panicles with increased spikelet number due to *qTSN4* thus resembled panicles receiving more resources during development, with specific structural effects being absent.

Hypothesis 3 – *qTSN4* effect on panicle size is related to pre-floral growth rate

Increases of spikelet number per panicle caused by *qTSN4* in the field, when they occurred, were associated with reduced panicle number, probably due to competition for resources, which were also associated with earlier cessation of tillering (Adriani et al. 2016). Such physiological trade-offs in the presence of *qTSN4* were apparently also involved in the results of Fujita et al. (2013) and Okami et al. (2015) and also those of Ohsumi et al. (2011) and Weng et al. (2014) who used different QTLs affecting panicle size. The phenotype generated by *qTSN4* can thus only be understood at the whole-plant or crop level. Complex pleiotropic effects were also reported for other QTLs or genes affecting the panicle, e.g., gene *OsSPL14 (WFP)* affecting shoot branching at vegetative stage and tiller final number (Miura et al. 2010), and gene *APO1* increasing leaf number, panicle primary branch number and harvest index (Ikeda et al. 2007; Terao et al. 2010).

Cereal crop models commonly predict spikelet number or post-floral sink capacity in function of shoot growth rate during panicle development (Bouman et al. 2001; Kouressy et al. 2008). At the individual (main) tiller level, we observed a significant correlation between panicle spikelet number and stem pre-floral growth rate, which explained at least partly the effects of *qTSN4*, environment and treatment. All factor effects were thus related to the internal availability of assimilates during panicle development, which in itself is not physiologically "expensive" but may be regulated by mechanisms such as sugar signaling (Rolland et al. 2006). Both spikelet number and stem growth rates between PI and flowering were larger in the field than in GH, and the correlation across factors was particularly strong for the NILs having IRRI146 background, which also showed the more consistent *qTSN4* effect. In addition, Sheehy et al. (2001), Takai et al. (2006) and Chen et al. (2008) reported that biomass accumulation in the stem (including leaf sheath) during panicle development contributes to panicle spikelet number. Also, spikelet number shares a common genetic basis with above-ground biomass (Zhang et al. 2004) and crop growth rate at 14 days before heading (Horie et al. 2003, Kamiji et al. 2011). Similarly, the largest panicles of a sorghum plant are carried by the most vigorous tillers of the plant (Lafarge et al. 2002), indicating that panicle size depends on phenologically earlier traits associated with tiller vigor.

The pleiotropic effects of *qTSN4* may thus have a common physiological basis, although a more direct genetic control of several traits by the QTL cannot be excluded. The QTL's primary effects shall be related to resource availability per tiller or to more upstream processes, also highlighted by Adriani et al. (2016), thereby affecting panicle size and, in some cases, yield. In this sense, according to Qi et al. (2008), the gene *Nal1* involved in leaf size, veining pattern and carboxylation is located within *qTSN4*. In addition, Takai et al. (2013) reported that *Nal1* has been identified as a QTL for GFS (Green for Photosynthesis) which changed leaf structure leading to pleiotropically enhanced photosynthesis rate. Therefore, leaf size *per se* can theoretically affect tiller number by compensation, which in turn would affect panicle number and ultimately, size.

Conclusions

In this study, *qTSN4* was confirmed to increase panicle sink size, particularly in IRRI146 background, while proportionally stimulating panicle branching and total branch length, under different light resources. Trade-offs between panicle size and panicle number were, however, observed in most cases in this study, with a weaker effect in the field where a systematic and positive effect of *qTSN4* on grain production could not be reported. Variation in panicle sink size was related to pre-floral stem dry-weight growth, explaining some of *qTSN4*, treatment and environment effects. We conclude that *qTSN4* probably increases primarily assimilate resources available at tiller level, followed by compensatory effects on tiller/panicle number, panicle size and structure that are environment and genetic-background dependent. While the role of *qTSN4* cannot be confirmed at crop level, its explicit effect reported here at plant level was confirmed at tissue level (Fabre et al. 2016).

Methods

Plant materials

Two pairs of parents/near-isogenic lines (P/NIL), IR64 vs IR64 NIL, and IRRI146 vs IRRI146 NIL, were used in this study. The polymorphic locus was *qTSN4* (for total spikelet number) syn. SPIKE, associated with the gene NAL1 (Fujita et al. 2013). The NIL with IR64 background was developed by self-pollination of a plant selected from BC4F2 population. In IRRI 146 background, a whole-genome survey of 96 BC3F1 plants using 116 polymorphic SSR markers that covered all chromosomes was conducted. One BC3F1 plant was selected and self-pollinated to develop the NIL with IRRI 146 background (Fujita et al. 2013). These NILs were identified carrying the high total spikelet number (TSN) QTL between SSR markers RM3423 and RM17492 on the long arm of chromosome 4 (Fujita et al. 2012). The isoline of IR64 received this QTL from the donor parent YP9, which was derived from a cross between indica cultivar Shennung 89–366 and tropical japonica landrace Daringan. The isoline of IRRI146 received the QTL from the donor parent YP4, which was derived from a cross between tropical japonica Bali Ontjer and IRRI146 (Fujita et al. 2013). The recipient lines (parents) were chosen because of their wide adaptability: IR64 is a mega variety grown in many parts of the world (Khush, 1987) and IRRI146 is a 2nd-generation New Plant Type (NPT) variety developed at IRRI and released in the Philippines in 2007 under the name NSIC Rc158, also known as a high-yielding *indica* cultivar (Brennan and Malabayabas, 2011). The NILs were morphologically distinguishable by their larger leaf size caused by the QTL, and the plant populations were visually homogenous (no off-types).

Design of experiment

Greenhouse (GH) experiment

The greenhouse experiment was performed from May to August 2013 at the National Center for Scientific Research (CNRS), Montpellier, France. The experiment adopted a two factors completely randomized design with three replications for panicle architecture, biomass and yield components measurements. The first factor was light treatment including two levels, C: natural daylight, S: shading from panicle initiation (PI) to heading. The second factor was rice genotype (G) including 4 genotypes: IR64 parent (IR64) and its NIL (*qTSN4.4* – YP9), IRRI146 parent (NSIC Rc158) and its NIL (*qTSN4.1* – YP4).

The seeds were grown in a germination chamber at 29 °C, then transplanted 4 days after germination in 3 l pots (3 seeds per pot) when seedlings were about 3 cm tall. The thinning of plant population to 1 plant per pot (downsizing to 45.65 plants m^{-2}) was conducted at 4-leaf stage. Pots contained about ¾ of their volume with EGOT 140 media (17 N-10P-14 K, pH of 5) and were placed side by side (corresponding to 14.8- cm spacing) on tables filled with 5 cm water depth. Basal fertilizer was applied using Basacot 6 M+ at 2 g l^{-1}, 11 N-9P-19 K +2 Mg incorporated before transplanting. A number of 104 pots were arranged on four aluminum tables, including border plants, at the beginning of the experiment. The tables were moved every week from two weeks after transplanting until maturity to avoid any bias due to the green house structure.

When the plants were at about PI (52 – 64 days after transplanting, depending on the genetic background), two layers of black net (90 × 140 cm^2) were set up at 30 cm above the canopy enclosing 50 plants (excluding border plants). The height of the net was then adjusted based on plant growth. Two tables were dedicated to the shading treatment. The light attenuation, 58 %, was then calculated from the difference of the average of daily PAR between control and shading divided by PAR under control. The plants were exposed back to normal conditions once the plants of the same genetic background under natural light were in heading stage.

Weather data were collected from the AWS (*Automatic Weather Station*) that was installed in the center of the tables measuring Photosynthetic Active Radiation (PAR), global radiation (Rg), air temperature (T) and relative humidity (RH). The average daily air temperature throughout the crop cycle was 27.3 ± 0.6 °C. The average daily PAR for the whole crop cycle was 24.7 ± 7.1 mol m^{-2} d^{-1} under full light and 10.3 ± 3.4 mol m^{-2} d^{-1} under shading, and the average relative humidity was 66.8 ± 7.7 %.

Field experiment

The field experiment was performed in split plot design with four replications during the dry season (December 2013 to April 2014) at the International Rice Research Institute (IRRI) experiment station in Los Baños, Philippines (14°11′N, 121°15′E, 21 m altitude). The soil was an Andaqueptic Haplaquoll with a topsoil of 61 % clay, 32 % silt, 7 % sand pH of 6.2. The main factor was plant density, as low plant density of 20 cm × 20 cm (LD, 25 plants m^{-2}), and high plant density of 10 cm × 10 cm (HD, 100 plants m^{-2}). The subsidiary factor was the same rice genotypes as used in the greenhouse experiment. The density levels were chosen to reflect the recommended standard population of transplanted rice in much of the tropics including the Philippines (25 plants m^{-2}) and a 4-fold greater population to approximate conditions under the increasingly popular direct seeding practice.

The seeds were soaked for 24 h, drained and incubated for another 24 h, then sown in the seeding trays in the greenhouse on December 5, 2013. The 2-week old seedlings were transplanted in the field at one plant per hill in a 2 × 2.4 m^2 plots. The field was initially flooded to hold two puddlings and two harrowings, standing water level of 3–5 cm, was maintained as the IRRI guide field standard. Phosphorus (30 kg P ha^{-1}), potassium (40 kg K ha^{-1}) and zinc (5 kg Zn ha^{-1}) were applied and incorporated into all the plots 2 days before transplanting. Nitrogen (60 kg N ha^{-1}) was applied 1 day before transplanting, then 40 kg N ha^{-1} and 60 kg N ha^{-1} were applied at mid-tillering and panicle initiation stage, respectively.

Weather data as radiation (MJ m^{-2}), daylight (hour), rainfall (mm), evaporation (mm), average temperature (°C), vapor pressure (kPa), relative humidity (%) and wind speed (m s^{-1}) were collected from the IRRI meteorological station. Average daily air temperature throughout crop cycle was 25.6 ± 1.5 °C. The average daily PAR for the whole crop cycle was 31.0 ± 11.3 mol m^{-2} d^{-1}, and the average relative humidity was 84.2 ± 4.8 %.

Plant measurements

All measurements described below were performed at physiological maturity. Plants were considered at physiological maturity when 75 % of the grains of the panicles had turned yellow and the texture was in dough stage.

Plant phenology, morphology and biomass accumulation

The whole plant was characterized with the stem length (distance from the soil to the highest collar) of main tiller and tiller 4 (the tiller that outgrew from the bud of phytomer 4 on the main stem), the number of leaves produced by the main tiller, the total green and dead (only for greenhouse experiment) leaf number, and the green tiller number. Three consecutive plants per treatment (as three replications chosen randomly) were tagged for phenological observations in the greenhouse and in the field. In this study, we considered the 'stem' as the addition of internode and sheath, and the 'tiller' as the addition of a stem, the leaf blades, and the panicle.

Panicle initiation (PI) was determined by dissecting and observing the main tiller of randomized collected plants (border plants for field experiment) from each single treatment every second day when PI was close. PI was considered to have occurred when the first row of floral primordial was visible on the shoot apex. Flowering (FLO) was considered to have occurred within each treatment when an average of 75 % spikelets per panicle of the main tiller exerted their anthers.

Biomass accumulation at maturity was measured in GH on the plants used for phenological observation (three plants) plus two other plants. In the field, three plants used for phenology observation in both densities were chosen for biomass measurement, plus 9 additional plants in HD treatment growing next to the three initial ones, in order to get the same soil area as the LD treatment. Leaves and stems of the main tiller and tiller 4 considered separately, and of the rest of the whole plant, were characterized by weighing the materials after having dried during 72 h in an oven at 70 °C. Panicle weight of the main tiller and tiller 4 was measured after sun drying. The dry weight fraction (panicle/stem) of a single tiller was determined by calculating the dry weight of the panicle divided by the corresponding stem dry weight. Main stem growth rate during panicle development was calculated between two key dates by dividing the increase in dry matter of the main stem by the number of days between the two key dates. In GH, the main stem growth rate was calculated from PI to heading and in the field from two weeks after PI to flowering. The description of the measured plant organs with their abbreviation are described in Table 1.

Table 1 Description of parameters with the unit of measurement

Traits	Abbreviation	Unit
Dry weight fraction (panicle/stem) of main tiller	$R_{DW\ (panicle)\ MT}$	
Dry weight fraction (panicle/stem) of tiller 4	$R_{DW\ (panicle)\ T4}$	
Panicle dry weight of main tiller	DW Panicle MT	g
Panicle dry weight of tiller 4	DW Panicle T4	g
Panicle architecture traits:		
Spikelet number	SN	
Panicle length	PL	cm
Total length of branches	TL	cm
Rachis length	RL	cm
Averaged primary branch length	PBL	cm
Averaged secondary branch length	SBL	cm
Primary branch number	PBN	
Secondary branch number	SBN	
Spikelet density	SD	

Panicle architecture

Panicle architecture was analyzed by spreading the panicle of the main tiller on a white background equipped with a scale. In GH, main-tiller panicles were sampled at physiological maturity from the 5 plants used for biomass accumulation and yield components for each genotype and each treatment. In the field, main-tiller panicles for each genotype and each treatment were sampled 10 days before physiological maturity to minimize spikelet losses and sun dried before being spread over the white background. Panicles of both experiments were systematically sampled by cutting the stem just below the neck node.

A picture was then taken for each spread panicles to analyze its structure by using P-TRAP, a Java-based standalone software. This software enables to automatize the extraction of parameters such as primary, secondary and tertiary axis length and number, and spikelet number. As an example (Fig. 5), yellow circles represent the points identified by P-TRAP to mark the starting and ending nodes of a panicle primary axis, bright green circles represent the primary node starting points, and blue circles represent the starting points of secondary axes, also named the primary branches, those directly attached to the primary axes. Green circles represent the starting points of tertiary axes, also named the secondary branches, those attached to the secondary axes. Red circles represent the end of axes (Al-Tam et al. 2013).

From the image analysis, many panicle structural traits and grain traits can be obtained. For this paper, we considered rachis length (RL), average primary branch length (PBL), average secondary branch length (SBL), primary branch number (PBN), secondary branch number (SBN), and one trait describing panicle grain that is spikelet number (SN). We also determined other traits

Fig. 5 P-TRAP analysis resulting panicle structure with the rachis length (**a**), primary branch length (**b**) and secondary branch length (**c**). Scale bar 5 cm

Table 2 P-values and changes of yield component values in the presence of the QTL (in comparison to the parent) and under low access to light (in comparison to the control) of the greenhouse experiment ($n = 5$)

Parameters	IR64 background				
	QTL	Increase or reduction	Treatment	Increase or reduction	QTL x T
Filled grain dry weight (g m^{-2})	0.0003	+50	<0.0001	-47	0.582
Panicle number m^{-2}	0.888	+1	0.778	+3	0.888
Panicle number plant $^{-1}$	0.888	+1	0.778	+3	0.888
Filled grain number panicle^{-1}	0.0002	+61	<0.0001	−46	0.810
1000 filled grain dry weight (g)	0.005	−6	0.001	−9	0.196
	IRRI146 background				
	QTL	Increase or reduction	Treatment	Increase or reduction	QTL x T
Filled grain dry weight (g m^{-2})	0.007	+20	<0.0001	−45	0.37
Panicle number m^{-2}	0.744	+3	0.648	+5	0.948
Panicle number plant^{-1}	0.744	+3	0.648	+5	0.948
Filled grain number panicle^{-1}	0.231	+12	<0.0001	−45	0.672
1000 filled grain dry weight (g)	0.038	+4	0.002	−6	0.456

related to panicle structure like panicle length (PL) (the length of panicle from the neck node up to the tip of the last spikelet), total length (TL) of the branches (RL + (PBL × PBN) + (SBL × SBN)) and spikelet density (SD), spikelet number divided by the sum of length of all branches. The abbreviations and units of measurements are summarized in Table 1.

Physiological maturity and yield components

The five plants harvested at maturity in GH were separated into panicles (after taken pictures for P-TRAP analysis), green leaf blades, senescent leaves, and productive stems (culms + sheaths). In the field, we harvested all the plants within a soil base area of 0.12 m^2 per plot that is 3 plants under LD and 12 plants under HD. They were then separated into panicles, green leaf blades, dead tissues and productive stems. The panicles were hand-threshed and then the filled spikelets were separated from the unfilled by a densitometric column (in GH) or submerging the spikelets in the water (in the field). Yield components were determined as panicle number per plant, 1000

Table 3 P-values and changes of yield component values in the presence of the QTL (in comparison to the parent) and under low access to light (in comparison to the high density) of the field experiment ($n = 4$)

	IR64 background				
Parameters	QTL	Increase or reduction	Treatment	Increase or reduction	QTL x T
Filled grain dry weight (g m^{-2})	0.612	−6	0.826	+3	0.741
Panicle number m^{-2}	0.050	−15	0.0003	+51	0.131
Panicle number plant $^{-1}$	0.216	−10	<0.0001	−62	0.74
Filled grain number panicle^{-1}	0.150	+16	0.002	−33	0.925
1000 filled grain dry weight	0.003	−6	0.802	0	0.25
Filling rate m^{-2}	0.018	+22	0.208	−9	0.769
	IRRI146 background				
	QTL	Increase or reduction	Treatment	Increase or reduction	QTL x T
Filled grain dry weight (g m^{-2})	0.032	−22	0.321	−10	0.992
Panicle number m^{-2}	0.010	−21	0.013	+25	0.946
Panicle number plant^{-1}	0.041	−22	<0.0001	−69	0.197
Filled grain number panicle^{-1}	0.774	−2	0.0003	−29	0.595
1000 filled grain dry weight	0.335	−4	0.213	+6	0.859
Filling rate m^{-2}	0.017	−10	0.774	−1	0.375

Table 4 P-values and changes of single tiller values in the presence of the QTL (in comparison to the parent) and under low access to light (in comparison to the control) of the greenhouse experiment ($n = 5$)

	IR64 background				
Parameters	QTL	Increase or reduction	Treatment	Increase or reduction	QTL x T
$R_{DW\ (panicle)\ MT}$	0.024	+31	0.0004	−39	0.849
$R_{DW\ (panicle)\ T4}$	0.037	+26	0.0001	−43	0.754
DW panicle MT	0.001	+57	<0.0001	−49	0.568
DW panicle T4	0.032	+29	<0.0001	−60	0.527
Filling rate in MT	<0.0001	+47	0.086	−12	0.085
	IRRI146 background				
	QTL	Increase or reduction	Treatment	Increase or reduction	QTL x T
$R_{DW\ (panicle)\ MT}$	0.043	+24	<0.0001	−55	0.58
$R_{DW\ (panicle)\ T4}$	0.058	+38	<0.0001	−62	0.302
DW panicle MT	0.031	+28	<0.0001	−50	0.546
DW panicle T4	0.023	+38	<0.0001	−52	0.36
Filling rate in MT	0.059	+6	0.068	−6	0.501

filled grain dry weight (the grain number was counted by a grain counter in GH and manually up to 100 grain number in the field), filled grain weight and filled grain number per panicle (considering 1000 grain dry weight), filled grain weight per square meter (considering plant density). In the field, the filling rate per square meter (ratio of filled number on total spikelet number) was also determined.

Statistical analyses

Yield components and single tiller parameters were tested for standard deviation. The graphs describing panicle architecture and main stem growth rate in relation to spikelet number were represented with standard error (standard deviation divided by square root of the number of samples). Data of Tables 2, 3, 4 and 5 and Figs. 2 and 3 were analyzed by an ANOVA procedure and mean comparison between parent vs NIL and between treatments for each pair of genotype by Duncan's multiple range test using Microsoft® Excel 2010/XLSTAT-PRO statistical software (version 2014, Addinsoft, Inc.,Brooklyn, NY, USA). The three-stages analysis of variance was used to analyze the QTL x T interactions effect on yield components using SAS version 9.1 software (SAS Institute Inc., Cary, NC, USA). SigmaPlot® Version 11.2 software (for Windows XP and below, copyright 2009–2010), Systat Software Inc. (Chicago, IL, USA) was used for plotting data and nonlinear regressions.

Table 5 P-values and changes of single tiller values in the presence of the QTL (in comparison to the parent) and under low access to light (in comparison to the high density) of the field experiment ($n = 4$)

	IR64 background				
Parameters	QTL	Increase or reduction	Treatment	Increase or reduction	QTL x T
$R_{DW\ (panicle)\ MT}$	0.013	−24	0.083	−23	0.972
$R_{DW\ (panicle)\ T4}$	0.050	−18	0.458	−9	0.161
DW panicle MT	0.983	0	0.005	−33	0.918
DW panicle T4	0.118	−13	<0.0001	−44	0.495
	IRRI146 background				
	QTL	Increase or reduction	Treatment	Increase or reduction	QTL x T
$R_{DW\ (panicle)\ MT}$	0.215	+9	0.036	+14	0.864
$R_{DW\ (panicle)\ T4}$	0.569	−5	0.824	+4	0.44
DW panicle MT	0.002	+48	0.545	−7	0.811
DW panicle T4	0.969	−6	0.016	−38	0.401

Competing interests

The authors declared that they have no competing interests.

Authors' contributions

DEA participated in performing the research, the data analyses and the writing of the manuscript; MD, DL and TL participated in designing and supervising the work and writing the manuscript; AD participated in collecting the data in the greenhouse; HA provided the P-TRAP software and helped in analyzing the P-TRAP results. All authors read and approved the final manuscript.

Acknowledgments

This work was a part of GRiSP (Global Rice Science Partnership) consortium and Yield Potential Project, in the scheme of Doctorate Double Degree program between Bogor Agricultural University, Indonesia and Montpellier SupAgro, France. The authors would like to gratitude to Faculty of Agriculture, University of Lambung Mangkurat for supporting the PhD program; and to the Directorate General of Higher Education of Indonesia, Agris Mundus (Agreenium) and CIRAD for providing a PhD scholarship. Lastly we thank the technicians Maximo Pelagio and Patrick Lumanglas and all contract workers for supporting field experiment in IRRI, Philippines.

Author details

[1]CIRAD, UMR AGAP, F-34398 Montpellier, France. [2]Faculty of Agriculture, University of Lambung Mangkurat, Banjarbaru, Indonesia. [3]Crop and Environment Science Division (CESD), International Rice Research Institute (IRRI), Los Baños, Philippines. [4]IRD, UMR DIADE, Montpellier, France.

References

Adriani DE, Lafarge T, Dardou A, Fabro A, Clément-Vidal A, Yahya S, Dingkuhn M, Luquet D (2016) The *qTSN* positive effect on panicle and flag leaf size of rice is associated with an early down-regulation of tillering. Front Plant Sci 6(1197):17

Al-Tam F, Adam H, dos Anjos A, Lorieux M, Larmande P, Ghesquiere A, Jouannic S, Shahbazkia HR (2013) P-TRAP: a panicle trait phenotyping tool. BMC Plant Biol 13:122

Bouman BAM, Kropff MJ, Tuong TP, Wopereis MCS, Ten Berge HFM, Van Laar HH (2001) ORYZA2000: Modeling Lowland Rice. International Rice Research Institute, and Wageningen: Wageningen University and Research Centre, Los Baños, p 235

Brennan JP, Malabayabas A (2011) Impacts of IRRI germplasm on the Philippines since 1985. International Rice Research Institute's contribution to rice varietal yield improvement in South-East Asia. ACIAR (Australian Center for International Agricultural Research) Impact Assesment Series Report 74:25–43

Chen S, Zeng F-R, Pao Z-Z, Zhang G-P (2008) Characterization of high-yield performance as affected by genotype and environment in rice. J Zhejiang Univ Sci B 9(5):363–370

Choi W-J, Lee M-S, Choi J-E, Yoon S, Kim H-Y (2013) How do weather extremes affect rice productivity in a changing climate? An answer to episodic lack of sunshine. Glob Chang Biol 19(4):1300–1310

Dingkuhn M, Penning de Vries FWT, De Datta SK, Van Laar HH (1991) Concepts for a new plant type for direct seeded flooded tropical rice. In: Direct Seeded Flooded Rice in the Tropics. International Rice Research Institute, Manila, pp 17–38

Dingkuhn M, Luquet D, Quilot B, de Reffye P (2005) Environmental and genetic control of morphogenesis in crops: towards models simulating phenotypic plasticity. Aust J Agric Res 56(11):1289–1302

Dingkuhn M, Laza MRC, Kumar U, Mendez KS, Collard B, Jagadish KSV, Kumar A, Singh RK, Padolina T, Malabayabas M, Torres E, Rebolledo MC, Manneh B, Sow A (2015) Improving yield potential of tropical rice: achieved levels and perspectives through improved ideotypes. Field Crop Res 182:43–59

Endo-Higashi N, Izawa T (2011) Flowering time genes *Heading date 1* and *early heading date 1* together control panicle development in rice. Plant Cell Physiol 52:1083–1094

Fabre D, Adriani DE, Dingkuhn M, Ishimaru T, Punzalan B, Lafarge T, Clément-Vidal A, Luquet D (2016) The *qTSN4* effect on flag leaf size, photosynthesis and panicle size, benefits to plant grain production in rice, depending on light availability. Front Plant Sci 7(623):12

Fujita D, Santos RE, Ebron LA, Telebanco-Yanoria MJ, Kato H, Kobayashi S, Uga Y, Araki E, Takai T, Tsunematsu H, Imbe T, Khush GS, Brar DS, Fukuta Y, Kobayashi N (2009) Development of introgression lines of an Indica-type rice variety, IR64, for unique agronomic traits and detection of the responsible chromosomal regions. Field Crop Res 114:244–254

Fujita D, Tagle AG, Ebron LA, Fukuta Y, Kobayashi N (2012) Characterization of near-isogenic lines carrying QTL for high spikelet number with the genetic background of an indica rice variety IR64 (*Oryza sativa* L.). Breed Sci 62:18–26

Fujita D, Trijatmiko KR, Tagle AG, Sapasap MV, Koide Y, Sasaki K, Tsakirpaloglou N, Gannaban RB, Nishimura T, Yanagihara S, Fukuta Y, Koshiba T, Slamet-Loedin IH, Ishimaru T, Kobayashi N (2013) *NAL1* allele from a rice landrace greatly increases yield in modern *indica* cultivars. Proc Natl Acad Sci U S A 110: 20431–20436

Hashida Y, Aoki N, Kawanishi H, Okamura M, Ebitani T, Hirose T, Yamagishi T, Ohsugi R (2013) A near isogenic line of rice carrying chromosome segments containing OsSPS1 of Kasalath in the genetic background of Koshihikari produces an increased spikelet number per panicle. Field Crop Res 149:56–62

Horie T, Lubis I, Takai T, Ohsumi A, Kuwasaki K, Katsura K, Nii A (2003) Physiological traits associated with high yield potential in rice. In: Mew TW, Brar DS, Peng S, Dawe D, Hardy B (eds) Rice Science: Innovations and Impact for Livelihood. IRRI, Los Baños, pp 117–145

Ikeda K, Ito M, Nagasawa M, Kyozuka J, Nagato Y (2007) Rice *ABERRANT PANICLE ORGANIZATION 1*, encoding an F-box protein, regulates meristem fate. Plant J 51(6):1030–1040

Itoh J, Nonomura K, Ikeda K, Yamaki S, Inukai Y, Yamagishi H, Kitano H, Nagato Y (2005) Rice plant development: from zygote to spikelet. Plant Cell Physiol 46:23–47

Kamiji Y, Yoshida H, Palta JA, Sakuratani T, Shiraiwa T (2011) N applications that increase plant N during panicle development are highly effective in increasing spikelet number in rice. Field Crop Res 122(3):242–247

Khush GS (1987) Rice breeding - past, present and future. J Genet 66(3):195–216

Kobata T, Sugawara M, Takatu S (2000) Shading during the early grain filling period does not affect potential grain dry matter increase in rice. Agron J 92:411–417

Kobayashi S, Fukuta Y, Yagi T, Sato T, Osaki M, Khush GS (2004) Identification and characterization of quantitative trait loci affecting spikelet number per panicle in rice (*Oryza sativa* L.). Field Crop Res 89(2–3):253–262

Kouressy M, Dingkuhn M, Vaksmann M, Heinemann AB (2008) Adaptation to diverse semi-arid environments of sorghum genotypes having different plant type and sensitivity to photoperiod. Agric For Meteorol 148:357–371

Lafarge TA, Broad IJ, Hammer GL (2002) Tillering in grain sorghum over a wide range of population densities: Identification of a common hierarchy for tiller emergence, leaf area development and fertility. Ann Bot 90(1):87–98

Lafarge T, Seassau C, Martin M, Bueno C, Clément-Vidal A, Shreck E, Luquet D (2010) Regulation and recovery of sink strength in rice plants grown under changes in light intensity. Funct Plant Biol 37(5):413–428

Li YH (1979) Morphology and Anatomy of Grass Family Crops. Shanghai science and Technology Press, Shanghai, pp 138–142

Li G, Zhang J, Yang C, Song Y, Zheng C, Wang S, Liu Z, Ding Y (2014) Optimal yield-related attributes of irrigated rice for high yield potential based on path analysis and stability analysis. Crop J 2:235–243

Luquet D, Zhang BG, Dingkuhn M, Dexet A, Clément-Vidal A (2005) Phenotypic plasticity of rice seedlings: case of phosphorus deficiency. Plant Prod Sci 8(2): 145–151

Luquet D, Clement-Vidal A, Fabre D, This D, Sonderegger N, Dingkuhn M (2008) Orchestration of transpiration, growth and carbohydrate dynamics in rice during a dry-down cycle. Funct Plant Biol 35(8):689–704

Mei HW, Xu JL, Li ZK, Yu XQ, Guo LB, Ying CS, Luo LJ (2006) QTLs influencing panicle size detected in two reciprocal introgressive line (IL) populations in rice (*Oryza sativa* L.). Theor Appl Genet 112(4):648–656

Miura K, Ikeda M, Matsubara A, Song XJ, Ito M, Asano K, Matsuoka M, Kitano H, Ashikari M (2010) *OsSPL14* promotes panicle branching and higher grain productivity in rice. Nat Genet 42(6):545–592

Ohsumi A, Takai T, Ida M, Yamamoto T, Arai-Sanoh Y, Masahiro Y, Tsuyu A, Motohiko K (2011) Evaluation of yield performance in rice near-isogenic lines with increased spikelet number. Field Crop Res 120(1):68–75

Okami M, Kato Y, Kobayashi N, Yamagishi J (2015) Morphological traits associated with vegetative growth of rice (*Oryza sativa* L.) during the recovery phase after early-season drought. Eur J Agron 64:58–66

Okawa S, Makino A, Mae T (2003) Effect of irradiance on the partitioning of assimilated carbon during the early phase of grain filling in rice. Ann Bot 92(3):357–364

Pallas B, Clement-Vidal A, Rebolledo MC, Soulié JC, Luquet D (2013) Using plant growth modeling to analyze C source–sink relations under drought: inter- and intra specific comparison. Front Plant Sci 4(437):1–13

Peng S, Khush GS, Cassman KG (1994) Evolution of the new plant ideotype for increased yield potential. In: Cassman KG (ed) Proceedings of a Workshop on Rice Yield Potential in Favorable Environments. International Rice Research Institute, Philippines, pp 5–20

Peng S, Khush GS, Virk P, Tang Q, Zou Y (2008) Progress in ideotype breeding to increase rice yield potential. Field Crop Res 108:32–38

Peng Y, Gao Z, Zhang B, Liu C, Xu J, Ruan B, Hu J, Dong G, Guo L, Liang G, Qian Q (2014) Fine mapping and candidate gene analysis of a major QTL for panicle structure in rice. Plant Cell Rep 33(11):1843–1850

Qi J, Qian Q, Bu Q, Li S, Chen Q, Sun J, Liang W, Zhou Y, Chu C, Li X, Ren F, Palme K, Zhao B, Chen J, Chen M, Li C (2008) Mutation of the rice *narrow leaf1* gene, which encodes a novel protein, affects vein patterning and polar auxin transport. Plant Physiol 147:1947–1959

Rolland F, Baena-Gonzalez E, Sheen J (2006) Sugar sensing and signaling in plants: conserved and novel mechanisms. Annu Rev Plant Biol 57:675–709

Sheehy JE, Dionora MJA, Mitchell PL (2001) Spikelet numbers, sink size and potential yield in rice. Field Crop Res 71(2):77–85

Shimizu H, Maruoka M, Ichikawa N, Baruah AR, Uwatoko N, Sano Y, Onishi K (2010) Genetic control of phenotypic plasticity in Asian cultivated and wild rice in response to nutrient and density changes. Genome 53:211–223

Shiratsuchi H, Ohdaira Y, Takanashi J-I (2007) Relationship between dry weight at heading and the number of spikelets on individual rice tillers. Plant Prod Sci 10:430–441

Sultan SE (2000) Phenotypic plasticity for plant development, function and life history. Trends Plant Sci 5(12):537–542

Takai T, Matsuura S, Nishio T, Ohsumi A, Shiraiwa T, Horie T (2006) Rice yield potential is closely related to crop growth rate during late reproductive period. Field Crop Res 96(2–3):328–335

Takai T, Adachi S, Taguchi-Shiobara F, Sanoh-Arai Y, Iwasawa N et al (2013) A natural variant of NAL1, selected in high-yield rice breeding programs, pleiotropically increases photosynthesis rate. Sci Rep 3:2149

Terao T, Nagata K, Morino K, Hirose T (2010) A gene controlling the number of primary rachis branches also controls the vascular bundle formation and hence is responsible to increase the harvest index and grain yield in rice. Theor Appl Genet 120(5):875–893

Wang Y, Li J (2005) The plant architecture of rice (*Oryza sativa*). Plant Mol Biol 59: 75–84

Weng X, Wang L, Wang J, Hu Y, Du H, Xu C, Xing Y, Li X, Xiao J, Zhang Q (2014) Grain number, plant height, and heading date7 is a central regulator of growth, development, and stress response. Plant Physiol 164(2):735–747

Xiong Y, Jing-Yong L, Wan-Jun R, Wen-Yu Y (2011) Primary study on the identification method of rice shade-endurance. Res J Agronomy 5(3–6): 32–37

Yamagishi J, Miyamoto N, Hirotsu S, Laza RC, Nemoto K (2004) QTLs for branching, floret formation, and pre-flowering floret abortion of rice panicle in a temperate japonica x tropical japonica cross. Theor Appl Genet 109: 1555–1561

Zhang ZH, Li P, Wang LX, Hu ZL, Zhu LH, Zhu YG (2004) Genetic dissection of the relationships of biomass production and partitioning with yield and yield related traits in rice. Plant Sci 167(1):1–8

Zhang G-H, Li S-Y, Wang L, Ye W-J, Zeng D-L, Rao Y-C, Peng Y-L, Hu J, Yang Y-L, Xu J, Ren D-Y, Gao Z-Y, Zhu L, Dong G-J, Hu X-M, Yan M-X, Guo L-B, Li C-Y, Qian Q (2014) LSCHL4 from Japonica cultivar, which is allelic to NAL1, increases yield of Indica Super Rice 93–11. Mol Plant 7:1350–1364

Zhuang JY, Lin HX, Lu J, Qian HR, Hittalmani S, Huang N, Zheng KL (1997) Analysis of QTL x environment interaction for yield components and plant height in rice. Theor Appl Genet 95(5–6):799–808

5

Diurnal Changes in Transcript and Metabolite Levels during the Iron Deficiency Response of Rice

Jamie Selby-Pham[1], Adrian Lutz[1,2], Laura T. Moreno-Moyano[1], Berin A. Boughton[1,2], Ute Roessner[1,2] and Alexander A. T. Johnson[1*]

Abstract

Background: Rice (*Oryza sativa* L.) is highly susceptible to iron (Fe) deficiency due to low secretion levels of the mugineic acid (MA) family phytosiderophore (PS) 2′-deoxymugineic acid (DMA) into the rhizosphere. The low levels of DMA secreted by rice have proved challenging to measure and, therefore, the pattern of DMA secretion under Fe deficiency has been less extensively studied relative to other graminaceous monocot species that secrete high levels of PS, such as barley (*Hordeum vulgare* L.).

Results: Gene expression and metabolite analyses were used to characterise diurnal changes occurring during the Fe deficiency response of rice. Iron deficiency inducible genes involved in root DMA biosynthesis and secretion followed a diurnal pattern with peak induction occurring 3–5 h after the onset of light; a result consistent with that of other Strategy II plant species such as barley and wheat. Furthermore, triple quadrupole mass spectrometry identified 3–5 h after the onset of light as peak time of DMA secretion from Fe-deficient rice roots. Metabolite profiling identified accumulation of amines associated with metal chelation, metal translocation and plant oxidative stress responses occurring with peak induction 10–12 h after the onset of light.

Conclusion: The results of this study confirmed that rice shares a similar peak time of Fe deficiency associated induction of DMA secretion compared to other Strategy II plant species but has less prominent daily fluctuations of DMA secretion. It also revealed metabolic changes associated with the remediation of Fe deficiency and mitigation of damage from resulting stress in rice roots. This study complements previous studies on the genetic changes in response to Fe deficiency in rice and constitutes an important advance towards our understanding of the molecular mechanisms underlying the rice Fe deficiency response.

Keywords: LC-MS, DMA, Strategy II, Plant stress response, Metal stress

Background

Iron (Fe) is utilized in a variety of essential biological processes such as the respiratory (Park et al. 1996) and photosynthetic (Hung et al. 2010) electron-transport chain reactions (Connolly and Guerinot 2002; Reddy 2006), due to its stable redox properties, specifically an ability to donate and accept electrons between both Fe^{+2} (ferrous) and Fe^{+3} (ferric) redox states. Although Fe is the fourth most abundant element in the earth's crust, it is not readily available to soil growing plants under aerobic conditions due to its tendency to form insoluble Fe^{3+} precipitates (Kim and Guerinot 2007). Insufficient Fe supply to plants leads to Fe depletion of both the mitochondria (Mori et al. 1991) and chloroplast (Stocking 1975; Terry and Low 1982), resulting in impaired efficiency of both cellular respiration and photosynthesis. Additionally, as Fe is required for the formation of the chlorophyll precursor δ-aminolevulinic acid (Marsh et al. 1963; Miller et al. 1984), Fe deficiency results in the characteristic yellowing or chlorosis of the youngest leaves (Terry and Low 1982). Furthermore, Fe deficiency is costly from an agronomic perspective (Hell and Stephan 2003) accounting for more

* Correspondence: johnsa@unimelb.edu.au
[1]School of BioSciences, The University of Melbourne, Parkville, Victoria, Australia
Full list of author information is available at the end of the article

than 25% of the crop yield losses of cereals, legumes and vegetables (Briat et al. 2015).

Iron is also the functional cofactor for the heme containing hemoprotein superfamily of enzymes involved in photosynthesis, respiration, DNA synthesis (Michel and Pistorius 2004) and antioxidants including peroxidases (Zaharieva and Abadía 2003). In cells, antioxidants quench reactive oxygen species (ROS) which are produced as an inevitable consequence of photosynthesis (Parida and Das 2005) and to a lesser extent due to respiration (Gill and Tuteja 2010). Cellular stress induced by Fe deficiency results in impaired antioxidant activity due to absence of the required catalytic Fe, leading to an increased abundance of cellular ROS (Becana et al. 1998; Iturbe-Ormaetxe et al. 1995). Accordingly Fe homeostasis must be tightly regulated in plants to prevent Fe-deficiency-induced oxidative damage to proteins and lipids (Sun et al. 2007).

Plants have evolved mechanisms to increase Fe uptake in response to conditions of limited Fe bioavailability. Dicots and non-graminaceous monocots employ Strategy I uptake, involving proton (H^+) secretion to the rhizosphere and reduction of soil Fe^{3+} to Fe^{2+}. Secretion of H^+ reduces soil pH and promotes the solubility of Fe ions and reduction of Fe^{3+} to the more bioavailable Fe^{2+} promotes increased uptake of Fe^{2+} through iron-regulated transporters (IRT) in the plasma membrane of plant root cells (Lee & An 2009; Scholz et al. 1992). Graminaceous plant species (grass family Gramineae) utilize Strategy II uptake involving secretion of mugineic acid (MA) family phytosiderophores (PS) into the rhizosphere to chelate and solubilise the soil Fe^{3+}, allowing for subsequent uptake of Fe^{3+}-PS complexes through yellow stripe 1-like (YSL) transporters in the plasma membrane of plant root cells (Inoue et al. 2009; Nagasaka et al. 2009).

Rice (*Oryza sativa* L.) is not only the model monocot species but also a staple food that contributes 35–59% of the daily caloric intake for approximately three billion people (Bhullar and Gruissem 2013). Rice is often grown in flooded paddies where microorganisms rapidly deplete dissolved oxygen levels, resulting in anoxic water where Fe has the tendency to be reduced to the more bioavailable Fe^{2+} form (Walker and Connolly 2008). Rice has therefore evolved the unusual ability amongst other Strategy II plant species to absorb Fe^{2+} via OsIRT transporters in a manner similar to Strategy I plant species (Bughio et al. 2002; Ishimaru et al. 2006). However, whilst genome sequencing demonstrates that the rice genome contains ferric reductase genes (Vasconcelos et al. 2004), rice does not show ferric reductase activity in roots nor the characteristic swelling of roots in response to Fe deficiency as seen in Strategy I plant species (Rong-li et al. 2012). The utilization of aspects from both Strategy I and Strategy II Fe uptake within the rice Fe deficiency response is commonly referred to as the 'Combined Strategy' (Ishimaru et al. 2006; Sperotto et al. 2012). The *OsIRT* homologous genes *HvIRT1* in barley and *ZmIRT1* in maize have been reported to be upregulated in response to Fe deficiency. Further, expression of these genes in Fe uptake-defective yeast mutants has been demonstrated to reverse growth defect associated with Fe uptake inability (Li et al. 2013; Pedas et al. 2008). However, *in planta* uptake of Fe^{2+} by Strategy II plants has so far only been confirmed in rice using tracer experiments (Ishimaru et al. 2006; Ricachenevsky and Sperotto 2014).

The amount of PS secreted by Strategy II plants corresponds with plant tolerance to growth on alkaline soils (soils with low Fe bioavailability) as identified by resistance to the onset of Fe-deficiency-associated chlorosis (Marschner et al. 1986). In response to Fe limiting conditions, barley (*Hordeum vulgare* L.), a plant species able to grow vigorously on high pH soils, secretes large amounts of PS including 2′-deoxymugineic acid (DMA) and the derivatives MA and 3-epihydroxymugineic acid (epiHMA). By contrast, rice, which is highly susceptible to Fe deficiency chlorosis on high pH soils, secretes only DMA in response to Fe deficiency and at much lower levels and shorter durations than barley (Higuchi et al. 1996; Mori et al. 1991). During Fe deficiency, barley (Marschner et al. 1986; Mori et al. 1987; Takagi et al. 1984; Walter et al. 1995), wheat (Oburger et al. 2014; Zhang et al. 1991), maize (Ueno et al. 2009), and red fescue (Ma et al. 2003) secrete PS following a diurnal pattern, with higher levels of PS secreted during day (light) periods and lower levels of PS secreted during night (dark) periods.

As Fe is a cofactor in enzymes involved in photosynthesis, this secretion pattern may enable plant Fe uptake to coincide with increased photosynthetic rates during light periods (Takagi et al. 1984). Additionally, as PS are readily degraded by soil microbes as a carbon and/or nitrogen source, the diurnal pattern of PS release may help reduce the degree of PS loss by microbial populations in the rhizosphere (Römheld and Marschner 1990). During Fe deficiency, rice induces expression of DMA biosynthetic genes following a diurnal pattern (Inoue et al. 2009; Nozoye et al. 2004, 2011). Additionally, rice secretes higher levels of DMA in the morning than afternoon or night periods (Nozoye et al. 2014). However, the diurnal pattern of transcript and metabolite levels within the rice Fe deficiency response has not been as extensively investigated as the Fe deficiency response of barley.

We studied the diurnal changes of DMA, NA, transcripts and metabolite levels within the rice Fe deficiency response. Rice roots exposed to Fe deficiency were used to investigate diurnal changes in expression levels of genes involved in Strategy II Fe uptake and translocation. A highly sensitive detection method was utilized to quantify *in planta* NA and DMA levels as well as DMA secretion levels by Fe-deficient rice roots. Additionally, metabolite profiling of Fe-deficient root tissues was performed to understand the molecular mechanisms utilized within the rice Fe deficiency response.

Results

Induction of Fe Deficiency Symptoms

Soil Plant Analysis Development (SPAD) measurements of leaf tissues identified that exposure of 5 week old hydroponically grown rice plants to one week of Fe deficiency induced a significant reduction in the youngest leaf 'greenness' ($p \leq 0.001$); however, no significant variation in plant fresh weight was detected ($p = 0.972$). Accordingly, this treatment was selected as a means of inducing Fe deficiency symptoms in rice without affecting biomass (Fig. 1).

Expression of Genes Involved in the Strategy II Response in Fe-deficient Rice Roots

Analysis of root transcripts (Additional file 1: Table S1A) by quantitative real time PCR (qRT-PCR) identified that Fe deficiency resulted in a significant increase in the expression of *OsNAS1* ($p \leq 0.001$), *OsNAS2* ($p \leq 0.001$), *OsNAS3* ($p \leq 0.001$), *OsNAAT1* ($p \leq 0.001$), *OsDMAS1* ($p \leq 0.001$), and *OsYSL15* ($p = 0.006$), but no significant change in the expression of *OsTOM1* ($p = 0.054$). Additionally, a statistically significant diurnal pattern of expression during Fe deficiency was detected for *OsNAS1* ($p = 0.016$), *OsNAS2* ($p = 0.017$), *OsNAAT1* ($p \leq 0.001$), *OsDMAS1* ($p = 0.049$) and *OsTOM1* ($p = 0.012$). Expression of *OsNAS1*, *OsNAS2*, *OsNAAT1*, *OsDMAS1* and *OsTOM1* were increased by Fe deficiency 1.5–2.0 fold during 6:00–8:00 and 2.3–2.9 fold during 11:00–13:00 (Additional file 1: Table S1B). In general, expression of these genes followed a diurnal pattern during Fe deficiency but not during Fe sufficiency (Fig. 2), wherein the peak of Fe deficiency induced expression occurred during 11:00–13:00. Accordingly, Fe deficiency promoted increased expression of genes involved in nicotianamine (NA) and DMA biosynthesis, DMA secretion, and uptake of Fe^{3+}-DMA complexes following a diurnal pattern with a peak time of induction 3–5 h after the onset of light.

Biosynthesis and Secretion of DMA in Fe-deficient Rice Roots

Targeted quantification of NA and DMA by liquid chromatography mass spectrometry (LC-MS) from root metabolite extracts collected after one week of Fe deficiency identified that concentrations of DMA (average 1.42-fold higher, $p \leq 0.001$), but not NA (average 1.05-fold higher, $p = 0.988$), were significantly higher in the Fe-deficient roots (Fig. 3a and b, Additional file 2: Table S2A). Root concentrations of NA and DMA in Fe-sufficient plants showed a similar diurnal pattern, with root NA ($p = 0.016$) and DMA ($p = 0.033$) concentrations significantly higher during dark periods than light periods. While NA and DMA concentrations in Fe-deficient roots did not follow a significant diurnal pattern, Fe deficiency induction of root NA and DMA concentrations were both highest during the 11:00–13:00 period (Fig. 3d).

Quantification of DMA by LC-MS from root secretions collected after one week of Fe deficiency, identified that Fe-deficient roots secreted on average 2.4-fold more DMA ($p = 0.009$) than Fe-sufficient roots across the day (Fig. 3c). Whilst no significant diurnal variation was observed in DMA secretion by Fe-sufficient roots, a diurnal

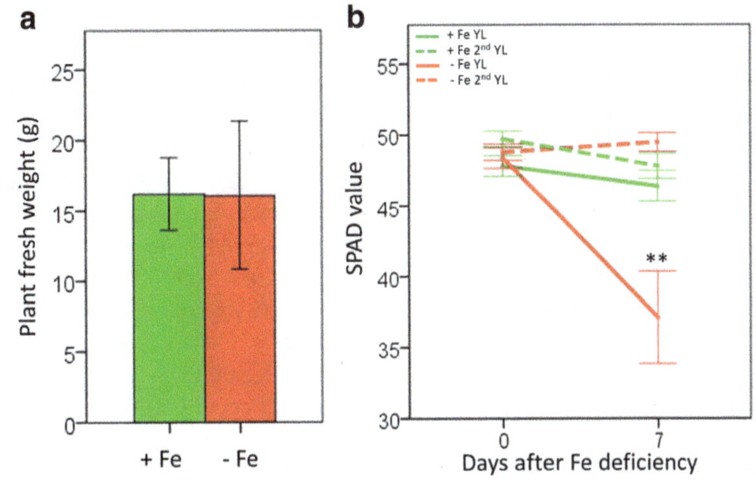

Fig. 1 Induction of Fe deficiency symptoms in rice plants. **a** Plant fresh weight ($n = 9$). **b** SPAD values of the youngest leaf (YL) and 2nd youngest leaf (2nd YL) ($n = 30$). Plant fresh weights and leaf SPAD values of Fe-sufficient (*green*) and Fe-deficient (*red*) plants were measured at 0 and 7 days after the Fe deficiency treatment. Values are presented as means ± SE of *n* biological replicates. Significant differences are indicated by ** for $p < 0.01$ by Student's t-tests performed between treatments within leaf types and days after Fe deficiency

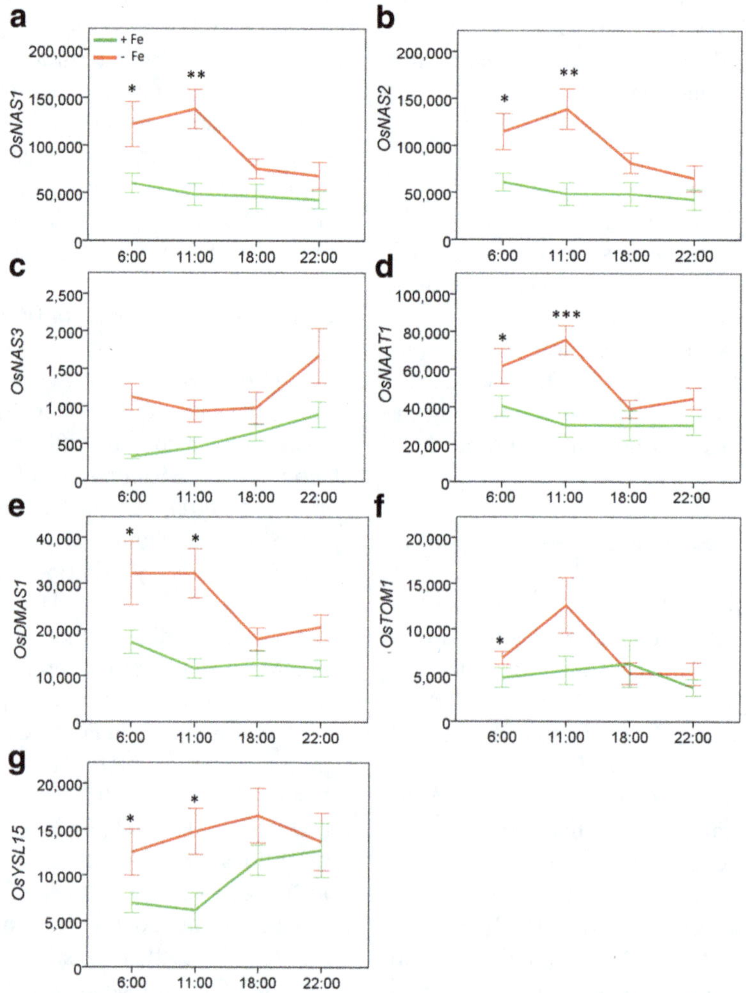

Fig. 2 Expression of genes associated with DMA biosynthesis in Fe-sufficient and Fe-deficient rice roots. Expression of **a** *OsNAS1*, **b** *OsNAS2*, **c** *OsNAS3*, **d** *OsNAAT1*, **e** *OsDMAS1*, **f** *OsTOM1* and **g** *OsYSL15* genes in the roots of Fe-sufficient (*green*) and Fe-deficient (*red*) rice plants. Data is presented as means of genes of interest (GOI) expression relative to the three gene normalization factor (3GNF) ± SE of 9 biological replicates. Significant differences are indicated by * for $p < 0.05$, ** for $p < 0.01$, *** for $p \leq 0.001$ by Student's t-tests performed between treatments within collection times

pattern with a peak during 11:00–13:00 ($p = 0.001$) was observed for DMA secretion by Fe-deficient roots. Accordingly, Fe deficiency promoted increased secretion of DMA following a diurnal pattern with a peak occurring 3–5 h after the onset of light.

Metabolite Profiles of Fe-deficient Rice Roots

Root tissue metabolite extracts were analysed by LC-MS to develop amines profiles (Additional file 2: Table S2A). PERMANOVA analyses showed highly significant differences in metabolite profiles between Fe treatments (Pseudo-$F = 14.197$, $p = 0.001$) and collection times (Pseudo-$F = 2.494$, $p = 0.003$). Principal component analysis (PCA) of root amines profiles (Fig. 4) identified clearly distinct 95% confidence regions between Fe-sufficient and Fe-deficient profiles for samples collected during the 11:00–13:00 time period only (Fig. 4b). However, significant differences in metabolite profiles between Fe treatments were observed for the collection times 6:00–8:00 ($t = 1.983$, $p = 0.011$), 11:00–13:00 ($t = 2.610$, $p = 0.002$), and 18:00–20:00 ($t = 2.521$, $p = 0.008$). No significant differences between metabolite profiles of different Fe treatments were observed during the 22:00–24:00 collection time ($t = 1.562$, $p = 0.059$). These findings indicate that the rice Fe deficiency response promoted changes to root metabolite profiles following a diurnal pattern with significant changes during 6:00–20:00 and a peak of Fe deficiency associated changes occurring during 11:00–13:00.

Biochemical Pathway Mapping of the Rice Fe Deficiency Response

Concentrations of metabolites in the following pathways were significantly increased in the roots of Fe-deficient rice: α–ketoglutarate family amino acids, 3-phosphoglycerates

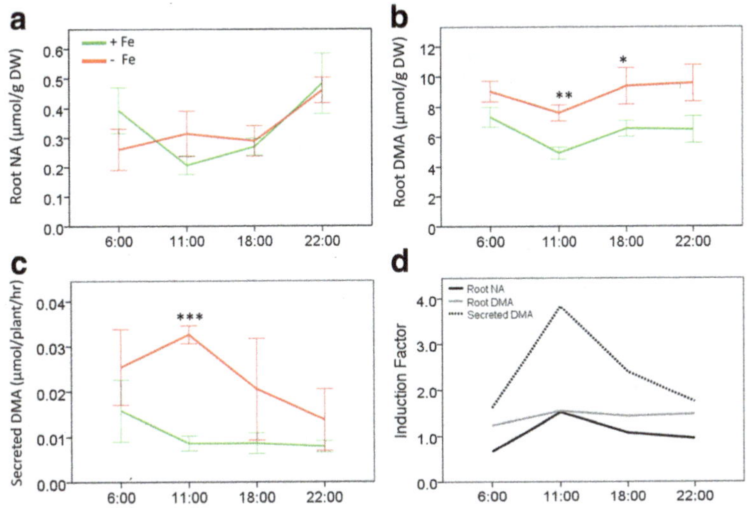

Fig. 3 Concentration of NA, DMA and secreted DMA of Fe-sufficient and Fe-deficient rice roots. Root concentrations of **a** NA ($n = 5$), **b** DMA ($n = 5$) and **c** DMA secretion ($n = 3$) from Fe-sufficient (*green*) and Fe-deficient (*red*) rice roots were quantified by LC-MS. **d** Induction factors of root NA, DMA and secreted DMA concentrations. Values are presented as means ± SE of n biological replicates. Significant differences are indicated by * for $p < 0.05$, ** for $p < 0.01$, *** for $p \leq 0.001$ as determined by Student's t-tests performed between treatments within collection times

(3-P-glycerates), pyruvate derivatives, aromatic amino acids, citrulline–nitric oxide (Cit-NO) cycle, oxaloacetate/aspartate family amino acids and the DMA biosynthetic pathway (Additional file 2: Table S2B). No metabolites were detected at significantly lower concentrations in the roots of the Fe-deficient rice.

Induction factors associated with Fe deficiency were mapped onto custom biochemical pathways (Fig. 5) using the Visualization and Analysis of NeTworks with related Experimental Data program (VANTED) v2.1.0 (Plant Bioinformatics Group, The Leibniz-Institute of Plant Genetics and Crop Plant Research, available at https://immersive-analytics.infotech.monash.edu/vanted/). Induced levels of the α–ketoglutarate family amino acids (proline, histidine and glutamine) were observed in Fe-deficient roots following a diurnal pattern with a peak of induction occurring during 18:00–20:00 (Fig. 5a). Induced levels of the 3-P-glycerate derivatives serine and glycine (Fig. 5b) and pyruvate derivatives: valine, leucine, alanine and β-alanine (Fig. 5c) were also observed in Fe-deficient roots, suggesting an increase in glycolysis during light periods (11:00–13:00 and 18:00–20:00). Increased activity of glycolysis due to Fe deficiency was also supported by a significant increase in concentration of the glucose-to-pyruvate intermediate 3-P-glycerate in Fe-deficient roots (Additional file 3: Figure S1). Induced levels of the aromatic amino acids: phenylalanine, serotonin, tyrosine, and tryptophan in Fe-deficient roots (Fig. 5d), suggest an increased activity of the shikimate pathway. Induced levels of arginine and citrulline but not ornithine in Fe-deficient roots (Fig. 5e), suggest an increased activity of the citrulline–nitric oxide (Cit-NO) cycle. Additionally, induced levels of the oxaloacetate/aspartate family amino acids: isoleucine, lysine, threonine (Fig. 5f) and methionine (Fig. 5g) were detected during 6:00–8:00, 11:00–13:00 and 18:00–20:00, but not 22:00–24:00. Activity of the DMA biosynthetic pathway was also increased in Fe-deficient roots based on induced levels of methionine (the dominant precursor of DMA biosynthesis), root DMA and secreted DMA (Fig. 5g). Additionally, the potential methionine precursors serine and cysteine (but not aspartate or homoserine) accumulated in response to Fe deficiency (Additional file 4: Figure S3).

Discussion

Iron Deficiency Induces Expression of Genes Involved in Strategy II Fe Uptake Following a Diurnal Pattern

Detailed analysis of rice gene expression related to NA and DMA biosynthesis, as well as DMA secretion under Fe deficiency, has been described in several previous investigations (Inoue et al. 2003; Nozoye et al. 2011; Wang et al. 2013). Here we have identified diurnal patterns of expression for the *OsNAS1*, *OsNAS2*, *OsNAAT1*, *OsDMAS1*, and *OsTOM1* genes in Fe-deficient rice roots with significant induction at the 6:00 and 11:00 sampling times (Fig. 2, Additional file 1: Table S1). These diurnal patterns are consistent with previous investigations (Nozoye et al. 2004, 2011), and may be summarised as a slight induction of expression prior to the onset of light, a peak of induction 3–5 h after the onset of light, and a return to expression levels similar to that of Fe-sufficient roots prior to dark. While the diurnal patterns reported in this study have been previously identified, other studies have reported much higher inductions ratios for

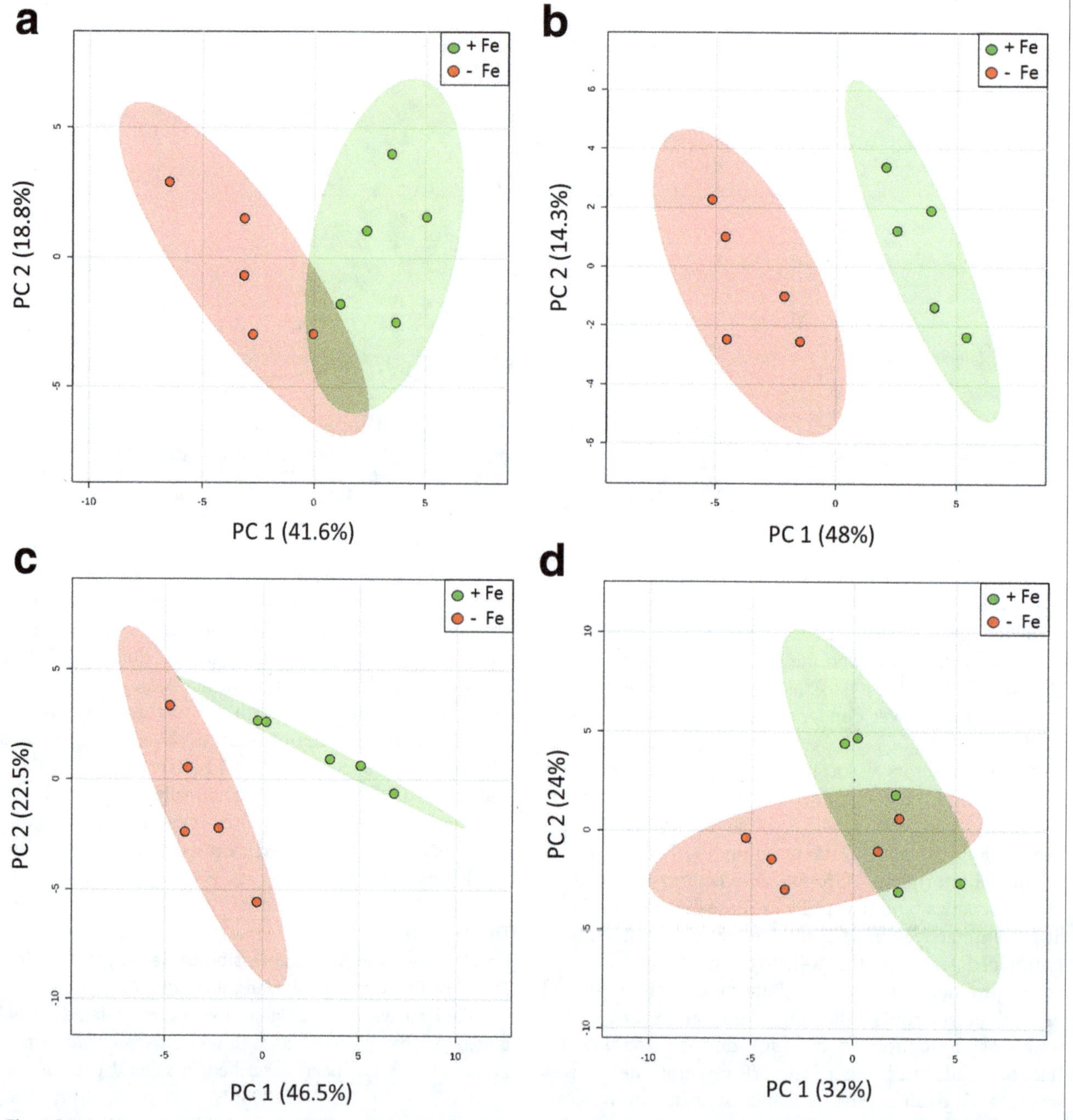

Fig. 4 Principal component analyses of root metabolite profiles from Fe-sufficient and Fe-deficient rice roots. Metabolite profiles from Fe-sufficient (*+Fe: green*) and Fe-deficient (*–Fe: red*) rice roots were compared using 95% confidence regions within the collection times **a** 6:00–8:00, **b** 11:00–13:00, **c** 18:00–20:00 and **d** 22:00–24:00

Fe-deficiency-inducible genes including *OsNAS1* and *OsNAS2* (Inoue et al. 2003; Kobayashi et al. 2009). This may have been due to differences in growing conditions, age of plants and length of exposure to the Fe deficiency treatment. In this study we cultivated rice plants in Fe-sufficient hydroponic solution for 5 weeks before transferring to Fe-deficient solution for a period of 1 week prior to sampling. Inoue et al. (2003) exposed very young seedlings to Fe-sufficient nutrient solution for 2 weeks, whereas Kobayashi et al. (2009) cultivated seedlings in Fe-sufficient nutrient solution for 3 weeks, before transferring plants to Fe-deficient solution for 1 week. Considering that our plants were more established than these studies prior to Fe deficiency, it is possible that our six week old plants were not fully Fe limited at the time of sampling and this may have reduced the induction ratios that we observed. We also found that expression of *OsYSL15* was significantly induced in response to Fe deficiency prior to the onset of

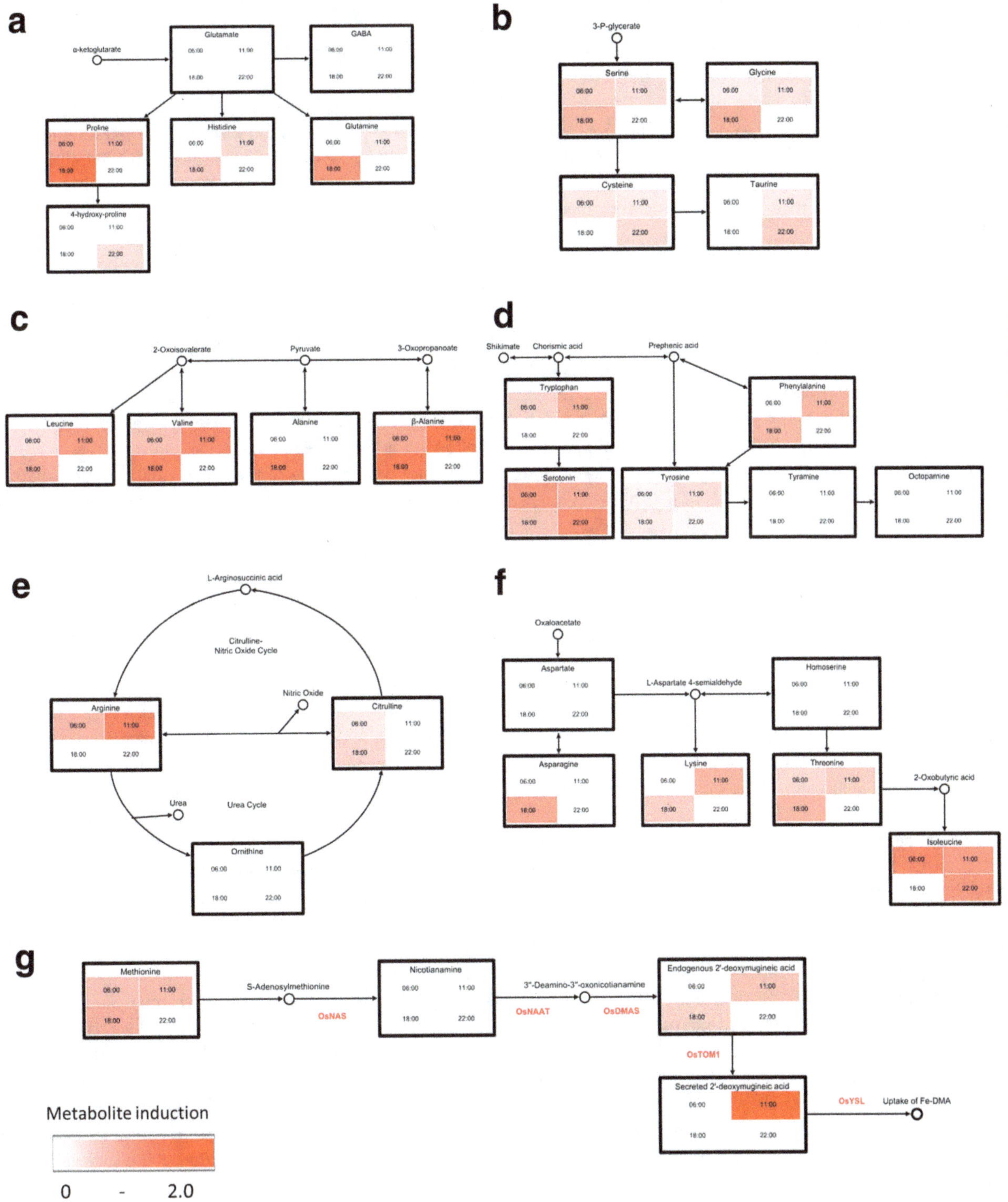

Fig. 5 Changes in metabolic pathways induced by Fe deficiency in rice roots. Heatmap (ranging from 0–2.0 as *white–red*) of log_2 transformed Fe deficiency associated induction of metabolite concentrations in biochemical pathways of **a** α-ketoglutarates, **b** 3-P-glycerates, **c** pyruvates, **d** aromatic amines, **e** Cit-NO/urea cycle, **f** aspartate family amino acids and **g** DMA biosynthesis and secretion

light and 3–5 h after the onset of light (Fig. 2g, Additional file 1: Table S1A). However, the diurnal pattern of *OsYSL15* in Fe-deficient roots differed from the diurnal pattern of *OsNAS1*, *OsNAS2*, *OsNAAT1*, *OsDMAS1* and *OsTOM1* by having a peak time of expression at 18:00–20:00 instead of 11:00–13:00 (Fig. 2). OsYSL15 is an

Fe-regulated Fe^{3+}-DMA transporter and is thought to be responsible for active uptake of Fe^{3+}-DMA complexes during daylight hours (Inoue et al. 2009). Thus, its expression may be induced by the preferential biosynthesis of root DMA at night time (Fig. 3b) and coordinated by diurnal secretion of DMA (Fig. 3c). Whilst increased expression of *OsYSL15* after the onset of light has been reported previously (Inoue et al. 2009), the diurnal pattern of *OsYSL15* that we observed is not consistent with the daily fluctuations of *OsYSL15* described by Inoue et al. (2009). This inconsistency could be explained by differences in rice genotype as well as experimental design. In this study we used rice cultivar Nipponbare whereas Inoue et al. (2009) utilized cultivar Tsukinohikari. These genotypic differences may influence *OsYSL15* expression under Fe deficiency. Inoue et al. (2009) also characterised the induction of expression of *OsYSL15* under Fe deficiency over a 27-h time period with tissue sampling intervals of 3 h, compared to our analysis over a 24-h time period with tissue sampling intervals of 4–7 h.

Iron Deficiency Promotes DMA Biosynthesis and Accumulation

In this study we observed significant upregulation of *OsNAS1, OsNAS2, OsNAAT1* and *OsDMAS1* under Fe-deficient conditions (Fig. 2), however, only DMA (not NA) concentrations were significantly increased in the Fe-deficient rice roots (Fig. 3a and b); a result that is consistent with previous studies in barley and rice (Higuchi et al. 2001). These results imply that *OsNAS* upregulation within the Strategy II Fe deficiency response is associated with increased NA production for DMA biosynthesis, rather than NA accumulation. These results also suggest that increased DMA, rather than NA, is utilized for the altered chelation of Fe in the roots of Fe-deficient rice. This is supported by predictive models indicating that at the rice root apoplastic pH of 4.9 (Nishiyama et al. 2008), DMA would outcompete NA as the predominant Fe complex (Hider et al. 2004). Recent speciation studies of metal ligands within chelation complexes in the xylem and phloem sap of rice have identified that DMA is a major chelator of Fe in the phloem sap (Nishiyama et al. 2012; Yoneyama et al. 2015), but neither DMA nor NA are major chelators of Fe in the xylem sap (Ariga et al. 2014). Accordingly DMA-mediated translocation of Fe within the rice Fe deficiency response may occur predominantly via the phloem.

Whilst NA and DMA concentrations in Fe-sufficient roots showed a distinct diurnal pattern with higher concentrations in dark periods than light periods, no significant diurnal pattern of NA or DMA concentrations were observed in Fe-deficient roots. In barley, a diurnal pattern is observed wherein root PS (DMA, MA and epiHMA) concentrations are reduced during periods of heightened PS secretions (Kawai et al. 1993; Nagasaka et al. 2009; Walter et al. 1995). In the current study, Fe deficiency induced a 1.5-fold increase in root DMA concentrations and a 2.4-fold increase in DMA secretions in rice plants (Fig. 3), accounting for root DMA concentrations changing minimally during periods of increased DMA secretion. By contrast in barley, the depletion of root PS concentrations may be explained by Fe deficiency inducing an 8.7-fold increase in root PS concentrations but up to a 250-fold increase in PS secretions (Walter et al. 1995). This difference may relate to the Combined Strategy utilized by rice grown in flooded conditions involving uptake of unbound Fe^{2+} via IRT transporters. As this mechanism of Fe uptake is not reliant on DMA secretion, rice may have evolved to retain a larger proportion of the increased DMA produced during Fe deficiency.

Iron Deficiency Promotes Increased Secretion of DMA Following a Diurnal Pattern

Successful modification of the DMA quantification method described in Kakei et al. (2009) from quadrupole time-of-flight mass spectrometry to the more sensitive triple quadrupole mass spectrometry provided a 3-fold improvement in sensitivity. Accordingly the developed DMA quantification method was more sensitive than DMA quantification methods described in previous publications including: Fe solubilisation assay (Rong-li et al. 2012), high performance liquid chromatography (Nozoye et al. 2014), LC-MS (Oburger et al. 2014), capillary electrophoresis mass spectrometry (Dell'mour et al. 2010) and nuclear magnetic resonance spectroscopy (Ma et al. 1995). This improved sensitivity allowed avoidance of the false-negative result wherein DMA within rice root secretions was deemed 'not detected' as the concentration was below the method's limit of detection as in previous publications utilizing nuclear magnetic resonance spectroscopy for analysis (Fan et al. 2001).

Previous studies have identified that rice secretes more DMA during the 6:00–13:00 'morning' period than the 13:00–20:00 'afternoon' or 20:00–6:00 'night' periods (Nozoye et al. 2014). Consistent with this result, our study identified that the peak time of DMA secretions by Fe-deficient rice occurred during the 6:00–13:00 'morning' period (Fig. 3). Additionally, the improved sensitivity of the developed method allowed for the analysis of root secretions collected during the 'morning' over two narrower time periods (6:00–8:00 and 11:00–13:00). The ability to analyse root secretion DMA concentrations collected over narrower time periods allowed for further characterisation of rice DMA secretions during this 'morning' period leading to the identification of 11:00–13:00 (3–5 h after the onset of light) as the peak time of DMA secretion from Fe-deficient rice roots (Fig. 3).

Whilst this diurnal pattern of the Fe deficiency response in rice is less distinct than that of other graminaceous plants, the peak time of DMA secretions by rice is consistent with the peak time of PS secretions observed under Fe deficiency in barley (Marschner et al. 1986; Mori et al. 1987; Takagi et al. 1984; Walter et al. 1995), wheat (Oburger et al. 2014; Zhang et al. 1991), maize (Ueno et al. 2009), and red fescue (Ma et al. 2003). Consistent with other graminaceous plant species, a major aspect of the rice Fe deficiency response appears to involve re-establishing Fe homeostasis via increased PS mediated Fe uptake and internal translocation with the time of peak secretions occurring 3–5 h after the onset of light.

Iron Deficiency in Rice Induces Changes to Carbon and Energy Metabolism

Previous proteome analyses have identified that the molecular mechanisms induced within plant Fe deficiency responses include changes to plant metabolism associated with: carbon and energy metabolism, S-adenosylmethionine (SAM) metabolism, ion transport, ROS scavenging and cell signalling (Kosová et al. 2011). Proteome analyses of the rice Fe deficiency response specifically identified increased levels of the glycolytic enzyme glyceraldehyde-3-phosphate dehydrogenase in the roots of Fe-deficient rice (Chen et al. 2015). Results of the current study indicated increased glycolysis and respiration within the Fe deficiency response during light periods (11:00–20:00) based on the induced levels of 3-P-glycreate derivatives, branched chain amino acids, pyruvate derivatives (Fig. 5), and 3-P-glycerate (Additional file 3: Figure S1). This is consistent with previous studies on plant stress responses (Di Martino et al. 2003; Wingler et al. 2000; Zoghlami et al. 2011). The rice Fe deficiency response therefore involves changes to carbon and energy metabolism potentially to accommodate the increased energy demand required to drive mechanisms induced within the Fe deficiency response.

Methionine is the dominant precursor of DMA biosynthesis (Mori and Nishizawa 1987) via the intermediate SAM, which is derived from methionine by SAM synthetase within the methionine cycle (Suzuki et al. 2006). Transcript analysis of barley suggested that methionine and SAM biosynthesis were induced in response to Fe deficiency (Nagasaka et al. 2009; Negishi et al. 2002). Similarly, transcript analyses of rice suggested that activity of the methionine cycle and DMA biosynthesis are induced 24 h after the onset of Fe deficiency (Itai et al. 2013). In the current study, Fe-deficient rice roots showed induced levels of methionine during 6:00–20:00 (Fig. 5g). However, unlike the diurnal patterns of gene expression detected (Fig. 2), a clear peak of methionine concentrations was not observed during 11:00–13:00. Serine (via cysteine) and/or aspartate (via homoserine) are used by plants to produce homocysteine, the immediate precursor to methionine (Droux et al. 1995). Serine and cysteine levels were induced in Fe-deficient rice roots whilst levels of aspartate and homoserine were unchanged (Fig. 5b and f; Additional file 4: Figure S4). The co-accumulation of serine, cysteine and methionine suggested that serine may be the dominant precursor to the increased methionine biosynthesis within the rice Fe deficiency response. To confirm this hypothesis, further investigations will be required such as radioisotope tracer experiments previously used to confirm that methionine is the dominant precursor of PS biosynthesis within the Fe deficiency responses of graminaceous plants (Mori and Nishizawa 1987; Nakanishi et al. 1999).

Iron Deficiency in Rice Induces Molecular Mechanisms Associated with Plant Metal-stress Responses

Ion transport, chelation and sequestration are key mechanisms involved in plant metal tolerance and reestablishment of metal homeostasis (Clemens 2001). In response to Fe deficiency, induced levels of histidine, proline and asparagine were detected during 18:00–20:00 (Fig. 5a and f). These amines have been identified as chelators of copper, nickel and zinc and facilitate tolerance to these metal stresses via complexation and xylem-mediated translocation (Rai 2002; Sharma and Dietz 2006). Considering that DMA is able to bind a range of metals in the rhizosphere (Scholz et al. 1992), and Fe-deficient rice has previously been shown to uptake increased concentrations of calcium, cadmium and zinc (da Silveira et al. 2007; Zhang et al. 1998), the increased DMA secretions during 11:00–13:00 within the rice Fe deficiency response may promote increased uptake of micronutrients other than Fe. The induced levels of chelators (histidine, proline, asparagine and DMA) during 18:00–20:00 within the rice Fe deficiency response may be associated with increased chelation and translocation of off-target metals (other than Fe) during the 11:00–13:00 period due to increased DMA secretions.

Iron Deficiency in Rice Induces Molecular Mechanisms Involved in Mediation of Oxidative Stress

The increased concentration of ROS from abiotic stresses, including Fe deficiency, results in an "oxidative burst" which in-turn causes damage to cell proteins, lipids, carbohydrates and DNA (Gill and Tuteja 2010; Spinelli et al. 2011). Accordingly, the plant Fe deficiency response includes the induction of molecular mechanisms associated with mediation of the damage due to oxidative stress. Accumulation of amino acids within abiotic stress responses play a protective role as antioxidants which detoxify ROS and prevent damage to the plant cell (Joshi et al. 2010). The antioxidant proline was induced in roots of our Fe-deficient rice plants with a peak of induction occurring during 18:00–20:00 (Fig. 5a). This peak time may be associated with off-target metal uptake as the localized increase to metal concentrations may cause oxidative stress via formation of ROS (Ercal et al. 2001). Iron-deficient rice roots also showed induced levels of phenylalanine, tyrosine, serotonin and tryptophan (Fig. 5d) which are antioxidant aromatic amines (Nimalaratne et al. 2011; Ramakrishna and Ravishankar 2011; Stadtman and Levine 2003). Particularly, phenylalanine is the main precursor to a suite of phenolic compounds which have very high antioxidant activity (Ghasemzadeh and Ghasemzadeh 2011; Michalak 2006).

Iron deficiency was associated with a significant decrease in root accumulation of ascorbic acid (Additional file 3: Figure S2), which is typically the primary polar antioxidant in plants (Weber et al. 2008). Previous proteomic studies have identified decreased levels of ascorbate peroxidase in the roots of Fe-deficient rice (Chen et al. 2015). As ascorbate peroxidase is a hemoprotein, Fe deficiency results in a decrease in ascorbate peroxidase activity (Becana et al. 1998; Iturbe-Ormaetxe et al. 1995; Sun et al. 2007). Additionally, previous analyses identified that in rice, but not in other graminaceous plants, Fe deficiency resulted in decreased root activity of glutathione reductase, an enzyme required for the ascorbic acid-glutathione cycle to efficiently mediate oxidative stress (Bashir et al. 2007). Accordingly, the rice Fe deficiency response includes accumulation of non-Fe-dependant antioxidants to scavenge ROS and minimise ROS associated damage.

Nitric oxide (NO) is a highly reactive radical that acts as a chemical signal (Gechev et al. 2006) in multiple plant stress responses including; drought, salt, metal toxicity (Spinelli et al. 2011) and Fe deficiency (Arnaud et al. 2006; Vigani et al. 2013). However, NO may also act as an antioxidant to quench superoxide anion radicals (Arasimowicz and Floryszak-Wieczorek 2007). In this study, Fe deficiency was associated with induced levels of citrulline and arginine (Fig. 5e). These results indicate that the Cit-NO cycle was induced which in turn suggests that NO production was increased (Mori 2007). Treatment of *Arabidopsis* with putrescine has previously been shown to alleviate Fe deficiency via the accumulation of NO (Zhu et al. 2016). Similarly, treatment with NO directly has been demonstrated to mitigate the chlorosis associated with Fe deficiency in rice (Lee et al. 2009) and maize (Sun et al. 2007) by acting as an antioxidant. Accordingly, increased NO production may occur as part of the rice Fe deficiency response to signal induction of gene expression and/or mitigate damage associated with oxidative stress.

Conclusion

The Strategy II Fe deficiency response of rice has been more difficult to study than that of wheat or barley due to very low levels of DMA secretion. The development of a highly sensitive method for the quantification of DMA and NA in this study allowed for the quantification of DMA in rice root secretions which for many previously published methodologies is below the limit of detection. This study reported a thorough characterization of the rice Fe deficiency response. Gene expression and metabolite profiling identified that Fe deficiency induced DMA biosynthesis and DMA secretions followed a diurnal pattern with a peak of induction occurring during 11:00–13:00. The identification of this peak time period, equating to 3–5 h after the onset of light, verified that the peak time of the rice Fe deficiency response is similar to that of other Strategy II plant species such as barley and wheat. However, the diurnal variation in DMA secretion observed within the rice Fe deficiency response was less prominent than that of other graminaceous species described in previous studies. The results presented herein indicate that the mechanisms utilized by rice in response to Fe deficiency include increased DMA-mediated Fe uptake and accumulation of metabolites associated with plant oxidative stress responses. These results constitute an important progression in our understanding of the rice Fe deficiency response.

Methods

Plant Materials and Growth Conditions

Rice (*Oryza sativa* L. cv. Nipponbare) grains were surface sterilized with 70% ethyl alcohol and 5% sodium hypochlorite and then germinated on filter paper wetted with 7 mL of reverse osmosis (RO) water in a controlled environment chamber with 12-h photoperiod at 24 °C. After 10 days, germinated rice seedlings were transferred to 13 L supported hydroponic tanks. The rice seedlings were grown for six weeks at a density of thirty plants per hydroponic tank in a growth room with 12-h photoperiod, and 28 °C/24 °C day/night temperature (8:00–20:00/20:00–8:00), with nutrient solution as described in Plett et al. (2010). Rice plants were initially acclimated for one week by growth in ¼ strength Fe-replete growth solution followed by ½ strength Fe-replete growth solution for an additional four days (Additional file 5: Table S3). Plants were grown for a further 4 weeks in full strength Fe-replete growth solution. Growth solutions were replaced every 3–4 days to maintain pH at 5.5. After the four week growth period, Fe-sufficient control plants were grown for an additional week in full strength Fe-replete growth solution while Fe-deficient treatment plants were grown in growth solution lacking $NaFe^{3+}$EDTA.

Measurement of Plant Size and Chlorophyll Score

Whole plant fresh weight was measured after one week of Fe deficiency treatment. Soil Plant Analysis Development (SPAD) measurements were taken on the youngest leaf (YL) and 2nd youngest leaf (2nd YL) at day 0 and 7 of the Fe deficiency treatment using a Minolta SPAD-502Plus

(Spectrum Technologies Inc., Plainfield, Illinois, USA) as an indicator of leaf chlorophyll content.

Tissue Sampling

Root tissue and root secretions were collected for gene expression and metabolite analyses during the time periods 6:00–8:00 (2–0 h prior to the onset of light), 11:00–13:00 (3–5 h after the onset of light), 18:00–20:00 (10–12 h after the onset of light) and 22:00–24:00 (2–4 h after the change to dark), after one week of Fe deficiency treatment.

Root Gene Expression Analyses

Extraction of RNA was performed from frozen tissue using the Qiagen RNeasy Mini Kit (Qiagen, Valencia, California, USA) according to manufacturer's instructions with the exception that 8 μL of the reducing agent dithiothreitol was substituted for 3.5 μL of β-mercaptoethanol (Sigma-Aldrich). Purified RNA was reverse transcribed using the iScript Select cDNA Synthesis Kit (Bio-Rad, Richmond, California, USA) according to manufacturer's instructions. The expression of *OsNAS1* (LOC_Os03g1942), *OsNAS2* (LOC_Os03g19420), *OsNAS3* (LOC_Os07g48980), *OsNAAT1* (LOC_Os02g20360), *OsDMAS1* (LOC_Os03g13390), *OsTOM1* (LOC_Os11g04020), and *OsYSL15* (LOC_Os02g43410) genes were analysed by quantitative real time PCR (qRT-PCR). The 30 μL reactions contained 6 μL MyTaq Reaction Buffer (Bioline), 7.5 pmol of each primer (Additional file 5: Table S4), 0.15 μL MyTaq polymerase, 25 ng of cDNA and 21.35 μL deionized MilliQ water (dH_2O). Reactants were initially denatured at 95 °C for 3 min, followed by 35 cycles of 95 °C for 30 s, 55 °C for 30 s, 72 °C for 1 min, and a final extension of 72 °C for 5 min. Expression analyses were performed relative to a three gene normalization factor (3GNF) from expression of the three housekeeping genes: *OsActin* (LOC_Os03g50885), *OsGAPDH* (LOC_Os02g38920), and *OsELF1* (LOC_Os03g29260).

Root Metabolite Extraction

Frozen root tissue was freeze dried using a Labconco Lyph-lock 6 freeze-dry system (Labconco Corp., Kansas City, Missouri, USA), transferred into 2 mL cryo-mill tubes (Lysing Matrix tube, MP Biomedicals, Solon, Ohio, USA), then 200 μL of methanol (MeOH) was added (containing the internal standards: ^{13}C-Sorbitol at 20 μg/mL, $^{13}C_5$-^{15}N Valine at 20 μg/mL, 2-aminobutyric acid at 10 μg/mL and pentafluorobenzoic acid at 10 μg/mL). The root tissue was then homogenized using a Precellys 24 cryo-mill coupled to a Cryolys (Bertin Technologies, Villeurbanne, France) using the following setting: 3 cycles involving 30 s of milling at 6,800 rpm followed by a 30 s rest period between cycles. The MeOH supernatant was removed and 200 μL of deionized MilliQ water was then added to the cryo-mill tube containing the homogenized root tissue pellet and vortexed for 30 s. The water supernatant was combined with the removed MeOH supernatant and 100 μL of dichloromethane (DCM) was added to the MeOH-water supernatant mixture.

Root Secretion Collection, Purification and Concentration

Root secretions were collected using the method described in Astolfi et al. (2012) with the following modifications: roots from three plants were briefly rinsed with RO water, transferred into 400 mL of autoclaved dH_2O in 1 L plastic containers, and then wrapped with aluminium foil to prevent exposure to light. After two hours of root secretion collection, 40 μL of Micropur Forte (Katadyn Products Inc.) was immediately added and filtered through Whatman filter discs. Root secretions were purified using cation exchange Amberlite IR120 H resin (Rohm and Haas Company, Philadelphia, Pennsylvania, USA) with a bed volume (BV) of 50 mL. Sample solutions were eluted through the column at two BV per hour, and this eluate was discarded. Recovery of exchangeable cations (including DMA) was performed by eluting 100 mL of 2 M ammonium hydroxide solution through the Amberlite resin at two BV per hour and stored at −80 °C. Concentration of root secretions was performed through centrifugal evaporation using a RVC 2–25 (Martin Christ Corporation, Osterode, Germany) at 30 °C.

LC-MS Quantification of Amines

Quantification of root metabolite concentrations was performed as described in Boughton et al. (2011) using an LC 1200 series binary pump with autosampler and heated column compartment coupled to a 6410 series Triple Quadrupole MS (Agilent Technologies Inc.). In brief, derivatization of samples involved combining 10 μL of sample, 70 μL of 200 mM borate buffer (containing the internal standard $^{13}C_5$-^{15}N-Valine), and 20 μL of fresh 2.85 mg/mL 6-aminoquinolyl-N-hydroxysuccinimidyl carbamate (Aqc) solution. The solution was briefly vortexed, centrifuged at 13,000 rpm for 1 min at 4 °C, then incubated in an Eppendorf thermomixer compact (Eppendorf AG) at 1,150 rpm for 10 min at 55 °C. Reverse phase (RP) chromatography was performed using a Zorbax Eclipse XDB-C18 Rapid Resolution HS 2.1x100 mm 1.8 Micron column (Agilent Technologies Inc.) with a flow rate of 300 μL/min and a temperature of 30 °C. The aqueous mobile phase, solvent 'A', was prepared as a 0.1% (v/v) formic acid (FA) in dH_2O, and the organic mobile phase, solvent 'B', was prepared as 0.1%

(v/v) FA in acetonitrile. Chromatography was performed using the following gradient:

Time (min)	A [%]	B [%]	Flow rate (μL/min)
0.00	99	1	300
2.00	99	1	300
9.00	85	15	300
14.00	70	30	300
14.10	99	1	300
19.00	99	1	300

During MS analyses, all metabolites were quantified based on formation of the m/z = 171 Amq product ion. Metabolites were quantified by multiple reaction monitoring (MRM) according to previously published transitions (Boughton et al. 2011; Callahan et al. 2007).

LC-MS Quantification of DMA and NA

Quantification of 9-fluorenyl methoxycarboxyl chloride (FMOC-Cl) derivatized DMA and NA by RP LC-MS was performed on an LC 1290 series infinity pump with autosampler and heated column compartment coupled to a 6490 series Triple Quadrupole MS (Agilent Technologies Inc.) using a modified version of the method described in Kakei et al. (2009). In short, derivatization of samples involved combining, 5 μL of sample, 10 μL of 1 M sodium borate buffer (pH = 8) containing 42 ng/mL of the internal standard 2-aminoanthracene (2-AA), 10 μL of 50 mM EDTA (pH = 8), and 40 μL of fresh 50 mM FMOC-Cl solution. The solution mixture was incubated at 60 °C with constant shaking at 700 rpm, for 15 min then the reaction was quenched via the addition of 8.9 μL of 5% FA solution (pH = 4). A Zorbax Eclipse XDB-C18 Rapid Resolution HS 2.1x100 mm 1.8 Micron column (Agilent Technologies Inc.) was used during chromatography. The aqueous mobile phase, solvent 'A', was prepared as a 0.1% (v/v) FA in dH_2O, and the organic mobile phase, solvent 'B', was prepared as 0.1% (v/v) FA in acetonitrile. Chromatography was performed using the following gradient:

Time (min)	A [%]	B [%]	Flow rate (μL/min)
0.00	98	2	500
0.50	98	2	500
1.00	40	60	500
4.00	15	85	500
4.10	0	100	500
7.00	0	100	500
7.10	98	2	500
10.00	98	2	500

DMA and NA were observed as $[M+(2 \times FMOC) + H]^+$ at retention times (RT) 2.88 min and 2.75 min, respectively. Additionally, 2-AA was observed as $[M+(FMOC) + H]^+$ at RT 5 min. Quantification of DMA, NA, and 2-AA by MRM used the following transitions:

	Transition	Precursor (m/z)	Product (m/z)	Collision energy (V)
$[DMA+(2 \times FMOC) + H]^+$	Quantifier	749.3	185.8	28
	Qualifier	749.3	139.8	40
$[NA+(2 \times FMOC) + H]^+$	Quantifier	748.3	526.2	28
	Qualifier	748.3	203.1	42
$[2\text{-}AA+(FMOC) + H]^+$	Quantifier	416.2	193.8	34
	Qualifier	416.2	237.7	30

Instrument parameters were set to: high pressure RF 210 V, low pressure RF 150 V, sheath gas temperature 400 °C, sheath gas flow 12 L/min, gas temperature 290 °C, gas flow 16 L/min, nebulizer 20 psi, capillary 4500 V, nozzle voltage 2000 V, cell accelerator voltage 2 V, and delta EMV 250 V.

Data Processing and Statistical Analyses

Spectra gathered during LC-MS analyses were initially processed using the MassHunter Workstation software package vB.06 (Agilent Technologies Inc.). For LC-MS analyses, compounds were identified by retention time and MRM match to an authentic standard, integrated peak areas were used for quantification relative to external calibration curves. General linear model and Tukey's test 95% confidence grouping analyses were performed in Minitab 16 statistical software (Minitab Inc., State College, Pennsylvania, USA). Pair-wise statistical analyses were performed using Student's t-tests between sets of biological replicates from different treatments collected during the same time period only. The metabolomics data analysis server Metaboanalyst 3.0 (Wishart Research Group, the University of Alberta, Edmonton, Alberta, Canada) was used to identify clustering of metabolites profiles between samples based on PCA. Autoscaling and cube-root normalization was performed on analyte levels prior to Metaboanalyst analyses. Permutational ANOVA (PERMANOVA) was used to test for significant differences ($p < 0.05$) in metabolite profiles between Fe treatments and collection times. PERMANOVA tests were run using unrestricted permutation of data and 999 permutations on a resemblance matrix of autoscaled metabolite profiles based on Euclidean distances (Anderson 2005). PERMANOVA tests were performed using PRIMER software (v.6.0) with the PERMANOVA+ add-on (v.1.0.6) (PRIMER-E, Plymouth Marine Laboratory, UK). The Visualization and Analysis of NeTworks with related Experimental Data program, VANTED v2.1.0 was used to map the log base 2 transformations of Fe deficiency associated induction factors of metabolite data on to custom made biochemical pathways maps developed from pathway maps available at the KEGG pathway database (available at http://www.genome.jp/kegg/pathway.html).

Additional files

Additional file 1: Table S1A. Iron deficiency induced gene expression detected by qRT-PCR. **Table S1B**. Diurnal pattern of Fe deficiency induced gene expression detected by qRT-PCR.

Additional file 2: Table S2A. Amines concentrations from Fe-sufficient and Fe-deficient rice roots. **Table S2B**. Iron deficiency associated induction of root amines concentrations.

Additional file 3: Figure S1. Levels of 3-P-glycerate in Fe-sufficient and Fe-deficient rice roots quantified by LC-qTOF-MS. **Figure S2**. Levels of ascorbic acid in Fe-sufficient and Fe-deficient rice roots quantified by LC-qTOF-MS.

Additional file 4: Figure S3. Induction of precursors to methionine and DMA biosynthesis within the rice Fe deficiency response.

Additional file 5: Table S3. Iron replete hydroponic growth solution. **Table S4**. Primers used for qRT-PCR analyses.

Abbreviations

2-AA: 2-aminoanthracene; 2nd YL: 2nd youngest leaf; 3GNF: 3 gene normalization factor; Aqc: 6-aminoquinolyl-N-hydroxysuccinimidyl carbamate; BV: Bed volume; Cit-NO cycle: Citrulline-nitric oxide cycle; CT: collection times; dH_2O: deionized MilliQ water; DMA: 2′-deoxymugineic acid; DMAS: 2′-deoxymugineic acid synthase; epiHMA: 3-epihydroxymugineic acid; FA: Formic acid; Fe: Iron; FMOC-Cl: 9-fluorenyl methoxycarboxyl chloride; GOI: Genes of interest; H^+: proton; IRT: Iron-regulated transporters; LC-MS: Liquid chromatography mass spectrometry; MA: Mugineic acids; MeOH: Methanol; MRM: Multiple reaction monitoring; NA: Nicotianamine; NAAT: Nicotianamine aminotransferase; NAS: Nicotianamine synthase; NO: Nitric oxide; PCA: Principal component analysis; PERMANOVA: Permutational ANOVA; ppm: parts per million; PS: Phytosiderophores; qRT-PCR: Quantitative real time PCR; RO water: Reverse osmosis water; ROS: Reactive oxygen species; RP chromatography: Reverse phase chromatography; RT: Retention times; SAM: S-adenosyl-methionine; SE: Standard error; SPAD: Soil plant analysis development; TOM: Transporter of mugineic acid family phytosiderophores; Tukey's HSD test: Tukey's honest significant difference test; VANTED: Visualization and analysis of networks with related experimental data program; YL: Youngest leaf; YSL: Yellow stripe-like

Acknowledgements

This research was supported by grants from the Australian Research Council (LP130100785) and the HarvestPlus Challenge Program. Metabolite analyses were conducted at Metabolomics Australia, a NCRIS initiative under Bioplatforms Australia Pty Ltd. The authors would like to thank Alice Chia Feen Ng, Priyanka Reddy, Dr. Thusitha Rupasinghe and Yuan Li for their assistance during metabolite and gene expression analyses.

Authors' contributions

JS, AJ, BB, and UR conceived and designed the experiments. JS, BB and AL developed and conducted the physiological and biochemical analyses. AJ and UR acquired the funds. JS, AJ and LM wrote the manuscript. All authors discussed, revised and approved the manuscript.

Competing interests

The authors declare that they have no competing interests.

Author details

[1]School of BioSciences, The University of Melbourne, Parkville, Victoria, Australia. [2]Metabolomics Australia, The University of Melbourne, Parkville, Victoria, Australia.

References

Anderson MJ (2005) Permutational multivariate analysis of variance. Department of Statistics, University of Auckland, Auckland 26:32–46.

Arasimowicz M, Floryszak-Wieczorek J (2007) Nitric oxide as a bioactive signalling molecule in plant stress responses. Plant Sci 172(5):876–87

Ariga T, Hazama K, Yanagisawa S, Yoneyama T (2014) Chemical forms of iron in xylem sap from graminaceous and non-graminaceous plants. Soil Sci Plant Nutr 60(4):460–9

Arnaud N, Murgia I, Boucherez J, Briat J-F, Cellier F, Gaymard F (2006) An iron-induced nitric oxide burst precedes ubiquitin-dependent protein degradation for *Arabidopsis AtFer1* ferritin gene expression. J Biol Chem 281(33):23579–88

Astolfi S, Zuchi S, Neumann G, Cesco S, Di Toppi LS, Pinton R (2012) Response of barley plants to Fe deficiency and Cd contamination as affected by S starvation. J Exp Bot 63(3):1241–50

Bashir K, Nagasaka S, Itai RN, Kobayashi T, Takahashi M, Nakanishi H, Mori S, Nishizawa NK (2007) Expression and enzyme activity of glutathione reductase is upregulated by Fe-deficiency in graminaceous plants. Plant Mol Biol 65(3):277–84

Becana M, Moran JF, Iturbe-Ormaetxe I, Escuredo PR (1998) Iron-dependent oxygen free radical generation in plants subjected to environmental stress: toxicity and antioxidant protection. Plant Soil 201(1):137–47

Bhullar NK, Gruissem W (2013) Nutritional enhancement of rice for human health: the contribution of biotechnology. Biotechnol Adv 31(1):50–7

Boughton BA, Callahan DL, Silva C, Bowne J, Nahid A, Rupasinghe T, Tull DL, McConville MJ, Bacic A, Roessner U (2011) Comprehensive profiling and quantitation of amine group containing metabolites. Anal Chem 83(19):7523–30

Briat J-F, Dubos C, Gaymard F (2015) Iron nutrition, biomass production, and plant product quality. Trends Plant Sci 20(1):33–40

Bughio N, Yamaguchi H, Nishizawa NK, Nakanishi H, Mori S (2002) Cloning an iron-regulated metal transporter from rice. J Exp Bot 53(374):1677–82

Callahan DL, Kolev SD, O'Hair RAJ, Salt DE, Baker AJM (2007) Relationships of nicotianamine and other amino acids with nickel, zinc and iron in *Thlaspi* hyperaccumulators. New Phytol 176(4):836–48

Chen L, Ding C, Zhao X, Xu J, Mohammad AA, Wang S, Ding Y (2015) Differential regulation of proteins in rice (*Oryza sativa* L.) under iron deficiency. Plant Cell Rep 34(1):83–96

Clemens S (2001) Molecular mechanisms of plant metal tolerance and homeostasis. Planta 212(4):475–86

Connolly EL, Guerinot ML (2002) Iron stress in plants. Genome Biol 3(8):1024.1–1024.4

da Silveira VC, de Oliveira AP, Sperotto RA, Espindola LS, Amaral L, Dias JF, da Cunha JB, Fett JP (2007) Influence of iron on mineral status of two rice (*Oryza sativa* L.) cultivars. Braz J Plant Physiol 19(2):127–39

Dell'mour M, Koellensperger G, Quirino JP, Haddad PR, Stanetty C, Oburger E, Puschenreiter M, Hann S (2010) Complexation of metals by phytosiderophores revealed by CE-ESI-MS and CE-ICP-MS. Electrophoresis 31(7):1201–7

Di Martino C, Delfine S, Pizzuto R, Loreto F, Fuggi A (2003) Free amino acids and glycine betaine in leaf osmoregulation of spinach responding to increasing salt stress. New Phytol 158(3):455–63

Droux M, Ravanel S, Douce R (1995) Methionine biosynthesis in higher plants. II. Purification and characterization of cystathionine β-lyase from spinach chloroplasts. Arch Biochem Biophys 316(1):585–95

Ercal N, Gurer-Orhan H, Aykin-Burns N (2001) Toxic metals and oxidative stress part I: mechanisms involved in metal-induced oxidative damage. Curr Top Med Chem 1(6):529–39

Fan TW-M, Lane AN, Shenker M, Bartley JP, Crowley D, Higashi RM (2001) Comperhensive chemical profiling of gramineous plant root exudates using high-resolution NMR and MS. Phytochemistry 57(2):209–21

Gechev TS, Breusegem FV, Stone JM, Denev I, Laloi C (2006) Reactive oxygen species as signals that modulate plant stress responses and programmed cell death. BioEssays 28(11):1091–101

Ghasemzadeh A, Ghasemzadeh N (2011) Flavonoids and phenolic acids: role and biochemical activity in plants and human. J Med Plants Res 5(31):6697–703

Gill SS, Tuteja N (2010) Reactive oxygen species and antioxidant machinery in abiotic stress tolerance in crop plants. Plant Physiol Biochem 48(12):909–30

Hell R, Stephan UW (2003) Iron uptake, trafficking and homeostasis in plants. Planta 216(4):541–51

Hider RC, Yoshimura E, Khodr H, von Wirén N (2004) Competition or complementation: the iron-chelating abilities of nicotianamine and phytosiderophores. New Phytol 164(2):204–8

Higuchi K, Kanazawa K, Nishizawa N-K, Mori S (1996) The role of nicotianamine synthase in response to Fe nutrition status in Gramineae. Plant Soil 178(2):171–7

Higuchi K, Watanabe S, Takahashi M, Kawasaki S, Nakanishi H, Nishizawa NK, Mori S (2001) Nicotianamine synthase gene expression differs in barley and rice under Fe-deficienct conditions. Plant J 25(2):159–67

Hung C-H, Hwang HJ, Chen Y-H, Chiu Y-F, Ke S-C, Burnap RL, Chu H-A (2010) Spectroscopic and functional characterizations of cyanobacterium *Synechocystis* PCC 6803 mutants on and near the heme axial ligand of cytochrome b559 in photosystem II. J Biol Chem 285(8):5563–663

Inoue H, Higuchi K, Takahashi M, Nakanishi H, Mori S, Nishizawa NK (2003) Three rice nicotianamine synthase genes, *OsNAS1*, *OsNAS2*, and *OsNAS3* are expressed in cells involved in long-distance transport of iron and differentially regulated by iron. Plant J 36(3):366–81

Inoue H, Kobayashi T, Nozoye T, Takahashi M, Kakei Y, Suzuki K, Nakazono M, Nakanishi H, Mori S, Nishizawa NK (2009) Rice OsYSL15 is an iron-regulated iron(III)-deoxymugineic acid transporter expressed in the roots and is essential for iron uptake in early growth of the seedlings. J Biol Chem 284(6):3470–9

Ishimaru Y, Suzuki M, Tsukamoto T, Suzuki K, Nakazono M, Kobayashi T, Wada Y, Watanabe S, Matsuhashi S, Takahashi M, Nakanishi H, Mori S, Nishizawa NK (2006) Rice plants take up iron as an Fe^{3+} phytosiderophore and as Fe^{2+}. Plant J 45(3):335–46

Itai RN, Ogo Y, Kobayashi T, Nakanishi H, Nishizawa NK (2013) Rice genes involved in phytosiderophore biosynthesis are synchronously regulated during the early stages of iron deficiency in roots. Rice 6(1):1–13

Iturbe-Ormaetxe I, Moran JF, Arrese-Igor C, Gogorcena Y, Klucas RV, Becana M (1995) Activated oxygen and antioxidant defenses in iron-deficient pea plants. Plant Cell Environ 18(4):421–9

Joshi V, Joung J-G, Fei Z, Jander G (2010) Interdependence of threonine, methionine and isoleucine metabolism in plants: accumulation and transcriptional regulation under abiotic stress. Amino Acids 39(4):933–47

Kakei Y, Yamaguchi I, Kobayashi T, Takahashi M, Nakanishi H, Yamakawa T, Nishizawa NK (2009) A highly sensitive, quick and simple quantification method for nicotianamine and 2′-deoxymugineic acid from minimum samples using LC/ESI-TOF-MS achieves functional analysis of these components in plants. Plant Cell Physiol 50(11):1988–93

Kawai S, Itoh K, Takagi SI (1993) Incorporation of ^{15}N and ^{14}C of methionine into the mugineic acid family of phytosiderophores in iron-deficient barley roots. Physiol Plant 88(4):668–74

Kim SA, Guerinot ML (2007) Mining iron: iron uptake and transport in plants. FEBS Lett 581(12):2273–80

Kobayashi T, Itai RN, Ogo Y, Kakei Y, Nakanishi H, Takahashi M, Nishizawa NK (2009) The rice transcription factor IDEF1 is essential for the early response to iron deficiency, and induces vegetative expression of late embryogenesis abundant genes. Plant J 60(6):948–61

Kosová K, Vítámvás P, Prášil IT, Renaut J (2011) Plant proteome changes under abiotic stress - Contribution of proteomics studies to understanding plant stress response. J Proteome 74(8):1301–22

Lee S, An G (2009) Over-expression of OsIRT1 leads to increased iron and zinc accumulations in rice. Plant, Cell and Environment 32(4):408–16.

Lee S, Chiecko JC, Kim SA, Walker EL, Lee Y, Guerinot ML, An G (2009) Disruption of *OsYSL15* leads to iron inefficiency in rice plants. Plant Physiol 150(2):786–800

Li S, Zhou X, Huang Y, Zhu L, Zhang S, Zhao Y, Guo J, Chen J, Chen R (2013) Identification and characterization of the zinc-regulated transporters, iron-regulated transporter-like protein (ZIP) gene family in maize. BMC Plant Biol 13(1):1

Ma JF, Shinada T, Matsuda C, Nomoto K (1995) Biosynthesis of phytosiderophores, mugineic acids, associated with methionine cycling. J Biol Chem 270(28):16549–54

Ma JF, Ueno H, Ueno D, Rombolà AD, Iwashita T (2003) Characterization of phytosiderophore secretion under Fe deficiency stress in *Festuca rubra*. Plant Soil 256(1):131–7

Marschner H, Römheld V, Kissel M (1986) Different strategies in higher plants in mobilization and uptake of iron. J Plant Nutr 9(3–7):695–713

Marsh HV Jr, Evans HJ, Matrone G (1963) Investigations of the role of iron in chlorophyll metabolism I. Effect of iron deficiency on chlorophyll and heme content and on the activities of certain enzymes in leaves. Plant Physiol 38(6):632–8

Michalak A (2006) Phenolic compounds and their antioxidant activity in plants growing under heavy metal stress. Pol J Environ Stud 15(4):523–30

Michel K-P, Pistorius EK (2004) Adaptation of the photosynthetic electron transport chain in cyanobacteria to iron deficiency: the function of IdiA and IsiA. Physiol Plant 120(1):36–50

Miller GW, Pushnik JC, Welkie GW (1984) Iron chlorosis, a world wide problem, the relation of chlorophyll biosynthesis to iron. J Plant Nutr 7(1–5):1–22

Mori M (2007) Regulation of nitric oxide synthesis and apoptosis by arginase and arginine recycling. J Nutr 137(6):1616S–20

Mori S, Nishizawa NK (1987) Methionine as a dominant precursor of phytosiderophores in *Graminaceae* plants. Plant Cell Physiol 28(6):1081–92

Mori S, Nishizawa N, Kawai S, Sato Y, Takagi S (1987) Dynamic state of mugineic acid and analogous phytosiderophores in Fe-deficient barley. J Plant Nutr 10(9–16):1003–11

Mori S, Nishizawa N, Hayashi H, Chino M, Yoshimura E, Ishihara J (1991) Why are young rice plants highly susceptible to iron deficiency? Plant Soil 130(1):143–56

Nagasaka S, Takahashi M, Nakanishi-Itai R, Bashir K, Nakanishi H, Mori S, Nishizawa NK (2009) Time course analysis of gene expression over 24 h in Fe-deficient barley roots. Plant Mol Biol 69(5):621–31

Nakanishi H, Bughio N, Matsuhashi S, Ishioka N-S, Uchida H, Tsuji A, Osa A, Sekine T, Kume T, Mori S (1999) Visualizing real time [^{11}C] methionine translocation in Fe-sufficient and Fe-deficient barley using a Positron Emitting Tracer Imaging System (PETIS). J Exp Bot 50(334):637–43

Negishi T, Nakanishi H, Yazaki J, Kishimoto N, Fujii F, Shimbo K, Yamamoto K, Sakata K, Sasaki T, Kikuchi S, Mori S, Nishizawa NK (2002) cDNA microarray analysis of gene expression during Fe-deficiency stress in barley suggests that polar transport of vesicles is implicated in phytosiderophore secretion in Fe-deficient barley roots. Plant J 30(1):83–94

Nimalaratne C, Lopes-Lutz D, Schieber A, Wu J (2011) Free aromatic amino acids in egg yolk show antioxidant properties. Food Chem 129(1):155–61

Nishiyama H, Ohya T, Tanoi K, Nakanishi TM (2008) A simple measurement of the pH of root apoplast by the fluorescence ratio method. Plant Root 2:3–6

Nishiyama R, Kato M, Nagata S, Yanagisawa S, Yoneyama T (2012) Identification of Zn–Nicotianamine and Fe–2′-Deoxymugineic acid in the phloem sap from rice plants (*Oryza sativa* L.). Plant Cell Physiol 53(2):381–90

Nozoye T, Itai RN, Nagasaka S, Takahashi M, Nakanishi H, Mori S, Nishizawa NK (2004) Diurnal changes in the expression of genes that participate in phytosiderophore synthesis in rice. Soil Sci Plant Nutr 50(7):1125–31

Nozoye T, Nagasaka S, Kobayashi T, Takahashi M, Sato Y, Sato Y, Nakanishi NUH, Nishizawa NK (2011) Phytosiderophore efflux transporters are crucial for iron acquisition in graminaceous plants. J Biol Chem 286(7):5446–54

Nozoye T, Nagasaka S, Bashir K, Takahashi M, Kobayashi T, Nakanishi H, Nishizawa NK (2014) Nicotianamine synthase 2 localizes to the vesicles of iron-deficient rice roots, and its mutation in the YXXφ or LL motif causes the disruption of vesicle formation or movement in rice. Plant J 77(2):246–60

Oburger E, Gruber B, Schindlegger Y, Schenkeveld WDC, Hann S, Kraemer SM, Wenzel WW, Puschenreiter M (2014) Root exudation of phytosiderophores from soil-grown wheat. New Phytol 203(4):1161–74

Parida AK, Das AB (2005) Salt tolerance and salinity effects on plants: a review. Ecotoxicol Environ Saf 60(3):324–49

Park S, Pan L-P, Chan SI, Alben JO (1996) Photoperturbation of the heme a_3-Cu_B binuclear center of cytochrome c oxidase CO complex observed by Fourier transform infrared spectroscopy. Biophys J 71(2):1036–47

Pedas P, Ytting CK, Fuglsang AT, Jahn TP, Schjoerring JK, Husted S (2008) Manganese efficiency in barley: identification and characterization of the metal ion transporter HvIRT1. Plant Physiol 148(1):455–66

Plett D, Safwat G, Gilliham M, Møller IS, Roy S, Shirley N, Jacobs A, Johnson A, Tester M (2010) Improved salinity tolerance of rice through cell type-specific expression of *AtHKT1;1*. PLoS ONE 5(9):e12571

Rai VK (2002) Role of amino acids in plant responses to stresses. Biol Plant 45(4):481–7

Ramakrishna A, Ravishankar GA (2011) Influence of abiotic stress signals on secondary metabolites in plants. Plant Signal Behav 6(11):1720–31

Reddy KJ (2006) Nutrient Stress. In: Rao KVM, Raghavendra AS, Reddy KJ (eds) Physiology and molecular biology of stress tolerance in plants. Springer Science & Business Media, Dordrecht, pp 187–219

Ricachenevsky FK, Sperotto RA (2014) There and back again, or always there? The evolution of rice combined strategy for Fe uptake. Front Plant Sci 5:189

Römheld V, Marschner H (1990) Genotypical differences among graminaceous species in release of phytosiderophores and uptake of iron phytosiderophores. Plant Soil 123(2):147–53

Rong-li S, Hong-mei H, Xiao-yun F, Karim MR, Fu-suo Z, Chun-qin Z (2012) Responses of aerobic rice (*Oryza sativa* L.) to iron deficiency. J Integr Agric 11(6):938–45

Scholz G, Becker R, Pich A, Stephan UW (1992) Nicotianamine - a common constituent of strategies I and II of iron acquisition by plants: a review. J Plant Nutr 15(10):1647–65

Sharma SS, Dietz K-J (2006) The significance of amino acids and amino acid-derived molecules in plant responses and adaptation to heavy metal stress. J Exp Bot 57(4):711–26

Sperotto RA, Ricachenevsky FK, De Abreu Waldow V, Fett JP (2012) Iron biofortification in rice: it's a long way to the top. Plant Sci 190:24–39

Spinelli F, Cellini A, Marchetti L, Nagesh KM, Piovene C (2011) Emission and function of volatile organic compounds in response to abiotic stress. In: Shanker A, Venkateswarlu B (eds) Abiotic stress in plants - mechanisms and adaptations. InTech, Rijeka

Stadtman ER, Levine RL (2003) Free radical-mediated oxidation of free amino acids and amino acid residues in proteins. Amino Acids 25(3):207–18

Stocking CR (1975) Iron deficiency and the structure and physiology of maize chloroplasts. Plant Physiol 55(4):626–31

Sun B, Jing Y, Chen K, Song L, Chen F, Zhang L (2007) Protective effect of nitric oxide on iron deficiency-induced oxidative stress in maize (*Zea mays*). J Plant Physiol 164(5):536–43

Suzuki M, Takahashi M, Tsukamoto T, Watanabe S, Matsuhashi S, Yazaki J, Kishimoto N, Kikuchi S, Nakanishi H, Mori S, Nishizawa NK (2006) Biosynthesis and secretion of mugineic acid family phytosiderophores in zinc-deficient barley. Plant J 48(1):85–97

Takagi SI, Nomoto K, Takemoto T (1984) Physiological aspect of mugineic acid, a possible phytosiderophore of graminaceous plants. J Plant Nutr 7(1–5):469–77

Terry N, Low G (1982) Leaf chlorophyll content and its relation to the intracellular localization of iron. J Plant Nutr 5(4–7):301–10

Ueno D, Yamaji N, Ma JF (2009) Further characterization of ferric—phytosiderophore transporters ZmYS1 and HvYS1 in maize and barley. J Exp Bot 60(12):3513–20

Vasconcelos M, Musetti V, Li C-M, Datta SK, Grusak MA (2004) Functional analysis of transgenic rice (*Oryza sativa* L.) transformed with an *Arabidopsis thaliana* ferric reductase (AtFR02). Soil Sci Plant Nutr 50(7):1151–7

Vigani G, Zocchi G, Bashir K, Philippar K, Briat J-F (2013) Signals from chloroplasts and mitochondria for iron homeostasis regulation. Trends Plant Sci 18(6):305–11

Walker EL, Connolly EL (2008) Time to pump iron: iron-deficiency-signaling mechanisms of higher plants. Curr Opin Plant Biol 11(5):530–5

Walter A, Pich A, Scholz G, Marschner H, Römheld V (1995) Diurnal variations in release of phytosiderophores and in concentrations of phytosiderophores and nicotianamine in roots and shoots of barley. J Plant Physiol 147(2):191–6

Wang L, Ying Y, Narsai R, Ye L, Zheng L, Tian J, Whelan J, Shou H (2013) Identification of OsbHLH133 as a regulator of iron distribution between roots and shoots in *Oryza sativa*. Plant Cell Environ 36(1):224–36

Weber G, von Wirén N, Hayen H (2008) Investigation of ascorbate-mediated iron release from ferric phytosiderophores in the presence of nicotianamine. BioMetals 21(5):503–13

Wingler A, Lea PJ, Quick WP, Leegood RC (2000) Photorespiration: metabolic pathways and their role in stress protection. Philos Trans R Soc Lond B Biol Sci 355(1402):1517–29

Yoneyama T, Ishikawa S, Fujimaki S (2015) Route and regulation of Zinc, Cadmium, and Iron transport in rice plants (*Oryza sativa* L.) during vegetative growth and grain filling: Metal transporters, metal speciation, grain Cd reduction and Zn and Fe biofortification. Int J Mol Sci 16(8):19111–29

Zaharieva TB, Abadía J (2003) Iron deficiency enhances the levels of ascorbate, glutathione, and related enzymes in sugar beet roots. Protoplasma 221(3–4):269–75

Zhang F-S, Römheld V, Marschner H (1991) Diurnal rhythm of release of phytosiderophores and uptake rate of zinc in iron-deficient wheat. Soil Sci Plant Nutr 37(4):671–8

Zhang X, Zhang F, Mao D (1998) Effect of iron plaque outside roots on nutrient uptake by rice (*Oryza sativa* L.). Zinc uptake by Fe-deficient rice. Plant Soil 202(1):33–9

Zhu XF, Wang B, Song WF, Zheng SJ, Shen RF (2016) Putrescine alleviates iron deficiency via NO-dependent reutilization of root cell-wall Fe in Arabidopsis. Plant Physiol 170(1):558–67

Zoghlami LB, Djebali W, Abbes Z, Hediji H, Maucourt M, Moing A, Brouquisse R, Chaïbi W (2011) Metabolite modifications in *Solanum lycopersicum* roots and leaves under cadmium stress. Afr J Biotechnol 10(4):567–79

6

Rice Bran Metabolome Contains Amino Acids, Vitamins & Cofactors, and Phytochemicals with Medicinal and Nutritional Properties

Iman Zarei[1,2], Dustin G. Brown[1], Nora Jean Nealon[1] and Elizabeth P. Ryan[1*]

Abstract

Background: Rice bran is a functional food that has shown protection against major chronic diseases (e.g. obesity, diabetes, cardiovascular disease and cancer) in animals and humans, and these health effects have been associated with the presence of bioactive phytochemicals. Food metabolomics uses multiple chromatography and mass spectrometry platforms to detect and identify a diverse range of small molecules with high sensitivity and precision, and has not been completed for rice bran.

Results: This study utilized global, non-targeted metabolomics to identify small molecules in rice bran, and conducted a comprehensive search of peer-reviewed literature to determine bioactive compounds. Three U.S. rice varieties (Calrose, Dixiebelle, and Neptune), that have been used for human dietary intervention trials, were assessed herein for bioactive compounds that have disease control and prevention properties. The profiling of rice bran by ultra-performance liquid chromatography-tandem mass spectrometry (UPLC-MS/MS) and gas chromatography–mass spectrometry (GC–MS) identified 453 distinct phytochemicals, 209 of which were classified as amino acids, cofactors & vitamins, and secondary metabolites, and were further assessed for bioactivity. A scientific literature search revealed 65 compounds with health properties, 16 of which had not been previously identified in rice bran. This suite of amino acids, cofactors & vitamins, and secondary metabolites comprised 46% of the identified rice bran metabolome, which substantially enhanced our knowledge of health-promoting rice bran compounds provided during dietary supplementation.

Conclusion: Rice bran metabolite profiling revealed a suite of biochemical molecules that can be further investigated and exploited for multiple nutritional therapies and medical food applications. These bioactive compounds may also be biomarkers of dietary rice bran intake. The medicinal compounds associated with rice bran can function as a network across metabolic pathways and this metabolite network may occur via additive and synergistic effects between compounds in the food matrix.

Keywords: Rice bran, Functional food, Metabolomics, Medicinal properties, Chronic diseases, Phytochemicals, Infectious diseases

* Correspondence: E.P.Ryan@colostate.edu
[1]Department of Environmental & Radiological Health Sciences, College of Veterinary Medicine and Biological Sciences, Colorado State University, 1680 Campus Delivery, Fort Collins, CO 80523, USA
Full list of author information is available at the end of the article

Background

Rice (*Oryza sativa* L.) is an essential staple food for more than half of the world's population (Hu et al. 2014; Qian et al. 2016) and is grown in more than 100 countries worldwide (Muthayya et al. 2014). Rice bran, the outer covering of the rice grain, contains a unique profile of phytochemicals with medicinal and nutritional properties that are beneficial to human health, some of which have been targeted for nutraceutical development for cancer (Henderson et al. 2012; Verschoyle et al. 2007), type 2 diabetes (Cheng et al. 2010; de Munter et al. 2007; Qureshi et al. 2002), lipid metabolism regulation (Kuriyan et al. 2005; Qureshi et al. 1997; Shibata et al. 2016; Wang et al. 2015), immune regulatory processes (Wang et al. 2015), and obesity (Ham et al. 2015). Furthermore, we recently showed whole rice bran can protect against enteric pathogens such as *Salmonella enterica serovar* Typhimurium, human rotavirus, and human norovirus (Goodyear et al. 2015; Kumar et al. 2012; Lei et al. 2016; Yang et al. 2015). Rice bran contains non-saponifiable lipids (i.e. gamma oryzanol), vitamin E (e.g. tocopherols and tocotriols), polyphenols (e.g. ferulic acid caffeic acid and salicylic acid), and phytosterols (e.g. beta-sitosterol) (Henderson et al. 2012) with reported health properties. Many of these compounds are available in the lipid fraction and also known as rice bran oil (Charoonratana et al. 2015; Iqbal et al. 2003; V. Panala, 2009). Additional compounds from other chemical classes in rice bran merit attention and can be identified via high throughput techniques, such as global, non-targeted metabolomics that can assess a large profile of small molecules present in the whole food. Given the emphasis in previous studies on rice bran lipids (Forster et al. 2013), this analysis focused on rice bran amino acids, cofactors & vitamins, and secondary metabolites that have medicinal and nutritional properties important to human health.

Despite the large body of scientific evidence on rice bran bioactivity, rice bran remains underutilized in human health and nutrition because it is considered an animal feed and is known to undergo hydrolytic rancidity after processing from whole grain rice (da Silva et al. 2006; Ramezanzadeh et al. 1999a, b). Thermal treatments applied to rice bran have helped to stabilize it and prevents rancidity by the inactivation of lipases and peroxidases. One major obstacle to achieving widespread human consumption and acceptance of rice bran is the global perception that rice bran is an animal feed (Ramezanzadeh et al. 1999a).

Food metabolomics, or "Foodomics", provides information on the presence and relative abundance of all compounds in a food matrix. Food metabolome studies have shown compounds across diverse chemical classes such as amino acids, lipids, sugars, peptides, organic acids, phenolic compounds and other secondary metabolites (Wishart, 2008). Entire metabolite profiles have been completed on several foods including cooked and uncooked rice grain (Heuberger et al. 2010; Hu et al. 2014; Kim et al. 2013a, 2013b), grape (*Vitis vinifera* L.) (Luca Narduzzi, 2015), human milk (Andreas et al. 2015; Wu et al. 2016), tomato (Moco et al. 2006), citrus juice (Arbona et al. 2015), and several other foods and crops (e.g., carrot, beer, wine, and coffee) (Johanningsmeier et al. 2016) through non-targeted screening methods. Nutritional metabolomics is an experimental approach that uses small molecule profiling to integrate the effects of diet on nutrition, and thus can be used to evaluate the health effects of foods at an individual level (Jones et al. 2012). Integrating food and nutritional metabolomic approaches can increase our knowledge on the bioactivity of food metabolites, and may increase evidence for metabolic mechanisms by which foods elicit important health effects (Capozzi and Bordoni, 2013; Herrero et al. 2012). Accurate food metabolite profiles in regards to food and nutritional metabolomics may also assist in the quantification of dietary intakes and specific food biomarkers.

The goal of the food metabolome approach applied herein was to obtain a complete characterization of the rice bran small molecule profile for bioactive components. This study used non-targeted metabolomics to investigate heat-stabilized rice bran from three U.S. rice cultivars for the identification of metabolites with medicinal and nutritional properties. These varieties were chosen for profiling based on human consumption in clinical trials, whereby rice bran intake improved intestinal health parameters by modulating gastrointestinal microbiota and host immunity (Borresen et al. 2016; Sheflin et al. 2016; Yang et al. 2015). The hypothesis was that rice bran contains a distinct stoichiometry of small molecules, covering multiple classes of phytochemicals, including but not limited to amino acids, cofactors & vitamins, and secondary metabolites that have medicinal properties and contribute to the nutritional benefits of rice bran as a whole food. A thorough examination of metabolites across chemical classes revealed a complex network of metabolic pathways that have not been previously examined for rice bran. A detailed analysis of rice bran functional food components allowed for a thorough understanding of how a suite of metabolites in a single food can exhibit therapeutic and preventive medicine properties.

Results

Non-Targeted Rice Bran Metabolomics

The metabolite profile composition of rice bran revealed 453 metabolites with known identity that were clustered into the following metabolic pathways: amino acid

(126 metabolites), carbohydrate (35 metabolites), cofactors & vitamins (28 metabolites), energy (11 metabolites), lipids (137 metabolites), nucleotides (40 metabolites), peptides (28 metabolites), secondary metabolites (55 metabolites), and xenobiotics (8 metabolites). Total metabolite numbers for each metabolic pathway are shown in Additional file 1: Table S1. Approximately 46% of total identified metabolites (209 metabolites) were classified as amino acids, cofactors & vitamins, and secondary metabolites and were interrogated for their potential human health-promoting properties. Medicinal and health promoting attributes were previously reported in the scientific literature for 65 rice bran metabolites from these three pathways. Table 1 lists the 29 amino acids, Table 2 lists the 13 cofactors & vitamins, and Table 3 lists the 23 secondary metabolites with previously reported health beneficial properties. A total of 16 out of 65 metabolites (noted by [1] in tables) with medicinal/nutritional properties were identified from this metabolome analysis that had not previously been reported from rice bran. These included 9 amino acids, 2 cofactors & vitamins, and 6 secondary metabolites.

Amino acids constituted ~28% of the rice bran metabolome from the three rice bran varieties tested. Of the 29 amino acids with medicinal properties in Table 1, 13 had reported antioxidant and/or anti-inflammatory activities, 6 compounds with known antimicrobial properties, 5 that were anti-hypertensive, 4 metabolites with lipid-lowering effects, 3 with cancer chemopreventive actions, one with evidence for prevention of obesity, and one metabolite with hypoglycemic, antidiabetic properties. Multiple metabolites, including taurine and betaine, varied in relative abundance across U.S. varieties, and exhibited more than one mechanism of action with nutritional benefits.

Figure 1 is a Cytoscape pathway classification network view specific to amino acids present in Calrose rice bran. The metabolite taurine is presented by a closed black node extending from methionine, cysteine, S-adenosylmethionine (SAM), & taurine pathway node, which connects to the central, hexagonal amino acid network node. Node size indicates the relative abundance Z-score for the metabolite. The median-scaled relative abundance for each metabolite in Table 1 can be inferred by the size of the node. For example, taurine (methionine, cysteine, S-adenosyl- methionine (SAM), & taurine pathway) has a bigger node size (i.e., bigger Z-score) when compared with betaine (glycine, serine and threonine metabolic pathway).

Cofactors and vitamins constituted ~ 6% of the rice bran metabolome. The 13 metabolites with established medicinal properties are listed in Table 2, with novel identifications of glucarate (median-scaled relative abundance of 0.89, 0.13 for Calrose and Dixiebelle respectively, and non-detectable in Neptune), and threonic acid (median-scaled relative abundance of 36.15, 20.99, and 16.44 for Calrose, Dixiebelle, and Neptune respectively) from rice bran. We found 6 cofactors & vitamins that had antioxidant and/or anti-inflammatory properties, 5 compounds had anti-hypertensive activity, 2 compounds had shown lipid-lowering effects, 4 compounds had evidence for cancer chemopreventive action, and one metabolite was a hypoglycemic agent and antidiabetic.

Figure 2 shows the Cytoscape pathway classification node network view for cofactors & vitamins metabolic pathways. The micronutrient richness of rice bran is depicted by inclusion of vitamin B6, ascorbate & aldarate, and nicotinate & nicotinamide sub-metabolic pathways with metabolite examples of pyridoxine, threonic acid, and nicotinate for each metabolic pathway, respectively.

The plant secondary metabolites from rice bran constitute more than 12% of the metabolome. From the 23 rice bran phytochemicals in Table 3 that have known medicinal properties, we identified 16 phytochemicals with antioxidant and/or anti-inflammatory properties, 9 that had antimicrobial effects, 2 components that were anti-hypertensive, 1 with lipid-lowering effects, 4 compounds with evidence of cancer chemoprevention, 1 had scientific reports for utility in obesity prevention, and 4 phytochemicals demonstrated hypoglycemic and antidiabetic properties. Figure 3 shows all of the secondary metabolites from the rice bran metabolome across three U.S. varieties with respect to their median-scaled relative abundances. Ferulate is an example of a secondary metabolite that has a relative abundance of 6.22, 10.28, and 19.69 in Calrose, Dixiebelle, and Neptune, respectively. Levels of this metabolite are in contrast to tartaric acid, which was 0.31, 0.32, and 0.26. The varied relative abundance detected across metabolites from diverse pathways supports the utility of a metabolome approach.

Figure 4 is the Cytoscape network view of the 16 newly identified metabolites within these metabolic pathways for Calrose (i.e. amino acids, cofactors & vitamins, and secondary metabolites) and their respective sub-metabolic pathways.

Calrose, Dixiebelle and Neptune have Similar Bran Metabolomes

To assess the potential for differences in the relative abundances of small molecules in Calrose, Dixiebelle, and Neptune, we performed a principal component analysis (PCA). Principal component 1 revealed 54% variance among these three rice bran cultivars (Fig. 5), yet this percent variation using PCA did not reach statistical significance, and a lower level of variation is expected between these three U.S varieties when compared to a

Table 1 Median-scaled relative abundance of rice bran amino acids metabolites with medicinal properties

Metabolite	HMDB ID	Calrose	Dixiebelle	Neptune	Function	References
4-guanidinobutanoate (Rhapontigenin)[a]	03464	99.84	194.48	153.27	• Anti-hyperlipidemic • Antifungal (*Candida albicans* in vitro) • Antioxidant (protection against cellular DNA damage caused by intracellular reactive oxygen species (ROS))	(Jo et al. 2014) (Kim et al. 2013a, b) (Zhang et al. 2007)
4-hydroxycinnamic acid	02035	3.61	6.62	9.49	• Antioxidant (protective against oxidative damage caused by ROS)	(Shang et al. 2015)
5-oxoproline (Pyroglutamic acid)	00267	10.80	22.81	41.36	• Increases the release of acetylcholine (Ach) and GABA from the cortical surface to improve the age associated memory impairment	(Antonelli et al. 1984) (Grioli et al. 1990)
3-(4-hydroxyphenyl) lactate	00755	0.35	0.57	1.09	• Antioxidant (decrease the ROS production in neutrophils and mitochondria) • Antifungal (against main genera: *Penicillum, Aspergillus* and *Fusarium*)	(Beloborodova et al. 2012) (Dallagnol et al. 2011)
α-hydroxyisocaproic acid (Leucic acid)[a]	00746	0.16	0.21	0.45	• Antifungal (against *Candida* and *Aspergillus* species) • Antibacterial (broad spectrum bacteriostatic properties) • Anti-catabolic (inhibitory effect on various matrix metalloproteinase enzymes, which are responsible for degradation of various connective and protein tissues → muscle gaining)	(Sakko et al. 2014) (Sakko et al. 2012) (Mero et al. 2010)
Agmatine[a]	01432	6.89	0.95	5.23	• Anti-nociceptive (without affecting morphine-induced gastrointestinal transit) • Anti-hypertensive • Cancer chemopreventive • Renal sodium regulation • Antihyperglycemic (increases insulin release from rat pancreatic islets of Langerhans cells)	(Raasch et al. 2001) (Piletz et al. 2013)
Betaine	00043	249.03	197.66	308.76	• Protects against atherosclerosis (helps reduce higher levels of homocysteine) • Prevents hepatic steatosis, prevents cirrhosis (prevents or reduces accumulation of fat in the liver) • Protects against genetic instability, senescence, and cancer through DNA methylation	(Craig, 2004)
Citrulline	00904	2.72	0.34	ND	• Anti-hypertensive (increases the arginine availability as its precursor and arginine sustains increase in nitrogen oxide (NO) production) • Anti-hyperlipidemic	(Kaore et al. 2013)
Cystathionine	00099	ND	0.75	1.05	• Anti-inflammatory	(Zhu et al. 2015)

Table 1 Median-scaled relative abundance of rice bran amino acids metabolites with medicinal properties *(Continued)*

Carboxyethyl-GABA	02201	11.50	12.42	16.56	• Immune modulator	(Cerino et al. 1988)
GABA (Gamma-aminobutyric acid)	00112	163.45	121.42	102.60	• Inhibitory neurotransmitter → Relaxant, anxiolytic and anti-convulsive (antiepileptic)	(Foster and Kemp, 2006) (Chapouthier and Venault, 2001)
Gentisate[a]	00152	ND	0.07	0.27	• Anti-inflammatory (via free radical scavenging) • Antirheumatic	(Carlin et al. 1985) (Clarke and Mosher, 1953; Kleinsorge and Pohl, 1953)
Glutamate	00148	777.00	581.99	571.80	• Required for central nervous system function and treating neuropsychological conditions • Immunomodulator (development of T-cell-mediated immunity by stimulating glutamate-specific receptors)	(Hettema et al. 2006) (Pacheco et al. 2007) (Pietersen et al. 1998)
Glutathione, reduced (GSH)	00125	25.76	21.87	14.02	• Antioxidant (as a nucleophile and a reductant, and can react with electrophilic or oxidizing species	(Pompella et al. 2003)
Hydrocinnamic acid	00764	0.07	0.09	ND	• Anti-inflammatory (control of the degranulation of mast cells, basophils and neutrophils)	(Panico et al. 2005)
Indoleacetate[a]	00197	1.85	1.73	0.98	• Cancer chemopreventive	(Folkes and Wardman, 2001)
N-Acetyl histidine (NAH)	32055	8.65	11.33	3.36	• Anti-cataract formation (it is one the two major constituents of the vertebrate brain and eye)	(Baslow, 1998; Baslow and Guilfoyle, 2015)
Serotonin	00259	4.71	2.50	14.11	• Neurotransmitter • Antiemetic	(Peroutka et al. 1981) (De-Miguel and Trueta, 2005)
N-Acetylserotonin	01238	ND	0.09	0.41	• Anti-insomnia (this is the immediate precursor of melatonin) • antidepressant & anxiolytic • Anti-hypertensive • Antioxidant (lowers resting levels of ROS in peripheral blood lymphocytes, and inhibits nitric oxide synthase) • Anti-inflammatory	(Touitou, 2001) (Oxenkrug, 1999; Oxenkrug et al. 2007) (Reiter et al. 1999) (Perianayagam et al. 2005)
N-acetylleucine[a]	11756	0.12	0.17	0.41	• Vertigo treatment	(Kanchan Rao Singh, 2012; Przybylski, 2008)
N-Acetyl-L-tyrosine	00866	0.10	0.20	0.44	• Treatment of neurotransmitter dysfunction (administration of this amino acid acts as a precursor of catecholamine, dopamine and serotonin)	(M. Hinz, 2003; M. C. Hinz, 2009)
N-acetyltryptophan	13713	ND	0.19	0.55	• Antioxidant (diminishes oxidation of human serum albumin)	(Anraku et al. 2004) (W. Li et al. 2015)

Table 1 Median-scaled relative abundance of rice bran amino acids metabolites with medicinal properties *(Continued)*

					• Neuro-protective (Treatment of neurodegenerative disease such as amyotrophic lateral sclerosis)	
N-methyltyrosine (Metyrosine)	14903	0.06	ND	ND	• Anti-hypertensive (inhibits tyrosine hydroxylase)	(DrugBank; Scriabine et al. 1978)
Norvaline	13716	0.13	ND	0.07	• Anti-inflammatory (via inhibition of ribosomal protein S6 kinase beta-1 (S6K1)) • Improves sperm motility, count and viability in diabetic rats (inhibits the arginase enzyme and increases arginine availability as substrate to interact with endothelial nitric oxide synthase (eNOS))	(Ming et al. 2009) (De et al. 2016)
Ornithine	03374	1.42	0.12	0.27	• Antifatigue (increases release of human growth hormone by stimulating pituitary gland)	(Sugino et al. 2008)
Phenyllactic acid[a]	00779	0.28	0.37	0.59	• Antifungal (against *Fusarium graminearum* IDM623, *Endomyces fibuliger* IDM3812, *Penicillium expansum IDM/FS2*, *Aspergillus niger* IDM1, and *Monilia sitophila* IDM/FS5) • Antibacterial (against *Klebsiella oxytoca* (g-), *Providencia stuartii* (g-), *Enterococcus faecalis* (g+), *Staphylococcus aureu* (g+) *and Listeria monocytogenes* (g+))	(Lavermicocca et al. 2000) (Dieuleveux et al. 1998)
Picolinic acid	02243	0.22	0.66	0.19	• Increases the bioavailability of elements such as zinc, iron, copper, manganese and molybdenum in the human body • Anti-inflammatory	(Grant et al. 2009) (Bosco et al. 2000)
Taurine[a]	00251	1.41	1.92	1.92	• Antioxidant (found in large quantities in the neutrophil and excitable tissues, is a powerful scavenger of hypochlorous acid) • Lowers lead and cadmium levels in blood and tissues • Obesity prevention (increases energy metabolism in white adipose tissue) • Anti-hypertensive • Neuroprotective against glutamate excitotoxicity • Anti-hyperlipidemic	(Christophersen, 2012) (Gurer et al. 2001) (Sinha et al. 2008) (Tsuboyama-Kasaoka et al. 2006) (Leon et al. 2009) (El Idrissi et al. 2003) (Yanagita et al. 2008) (Huxtable, 1992)
Trans-urocanate (t-Urocanic acid)	00301	0.58	0.77	2.24	• Ultraviolet (UV) protectant (natural sunscreen)	(Egawa et al. 2010)

ND Not Detected
[a]Newly identified for expression in rice bran

Table 2 Median-scaled relative abundance of rice bran cofactors & vitamins metabolites with medicinal properties

Metabolite	HMDB ID	Calrose	Dixiebelle	Neptune	Function	References
Alpha-tocopherol	01893	0.02	0.05	0.03	• Antioxidant (antioxidant activity against (Fe^{2+} + ascorbate) and (Fe^{2+} + NADPH)-induced lipid peroxidation, decreases plasma and low-density lipoprotein (LDL) oxidizability) • Anti-hypertensive (inhibits smooth muscle cell proliferation by inhibiting protein kinase C activity)	(Serbinova et al. 1991) (Kontush et al. 1996) (Chatelain et al. 1993)
Alpha-tocopherol acetate	34227	0.02	0.04	ND	• Boosts alpha-tocopherol antioxidant bioactivity	(Brigelius-Flohe and Traber, 1999)
Alpha-tocotrienol	06327	0.01	0.08	0.05	• Antioxidant	(Serbinova et al. 1991)
Beta-tocopherol	06335	0.001	ND	ND	• Antioxidant (free radical scavenging)	(Kadoma et al. 2006)
Delta-tocopherol	02902	ND	0.01	ND	• Antioxidant (free radical scavenging) • Anti-hypertensive (inhibits smooth muscle cell proliferation by inhibiting protein kinase C activity)	(Kadoma et al. 2006) (Chatelain et al. 1993)
Gamma-tocopherol	01492	ND	0.07	0.03	• Anti-inflammatory • Anti-hypertensive (inhibits smooth muscle cell proliferation by inhibiting protein kinase C activity)	(Jiang et al. 2000) (Chatelain et al. 1993)
Gamma-tocotrienol	12958	0.03	0.17	0.12	• Radio-protector • Anti-hypertensive	(Ghosh et al. 2009)
Glucarate (saccharate)[a]	00663	0.89	0.13	ND	• Cancer chemopreventive (by inhibiting serum β-glucuronidase)	(Lampe et al. 2002) (Hanausek et al. 2003)
Nicotinamide (vitamin B3)	01406	2.00	2.04	4.28	• Anti-inflammatory (treatment of skin disorders such as acne vulgaris) • Cancer chemopreventive (treatment of non-melanoma skin cancer) • Treatment of Alzheimer's disease (by inhibiting of poly(ADP-ribose) polymerase-1 (PARP-1) and neuro-inflammatory enzyme)	(Khodaeiani et al. 2013; Niren, 2006) (A. C. Chen et al. 2015; Surjana et al. 2012) (Turunc Bayrakdar et al. 2014)
Nicotinate	01488	12.11	12.11	21.17	• Anti-hyperlipidemic	(Duggal et al. 2010; Figueroa et al. 2015; Zema, 2000)
Pyridoxine (Vitamin B6)	02075	2.67	4.22	5.03	• Required for biosynthesis of amino acids, sugars and fatty acids, and major centrally-acting neurotransmitters (e.g. serotonin, dopamine, and GABA) • Anti-hypertensive • Anti-hyperlipidemic • Anti-hyperglycemic • Cancer chemopreventive (colon cancer)	(Percudani and Peracchi, 2009) (Yarlagadda and Clayton, 2007) (Hellmann and Mooney, 2010) (Larsson et al. 2010)
Threonic acid[a]	00943	36.15	20.99	16.44	• Prevention of androgen-driven balding • Increases bioavailability of vitamin C for T-cells (calcium L-threonate) • Increases bioavailability of iron (is used for treatment of iron deficiency anemia	(Kwack et al. 2010) (Fay and Verlangieri, 1991) (X. X. Li et al. 2005)
Trigonelline	00875	608.11	0.77	11.59	• Cancer chemopreventive	(Hirakawa et al. 2005) (Liao et al. 2015)

ND: Not Detected
[a]Newly identified for expression in rice bran

Table 3 Median-scaled relative abundance of rice bran secondary metabolites with medicinal properties

Metabolite	HMDB ID	Calrose	Dixiebelle	Neptune	Function	References
4-hydroxybenzoate[a]	00500	1.18	1.87	1.04	• Antimicrobial (broad spectrum bacteriostatic and fungistatic properties • Antioxidant (scavenges free radicals to reduce skin damage)	(Kosova et al. 2015) (Barreca et al. 2016)
Abscisate	35140	0.62	0.31	0.40	• Anti-inflammatory (type II diabetes and obesity-related inflammation)	(Guri et al. 2007)
Alpha-amyrin[a]	36657	0.06	0.08	0.12	• Anti-inflammatory • Antimicrobial (broad spectrum bacteriostatic properties)	(Liliana Hernández Vázquez, 2012)
Apigenin	02124	ND	0.08	0.46	• Cancer chemopreventive • Anti-inflammatory and anti-metastatic agent	(Balasubramanian et al. 2006) (Vargo et al. 2006)
Astragalin	37429	ND	ND	2.89	• Anti-histaminergic (reduces IgE production) • Anti-parasitic (against *Fasciolopsis buski*) • Anti-inflammatory (reduces inflammation through nuclear factor NF-kB inhibition) • Neuroprotective (protects neurons from hydrogen peroxide-mediated cell death during neuroblastoma)	(Kotani et al. 2000) (Ananta Swargiary 2015) (Chung et al. 2016)
Benzoate	01870	4.38	3.50	3.36	• Antimicrobial (broad spectrum bacteriostatic and fungistatic properties)	(Nascimento et al. 2000)
Caffeate	01964	ND	0.20	1.23	• Antioxidant • Cancer chemopreventive (functions as an antiproliferative chemotherapeutic agent)	(Olthof et al. 2001) (Rajendra Prasad et al. 2011)
Chlorogenic acid	03164	ND	ND	0.03	• Antioxidant • Anti-diabetic (influences glucose metabolism by inhibiting alpha-glucosidase) • Anti-septic arthritis caused by *Candida albicans*. • Sleep-wake cycle modulator	(Olthof et al. 2001) (Upadhyay and Mohan Rao, 2013)
Chrysoeriol	30667	0.49	0.16	0.92	• Antioxidant • Anti-inflammatory (Inhibits lipid peroxidation, inhibits production of superoxide anion by xanthine/xanthine oxidase system, inhibits the lipopolysaccharide induction of the nitric oxide synthase (iNOS) gene.)	(Mishra et al. 2003) (Choi et al. 2005)
Cinnamate	00930	0.31	0.12	0.18	• Antimicrobial (anti-mycobacterial agent that can be developed against tuberculosis) • Cancer chemopreventive (anti-proliferative activity against melanoma cells and lung carcinoma cells, inhibition of histone deacetylases in colon cancer cells) • Anti-obesogenic • Anti-hypertensive (inhibits serum lipase and angiotensin-converting enzyme) • Cardio-protective (In electrocardiography, it decreases the ST segment elevation induced by acute myocardial ischemia) • Antioxidant • Anti-inflammatory	(Y. L. Chen et al. 2011) (Zhu et al. 2016) (Mnafgui et al. 2015) (Song et al. 2013)

Table 3 Median-scaled relative abundance of rice bran secondary metabolites with medicinal properties *(Continued)*

Ergothioneine[a]	03045	ND	0.79	0.99	• Antioxidant	(Aruoma et al. 2012)
Ferulate	00954	6.22	10.28	19.69	• Antioxidant	(Kanski et al. 2002 2002)
Indolin-2-one	-	1.19	0.30	3.80	• Cancer chemopreventive (anti-proliferative and inhibits thioredoxin reductase (TrxR))	(Kaminska et al. 2016)
Luteolin	05800	ND	0.04	1.67	• Antioxidant (scavenges ROS) • Antimicrobial (broad spectrum gram positive bacteriostatic properties and yeast)	(Lin et al. 2008) (Singh et al. 2016)
Piperidine[a]	34301	0.76	0.69	0.91	• Anti-hypertensive	(Aisaka et al. 1985; Patel et al. 2006)
Quinate[a]	03072	8.86	1.65	7.27	• Anti-inflammatory • Antioxidant	(Pero et al. 2009)
Salicylate	01895	6.67	0.48	2.12	• Anti-diabetic (reduces blood glucose via activation of adenosine monophosphate-activated protein kinase) • Anti-inflammatory	(Hawley et al. 2012)
Sinapic acid	32616	0.15	1.99	2.22	• Anti-inflammatory • Anti-diabetic (increases expression of glucose transporter type 4)	(Yun et al. 2008) (Cherng et al. 2013)
Sitostanol (Stigmastanol)	00494	ND	0.10	0.06	• Anti-hyperlipidemic	(Batta et al. 2006)
Syringic acid	02085	0.32	0.36	0.49	• Antioxidant (prevents oxidative stress, inhibits the production of free radicals and reduces lipid peroxidation) • Antimicrobial (against *Escherichia coli* LY01 (g-), *Cronobacter sakazakii* (g-) and, *Oenococcus oeni* (g+)) • Anti-diabetic (enhancement of insulin action and C-peptide)	(Cikman et al. 2015) (Shi et al. 2016) (Muthukumaran et al. 2013)
Tartaric acid[a]	00956	0.31	0.32	0.26	• Antimicrobial (against *E. coli, Bacillus subtilis* and *Streptococcus suis*)	(Zhihong Gao, 2012)
Vanillate	00484	1.21	2.04	2.20	• Antibacterial (against *Cronobacter* spp) • Anticoagulant (against snake venom)	(Yemis et al. 2011) (Dhananjaya et al. 2006)
Vanillin	12308	0.69	1.18	0.83	• Antioxidant • Anti-inflammatory • Antibacterial (agains *Cronobacter* spp) • Treatment of sickle cell disease	(Makni et al. 2012; Makni et al. 2011) (Yemis et al. 2011) (Abdulmalik et al. 2011)

ND Not Detected
[a]Newly identified for expression in rice bran

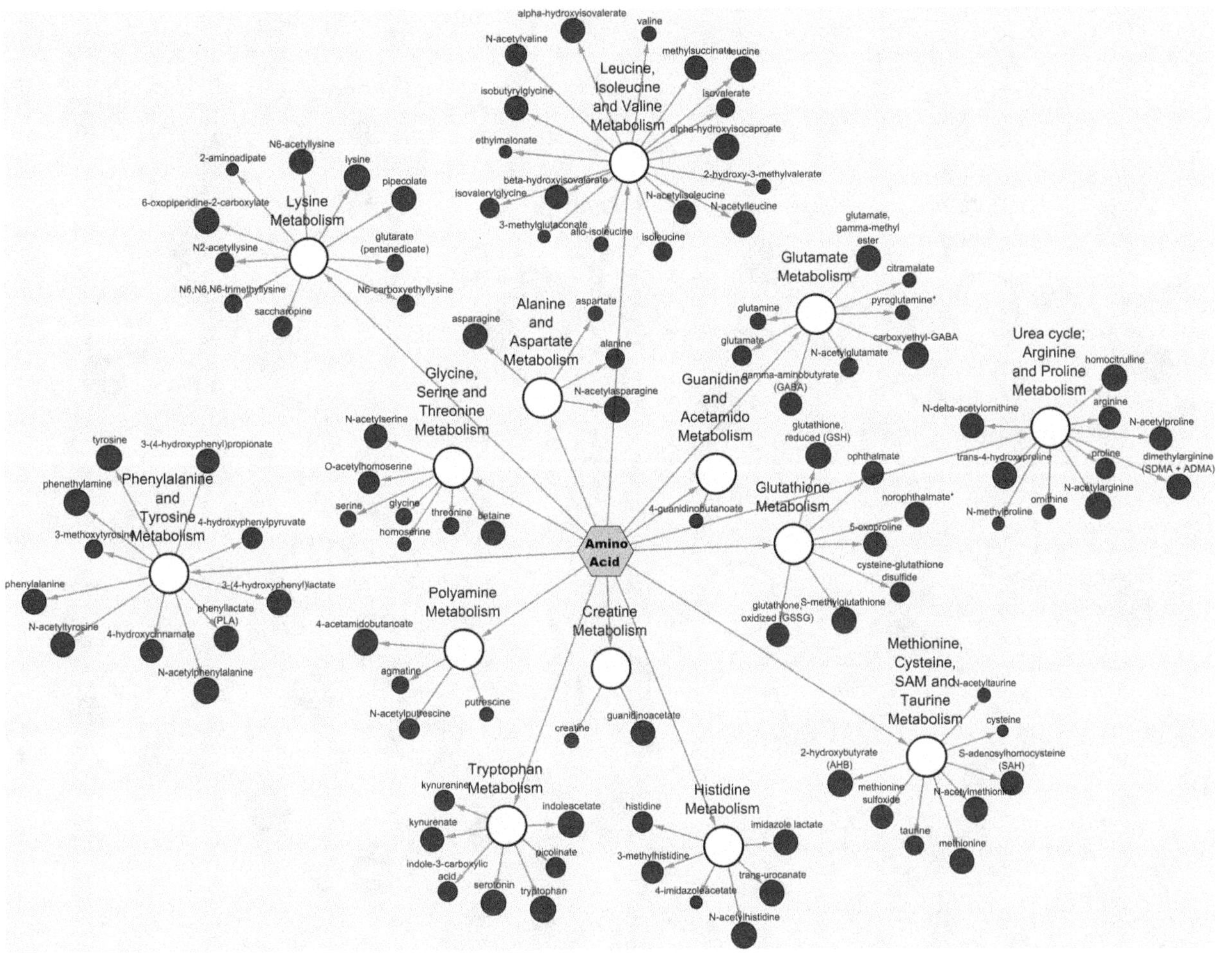

Fig. 1 Cytoscape network analysis of rice bran amino acid metabolic pathways. Pathway specific network visualization is shown for Calrose rice bran. Each metabolite is represented as a node (circle), extending from a central sub-metabolic pathway node. The central hexagon represents the super metabolic pathway. Node size corresponds to the Z-score using the relative abundance mean value for all three varieties

larger sample set of global varieties with established genome diversity.

Discussion

The rice bran metabolome analysis herein focused on amino acid, cofactor & vitamin, and secondary metabolite compounds that exhibited medicinal and nutritional properties with an emphasis on chronic and infectious disease control and prevention. The three selected classes of metabolites represented ~46% of total rice bran metabolite profile. Metabolites were described as antioxidative and anti-inflammatory (35 metabolites), antimicrobial (15 metabolites), anti-hypertensive (12 metabolites), cancer chemopreventive (11 compounds). anti-hyperlipidemic (8 metabolites), anti-hyperglycemic (6 compounds), and anti-obesogenic (2 compounds).

Antioxidants represented a broad class of compounds available from many different foodstuffs (Carlsen et al. 2010). Rice bran is a promising candidate for dietary supplementation and nutritional therapy for prevention of chronic and infectious disease via its antioxidant composition. A majority of the rice bran antioxidants (e.g. 4-guanidinobutanoate and taurine from amino acids, tocopherols and tocotrienols from cofactors & vitamins, and ergothioneine and quinate from secondary-metabolites) work through different mechanisms to combat lipid peroxidation, DNA damage, protein modification, and enzyme inactivation caused by free radicals, in particular reactive oxygen species (ROS) (Lü et al. 2010; Nimse and Pal, 2015). Oxidative stress caused by free radicals damages host cells and may initiate early stage development of chronic diseases such as cancer, heart disease, Alzheimer's disease, arthritis, cataracts, diabetes, and kidney disease (Morales-González, 2013). Antioxidants from rice bran can safely interact with and detoxify free radicals to stop the chain of damaging reactions for disease prevention (Iqbal et al. 2005; Jun et al. 2012; Parrado et al. 2006; Parrado et al. 2003). For

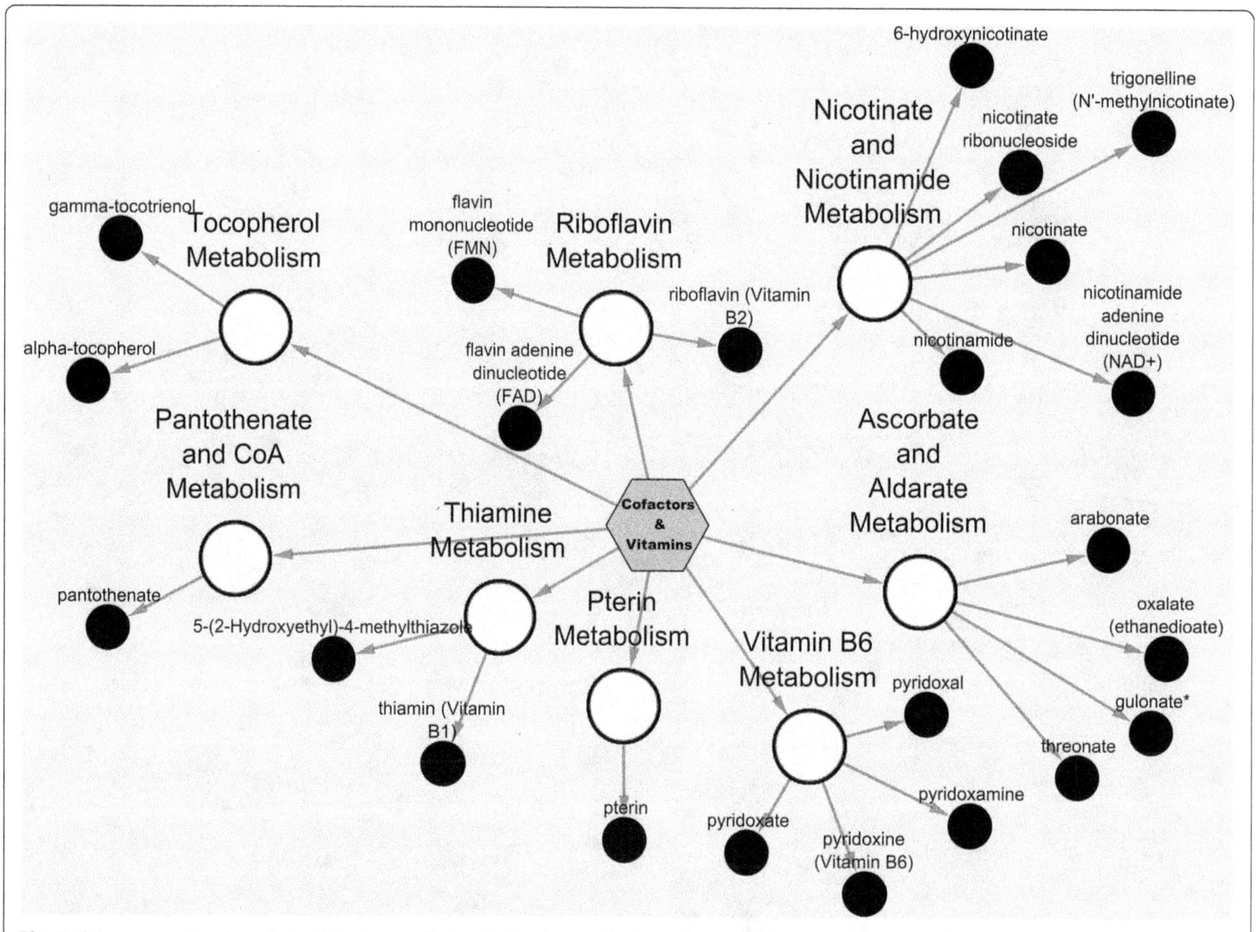

Fig. 2 Cytoscape network analysis of rice bran cofactor & vitamin metabolic pathways. Pathway specific network visualization is shown for Calrose rice bran. Each metabolite is represented as a node (circle), extending from a central sub-metabolic pathway node., The central hexagon node represents the super metabolic pathway. Node size corresponds to the Z-score using the relative abundance mean value for all three varieties

example, quinate, an antioxidant that is naturally synthesized in plants and microorganisms is now described from rice bran via metabolomics (Fig. 4). It was shown that consumption of 3000 milligrams of quinic acid ammonium chelate per day can regulate activation of NF-kB (nuclear factor kappa-light-chain-enhancer of activated B cells) and enhances DNA repair by increasing serum thiol levels (Pero et al. 2009). Rice bran derived quinate merits further evaluation for similar antioxidant activities.

The antimicrobial activity of dietary rice bran can be attributed to at least 15 metabolites across amino acids and secondary metabolites. Understanding the relative contribution of rice bran compounds and the mechanisms of antimicrobial action could be helpful in combating emerging and existing problems associated with resistance to antibiotics. Hence, treatment strategies using natural food molecules from rice bran may prevent progression of infection and associated symptoms as a sustainable, globally available long-term solution (Cowan, 1999; Kondo et al. 2011; Srivastava et al. 2014).

For instance, luteolin is a rice bran flavonoid (shown in Table 3) that reduced the growth of a variety of gram-positive bacteria and yeast (Singh et al. 2016; Srivastava et al. 2014). We, and others, have previously shown that dietary rice bran has antimicrobial activity in animals and on isolated bacterial strains (Goodyear et al. 2015; Irfan A Ghazi et al. 2016; Kim et al. 2014; Kondo et al. 2011; Kumar et al. 2012; Nealon et al. 2017; Yang et al. 2015; Yang et al. 2014). Our study revealed two newly identified rice bran amino acids (out of 15) and three newly identified rice bran secondary metabolites with antimicrobial properties; Phenyllactic acid and α-hydroxyisocaproic acid (leucic acid) from the amino acid metabolic pathway, and 4-hydroxybenzoate, alpha-amyrin, and tartaric acid from the secondary metabolite metabolic pathway. Phenyllactic acid is found in many bacteria as a metabolic byproduct (e.g. *Lactobacillus* spp.) but not previously identified in any plant sources (Valerio et al. 2004). Leucic acid has been identified in fermented foods, including certain cheeses, wines, and soy sauce (Mero et al. 2010). 4-hydroxybenzoate has been previously found in pistachio

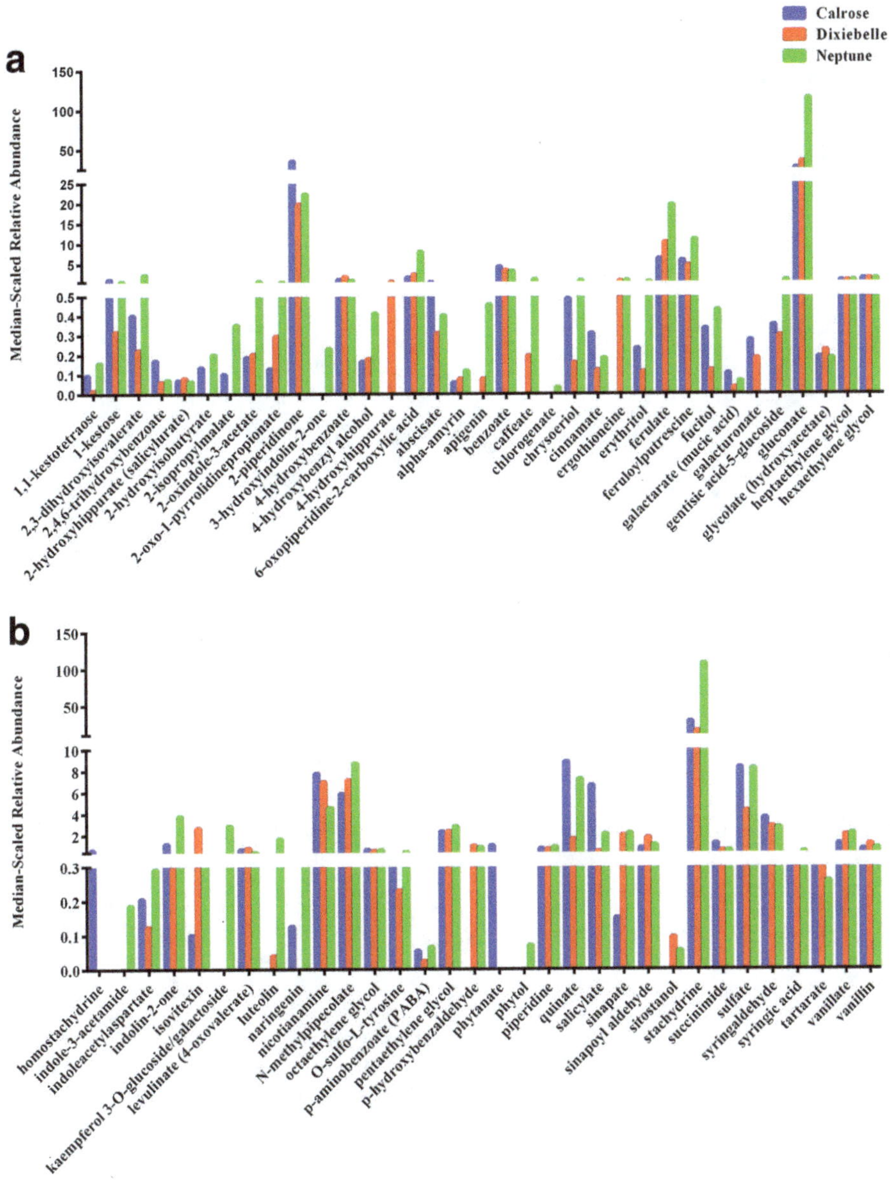

Fig. 3 Secondary metabolites detected in rice bran across 3 U.S. varieties. Rice bran metabolites were normalized by median of relative abundance for the entire dataset. **a** shows half of metabolites identified within secondary metabolites metabolic pathways and their difference in median-scaled relative abundance across three varieties. **b** shows another half of identified secondary metabolites and their difference in medin-scaled relative abundance across three varieties

hulls (Barreca et al. 2016). Alpha-amyrin is found in Carissa carandas (karanda fruit) (Akansha Singh, 2015). Additionally, tartaric acid has been found in Hibiscus sabdariffa flower (Da-Costa-Rocha et al. 2014). Our metabolomics analysis results suggest that nutritional therapy through rice bran's multi-faceted antimicrobial actions merits testing in medical clinical applications to mitigate microbial resistance.

Rice bran merits attention for being of considerably high nutritional value. These metabolome analyses confirm that it is a rich source of proteins, fats, minerals and micronutrients, such as B vitamins and trace elements. For example, at 12 – 15% protein content and with protein digestibility that is comparable to casein, the macro-nutritional value of rice bran which also contains healthy fats and fibers warrants greater pubic health attention (Saunders, 1990; Wang et al. 1999). Rice bran is also a rich source of B-complex vitamins, particularly thiamine and nicotinic acid, riboflavin and vitamin B_6. A single serving of rice bran (28 grams in accordance to USDA) delivers more than half of the daily nutritional requirements for thiamine, niacin and

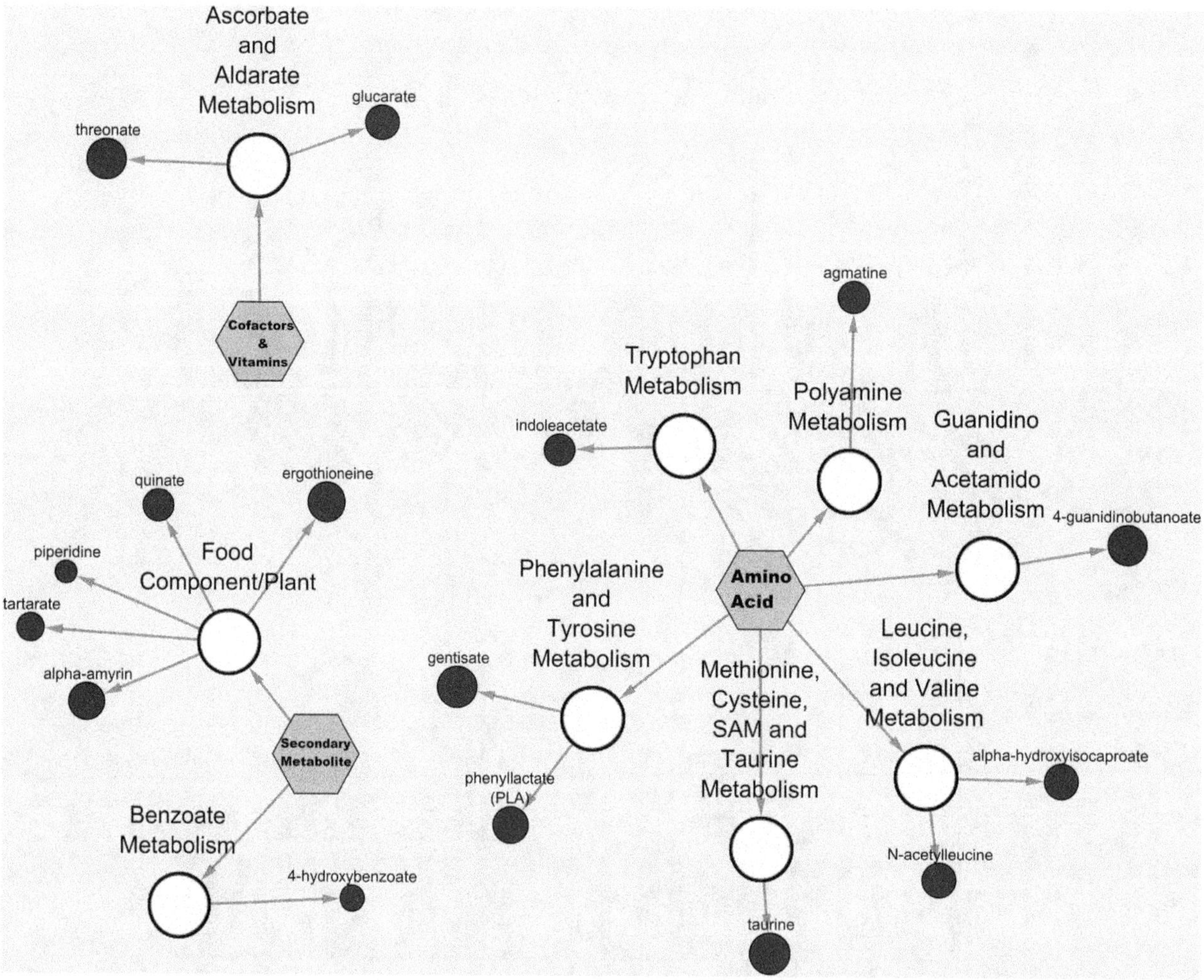

Fig. 4 Cytoscape visualisation of 16 newly identified metabolites from the rice bran metabolome. Medicinal and nutritional value for these bioactive compounds are described as deonted by [1] in Tables 1, 2 and 3. Pathway specific network visualization is shown for Calrose rice bran. Each metabolite is represented as a node (circle), extending from a central sub-metabolic pathway node. The central hexagon node represents the super metabolic pathway. Node size corresponds to the Z score using the relative abundance mean value for all three varieties

vitamin B6 (based on a 2,000 calorie reference diet) (SELFNutritionDada; United States Department of Agriculture, 2016). Vitamins cannot be synthesized by the body and must be ingested, as such inadequate intake or subtle deficiencies in vitamins are risk factors for multiple chronic diseases (Fairfield and Fletcher, 2002). Recent evidence showed intake levels of thiamin, niacin, vitamin B_6, total folate, and alpha-tocopherol was improved in colorectal cancer survivors consuming rice bran and suggests that foods with multiple bioactive components and nutrients can play a pivotal role in the prevention of chronic diseases such as cancer and cardiovascular disease (Borresen et al. 2016; Borresen EC, 2016).

A major strength of the non-targeted metabolomics approach herein was the identification of novel compounds from rice bran with medicinal properties (Fig. 4). The limitations of non-targeted metabolomics in dietary exposure biomarker discovery platforms arise from metabolite concentrations that can vary across cultivars, and inconsistencies in extraction methods or instrument detection limits. Additional limitations for results interpretations from this study involve the limited information for bioavailability of rice bran compounds. The biological properties for rice bran will be dependent on host bioavailability and bioaccessiblity following ingestion, and thus this rice bran food metabolome investigation will assist to identify rice bran exposure biomarkers of intake in people. The variation in gut microbiota composition is another major factor that can influence bioavailability of food metabolites as well as the biological activities (Conlon and Bird, 2015; Krajmalnik-Brown et al. 2012). This is the first non-targeted whole food metabolome study of rice bran with an investigative focus towards the suites of amino acids, cofactors & vitamins, and secondary metabolites.

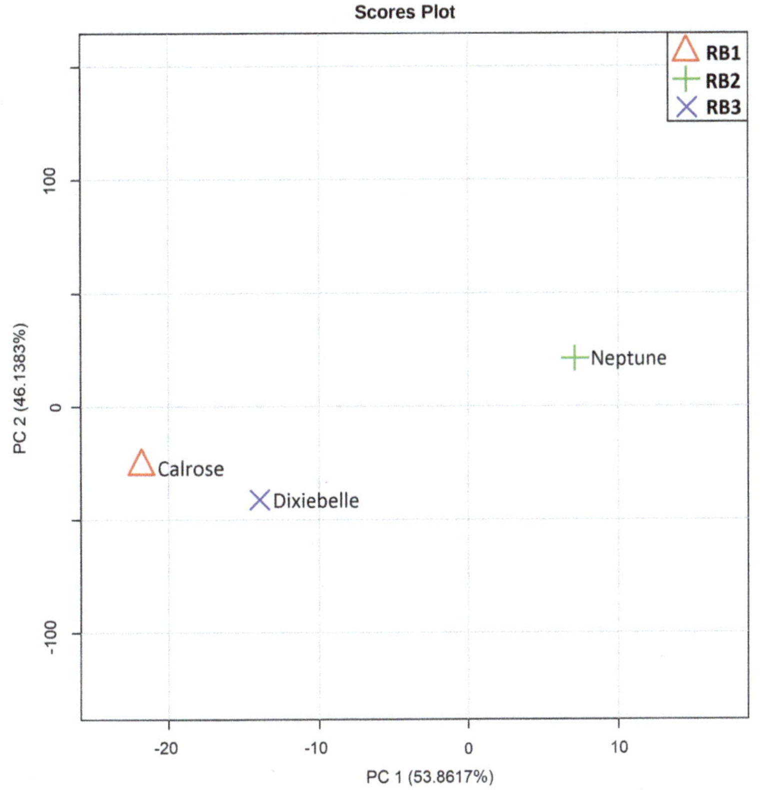

Fig. 5 Principal Component Analysis of three U.S. varieties. Principal component analysis (PCA) revealed that the first component described 54% variation herein and between separation of Calrose, Dixiebelle and Neptune rice bran cultivars. The second component describing 46% of the variance did not separate the varieties. No statistical significance was established between the metabolites detected from each of the three U.S rice varieties analyzed

Additional metabolic pathways and chemical classes of metabolites from this analysis (listed in Additional file 1: Table S1) merit continued investigation for medicinal properties and nutritional value.

Conclusions

This study identified approximately 453 metabolites from the rice bran metabolome, many of which are described herein as cofactors & vitamins, amino acids and secondary metabolites. These metabolic pathways, among others found in rice bran, have shown positive health effects in animals and humans. The wide range of phytochemicals found in rice bran are likely working synergystically to contribute to rice bran's functional food properties. The ability of rice bran to fight both infectious and chronic diseases may be in part due to synergistic combinations of phytochemicals, and alongside metabolism by the gut microbiota (Borresen et al. 2016; Sheflin et al. 2016; Sheflin et al. 2015). Rice bran biochemical composition merits further investigation for multiple nutritional therapies and medical food applications.

Methods

Rice: Milling and Heat Stabilization of Bran

Rice bran from 3 U.S. rice varieties (Calrose, Dixiebelle, and Neptune), representive of rice production in the southeastern U.S. and California, were chosen for this study (Additional file 2: Table S2). Bran was collected and heat stabilized by the United States Department of Agriculture-Agricultural Research Service (USDA-ARS; Stuttgart, AR) as previously reported and utilized in animal and human studies (Borresen EC, 2016; Goodyear et al. 2015; Sheflin et al. 2016; Sheflin et al. 2015).

Rice was milled using a Yamamoto test whitening machine Rice pal VP-31 T grinder and milling system. This laboratory-based instrument has specifications of high accuracy recovery rate (defined as a maximum of 12% bran removal from the whole rice grain). Rice samples were milled at room temperature prior to bran preparation. Testing sieve No. 20 was used to separate bran from broken rice and hulls into a clean container. Once separated, milled bran was heat stabilized at 110 °C for 6 min to prevent rancidity during storage. Rice bran was stored at −20 °C until further processing for metabolite analysis.

Rice Bran Extraction

Metabolomics analysis was performed by Metabolon Inc. (Durham, NC). Prior to the first step of extraction, several recovery standards (also called quality control standards) were added into the samples for quality control purposes. Quality control standards, that were carefully chosen not to interfere with the measurement of endogenous compounds, were spiked into every analyzed sample to allow instrument performance monitoring and aid chromatographic alignment. The purpose of adding the standards was to assess variability and verify performance of extraction and instrumentation. Values for instrument and process variability met Metabolon's acceptance criteria. Rice bran samples were mixed with 80% ice-cold methanol under vigorous shaking for 2 min (Glen Mills GenoGrinder 2000) and then were centrifuged to precipitate protein, and free small molecules bound to other macromolecules. The supernatant fraction, i.e., rice bran extract (RBE), was used for further analysis and was divided into four portions: three portions for different mode of analysis by ultra-performance liquid chromatography-tandem mass spectrometry (UPLC-MS/MS) (i.e., one portion for analysis with positive ion mode electrospray ionization, one for negative ion mode electrospray ionization, and one portion for analysis by UPLC-MS/MS polar platform (negative ionization), and one portion for analysis by gas chromatography–mass spectrometry (GC-MS)). Samples were placed briefly on a TurboVap® (Zymark) concentration evaporator to remove the organic solvent. For UPLC, the samples were stored overnight under nitrogen before preparation for analysis. For GC, each sample was dried under vacuum overnight before preparation for analysis.

Ultra-Performance Liquid Chromatography-Tandem Mass Spectroscopy (UPLC-MS/MS)

The UPLC-MS/MS was performed for a non-targeted metabolomics analysis based on a Waters ACQUITY ultra-performance liquid chromatography (UPLC) and a Thermo Scientific Q-Exactive high resolution/accurate mass spectrometer interfaced with a heated electrospray ionization (HESI-II) source and Orbitrap mass analyzer operated at 35,000 mass resolution. The dried RBE was re-suspended in acidic or basic UPLC-compatible solvents. Each sample contained 8 injection quality control standards at fixed concentrations to ensure injection and chromatographic consistency. These standards, in concert with experimental samples, generated a pooled matrix served as technical replicates throughout the data set across all samples and varieties. The acidic solution was analyzed using acidic positive ion optimized conditions, and the basic solution was analyzed using basic negative ion optimized conditions under two identical separate dedicated columns independently (Waters UPLC BEH C18-2.1x100 mm, 1.7 μm). For acidic conditions, RBE was eluted at 350 μl/min from a C18 column using (A) 0.1% formic acid in water and (B) 0.1% formic acid in methanol (0% B to 70% B in 4 min, 70-98% B in 0.5 min, 98% B for 0.9 min). Likewise, the basic extracts were eluted from C18 using ammonium bicarbonate instead of formic acid. Another portion of RBE was also analyzed through negative ionization using bicarbonate used (A) 6.5 mM ammonium bicarbonate in water, pH 8, and (B) 6.5 mM ammonium bicarbonate in 95/5 methanol/water (same gradient profile as above) at 350 μL/min, followed by elution from a hydrophilic interaction liquid chromatography (HILIC) column (Waters UPLC BEH Amide 2.1x150 mm, 1.7 μm). 10 mM ammonium formate were gradient eluted at 500 μL/min using (A) 10 mM ammonium formate in water and (B) 10 mM ammonium formate in acetonitrile (5% B to 50% B in 3.5 min, 50% B to 95% B in 2 min and 95% B for 1 min). The MS analysis alternated between MS and data-dependent MS2 scans using dynamic exclusion, and the scan range was from 80–1000 m/z (Brown et al. 2016).

Gas Chromatography-Mass Spectroscopy (GC-MS)

The RBEs assigned for GC-MS analysis were dried under vacuum overnight (18 h). They were further derivatized with bistrimethyl-silyltrifluoroacetamide under dried nitrogen. Derivatized RBEs were separated on a 5% diphenyl/95% dimethyl polysiloxane fused silica column (20 m x 0.18 mm ID; 0.18 um film thickness) with an appropriate carrier gas. Samples were analyzed on a Thermo-Finnigan Trace DSQ™ fast-scanning single-quadrupole mass spectrometer using electron impact ionization (EI) and operated at unit mass resolving power. The scan range was from 50–750 m/z (Brown et al. 2016).

Metabolite Data Extraction and Compound Identification

Raw instrument data was extracted and then processed through Metabolon in-house developed peak detection and integration software (quantitation is based on area under the curve from MS data). This software uses standard industry approaches for MS peak detection, including using minimum height, signal-to-noise, width and area criteria. These systems are built on a web-service platform utilizing Microsoft's .NET technologies, which run on high-performance application servers and fiber-channel storage arrays in clusters to provide active failover and load-balancing. Compounds were identified by comparison to library entries of purified standards or recurrent unknown entities. Metabolon maintains a library based on authenticated standards that contains the retention time/index (RI), mass to charge ratio (m/z), and chromatographic data (including MS/MS spectral data) on all molecules present in the library. Biochemical identifications were based on three criteria: the experimentally detected signature matching the accurate mass of the authentic standard within 8 ppm, i.e. match to the NIST (National Institute of Standards and Technology) library within +/− 0.005 atomic mass units), retention index match within a defined window (approximately 5 s), and the Tandem mass spectrometry (MS/MS) forward and reverse scores between the experimental data and high quality standards. The MS/MS scores are based on a comparison of the ions present in the experimental spectrum to the ions present in the library spectrum. There are more than 3300 commercially purified standard compounds that have been registered in Metabolon Laboratory Information Management System for distribution to both the LC-MS and GC-MS platforms for determination of their analytical characteristics (Evans AM, 2014; Evans et al. 2009).

Statistical Analysis

A data normalization step was performed to correct variation between instrument inter-day tuning differences. Each compound was corrected in run-day blocks by registering the medians to equal 1.00 and normalizing each data point proportionally. The relative abundance of each metabolite was also normalized by median of the metabolite across the entire dataset (i.e. median-scaled). Median-scaled relative abundance is calculated as: raw abundance of a metabolite/median raw abundance of that metabolite across the entire dataset. Median-scaled relative abundance Z-score was further used as a basis for metabolic pathway network visualization. Z-scores are expressed as standard deviations from the mean and were calculated using the following formula: $z = (x- \mu)/\sigma$ where "x" is median-scaled relative abundance of the metabolite, "μ" is mean of median-scaled relative abundance for the metabolite across three rice bran varieties, and "σ" is the median-scaled relative abundance standard deviation of same metabolite across three varieties. Metabolic pathway and the graphical presentation of metabolite interaction network were composed with Cytoscape version 3.4.0.

Acknowledgements
This work was supported by a Grand Challenges Explorations in Global Health grant (OPP1015267) from the Bill and Melinda Gates Foundation and a grant from the National Institute of Food and Agriculture (NIFA) and United States Department of Agriculture (USDA) (004228–00002).

Authors' Contributions
EPR & IZ conceived and designed the study. IZ performed metabolomics analysis. IZ, DGB, NJN & EPR interpreted the results and wrote the manuscript. All authors made editorial comments, read and approved the final version of the article.

Competing Interests
The authors declare that they have no competing interests.

Author details
[1]Department of Environmental & Radiological Health Sciences, College of Veterinary Medicine and Biological Sciences, Colorado State University, 1680 Campus Delivery, Fort Collins, CO 80523, USA. [2]Institute of Human Nutrition and Food, College of Human Ecology, University of the Philippines Los Baños, Los Baños 4031, Laguna, Philippines.

References

Abdulmalik O, Ghatge MS, Musayev FN, Parikh A, Chen Q, Yang J et al (2011) Crystallographic analysis of human hemoglobin elucidates the structural basis of the potent and dual antisickling activity of pyridyl derivatives of vanillin. Acta Crystallogr D Biol Crystallogr 67(Pt 11):920–928. doi:10.1107/s0907444911036353

Aisaka K, Hattori Y, Ishihara T, Morita M, Kase Y, Miyata T (1985) The effects of piperidine and its related substances on blood vessels. Jpn J Pharmacol 37(4):345–353

Andreas NJ, Hyde MJ, Gomez-Romero M, Lopez-Gonzalvez MA, Villasenor A, Wijeyesekera A et al (2015). Multiplatform characterization of dynamic changes in breast milk during lactation. Electrophoresis. doi:10.1002/elps.201500011.

Anraku M, Tsurusaki Y, Watanabe H, Maruyama T, Kragh-Hansen U, Otagiri M (2004) Stabilizing mechanisms in commercial albumin preparations: octanoate and N-acetyl-L-tryptophanate protect human serum albumin against heat and oxidative stress. Biochim Biophys Acta 1702(1):9–17. doi:10.1016/j.bbapap.2004.07.002

Antonelli T, Carla V, Lambertini L, Moroni F, Bianchi C (1984) Pyroglutamic acid administration modifies the electrocorticogram and increases the release of acetylcholine and GABA from the guinea-pig cerebral cortex. Pharmacol Res Commun 16(2):189–197

Arbona V, Iglesias DJ, Gomez-Cadenas A (2015) Non-targeted metabolite profiling of citrus juices as a tool for variety discrimination and metabolite flow analysis. BMC Plant Biol 15:38. doi:10.1186/s12870-015-0430-8

Aruoma OI, Coles LS, Landes B, Repine JE (2012) Functional benefits of ergothioneine and fruit- and vegetable-derived nutraceuticals: overview of the supplemental issue contents. Prev Med 54(Suppl):S4–8. doi:10.1016/j.ypmed.2012.04.001

Balasubramanian S, Zhu L, Eckert RL (2006) Apigenin inhibition of involucrin gene expression is associated with a specific reduction in phosphorylation of protein kinase Cdelta Tyr311. J Biol Chem 281(47):36162–36172. doi:10.1074/jbc.M605368200

Barreca D, Lagana G, Leuzzi U, Smeriglio A, Trombetta D, Bellocco E (2016) Evaluation of the nutraceutical, antioxidant and cytoprotective properties of ripe pistachio (Pistacia vera L., variety Bronte) hulls. Food Chem 196:493–502. doi:10.1016/j.foodchem.2015.09.077

Baslow MH (1998) Function of the N-acetyl-L-histidine system in the vertebrate eye. Evidence in support of a role as a molecular water pump. J Mol Neurosci 10(3):193–208. doi:10.1007/bf02761774

Baslow MH, Guilfoyle DN (2015) N-acetyl-L-histidine, a Prominent Biomolecule in Brain and Eye of Poikilothermic Vertebrates. Biomolecules 5(2):635–646. doi:10.3390/biom5020635

Batta AK, Xu G, Honda A, Miyazaki T, Salen G (2006) Stigmasterol reduces plasma cholesterol levels and inhibits hepatic synthesis and intestinal absorption in the rat. Metabolism 55(3):292–299. doi:10.1016/j.metabol.2005.08.024

Beloborodova N, Bairamov I, Olenin A, Shubina V, Teplova V, Fedotcheva N (2012) Effect of phenolic acids of microbial origin on production of reactive oxygen species in mitochondria and neutrophils. J Biomed Sci 19:89. doi:10.1186/1423-0127-19-89

Borresen BC, Jenkins-Puccetti N, Schmitz K, Brown DG, Pollack A, Fairbanks A, . . . Ryan EP (2017) A Pilot Randomized Controlled Clinical Trial to Assess Tolerance and Efficacy of Navy Bean and Rice Bran Supplementation for Lowering Cholesterol in Children. Glob Pediatr Health, 4, 2333794X17694231. doi: 10.1177/2333794x17694231

Borresen EC, Brown DG, Harbison G, Taylor L, Fairbanks A, O'Malia J et al (2016) A randomized controlled trial to increase navy bean or rice bran consumption in colorectal cancer survivors. Nutr Cancer, 1–15. doi: 10.1080/01635581.2016.1224370

Bosco MC, Rapisarda A, Massazza S, Melillo G, Young H, Varesio L (2000) The tryptophan catabolite picolinic acid selectively induces the chemokines macrophage inflammatory protein-1 alpha and –1 beta in macrophages. J Immunol 164(6):3283–3291

Brigelius-Flohe R, Traber MG (1999) Vitamin E: function and metabolism. FASEB J 13(10):1145–1155

Brown DG, Rao S, Weir TL, O'Malia J, Bazan M, Brown RJ, Ryan EP (2016) Metabolomics and metabolic pathway networks from human colorectal cancers, adjacent mucosa, and stool. Cancer Metab 4:11. doi:10.1186/s40170-016-0151-y

Capozzi F, Bordoni A (2013) Foodomics: a new comprehensive approach to food and nutrition. Genes Nutr 8(1):1–4. doi:10.1007/s12263-012-0310-x

Carlin G, Djursäter R, Smedegård G, Gerdin B (1985) Effect of anti-inflammatory drugs on xanthine oxidase and xanthine oxidase induced depolymerization of hyaluronic acid. Agents Actions 16(5):377–384. doi:10.1007/bf01982876

Carlsen MH, Halvorsen BL, Holte K, Bøhn SK, Dragland S, Sampson L et al (2010) The total antioxidant content of more than 3100 foods, beverages, spices, herbs and supplements used worldwide. Nutr J 9:3–3. doi:10.1186/1475-2891-9-3

Cerino A, Bestagno M, Colonna M, Fussi F, Astaldi Ricotti GC (1988) Carboxyethyl γ - Aminobutyric Acid, a polyamine derivative, improves the recovery of EBV-transformed lymphocytes. Biochem Biophys Res Commun 150(3):931–936. doi:10.1016/0006-291X(88)90718-8

Chapouthier G, Venault P (2001) A pharmacological link between epilepsy and anxiety? Trends Pharmacol Sci 22(10):491–493

Charoonratana T, Songsak T, Sakunpak A, Pathompak P, Charoenchai L (2015) Using liquid chromatography-mass spectrometry based metabolomics to discriminate between cold pressed rice bran oils produced from two different cultivars of Oryza sativa L. ssp. indica in Thailand. Se Pu 33(9):966–973

Chatelain E, Boscoboinik DO, Bartoli GM, Kagan VE, Ge
y FK, Packer L, Azzi A (1993) Inhibition of smooth muscle cell proliferation and protein kinase C activity by tocopherols and tocotrienols. Biochim Biophys Acta 1176(1–2):83–89

Chen YL, Huang ST, Sun FM, Chiang YL, Chiang CJ, Tsai CM, Weng CJ (2011) Transformation of cinnamic acid from trans- to cis-form raises a notable bactericidal and synergistic activity against multiple-drug resistant Mycobacterium tuberculosis. Eur J Pharm Sci 43(3):188–194. doi:10.1016/j.ejps.2011.04.012

Chen AC, Martin AJ, Choy B, Fernandez-Penas P, Dalziell RA, McKenzie CA et al (2015) A Phase 3 Randomized Trial of Nicotinamide for Skin-Cancer Chemoprevention. N Engl J Med 373(17):1618–1626. doi:10.1056/NEJMoa1506197

Cheng HH, Huang HY, Chen YY, Huang CL, Chang CJ, Chen HL, Lai MH (2010) Ameliorative effects of stabilized rice bran on type 2 diabetes patients. Ann Nutr Metab 56(1):45–51. doi:10.1159/000265850

Cherng YG, Tsai CC, Chung HH, Lai YW, Kuo SC, Cheng JT (2013) Antihyperglycemic action of sinapic acid in diabetic rats. J Agric Food Chem 61(49):12053–12059. doi:10.1021/jf403092b

Choi DY, Lee JY, Kim MR, Woo ER, Kim YG, Kang KW (2005) Chrysoeriol potently inhibits the induction of nitric oxide synthase by blocking AP-1 activation. J Biomed Sci 12(6):949–959. doi:10.1007/s11373-005-9028-8

Christophersen, O. A. (2012). Radiation protection following nuclear power accidents: a survey of putative mechanisms involved in the radioprotective actions of taurine during and after radiation exposure. 2012, 23. doi:10.3402/mehd.v23i0.14787.

Chung MJ, Lee S, Park YI, Lee J, Kwon KH (2016) Neuroprotective effects of phytosterols and flavonoids from Cirsium setidens and Aster scaber in human brain neuroblastoma SK-N-SH cells. Life Sci 148:173–182. doi:10.1016/j.lfs.2016.02.035

Cikman O, Soylemez O, Ozkan OF, Kiraz HA, Sayar I, Ademoglu S et al (2015) Antioxidant Activity of Syringic Acid Prevents Oxidative Stress in l-arginine-Induced Acute Pancreatitis: An Experimental Study on Rats. Int Surg 100(5):891–896. doi:10.9738/intsurg-d-14-00170.1

Clarke NE, Mosher RE (1953) Phenolic compounds in the treatment of rheumatic fever. II. The metabolism of gentisic acid and the ethanolamide of gentisic acid. Circulation 7(3):337–344

Conlon MA, Bird AR (2015) The Impact of Diet and Lifestyle on Gut Microbiota and Human Health. Nutrients 7(1):17–44. doi:10.3390/nu7010017

Cowan MM (1999) Plant products as antimicrobial agents. Clin Microbiol Rev 12(4):564–582

Craig SA (2004) Betaine in human nutrition. Am J Clin Nutr 80(3):539–549

da Silva MA, Sanches C, Amante ER (2006) Prevention of hydrolytic rancidity in rice bran. J Food Eng 75(4):487–491. doi:10.1016/j.jfoodeng.2005.03.066

Da-Costa-Rocha, I., Bonnlaender, B., Sievers, H., Pischel, I., & Heinrich, M. (2014). Hibiscus sabdariffa L. – A phytochemical and pharmacological review. Food Chem, 165, 424–443. doi:http://dx.doi.org/10.1016/j.foodchem.2014.05.002.

Dallagnol AM, Catalan CA, Mercado MI, Font de Valdez G, Rollan GC (2011) Effect of biosynthetic intermediates and citrate on the phenyllactic and hydroxyphenyllactic acids production by Lactobacillus plantarum CRL 778. J Appl Microbiol 111(6):1447–1455. doi:10.1111/j.1365-2672.2011.05159.x

de Munter JS, Hu FB, Spiegelman D, Franz M, van Dam RM (2007) Whole grain, bran, and germ intake and risk of type 2 diabetes: a prospective cohort study and systematic review. PLoS Med 4(8):e261. doi:10.1371/journal.pmed.0040261

De A, Singh MF, Singh V, Ram V, Bisht S (2016) Treatment effect of l-Norvaline on the sexual performance of male rats with streptozotocin induced diabetes. Eur J Pharmacol 771:247–254. doi:10.1016/j.ejphar.2015.12.008

De-Miguel FF, Trueta C (2005) Synaptic and extrasynaptic secretion of serotonin. Cell Mol Neurobiol 25(2):297–312

Dhananjaya BL, Nataraju A, Rajesh R, Raghavendra Gowda CD, Sharath BK, Vishwanath BS, D'Souza CJ (2006) Anticoagulant effect of Naja naja venom 5'nucleotidase: demonstration through the use of novel specific inhibitor, vanillic acid. Toxicon 48(4):411–421. doi:10.1016/j.toxicon.2006.06.017

Dieuleveux V, Lemarinier S, Gueguen M (1998) Antimicrobial spectrum and target site of D-3-phenyllactic acid. Int J Food Microbiol 40(3):177–183

DrugBank. http://www.drugbank.ca/drugs/DB00765.

Duggal JK, Singh M, Attri N, Singh PP, Ahmed N, Pahwa S et al (2010) Effect of niacin therapy on cardiovascular outcomes in patients with coronary artery disease. J Cardiovasc Pharmacol Ther 15(2):158–166. doi:10.1177/1074248410361337

Egawa M, Nomura J, Iwaki H (2010) The evaluation of the amount of cis- and trans-urocanic acid in the stratum corneum by Raman spectroscopy. Photochem Photobiol Sci 9(5):730–733. doi:10.1039/b9pp00143c

El Idrissi A, Messing J, Scalia J, Trenkner E (2003) Prevention of epileptic seizures by taurine. Adv Exp Med Biol 526:515–525

Evans AM, B. B., Liu Q, Mitchell MW, Robinson RJ, Dai H, Stewart SJ, DeHaven CD and Miller LAD. (2014). High Resolution Mass Spectrometry Improves Data Quantity and Quality as Compared to Unit Mass Resolution Mass Spectrometry in High- Throughput Profiling Metabolomics. Metabolomics, 4(2). doi:http://dx.doi.org/10.4172/2153-0769.1000132

Evans AM, DeHaven CD, Barrett T, Mitchell M, Milgram E (2009) Integrated, nontargeted ultrahigh performance liquid chromatography/electrospray ionization tandem mass spectrometry platform for the identification and relative quantification of the small-molecule complement of biological systems. Anal Chem 81(16):6656–6667. doi:10.1021/ac901536h

Fairfield KM, Fletcher RH (2002) Vitamins for chronic disease prevention in adults. Scientific review. JAMA 287(23):3116–3126. doi:10.1001/jama.287.23.3116

Fay MJ, Verlangieri AJ (1991) Stimulatory action of calcium L-threonate on ascorbic acid uptake by a human T-lymphoma cell line. Life Sci 49(19):1377–1381

Figueroa C, Droppelmann K, Quinones V, Amigo L, Mendoza C, Serrano V et al (2015) Nicotinic acid increases cellular transport of high density lipoprotein cholesterol in patients with hypoalphalipoproteinemia. Rev Med Chil 143(9):1097–1104. doi:10.4067/s0034-98872015000900001

Folkes LK, Wardman P (2001) Oxidative activation of indole-3-acetic acids to cytotoxic species- a potential new role for plant auxins in cancer therapy. Biochem Pharmacol 61(2):129–136

Forster GM, Raina K, Kumar A, Kumar S, Agarwal R, Chen MH et al (2013) Rice varietal differences in bioactive bran components for inhibition of colorectal cancer cell growth. Food Chem 141(2):1545–1552. doi:10.1016/j.foodchem.2013.04.020

Foster AC, Kemp JA (2006) Glutamate- and GABA-based CNS therapeutics. Curr Opin Pharmacol 6(1):7–17. doi:10.1016/j.coph.2005.11.005

Irfan A Ghazi, Iman Zarei, Job O Mapesa, Jessie R Wilburn, Jan E Leach, Sangeeta Rao, et al (2016). Rice Bran Extracts Inhibit Invasion and Intracellular Replication of Salmonella typhimurium in Mouse and Porcine Intestinal Epithelial Cells. Med Aromatic Plants (Loa Angel), 5(5), 10. doi:10.4172/2167-0412.1000271.

Ghosh SP, Kulkarni S, Hieber K, Toles R, Romanyukha L, Kao TC et al (2009) Gamma-tocotrienol, a tocol antioxidant as a potent radioprotector. Int J Radiat Biol 85(7):598–606. doi:10.1080/09553000902985128

Goodyear A, Kumar A, Ehrhart EJ, Swanson KS, Grusak MA, Leach JE et al (2015) Dietary rice bran supplementation prevents Salmonella colonization differentially across varieties and by priming intestinal immunity. J Funct Foods 18(Part A):653–664. doi:10.1016/j.jff.2015.08.027

Grant RS, Coggan SE, Smythe GA (2009) The physiological action of picolinic Acid in the human brain. Int J Tryptophan Res 2:71–79

Grioli S, Lomeo C, Quattropani MC, Spignoli G, Villardita C (1990) Pyroglutamic acid improves the age associated memory impairment. Fundam Clin Pharmacol 4(2):169–173

Gurer H, Ozgunes H, Saygin E, Ercal N (2001) Antioxidant effect of taurine against lead-induced oxidative stress. Arch Environ Contam Toxicol 41(4):397–402. doi:10.1007/s002440010265

Guri AJ, Hontecillas R, Si H, Liu D, Bassaganya-Riera J (2007) Dietary abscisic acid ameliorates glucose tolerance and obesity-related inflammation in db/db mice fed high-fat diets. Clin Nutr 26(1):107–116. doi:10.1016/j.clnu.2006.07.008

Ham H, Sung J, Lee J (2015) Effect of Rice Bran Unsaponifiables on High-Fat Diet-Induced Obesity in Mice. J Food Biochem 39(6):673–681. doi:10.1111/jfbc.12170

Hanausek M, Walaszek Z, Slaga TJ (2003) Detoxifying cancer causing agents to prevent cancer. Integr Cancer Ther 2(2):139–144. doi:10.1177/1534735403002002005

Hawley SA, Fullerton MD, Ross FA, Schertzer JD, Chevtzoff C, Walker KJ et al (2012) The ancient drug salicylate directly activates AMP-activated protein kinase. Science 336(6083):918–922. doi:10.1126/science.1215327

Hellmann H, Mooney S (2010) Vitamin B6: a molecule for human health? Molecules 15(1):442–459. doi:10.3390/molecules15010442

Henderson AJ, Ollila CA, Kumar A, Borresen EC, Raina K, Agarwal R, Ryan EP (2012) Chemopreventive properties of dietary rice bran: current status and future prospects. Adv Nutr 3(5):643–653. doi:10.3945/an.112.002303

Herrero M, Simó C, García-Cañas V, Ibáñez E, Cifuentes A (2012) Foodomics: MS-based strategies in modern food science and nutrition. Mass Spectrom Rev 31(1):49–69. doi:10.1002/mas.20335

Hettema JM, An SS, Neale MC, Bukszar J, van den Oord EJ, Kendler KS, Chen X (2006) Association between glutamic acid decarboxylase genes and anxiety disorders, major depression, and neuroticism. Mol Psychiatry 11(8):752–762. doi:10.1038/sj.mp.4001845

Heuberger AL, Lewis MR, Chen MH, Brick MA, Leach JE, Ryan EP (2010) Metabolomic and functional genomic analyses reveal varietal differences in bioactive compounds of cooked rice. PLoS ONE 5(9):e12915. doi:10.1371/journal.pone.0012915

Hinz MC (2003) Serotonin and catecholamine system segment optimization technology. U.S. Patent Application No. 10/394,597, https://www.google.com/patents/US20030181509

Hinz MC (2009) Administration of dopa precursors with sources of dopa to effectuate optimal catecholamine neurotransmitter outcomes. U.S. Patent Application No. 12/400,291, https://www.google.com/patents/US20090234012

Hirakawa N, Okauchi R, Miura Y, Yagasaki K (2005) Anti-invasive activity of niacin and trigonelline against cancer cells. Biosci Biotechnol Biochem 69(3):653–658

Hu C, Shi J, Quan S, Cui B, Kleessen S, Nikoloski Z et al (2014) Metabolic variation between japonica and indica rice cultivars as revealed by non-targeted metabolomics. Sci Rep 4:5067. doi:10.1038/srep0506, http://www.nature.com/articles/srep05067#supplementary-information

Huxtable RJ (1992) Physiological actions of taurine. Physiol Rev 72(1):101

Iqbal J, Minhajuddin M, Beg ZH (2003) Suppression of 7,12-dimethylbenz[alpha]anthracene-induced carcinogenesis and hypercholesterolaemia in rats by tocotrienol-rich fraction isolated from rice bran oil. Eur J Cancer Prev 12(6):447–453. doi:10.1097/01.cej.0000102802.33147.4b

Iqbal S, Bhanger MI, Anwar F (2005) Antioxidant properties and components of some commercially available varieties of rice bran in Pakistan. Food Chem 93(2):265–272. doi:10.1016/j.foodchem.2004.09.024

Jiang Q, Elson-Schwab I, Courtemanche C, Ames BN (2000) gamma-tocopherol and its major metabolite, in contrast to alpha-tocopherol, inhibit cyclooxygenase activity in macrophages and epithelial cells. Proc Natl Acad Sci U S A 97(21):11494–11499. doi:10.1073/pnas.200357097

Jo SP, Kim JK, Lim YH (2014) Antihyperlipidemic effects of rhapontin and rhapontigenin from rheum undulatum in rats fed a high-cholesterol diet. Planta Med 80(13):1067–1071. doi:10.1055/s-0034-1382999

Johanningsmeier SD, Harris GK, Klevorn CM (2016) Metabolomic Technologies for Improving the Quality of Food: Practice and Promise. Annu Rev Food Sci Technol 7:413–438. doi:10.1146/annurev-food-022814-015721

Jones DP, Park Y, Ziegler TR (2012) Nutritional metabolomics: progress in addressing complexity in diet and health. Annu Rev Nutr 32:183–202. doi:10.1146/annurev-nutr-072610-145159

Jun HI, Song GS, Yang EI, Youn Y, Kim YS (2012) Antioxidant activities and phenolic compounds of pigmented rice bran extracts. J Food Sci 77(7):C759–764. doi:10.1111/j.1750-3841.2012.02763.x

Kadoma Y, Ishihara M, Okada N, Fujisawa S (2006) Free radical interaction between vitamin E (alpha-, beta-, gamma- and delta-tocopherol), ascorbate and flavonoids. Vivo 20(6B):823–827

Kaminska KK, Bertrand HC, Tajima H, Stafford WC, Cheng Q, Chen W et al (2016) Indolin-2-one compounds targeting thioredoxin reductase as potential anticancer drug leads. Oncotarget 7(26):40233–40251. doi:10.18632/oncotarget.9579

Kanchan Rao Singh MS (2012) Current Perspectives in the Pharmacotherapy of Vertigo. Otorhinolaryngology Clin Int J 4(2):5. doi:10.5005/jp-journals-10003-1091

Kanski J, Aksenova M, Stoyanova A, Butterfield DA (2002) Ferulic acid antioxidant protection against hydroxyl and peroxyl radical oxidation in synaptosomal and neuronal cell culture systems in vitro: structure-activity studies. J Nutr Biochem 13(5):273–281

Kaore SN, Amane HS, Kaore NM (2013) Citrulline: pharmacological perspectives and its role as an emerging biomarker in future. Fundam Clin Pharmacol 27(1):35–50. doi:10.1111/j.1472-8206.2012.01059.x

Khodaeiani E, Fouladi RF, Amirnia M, Saeidi M, Karimi ER (2013) Topical 4% nicotinamide vs. 1% clindamycin in moderate inflammatory acne vulgaris. Int J Dermatol 52(8):999–1004. doi:10.1111/ijd.12002

Kim JK, Park S-Y, Lim S-H, Yeo Y, Cho HS, Ha S-H (2013a) Comparative metabolic profiling of pigmented rice (Oryza sativa L.) cultivars reveals primary metabolites are correlated with secondary metabolites. J Cereal Sci 57(1):14–20. doi:10.1016/j.jcs.2012.09.012

Kim N, Kim JK, Hwang D, Lim YH (2013b) The possible mechanism of rhapontigenin influencing antifungal activity on Candida albicans. Med Mycol 51(1):45–52. doi:10.3109/13693786.2012.689021

Kim SP, Park SO, Lee SJ, Nam SH, Friedman M (2014) A Polysaccharide isolated from the liquid culture of Lentinus edodes (Shiitake) mushroom mycelia containing black rice bran protects mice against salmonellosis through upregulation of the Th1 immune reaction. J Agric Food Chem 62(11):2384–2391. doi:10.1021/jf405223q

Kleinsorge H, Pohl W (1953) Gentisic acid in therapy of rheumatic polyarthritis. Med Klin 48(29):1038–1040

Kondo S, Teongtip R, Srichana D, Itharat A (2011) Antimicrobial activity of rice bran extracts for diarrheal disease. J Med Assoc Thai 94(Suppl 7):S117–121

Kontush A, Finckh B, Karten B, Kohlschutter A, Beisiegel U (1996) Antioxidant and prooxidant activity of alpha-tocopherol in human plasma and low density lipoprotein. J Lipid Res 37(7):1436–1448

Kosova M, Hradkova I, Matlova V, Kadlec D, Smidrkal J, Filip V (2015) Antimicrobial effect of 4-hydroxybenzoic acid ester with glycerol. J Clin Pharm Ther 40(4):436–440. doi:10.1111/jcpt.12285

Kotani M, Matsumoto M, Fujita A, Higa S, Wang W, Suemura M et al (2000) Persimmon leaf extract and astragalin inhibit development of dermatitis and IgE elevation in NC/Nga mice. J Allergy Clin Immunol 106(1, Part 1):159–166. doi:10.1067/mai.2000.107194

Krajmalnik-Brown R, Ilhan Z-E, Kang D-W, DiBaise JK (2012) Effects of Gut Microbes on Nutrient Absorption and Energy Regulation. Nutr Clin Pract 27(2):201–214. doi:10.1177/0884533611436116

Kumar A, Henderson A, Forster GM, Goodyear AW, Weir TL, Leach JE et al (2012) Dietary rice bran promotes resistance to Salmonella enterica serovar Typhimurium colonization in mice. BMC Microbiol 12:71. doi:10.1186/1471-2180-12-71

Kuriyan R, Gopinath N, Vaz M, Kurpad AV (2005) Use of rice bran oil in patients with hyperlipidaemia. Natl Med J India 18(6):292–296

Kwack MH, Ahn JS, Kim MK, Kim JC, Sung YK (2010) Preventable effect of L-threonate, an ascorbate metabolite, on androgen-driven balding via repression of dihydrotestosterone-induced dickkopf-1 expression in human hair dermal papilla cells. BMB Rep 43(10):688–692. doi:10.5483/BMBRep.2010.43.10.688

Lampe JW, Li SS, Potter JD, King IB (2002) Serum beta-glucuronidase activity is inversely associated with plant-food intakes in humans. J Nutr 132(6):1341–1344

Larsson SC, Orsini N, Wolk A (2010) Vitamin B6 and risk of colorectal cancer: a meta-analysis of prospective studies. JAMA 303(11):1077–1083. doi:10.1001/jama.2010.263

Lavermicocca P, Valerio F, Evidente A, Lazzaroni S, Corsetti A, Gobbetti M (2000) Purification and characterization of novel antifungal compounds from the sourdough Lactobacillus plantarum strain 21B. Appl Environ Microbiol 66(9):4084–4090

Lei S, Ramesh A, Twitchell E, Wen K, Bui T, Weiss M et al (2016) High Protective Efficacy of Probiotics and Rice Bran against Human Norovirus Infection and Diarrhea in Gnotobiotic Pigs. Front Microbiol 7:1699. doi:10.3389/fmicb.2016.01699

Leon R, Wu H, Jin Y, Wei J, Buddhala C, Prentice H, Wu JY (2009) Protective function of taurine in glutamate-induced apoptosis in cultured neurons. J Neurosci Res 87(5):1185–1194. doi:10.1002/jnr.21926

Li XX, Chen XL, Zhang MH, Wang YH, Da WM, Li JY (2005) A randomized controlled and multicenter clinical study of ferrous L-threonate in treatment of iron deficiency anemia. Zhonghua Nei Ke Za Zhi 44(11):844–847

Li W, Fotinos A, Wu Q, Chen Y, Zhu Y, Baranov S et al (2015) N-acetyl-L-tryptophan delays disease onset and extends survival in an amyotrophic lateral sclerosis transgenic mouse model. Neurobiol Dis 80:93–103. doi:10.1016/j.nbd.2015.05.002

Liao JC, Lee KT, You BJ, Lee CL, Chang WT, Wu YC, Lee HZ (2015) Raf/ERK/Nrf2 signaling pathway and MMP-7 expression involvement in the trigonelline-mediated inhibition of hepatocarcinoma cell migration. Food Nutr Res 59:29884. doi:10.3402/fnr.v59.29884

Liliana Hernández Vázquez, J. P. a. A. N.-O. (2012). The Pentacyclic Triterpenes α, β-amyrins: A Review of Sources and Biological Activities Phytochemicals A Global Perspect Role Nutr Health, 17. doi:10.5772/27253.

Lin Y, Shi R, Wang X, Shen H-M (2008) Luteolin, a flavonoid with potentials for cancer prevention and therapy. Curr Cancer Drug Targets 8(7):634–646

Lü J-M, Lin PH, Yao Q, Chen C (2010) Chemical and molecular mechanisms of antioxidants: experimental approaches and model systems. J Cell Mol Med 14(4):840–860. doi:10.1111/j.1582-4934.2009.00897.x

Luca Narduzzi JS, Mattivi F (2015) Comparing Wild American Grapes with Vitis vinifera: A Metabolomics Study of Grape Composition. J Agric Food Chem 63(30):12. doi:10.1021/acs.jafc.5b01999

Makni M, Chtourou Y, Fetoui H, el Garoui M, Boudawara T, Zeghal N (2011) Evaluation of the antioxidant, anti-inflammatory and hepatoprotective properties of vanillin in carbon tetrachloride-treated rats. Eur J Pharmacol 668(1–2):133–139. doi:10.1016/j.ejphar.2011.07.001

Makni M, Chtourou Y, Barkallah M, Fetoui H (2012) Protective effect of vanillin against carbon tetrachloride (CCl(4))-induced oxidative brain injury in rats. Toxicol Ind Health 28(7):655–662. doi:10.1177/0748233711420472

Mero AA, Ojala T, Hulmi JJ, Puurtinen R, Karila TA, Seppala T (2010) Effects of alfa-hydroxy-isocaproic acid on body composition, DOMS and performance in athletes. J Int Soc Sports Nutr 7:1. doi:10.1186/1550-2783-7-1

Ming X-F, Rajapakse AG, Carvas JM, Ruffieux J, Yang Z (2009) Inhibition of S6K1 accounts partially for the anti-inflammatory effects of the arginase inhibitor L-norvaline. BMC Cardiovasc Disord 9:12–12. doi:10.1186/1471-2261-9-12

Mishra B, Priyadarsini KI, Kumar MS, Unnikrishnan MK, Mohan H (2003) Effect of O-glycosilation on the antioxidant activity and free radical reactions of a plant flavonoid, chrysoeriol. Bioorg Med Chem 11(13):2677–2685

Mnafgui K, Derbali A, Sayadi S, Gharsallah N, Elfeki A, Allouche N (2015) Anti-obesity and cardioprotective effects of cinnamic acid in high fat diet- induced obese rats. J Food Sci Technol 52(7):4369–4377. doi:10.1007/s13197-014-1488-2

Moco S, Bino RJ, Vorst O, Verhoeven HA, de Groot J, van Beek TA et al (2006) A liquid chromatography-mass spectrometry-based metabolome database for tomato. Plant Physiol 141(4):1205–1218. doi:10.1104/pp.106.078428

Morales-González JA (2013) Oxidative stress and chronic degenerative diseases-a role for antioxidants. Rijeka: Croatia InTech, 500. https://www.intechopen.com/books/oxidative-stress-and-chronic-degenerative-diseases-a-role-for-antioxidants

Muthayya S, Sugimoto JD, Montgomery S, Maberly GF (2014) An overview of global rice production, supply, trade, and consumption. Ann N Y Acad Sci 1324:7–14. doi:10.1111/nyas.12540

Muthukumaran J, Srinivasan S, Venkatesan RS, Ramachandran V, Muruganathan U (2013) Syringic acid, a novel natural phenolic acid, normalizes hyperglycemia with special reference to glycoprotein components in experimental diabetic rats. J Acute Dis 2(4):304–309. doi:10.1016/S2221-6189(13)60149-3

Nascimento GGF, Locatelli J, Freitas PC, Silva GL (2000) Antibacterial activity of plant extracts and phytochemicals on antibiotic-resistant bacteria. Braz J Microbiol 31:247–256

Nealon NJ, Worcester CR, & Ryan EP (2017) Lactobacillus paracasei metabolism of rice bran reveals metabolome associated with salmonella Typhimurium growth reduction. J Appl Microbiol. doi: 10.1111/jam.13459

Nimse SB, Pal D (2015) Free radicals, natural antioxidants, and their reaction mechanisms. RSC Adv 5(35):27986–28006. doi:10.1039/C4RA13315C

Niren NM (2006) Pharmacologic doses of nicotinamide in the treatment of inflammatory skin conditions: a review. Cutis 77(1 Suppl):11–16

Olthof MR, Hollman PC, Katan MB (2001) Chlorogenic acid and caffeic acid are absorbed in humans. J Nutr 131(1):66–71

Oxenkrug GF (1999) Antidepressive and antihypertensive effects of MAO-A inhibition: role of N-acetylserotonin. A review. Neurobiology (Bp) 7(2):213–224

Oxenkrug GF, Sablin SO, Requintina PJ (2007) Effect of methylene blue and related redox dyes on monoamine oxidase activity; rat pineal content of N-acetylserotonin, melatonin, and related indoles; and righting reflex in melatonin-primed frogs. Ann N Y Acad Sci 1122:245–252. doi:10.1196/annals.1403.017

Pacheco R, Gallart T, Lluis C, Franco R (2007) Role of glutamate on T-cell mediated immunity. J Neuroimmunol 185(1–2):9–19. doi:10.1016/j.jneuroim.2007.01.003

Panala V, M. V, Boateng J, Field R, Shackelford L, Walker LT (2009) A Comparison of Rice Bran, Corn Oil and Soybean Oil Against Azoxymethane Induced Colon Cancer in a Fisher 344 Rat Model. Int J Cancer Res 5(1):11. doi:10.3923/ijcr.2009.25.35

Panico AM, Cardile V, Garufi F, Puglia C, Bonina F, Ronsisvalle G (2005) Protective effect of Capparis spinosa on chondrocytes. Life Sci 77(20):2479–2488. doi:10.1016/j.lfs.2004.12.051

Parrado J, Miramontes E, Jover M, Márquez JC, AngelesMejias M, CollantesdeTeran L et al (2003) Prevention of brain protein andlipid oxidation elicited by a water-soluble oryzanol enzymaticextract derived from rice bran. Eur J Nutr 42(6):307–314. doi:10.1007/s00394-003-0424-4

Parrado J, Miramontes E, Jover M, Gutierrez JF, Collantes de Terán L, Bautista J (2006) Preparation of a rice bran enzymatic extract with potential use as functional food. Food Chem 98(4):742–748. doi:10.1016/j.foodchem.2005.07.016

Patel K, Chen Y, Dennehy K, Blau J, Connors S, Mendonca M et al (2006) Acute antihypertensive action of nitroxides in the spontaneously hypertensive rat. Am J Physiol Regul Integr Comp Physiol 290(1):R37–43. doi:10.1152/ajpregu.00469.2005

Percudani R, Peracchi A (2009) The B6 database: a tool for the description and classification of vitamin B6-dependent enzymatic activities and of the corresponding protein families. BMC Bioinformatics 10(1):273. doi:10.1186/1471-2105-10-273

Perianayagam MC, Oxenkrug GF, Jaber BL (2005) Immune-Modulating Effects of Melatonin, N-Acetylserotonin, and N-Acetyldopamine. Ann N Y Acad Sci 1053(1):386–393. doi:10.1111/j.1749-6632.2005.tb00046.x

Pero RW, Lund H, Leanderson T (2009) Antioxidant metabolism induced by quinic acid. Increased urinary excretion of tryptophan and nicotinamide. Phytother Res 23(3):335–346. doi:10.1002/ptr.2628

Peroutka SJ, Lebovitz RM, Snyder SH (1981) Two distinct central serotonin receptors with different physiological functions. Science 212(4496):827–829

Pietersen HG, Langenberg CJ, Geskes G, Soeters PB, Wagenmakers AJ (1998) Glutamate metabolism of the heart during coronary artery bypass grafting. Clin Nutr 17(2):73–75

Piletz JE, Aricioglu F, Cheng J-T, Fairbanks CA, Gilad VH, Haenisch B et al (2013) Agmatine: clinical applications after 100 years in translation. Drug Discov Today 18(17–18):880–893. doi:10.1016/j.drudis.2013.05.017

Pompella A, Visvikis A, Paolicchi A, De Tata V, Casini AF (2003) The changing faces of glutathione, a cellular protagonist. Biochem Pharmacol 66(8):1499–1503

Przybylski C (2008) Use of acetyl-leucine for preparing a drug for treating balance disorders. U.S. Patent Application No. 11/886,419. https://www.google.com/patents/US20080108702

Qian Q, Guo L, Smith SM, & Li J. (2016). Breeding high-yield superior quality hybrid super rice by rational design. Natl Sci Rev. doi:10.1093/nsr/nww006.

Qureshi AA, Bradlow BA, Salser WA, Brace LD (1997) Novel tocotrienols of rice bran modulate cardiovascular disease risk parameters of hypercholesterolemic humans. J Nutr Biochem 8(5):290–298. doi:10.1016/S0955-2863(97)89667-2

Qureshi AA, Sami SA, Khan FA (2002) Effects of stabilized rice bran, its soluble and fiber fractions on blood glucose levels and serum lipid parameters in humans with diabetes mellitus Types I and II. J Nutr Biochem 13(3):175–187

Raasch W, Schäfer U, Chun J, Dominiak P (2001) Biological significance of agmatine, an endogenous ligand at imidazoline binding sites. Br J Pharmacol 133(6):755–780. doi:10.1038/sj.bjp.0704153

Rajendra Prasad N, Karthikeyan A, Karthikeyan S, Venkata Reddy B (2011) Inhibitory effect of caffeic acid on cancer cell proliferation by oxidative mechanism in human HT-1080 fibrosarcoma cell line. Mol Cell Biochem 349(1):11–19. doi:10.1007/s11010-010-0655-7

Ramezanzadeh FM, Rao RM, Windhauser M, Prinyawiwatkul W, Marshall WE (1999a) Prevention of oxidative rancidity in rice bran during storage. J Agric Food Chem 47(8):2997–3000

Ramezanzadeh FM, Rao RM, Windhauser M, Prinyawiwatkul W, Tulley R, Marshall WE (1999b) Prevention of hydrolytic rancidity in rice bran during storage. J Agric Food Chem 47(8):3050–3052

Reiter RJ, Tan DX, Cabrera J, D'Arpa D (1999) Melatonin and tryptophan derivatives as free radical scavengers and antioxidants. Adv Exp Med Biol 467:379–387

Sakko M, Tjaderhane L, Sorsa T, Hietala P, Jarvinen A, Bowyer P, Rautemaa R (2012) 2-Hydroxyisocaproic acid (HICA): a new potential topical antibacterial agent. Int J Antimicrob Agents 39(6):539–540. doi:10.1016/j.ijantimicag.2012.02.006

Sakko M, Moore C, Novak-Frazer L, Rautemaa V, Sorsa T, Hietala P et al (2014) 2-hydroxyisocaproic acid is fungicidal for Candida and Aspergillus species. Mycoses 57(4):214–221. doi:10.1111/myc.12145

Saunders RM (1990) The properties of rice bran as a foodstuff. Cereal Foods World 35(7):5

Scriabine A, Ludden CT, Sweet CS, Porter CC, Ulm EH, Stone CA et al (1978) Antihypertensive activity of metyrosine in spontaneously hypertensive rats and its enhancement by carbidopa. Clin Sci Mol Med Suppl, 4:255s–257s

SELFNutritionDada. Rice bran, crude Nutritional Facts & Calories. http://nutritiondata.self.com/facts/cereal-grains-and-pasta/5725/2

Serbinova E, Kagan V, Han D, Packer L (1991) Free radical recycling and intramembrane mobility in the antioxidant properties of alpha-tocopherol and alpha-tocotrienol. Free Radic Biol Med 10(5):263–275. doi:10.1016/0891-5849(91)90033-Y

Shang YJ, B Y L, M M Z (2015) Details of the Antioxidant Mechanism of Hydroxycinnamic Acids. Czech J Food Sci 33(3):7. doi:10.17221/611/2014-CJFS

Sheflin AM, Borresen EC, Wdowik MJ, Rao S, Brown RJ, Heuberger AL et al (2015) Pilot dietary intervention with heat-stabilized rice bran modulates stool microbiota and metabolites in healthy adults. Nutrients 7(2):1282–1300. doi:10.3390/nu7021282

Sheflin, A. M., Borresen, E. C., Kirkwood, J. S., Boot, C. M., Whitney, A. K., Lu, S., et al. (2016). Dietary supplementation with rice bran or navy bean alters gut bacterial metabolism in colorectal cancer survivors. Mol Nutr Food Res. doi:10.1002/mnfr.201500905.

Shi C, Sun Y, Zheng Z, Zhang X, Song K, Jia Z et al (2016) Antimicrobial activity of syringic acid against Cronobacter sakazakii and its effect on cell membrane. Food Chem 197(Pt A):100–106. doi:10.1016/j.foodchem.2015.10.100

Shibata, A., Kawakami, Y., Kimura, T., Miyazawa, T., & Nakagawa, K. (2016). alpha-Tocopherol Attenuates the Triglyceride- and Cholesterol-Lowering Effects of Rice Bran Tocotrienol in Rats Fed a Western Diet. J Agric Food Chem. doi:10.1021/acs.jafc.6b02228.

Singh A, G. K. U (2015) A review on Carissa carandas - phytochemistry, ethno- pharmacology, and micropropagation as conservation strategy. Asian J Pharm Clin Res 8(3):5

Singh G, Passsari AK, Leo VV, Mishra VK, Subbarayan S, Singh BP et al (2016) Evaluation of Phenolic Content Variability along with Antioxidant, Antimicrobial, and Cytotoxic Potential of Selected Traditional Medicinal Plants from India. Front Plant Sci 7:407. doi:10.3389/fpls.2016.00407

Sinha M, Manna P, Sil PC (2008) Taurine protects the antioxidant defense system in the erythrocytes of cadmium treated mice. BMB Rep 41(9):657–663

Song F, Li H, Sun J, Wang S (2013) Protective effects of cinnamic acid and cinnamic aldehyde on isoproterenol-induced acute myocardial ischemia in rats. J Ethnopharmacol 150(1):125–130. doi:10.1016/j.jep.2013.08.019

Srivastava J, Chandra H, Nautiyal AR, Kalra SJS (2014) Antimicrobial resistance (AMR) and plant-derived antimicrobials (PDA(m)s) as an alternative drug line to control infections. 3 Biotech 4(5):451–460. doi:10.1007/s13205-013-0180-y

Sugino T, Shirai T, Kajimoto Y, Kajimoto O (2008) L-ornithine supplementation attenuates physical fatigue in healthy volunteers by modulating lipid and amino acid metabolism. Nutr Res 28(11):738–743. doi:10.1016/j.nutres.2008.08.008

Surjana D, Halliday GM, Martin AJ, Moloney FJ, Damian DL (2012) Oral Nicotinamide Reduces Actinic Keratoses in Phase II Double-Blinded Randomized Controlled Trials. J Investig Dermatol 132(5):1497–1500. doi:10.1038/jid.2011.459

Swargiary A, B. R (2015) in vitro anthelmintic efficacy of alpinia nigra and its bioactive compound, astragalin against fasciolopsis buski. Int J Pharm Pharm Sci 7(10):6

Touitou Y (2001) Human aging and melatonin. Clinical relevance. Exp Gerontol 36(7):1083–1100

Tsuboyama-Kasaoka N, Shozawa C, Sano K, Kamei Y, Kasaoka S, Hosokawa Y, Ezaki O (2006) Taurine (2-aminoethanesulfonic acid) deficiency creates a vicious circle promoting obesity. Endocrinology 147(7):3276–3284. doi:10.1210/en.2005-1007

Turunc Bayrakdar E, Uyanikgil Y, Kanit L, Koylu E, Yalcin A (2014) Nicotinamide treatment reduces the levels of oxidative stress, apoptosis, and PARP-1 activity in Aβ(1–42)-induced rat model of Alzheimer's disease. Free Radic Res 48(2):146–158. doi:10.3109/10715762.2013.857018

US Department of Agriculture, Agricultural research service, Nutrient data Laboratory. USDA National Nutrient Database for Standard Reference, Release 28. Version Current: September 2015, slightly revised May 2016. Internet: /nea/bhnrc/ndl

Upadhyay R, Mohan Rao LJ (2013) An Outlook on Chlorogenic Acids—Occurrence, Chemistry, Technology, and Biological Activities. Crit Rev Food Sci Nutr 53(9):968–984. doi:10.1080/10408398.2011.576319

Valerio F, Lavermicocca P, Pascale M, Visconti A (2004) Production of phenyllactic acid by lactic acid bacteria: an approach to the selection of strains contributing to food quality and preservation. FEMS Microbiol Lett 233(2):289–295. doi:10.1016/j.femsle.2004.02.020

Vargo MA, Voss OH, Poustka F, Cardounel AJ, Grotewold E, Doseff AI (2006) Apigenin-induced-apoptosis is mediated by the activation of PKCdelta and caspases in leukemia cells. Biochem Pharmacol 72(6):681–692. doi:10.1016/j.bcp.2006.06.010

Verschoyle RD, Greaves P, Cai H, Edwards RE, Steward WP, Gescher AJ (2007) Evaluation of the cancer chemopreventive efficacy of rice bran in genetic mouse models of breast, prostate and intestinal carcinogenesis. Br J Cancer 96(2):248–254. doi:10.1038/sj.bjc.6603539

Wang M, Hettiarachchy NS, Qi M, Burks W, Siebenmorgen T (1999) Preparation and functional properties of rice bran protein isolate. J Agric Food Chem 47(2):411–416

Wang J, Shimada M, Kato Y, Kusada M, Nagaoka S (2015) Cholesterol-lowering effect of rice bran protein containing bile acid-binding proteins. Biosci Biotechnol Biochem 79(3):456–461. doi:10.1080/09168451.2014.978260

Wishart DS (2008) Metabolomics: applications to food science and nutrition research. Trends Food Sci Technol 19(9):482–493. doi:10.1016/j.tifs.2008.03.003

Wu J, Domellof M, Zivkovic AM, Larsson G, Ohman A, Nording ML (2016) NMR-based metabolite profiling of human milk: A pilot study of methods for investigating compositional changes during lactation. Biochem Biophys Res Commun 469(3):626–632. doi:10.1016/j.bbrc.2015.11.114

Yanagita T, Han S-Y, Hu Y, Nagao K, Kitajima H, Murakami S (2008) Taurine reduces the secretion of apolipoprotein B100 and lipids in HepG2 cells. Lipids Health Dis 7(1):38. doi:10.1186/1476-511x-7-38

Yang X, Wen K, Tin C, Li G, Wang H, Kocher J et al (2014) Dietary rice bran protects against rotavirus diarrhea and promotes Th1-type immune responses to human rotavirus vaccine in gnotobiotic pigs. Clin Vaccine Immunol 21(10):1396–1403. doi:10.1128/cvi.00210-14

Yang X, Twitchell E, Li G, Wen K, Weiss M, Kocher J et al (2015) High protective efficacy of rice bran against human rotavirus diarrhea via enhancing probiotic growth, gut barrier function, and innate immunity. Sci Rep 5:15004. doi:10.1038/srep15004

Yarlagadda A, Clayton AH (2007) Blood Brain Barrier: The Role of Pyridoxine. Psychiatry (Edgmont) 4(8):58–60

Yemis GP, Pagotto F, Bach S, Delaquis P (2011) Effect of vanillin, ethyl vanillin, and vanillic acid on the growth and heat resistance of Cronobacter species. J Food Prot 74(12):2062–2069. doi:10.4315/0362-028x.jfp-11-230

Yun KJ, Koh DJ, Kim SH, Park SJ, Ryu JH, Kim DG et al (2008) Anti-inflammatory effects of sinapic acid through the suppression of inducible nitric oxide synthase, cyclooxygase-2, and proinflammatory cytokines expressions via nuclear factor-kappaB inactivation. J Agric Food Chem 56(21):10265–10272. doi:10.1021/jf802095g

Zema MJ (2000) Gemfibrozil, nicotinic acid and combination therapy in patients with isolated hypoalphalipoproteinemia: a randomized, open-label, crossover study. J Am Coll Cardiol 35(3):640–646. doi:10.1016/S0735-1097(99)00585-9

Zhang R, Kang KA, Piao MJ, Lee KH, Jang HS, Park MJ et al (2007) Rhapontigenin from Rheum undulatum protects against oxidative-stress-induced cell damage through antioxidant activity. J Toxicol Environ Health A 70(13):1155–1166. doi:10.1080/15287390701252766

Zhihong Gao JS, Sun H, Zhong W, Zhuang W, Zhang Z (2012) Evaluation of different kinds of organic acids and their antibacterial activity in Japanese Apricot fruits. Afr J Agric Res 7(35):8. doi:10.5897/AJAR12.1347

Zhu M, Du J, Liu AD, Holmberg L, Chen SY, Bu D et al (2015) L-cystathionine inhibits oxidized low density lipoprotein-induced THP-1-derived macrophage inflammatory cytokine monocyte chemoattractant protein-1 generation via the NF-kappaB pathway. Sci Rep 5:10453. doi:10.1038/srep10453

Zhu B, Shang B, Li Y, Zhen Y (2016) Inhibition of histone deacetylases by trans-cinnamic acid and its antitumor effect against colon cancer xenografts in athymic mice. Mol Med Rep 13(5):4159–4166. doi:10.3892/mmr.2016.5041

7

Overexpression of Rice Auxilin-Like Protein, XB21, Induces Necrotic Lesions, up-Regulates Endocytosis-Related Genes, and Confers Enhanced Resistance to *Xanthomonas oryzae* pv. *oryzae*

Chang-Jin Park[1,2], Tong Wei[1,4,5], Rita Sharma[1,3] and Pamela C. Ronald[1,4,5*]

Abstract

Background: The rice immune receptor XA21 confers resistance to the bacterial pathogen, *Xanthomonas oryzae* pv. *oryzae* (*Xoo*). To elucidate the mechanism of XA21-mediated immunity, we previously performed a yeast two-hybrid screening for XA21 interactors and identified XA21 binding protein 21 (XB21).

Results: Here, we report that XB21 is an auxilin-like protein predicted to function in clathrin-mediated endocytosis. We demonstrate an XA21/XB21 in vivo interaction using co-immunoprecipitation in rice. Overexpression of *XB21* in rice variety Kitaake and a Kitaake transgenic line expressing XA21 confers a necrotic lesion phenotype and enhances resistance to *Xoo*. RNA sequencing reveals that *XB21* overexpression results in the differential expression of 8735 genes (4939 genes up- and 3846 genes down-regulated) (≥2-folds, FDR ≤0.01). The up-regulated genes include those predicted to be involved in 'cell death' and 'vesicle-mediated transport'.

Conclusion: These results indicate that XB21 plays a role in the plant immune response and in regulation of cell death. The up-regulation of genes controlling 'vesicle-mediated transport' in XB21 overexpression lines is consistent with a functional role for XB21 as an auxilin.

Keywords: Auxilin, Auxilin-Like Protein, Disease Resistance, Heat Shock Cognate Protein, Heat Shock Protein, XA21 Immune Receptor, XA21 Binding Protein 21, *Xanthomonas oryzae* pv. *oryzae*

Background

Protein folding, unfolding, and turnover is central to cell function and is regulated by molecular chaperones, such as heat shock proteins (Hsps) (Saibil 2013). Dysregulation of protein folding can cause cell death and other protein misfolding diseases. Among the major families of HSPs, heat shock cognate protein 70s (Hsc70s) play important roles in protein folding, disaggregation, and transportation. Hsp40, which acts as a co-chaperone of these processes, binds to Hsc70 and stimulates the Hsc70 ATPase activity to stabilize the interaction with substrates (Kampinga and Craig 2010). Hsp40 carries a 70 amino acid signature region, called the J domain, thus Hsp40s are also called J proteins.

J-proteins are classified into three types based on the domain arrangement (Cheetham and Caplan 1998). Type I J-proteins contain an N-terminal J-domain, a glycine-rich region, a zinc-finger domain, and a C-terminal domain. Type II J-proteins are like type I J-proteins except that the zinc-finger domain is missing. Type III J-proteins contain the characteristic J-domain but are structurally divergent outside the J-domain (Koutras and Braun 2014). Type I and II J-proteins are functionally similar in substrates binding, whereas type III J-proteins have distinct roles in stimulating Hsc70 activity due to the flexibility in

* Correspondence: pcronald@ucdavis.edu
[1]Department of Plant Pathology and the Genome Center, University of California Davis, Davis, CA 95616, USA
[4]Feedstocks Division, Joint BioEnergy Institute, Lawrence Berkeley National Laboratory, Berkeley, CA 94720, USA
Full list of author information is available at the end of the article

the structures. One class of type III J-proteins, auxilin and auxilin-like protein (ALP), plays an important role in clathrin-mediated endocytosis. During this process, vesicles are coated by clathrin, a critical component for endocytosis. The clathrin-coated vesicles then transport receptor proteins and their ligands from the cell surface and the trans-Golgi network to the endosomal system (Ungewickell et al. 1995). The J-domain of auxilin recruits Hsc70 to newly budded clathrin-coated vesicles and the Hsc70 and auxilin cooperatively remove the clathrin coat uncoating the vesicles (Kampinga and Craig 2010; Lemmon 2001).

Plant genomes contain more J-proteins than animals. For example, the *Arabidopsis* and rice genomes contains 120 and 125 potential J-proteins respectively, whereas the human genome contains only 41 J-proteins (Sarkar et al. 2013; Rajan and D'Silva 2009). *Arabidopsis*, rice and human encode 92, 83 and 23, type III J-proteins, respectively (Sarkar et al. 2013; Rajan and D'Silva 2009; Qiu et al. 2006), which are the most abundant type of J-proteins. Recent studies of several plant J-proteins suggest that they function in both plant development and the immune response (examples in Sarkar et al. 2013). However, most of the J-proteins have not been characterized and the mechanism of action is unknown.

The rice XA21 immune receptor is representative of a large class of receptor kinases involved in plant innate immunity (Pruitt et al. 2015; Schwessinger and Ronald 2012; Song et al. 1995; Dardick and Ronald 2006). A tyrosine-sulfated peptide from *Xoo*, called RaxX (required for activation of XA21-mediated immunity X), is recognized by XA21 and triggers XA21-mediated immune responses (Pruitt et al. 2015). To elucidate the mechanism of XA21-mediated resistance, we previously performed a yeast two hybrid screening for XA21 interaction partners using a rice cDNA library (Park et al. 2008; Seo et al. 2011). We previously reported the characterization of four of these XA21 binding proteins including a WRKY transcription factor (Peng et al. 2008), a ubiquitin ligase (Wang et al. 2006), an ATPase (Chen et al. 2010b) and a protein phosphatase 2C (Park et al. 2008).

Here we report the characterization of XA21 binding protein 21 (XB21), predicted to encode a type III J-protein. To determine the in vivo biological function of XB21, we carried out biochemical and transgenic analysis. Co-immunoprecipitation indicates that XB21 directly interacts with the XA21 immune receptor in vivo. Overexpression of *XB21* in rice confers enhanced resistance to *Xoo*, sometimes accompanied with a cell death phenotype. RNA sequencing (RNAseq) analysis of XA21 transgenic plants overexpressing *XB21* indicates that genes related to 'vesicle-mediated transport' (which includes genes controlling clathrin-mediated endocytosis) and 'cell death' are significantly up-regulated. Taken together, these results indicate that XB21 functions in plant immunity and cell death regulation and suggest that XB21 functions as an auxilin.

Results

XB21 Encodes a Type III J-Protein

XB21 contains a predicted 2778 bp open reading frame that encodes a 926 amino acid protein with a molecular weight of 102.0 kDa (Fig. 1a). Phylogenetic analysis with closely related rice and *Arabidopsis* type III J-proteins demonstrates that At4g12770, At4g12780, Os11g43950, Os12g36180-XB21, and Os01g25320 cluster in one group (Fig. 1b). XB21 shows relatively high identity to one rice type III J-proteins (Os11g43950, 63.9% identity) and two *Arabidopsis* type III J-proteins (At4g12770, 50.7% identity and At4g12780, 50.4% identity) (Fig. 1a; Additional file 1: Figure S1).

XB21 contains a conserved J-domain at the C-terminus (837–925 amino acids, underlined in Fig. 1a), and possesses a conserved histidine-proline-aspartic acid (HPD) motif known to interact with Hsc70 (Fig. 1a; Additional file 1: Figure S1). The N-terminal region of XB21 does not contain other recognizable domains. XB21 carries an Asp-Pro-Phe/Trp (DPF/W) adaptor protein-binding motif (bold and italic) (Brett et al. 2002). This motif is predicted to bind to the appendage domain of α- or β-subunits of an adaptor protein complex AP-2, a key component in clathrin-mediated endocytosis. XB21 also contains another Phe-x-Asp/Asn-x-Phe (FxD/NxF; x is any amino acid) motif within the short conserved segment at the N terminus (bold and underlined) (Fig. 1a; Additional file 1: Figure S1). This FxD/NxF motif is predicted to serve as an AP-2 α appendage-binding motif and is found in several accessory proteins indispensable for assembly of clathrin-coated vesicles (Brett et al. 2002). These results indicate that XB21 is a type III J-protein with motifs consistent with a role in clathrin-mediated endocytosis.

XB21 Interacts With XA21 in Vivo

We previously demonstrated that XB21 and XA21 interact in yeast (Seo et al. 2011). To assess if the proteins interact in vivo, we carried out a co-immunoprecipitation assay. Proteins extracted from Kitaake transgenic rice plants expressing N-terminal Myc-tagged XA21 under the control of its native promoter (Myc-XA21, line T330-16-1) (Park et al. 2008; Park et al. 2010) and the Kitaake control not carrying *XA21* were immunoprecipitated with agarose beads conjugated with anti-Myc antibodies. Western blot analysis of the immunoprecipitated proteins revealed the presence of a 140 kDa polypeptide corresponding to the size of Myc-XA21 in mock treated and *Xoo*-infected Myc-XA21 plants but not from the Kitaake

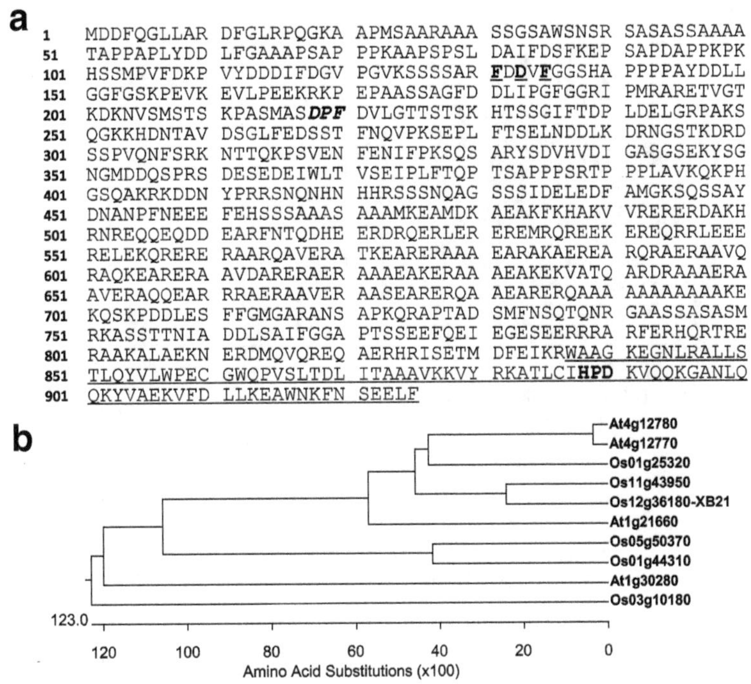

Fig. 1 Deduced amino acid sequence of XB21 and phylogenetic analysis of XB21 and type III J-proteins from rice and *Arabidopsis*. **a** The XB21 protein has a C-terminal J-domain similar to auxilin. The auxilin-like-C-terminal domain including a J-domain is underlined. Underlined bolds and bold italic indicate an FxD/NxF motif and a DPF AP-2 binding motif, respectively. **b** Phylogenetic relationships among XB21 and type III J-proteins from rice and *Arabidopsis*. For determining phylogenetic relationships, protein sequences of putative plant proteins related to type III J-proteins were aligned and then used in ClustalW. Ten thousand bootstrap replicates were performed. At4g12780, At4g12770, At1g21660, and At1g30280 from *Arabidopsis*; and Os01g25320, Os11g43950, Os05g50370, Os01g44310, Os03g10180, and Os12g36180 (XB21) from rice

control (Fig. 2a). In the protein sample prepared from the Myc-XA21 infected with *Xoo*, a band of approximately 120 kDa polypeptide was detected with an anti-XB21 antibody (Fig. 2b). The band was not observed in Kitaake and was barely detectable in the mock-treated Myc-XA21 sample. These results demonstrate the *in planta* interaction between XB21 and XA21.

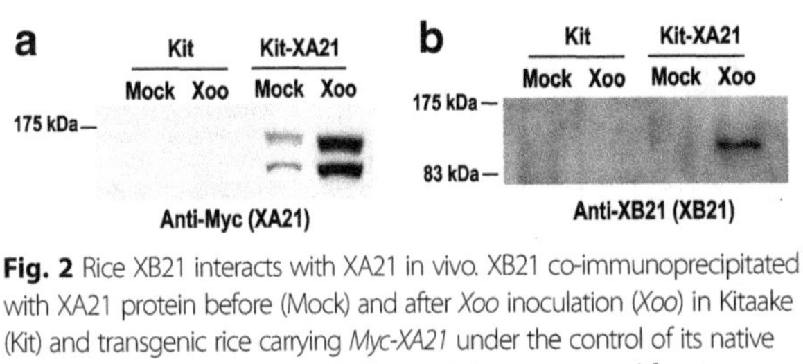

Fig. 2 Rice XB21 interacts with XA21 in vivo. XB21 co-immunoprecipitated with XA21 protein before (Mock) and after *Xoo* inoculation (*Xoo*) in Kitaake (Kit) and transgenic rice carrying *Myc-XA21* under the control of its native promoter (Myc-XA21, T330-16-1). The precipitates were used for western gel blot analysis using anti-Myc antibody (**a**) or anti-XB21 antibody (**b**). Myc-XA21 and its cleavage product displayed bands at approximately 140 and 100 kDa, respectively, and XB21 was detected as approximately 110 kDa band. Experimental results were repeated three times, with similar results

Ectopic Expression of *XB21* Confers Enhanced Resistance to *Xoo* in Both Kitaake and *XA21* Transgenic Plants

To investigate the biological relevance of *XB21*, we generated transgenic rice lines overexpressing *XB21* (XB21ox/Kit) under the control of the maize ubiquitin promoter in the Kitaake genetic background. Eighteen independent T_0 transgenic lines overexpressing *XB21* were inoculated with *Xoo*. Six lines (4A, 5A, 7A, 8A, 10A, and 12A) displayed enhanced resistance as compared with the Kitaake control (Additional file 1: Figure S2). We further analyzed XB21ox/Kit T_1 progenies from 4A, 5A, 7A, 10A, 10B, and 15A, and found that the transgenic plants developed necrotic lesions at the 4 to 5-week stage and became severe at 6-week stage regardless of *Xoo* inoculation (Additional file 1: Figure S3). Six-week-old XB21ox/Kit plants were then inoculated *Xoo*.

They exhibited enhanced resistance to *Xoo*, with lesions of approximately 3–5 cm, much shorter than the infected Kitaake plants, which display lesion lengths of approximately 15–20 cm (Fig. 3; Additional file 1: Figure S4). Overexpression of *XB21* was assessed in XB21ox/Kit T_1 progeny from 4A, 5A and 10B using RT-PCR with *XB21* specific primers (Additional file 1: Figure S5). Transgenic progenies from 4A, 5A, and 10B were

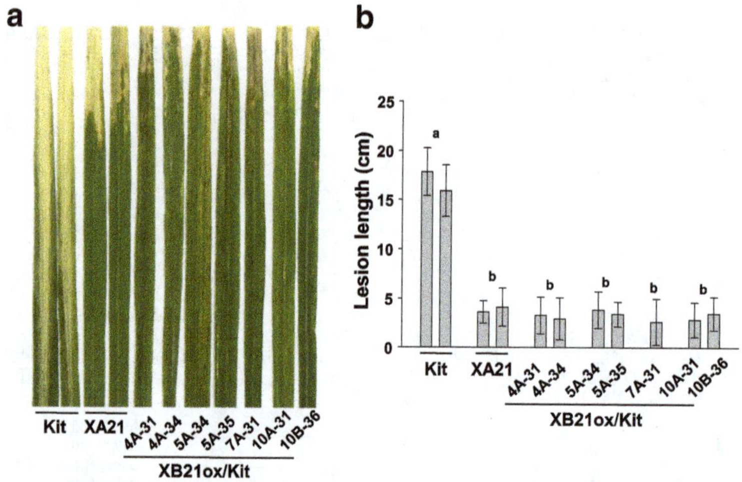

Fig. 3 Kitaake wild type rice plants overexpressing *XB21* (XB21ox/Kit) exhibit enhanced resistance to *Xoo*. **a** Rice plants 14 days after inoculation with *Xoo*. From left to right: Kitaake wild type (Kit), transgenic line (XA21, 23A-1-14) carrying *XA21* driven from its native promoter, and transgenic Kitaake lines carrying *XB21ox* construct (XB21ox/Kit). **b** Lesion lengths of *Xoo* inoculated plants of Kitaake wild type (Kit), transgenic line (XA21) carrying *XA21* driven from its native promoter, and transgenic Kitaake lines carrying *XB21ox* construct (XB21ox/Kit)

displayed high levels of *XB21* transcript compared with Kitaake plants and progenies not carrying *XB21ox* construct. To determine if the enhanced resistance and cell death phenotypes observed in XB21ox/Kit T_1 are stably transmitted to a third generation, 6-week old XB21ox/Kit plants T_2 (progeny from 5A-7 and 10A-1) were inoculated with *Xoo*. All T_2 progenies carrying the XB21ox construct displayed enhanced resistance and necrotic lesions (Additional file 1: Figure S6). These results demonstrate that *XB21* overexpression correlates with the presence of a necrotic lesion phenotype and enhanced resistance to *Xoo*.

To determine if *XB21* overexpression also enhances resistance in the XA21 genetic background, we introduced the XB21ox construct into the transgenic Kitaake carrying *XA21* driven by its native promoter (XA21, line 23A-1-14; described in Chern et al. 2005). We generated 14 independently transformed XB21ox/XA21 T_0 lines. Of these, six lines (2A, 3A, 5A, 6A, 15A, and 23A) developed necrotic lesions that became more severe as the plants grew older (after 4-weeks) under greenhouse condition. Many T_1 progenies derived from XB21ox/XA21 transgenic lines 2A, 3A, 5A and 6A displayed necrotic lesions (Additional file 1: Figures S7 and S8) and, exhibited moderate but significant enhanced resistance compared with the XA21 control plants after *Xoo* inoculation, (Fig. 4a and b). The necrotic lesions by XB21 overexpression, however, were not consistently observed in our greenhouse and chamber

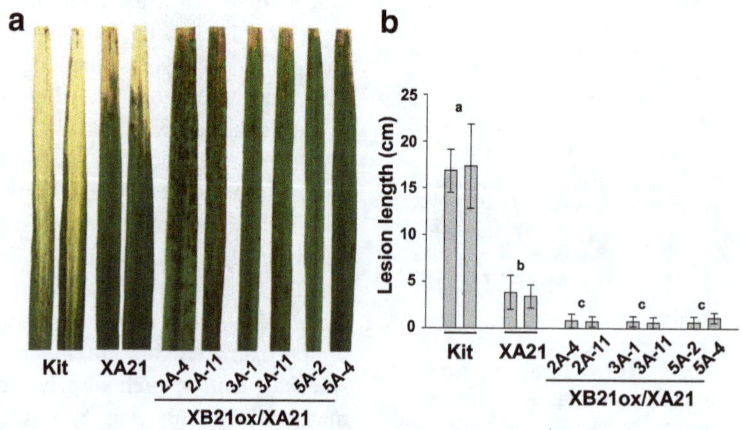

Fig. 4 XA21 transgenic rice plants overexpressing *XB21* (XB21ox/XA21) confer enhanced resistance to *Xoo*. **a** Rice plants 12 days after inoculation with *Xoo*. From left to right: Transgenic lines carrying *XB21ox* construct (XB21ox/XA21), transgenic line (XA21, 23A-1-14) carrying *XA21* driven from its native promoter, and Kitaake wild type (Kit). **b** Lesion lengths of *Xoo* inoculated plants of transgenic lines carrying *XB21ox* construct (XB21ox/XA21), transgenic line (XA21) carrying *XA21* driven from its native promoter, and Kitaake wild type (Kit)

conditions. The T_1 progenies from these lines segregated for expression of necrotic lesions and enhanced resistance to *Xoo*. Only the XB21ox/XA21 transgenic plants exhibiting necrotic lesions (black bars in Additional file 1: Figure S8) displayed enhanced resistance to *Xoo*. To investigate whether the necrotic lesions are associated with the expression levels of *XB21*, overexpression of *XB21* was assessed in XB21ox/XA21 T_1 progeny from 2A, 3A, and 5A using RT-PCR with *XB21* specific primers (Additional file 1: Figure S9). We found no significant difference in *XB21* expression between segregants displaying and not displaying necrotic lesions. Thus, *XB21* overexpression does not correlate with the expression of necrotic lesion in the XA21 genetic background. However, expression of necrotic lesions does correlate with enhanced resistance to *Xoo*.

Silencing of *XB21* Does not Affect Resistance to *Xoo*

Based on the observation that many of the XB21 overexpressing lines display necrotic lesions and enhanced resistance to *Xoo*, we next investigated whether *XB21* silencing affects resistance in the presence and absence of XA21. For this purpose, a DNA sequence specific to *XB21* that diverges from its closely related Os11g43950 was chosen for *XB21* RNAi construction (Additional file 1: Figure S10). The resulting XB21RNAi construct was introduced into the Kitaake and XA21 genetic backgrounds to generate XB21RNAi/Kit and XB21RNAi/XA21 lines, respectively.

Thirteen independently transformed XB21RNAi/Kit T_0 lines (Additional file 1: Figure S11) were inoculated with *Xoo* and showed no obvious difference in lesion lengths compared with Kitaake control plants inoculated with *Xoo*. Silencing of *XB21* was confirmed in the T_1 progeny from four T_0 lines (9B, 12A, 13A, and 14A) by qRT-PCR (Fig. 5a). The progeny from these lines also did not display any significant difference in lesion lengths or bacterial multiplication compared with the Kitaake control (Fig. 5b and c). Additional inoculation experiments on the T_1 progeny of seven T_0 lines (3B, 5B, 7A, 8A, 12A, 14A, and 16A) revealed similar results (Additional file 1: Figure S12). These results indicate that silencing of *XB21* expression does not affect resistance to *Xoo* in the Kitaake genetic background.

We next generated nine independent lines expressing the XB21 RNAi construct in the XA21 genetic background (XB21RNAi/XA21). Silencing of *XB21* was confirmed in the T_1 progeny from five representative T_0 lines, 1B, 2B, 4B, 6A and 7A (Fig. 6a). To assess their

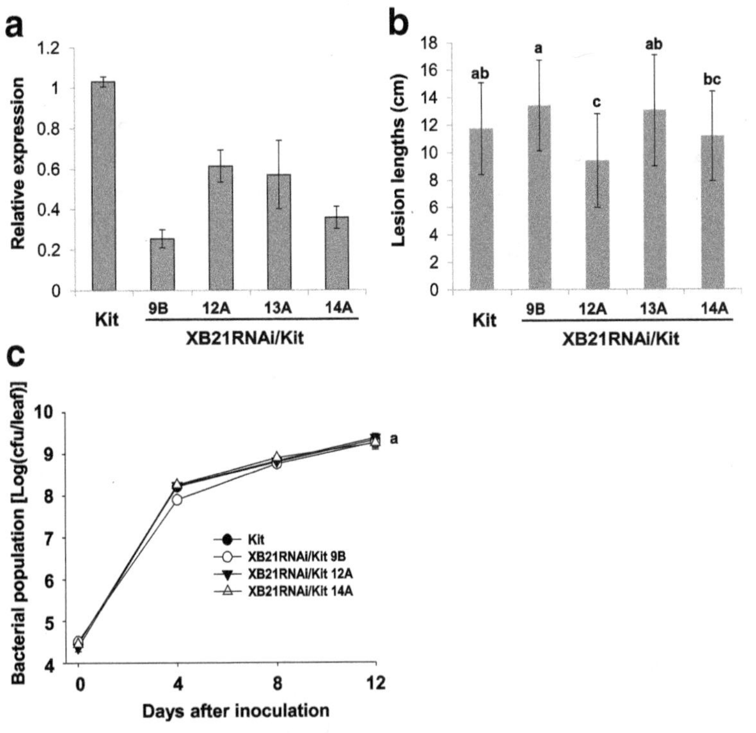

Fig. 5 Silencing of XB21 in Kitaake plants does not affect resistance to *Xoo*. **a** Relative expression of *XB21* in Kitaake plants silencing *XB21*. qRT-PCR was performed with specific primers for each genes. Gene expression level was normalized using *ubiquitin* as an internal reference. **b** Lesion lengths measured 14 days after inoculation from Kitaake plants silencing *XB21* (XB21RNAi/Kit 9B, 12A, 13A, and 14A) and Kitaake plants. The error bars represent standard error between three biological replicates. **c** *Xoo* populations were monitored over 12 days in XB21RNAi/Kit (9B, 12A-2, and 14A-2) and Kitaake wild type plants. For each time point, bacterial populations were obtained in three separate leaves for each genotype. The error bars represent standard deviation values obtained from the three samples

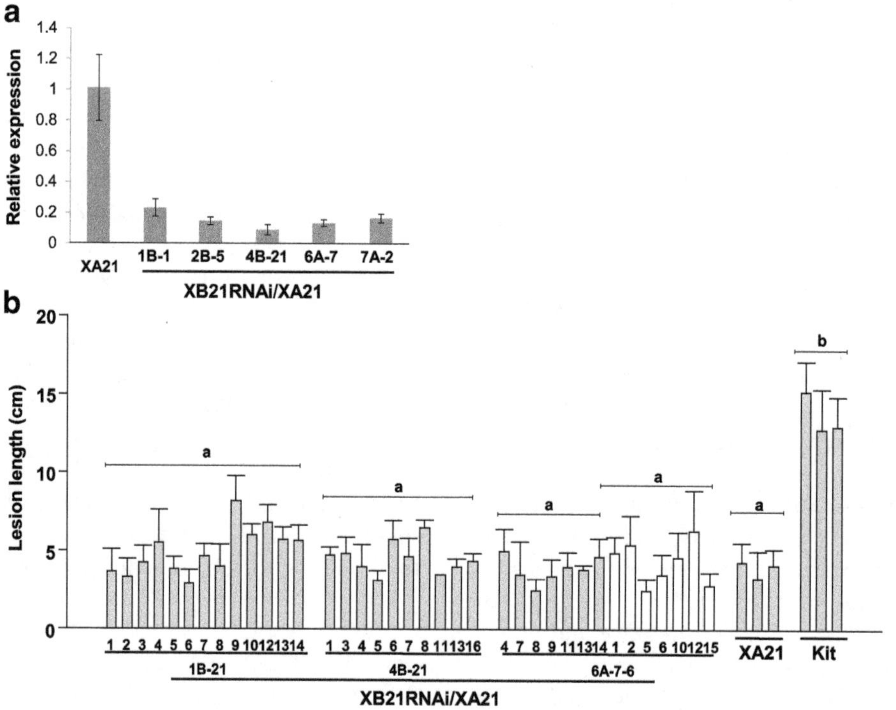

Fig. 6 Silencing of XB21 in XA21 transgenic plants does not affect resistance to *Xoo*. **a** Relative expression of *XB21* in XA21 transgenic plants silencing *XB21*. qRT-PCR was performed with specific primers for each genes. Gene expression level was normalized using *ubiquitin* as an internal reference. **b** Lesion lengths measured 12 days after inoculation from XA21 transgenic plants silencing *XB21* (XB21RNAi/XA21 1B-21, 4B-21, and 6A-7-6) and XA21 transgenic plants (23A-1-14). Progeny not carrying *XB21RNAi* construct are labeled with a *white bar*. The error bars represent standard deviation values obtained from the two to seven samples from each plant

resistance to *Xoo*, 6-week T_2 or T_3 progeny from three XB21RNAi/XA21 lines were inoculated with *Xoo* along with Kitaake and XA21 control plants (Fig. 6b). The progeny do not display significant difference in lesion lengths compared with the XA21 control. In addition, the presence of *XB21ox* construct does not correlated with the lesion lengths in line 6A-7-6. These results suggest that silencing of *XB21* expression does not affect resistance to *Xoo* in the Xa21 genetic background.

Transcriptomic Analysis Reveals the up-Regulation of Genes Involved in 'Cell death' and 'Vesicle-Mediated transport' in the XB21ox/XA21 Transgenic Plants

To further investigate the mechanism of *XB21* function, we performed a transcriptome analysis of the XB21ox/XA21 line (homozygous progeny derived from the 3A-11 line), the XA21 transgenic line (23A-1-14), and Kitaake controls via RNA-sequencing. Leaf tissue was harvested from five-week-old plants at the point where the XB21ox/XA21 lines began to develop the necrotic lesions. The genes, exhibiting ≥2-fold change in expression with the false discover rate (FDR) ≤0.01, were considered as '*differentially expressed*' and further examined to identify affected pathways. Ninety genes were differentially expressed (≥2-folds, FDR ≤0.01) in the XA21 transgenic plants compared with Kitaake control plants (Fig. 7a; Additional file 2: Table S1A and B). 8735 genes (4939 genes up- and 3846 genes down-regulated) were '*differentially expressed*' in the XB21ox/XA21 plants compared with XA21 plants (Fig. 7a; Additional file 2: Table S1C and D). None of these genes were up-regulated in XA21 plants compared with Kitaake plants indicating that the differential regulation observed in XB21ox/XA21 is dues to XB21 overexpression.

The '*differentially expressed*' genes were used for the Gene Ontology (GO) enrichment analysis (Fig. 7b). The P-value of this analysis revealed that GO terms associated with 'protein transport', 'vesicle-mediated transport' and 'protein targeting' are significantly enriched among the up-regulated genes. Those related with 'response to biotic stimulus' and 'cell death' are also significantly enriched. Whereas, those regulating 'photosynthesis' and 'response to abiotic stimulus' were enriched in the down-regulated genes. Because of the necrotic phenotype observed in the XB21ox/XA21 plants and the predicted function of XB21 as an auxilin involved in clathrin-mediated endocytosis based on sequence analysis, we further analyzed genes associated with the GO terms, 'cell death' and 'vesicle-mediated transport', (Additional file 3: Table S2A and B).

a

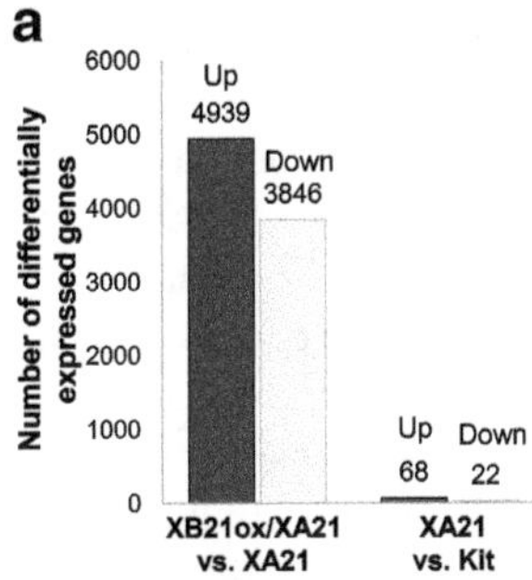

b

GO ID	GO Term	Observed Freq. (%)	Expected Freq. (%)	P-value
Significant GO Terms in Up-regulated genes:				
GO:0015031	Protein transport	3.00	1.20	1.09E-10
GO:0016192	Vesicle-mediated transport	1.50	0.60	4.54E-06
GO:0008104	Protein Targeting	3.40	1.50	2.90E-10
GO:0006468	Protein amino acid phosphorylation	12.60	6.60	7.89E-28
GO:0009607	Response to biotic stimulus	11.20	6.80	3.74E-15
GO:0008219	Cell death	4.50	2.80	4.65E-06
GO:0030154	Cell differentiation	8.50	5.30	1.02E-10
GO:0007165	Signal transduction	20.70	14.30	6.03E-18
Significant GO Terms in Down-regulated genes:				
GO:0015979	Photosynthesis	2.40	1.20	1.35E-04
GO:0009628	Response to abiotic stimulus	19.20	14.30	1.98E-07

Fig. 7 Transcriptional dynamics in response to overexpression of *XB21* in transgenic rice plants carrying *XA21*. **a** Results of differential expression analysis. Numbers of up- and down-regulated genes (≥2-fold change, FDR ≤0.01) in XA21 plants overexpressing *XB21* vs. XA21 plants (XB21ox/XA21 vs. XA21) and XA21 plants vs. Kitaake wild type (XA21 vs. Kit) are presented. **b** Gene Ontology (GO) enrichment analysis for differentially expressed genes (≥2-fold with FDR ≤0.01) performed using Virtual Plant database. Only the top ranked GO terms for "biological process" category are listed. The FDR, calculated using hypergeometric distribution analysis, are provided

The GO term 'cell death' contains 143 differentially expressed genes (Additional file 3: Table S2A). These include up-regulation of putative immune receptors including RGAs (resistance gene analogs) (Sekhwal et al. 2015), RPSs (resistance to *Pseudomonas syringae*) (Axtell and Staskawicz 2003), RPPs (Recognition of *Peronospora parasitica*) (Rehmany et al. 2005), MLOs (powdery mildew locus O) (Piffanelli et al. 2002), and Yr10s (yellow rust resistance protein 10) (Spielmeyer and Lagudah 2003). The differentially regulated genes in the "cell death" set also include Bcl-2 associated athanogene (BAG) domain-containing proteins, which are associated with stress responses and modulation of cell death in humans and *Arabidopsis* (Kabbage and Dickman 2008), chitinases (Jwa et al. 2006), defender against death (DAD family) protein (Danon et al. 2004), BAX inhibitor motif-containing protein, anti-apoptotic protein (Lisak et al. 2015). The up-regulation of the 'cell death'-related genes in the XB21 overexpression lines correlates with the presence of the necrotic lesions and enhanced resistance observed in the XB21ox/XA21 lines.

We also identified 48 genes with the GO term 'vesicle-mediated transport' that are differentially expressed in XB21ox/XA21 (Additional file 3: Table S2B). This set of up-regulated genes include those annotated as exo70 exocyst complex subunits (Wu and Guo 2015), vesicle-associated membrane proteins (Yun et al. 2013), coatomer alpha subunit (Kaur and Subramanian 2015), Sec1 family transport proteins (Karnik et al. 2013), vesicle transport v-soluble N-ethylmaleimide-sensitive factor attachment protein receptor (SNARE) proteins (Bao et al. 2008), adaptins (Park and Jurgens 2011), and component proteins of the clathrin adaptor complex (Qiao et al. 2010). We also identified in this gene set, many clathrin-related protein genes including many clathrin adaptor complex proteins, clathrin assembly proteins, and dynamins (Additional file 2: Table S1 and Additional file 3: Table S2). The up-regulation of genes involved in 'vesicle-mediated transport' suggests that XB21 may exert its influence in clathrin-mediated endocytosis.

Discussion

Here, we characterize a novel XA21-binding protein that carries motifs typical of type III J-protein and is involved in cell death and resistance to *Xoo*.

XB21 is a Putative Auxilin Protein

XB21 is predicted to encode an auxilin, based on the presence of the C-terminal J domain and the FxD/NxF and DPF/W motifs, which have been found in accessory proteins indispensable for assembly of clathrin-coated vesicles (Lemmon 2001; Brett et al. 2002).

In mammals, there are two highly homologous auxilin-related proteins, auxilin 1 and cyclin-G-dependent kinase/auxilin2 (GAK/auxilin 2). The main structural difference is that GAK/auxilin 2 has an additional N-terminal kinase domain. Despite this structural difference, both auxilin 1 and GAK/auxilin 2 are required for uncoating of clathrin-coated vesicles and for endocytic membrane trafficking in HeLa cells, zebrafish, and *Drosophila* (Hirst et al. 2008; Bai et al. 2010; Kandachar et al. 2008). In *Arabidopsis*, an auxilin-like protein (At4g12780, orthologous gene of rice *XB21*) interacts with plant clathrin and increases clathrin uncoating from microsomal membranes in presence of animal Hsc70 (Lam et al. 2001). These similarities suggest that XB21 also interacts with plant clathrin. In support of this hypothesis, we found that overexpression of XB21 in XA21 rice plants, led to significant induction of genes involved in 'vesicle-mediated transport'. One of the up-regulated genes, *OsSYP71*, has previously been demonstrated to confer resistance to *Magnaporthe oryzae*, the fungus causing rice blast disease (Bao et al. 2012).

OsSYP71 belongs to the SNARE protein family that functions in vesicle trafficking in eukaryotic cells (Heese et al. 2001; Wick et al. 2003). Taken together, these results suggest that XB21, like mammalian auxilins and *Arabidopsis* auxilin-like proteins, plays a role in uncoating clathrin-coated vesicles.

In animals, a physical interaction between auxilin and cell-surface receptors has not been reported. There are, however, an increasing number of reports showing that auxilin regulates receptor-mediated signaling pathways. For example, *Drosophila* auxilin exhibits specific genetic interactions with Notch receptors, mediating cell-cell interactions during development (Hagedorn et al. 2006). Down-regulation of GAK/auxilin 2 in HeLa cells showed 50-folds *increase* in expression of epidermal growth factor receptor (EGFR), as well as significant changes in downstream EGFR trafficking (Zhang et al. 2004). Depletion of GAK/auxilin 2 in in HeLa cells inhibited receptor-mediated endocytosis and recruitment of clathrin (Lee et al. 2005).

In plants, several reports describing cell-surface receptor endocytosis have been recently published. For example, it has been reported that resistance mediated by *Arabidopsis* Flagellin-sensing 2 (FLS2) immune receptor depends on clathrin-mediated endocytosis (Mbengue et al. 2016). In *Nicotiana benthamiana* silencing *clathrin heavy chain* (*NbCHC*), the endosomal location of heterologously expressed FLS2 was decreased by approximately 75%, indicating defects in the normal endosomal trafficking of FLS2-GFP. XA21 is also reported to be endocytosed and probably transported via trans-Goli network and early endosome compartment (Chen et al. 2010a). Together with these reports, our results suggest that the auxilin-like protein XB21 may function as clathrin uncoating factor to mediate XA21 endocytosis. It is also possible that, considering the general mechanism of clathrin uncoating in the endocytosis process, XB21 could be involved in the endocytosis of other unknown receptors.

Overexpression of XB21 Causes a Cell Death Phenotype

In animals and plants, cell death occurs during growth and development to remove unnecessary or unwanted cells. In a few cases, mouse and zebrafish, this cell death is mediated by an auxilin (Lee et al. 2008; Bai et al. 2010). In soybean, overexpression of a type III J-protein (GmHsp40.1) caused necrotic cell death (Liu and Whitham 2013). Although, a relationship between auxilin-like proteins and cell death has not been previously observed in plants, the development of cell death lesions in many of the *XB21*-overexpressing plants suggests that XB21 is at least partially involved in this cell death phenotype. We have noted, however, that the XB21-induced cell death is not consistently observed. For example during winter greenhouse trials of the T_1 progeny of XB21ox/XA21 plants, more than half of the segregating progeny carrying *XB21ox* construct did not display a necrotic lesion phenotype. These results suggest that the developmentally regulated cell death observed in some *XB21* overexpressing plants is also regulated by additional, unidentified abiotic factors.

Several studies have reported a role of vesicular trafficking in plant cell death. For example, rice spotted leaf 28 (*SPL28*) encoding a clathrin-associated adaptor protein complex 1, medium subunit μ1 (AP1M1), which is involved in the post-Golgi trafficking pathway (Qiao et al. 2010). The *spl28* mutants displayed a necrotic lesion phenotype spontaneously in the absence of pathogen attack. Another example is the discovery of the μ subunit of *Arabidopsis* adaptor protein-2 (AP2M), mediating clathrin-coated vesicle formation, is also involved in the necrotic lesion formation. An *ap2m* mutant displayed reduced amount of cell death after *Pseudomonas syringae* pv. *tomato* DC3000 (Hatsugai et al. 2016). A third example is *Arabidopsis* enhanced disease resistance 4 (EDR4) negatively regulating resistance to powdery mildew. The EDR4 interacts with clathrin heavy chain 2 (CHC2) (Wu et al. 2015). The *edr4* mutants displayed cell death and reduced endocytosis rates, indicating that clathrin-mediated membrane trafficking is involved in cell deaths. The molecular mechanisms of how these proteins regulate cell death have not yet been elucidated.

Considering that XB21 possesses a DPF/W motif and a FxD/NxF motif predicted to interact with adaptor protein complex indispensable for clathrin-coated vesicle formation, we hypothesize that XB21 cooperatively works with adaptor proteins and/or clathrins. Supporting this hypothesis, RNAseq analysis in XB21ox/XA21 transgenic plants displayed up-regulation of many clathrin adaptor complex proteins, clathrin assembly proteins and dynamins as well as other genes involved in 'vesicle-mediated transport'.

Conclusions

XB21 was originally isolated as an XA21 binding protein in yeast. Here we demonstrate an in vivo interaction between XB21 and XA21 using co-immunoprecipitation assays in rice plants. RNAseq analysis of *XB21*-overexpressing rice plants revealed up-regulation of genes related to diverse biological processes including those regulating 'vesicle-mediated transport' and 'cell death'. These results together with the predicted domain structure of XB21 support the hypothesis that XB21 functions as an auxilin.

Methods

Plant Materials and Growth Conditions

Oryza sativa L. spp. *japonica* rice variety Kitaake plants (lacking *XA21*) and transgenic plants were maintained in the greenhouse facility at University of California, Davis. The growth chamber for inoculation experiments was set on a 16-h light and 8-h dark period, 28/26 °C temperature cycle and 90% humidity. Healthy and well-expanded leaves from six-week-old plants were used for *Xoo* inoculation as well as nucleic acid and protein extractions.

In Vivo Co-Immunoprecipitation

To co-immunoprecipitate Myc-XA21 and XB21, total proteins were extracted from 5 g of rice leaf tissue in 25 ml of ice-cold Extraction Buffer [0.15 M NaCl, 0.01 M Na-phosphate pH 7.2, 2 mM EDTA, 10 mM β-mercaptoethanol, 1% Triton X-100, 1 mM PMSF, 20 mM NaF, 1% Protease cocktail (Sigma), 2 μg/ml antipain, 2 μg/ml leupeptin, and 2 μg/ml aprotinin] as described earlier (Park et al. 2008). After filtering through Miracloth (Calbiochem) followed by centrifugation at 13,000 g for 20 min at 4 °C, 50 μl of agarose conjugated anti-Myc antibody (Santa Cruz) was added to the supernatant and incubated at 4 °C for 2 h. The agarose beads were then washed three times in 1.5 ml of extraction buffer without proteinase inhibitors. The immunoprecipitated proteins were eluted with 4× Laemmli sample buffer. Western blot analyses were performed as previously (Park et al. 2008). The anti-XB21 antibody was generated by Pacific Immunology based on a synthetic peptide of XB21. Detailed information about their methods can be obtained from their website (http://www.pacificimmunology.com/). In brief, a short peptide (KPEVKEVLPEEKRKPEPA, 156–173 a.a of XB21) was synthesized and conjugated to an immune-stimulating carrier protein, keyhole limpet hemocyanin. Anti-XB21 antibody was affinity-purified using the synthetic peptides bound to C3-SEP-PAK cartridges.

Xoo Inoculations and Determination of Bacterial Populations

Xanthomonas oryzae pv. *oryzae* (*Xoo*) Philippine race 6 (PR6) strain PXO99Az (generously provided by Jan Leach; referred to as *Xoo* throughout text) was used in this study. For *Xoo* inoculations, 6-week old rice plants were transferred to the growth chamber. *Xoo* suspensions (OD600 of 0.5) were used to inoculate rice plants by the scissors-dip method (Song et al. 1995; Chern et al. 2005). Only the top two to three fully expanded leaves of each tiller were inoculated. Lesion lengths were measured at the indicated days after inoculation. Statistical analysis was performed using the Tukey's honest significant difference test (Tukey's HSD test). For bacterial colony counts, the inoculated leaves, including lesions and tissue showing no lesions, were cut into approximately 1 mm pieces with sterile scissors and incubated in 10 mL water for 1 h to harvest bacteria. The extract was diluted accordingly and plated on peptone sucrose agar (PSA) plates containing 20 mg/L of cephalexin. The plates were scored after incubation at 37 °C for two days.

Plasmid Construction for *XB21* Overexpression and Silencing

A 2778 nt cDNA fragment encoding full-length XB21 protein was amplified from rice cDNA using primers, 5′-CACCATGGACGACTTCCAGGGCCTCCTGGCC-3′ and 5′-TTAGAAGAGTTCCTCTGAGTTGAATTTG-3′. The PCR product was cloned in pENTR™/D-TOPO® according to the manufacturer's instruction (Invitrogen) and the insert was confirmed by Sanger sequencing. For over-expression in rice, the *XB21* cDNA in pENTR™/D-TOPO® was recombined into an Ubi-1300 vector using Gateway® LR Clonase (Invitrogen). Ubi-1300 is a pCAMBIA-1300 (AF234296) derivative containing an additional expression cassette with the maize ubiquitin promoter (Ubi) and a nopaline synthase 3′-polyadenylation region (NOS), to which the Gateway® cassette was added (Rohila et al. 2006).

For *XB21* silencing, a 427 nt cDNA fragment from 529 to 955 bp of *XB21* cDNA was amplified from rice cDNA using *XB21* specific primers, 5′-CACCTCGGCTGGGTTTGATGATTT-3′ and 5′-CAGAAGGTTTTTGGGTTGTATTTT-3′. The cDNA fragment was chosen because of its high diversity with other *XB21* paralogs. The PCR product was cloned in pENTR™/D-TOPO® (Invitrogen) and the insert was confirmed by sequencing. For silencing in rice, the partial *XB21* cDNA fragment in pENTR™/D-TOPO® was recombined into the pANDA silencing vector (Miki and Shimamoto 2004) using gateway LR Clonase (Invitrogen).

Rice Transformations

Rice transformations were carried out as described previously (Chern et al. 2005). *Agrobacterium tumefaciens* strain EHA105 was used to infect rice callus. Transformants of rice cultivar Kitaake and transgenic XA21 plants (23A-1-14) (referred to as XA21 in this paper) (Park et al. 2010; Chern et al. 2005) carrying *XB21ox* or *XB21RNAi* construct were selected using hygromycin and confirmed by PCR using gene-specific primers.

Quantitative Real-Time PCR Analysis

For quantitative real time PCR (qRT-PCR) analysis, we harvested fully expended leaves from wild type and transgenic plants and extracted RNA using TRIzol reagent (Invitogen) as per manufacturer's instructions. Total RNA was treated with DNase I and purified using Macherey-Nagel NucleoSpin RNA II kit. We quantified RNA using Nanodrop (ND-1000 spectrophotomter from Thermo Scientific) and performed cDNA synthesis using superscript VILO cDNA synthesis kit (Invitrogen).

About 100 ng of cDNA was used as template for each qRT-PCR reaction using SsoFast EvaGreen Supermix (Bios-Rad). All PCR reactions were performed in 96-well plates on a Bio-Rad CFX96 Real-Time System coupled with a C1000 Thermal Cycler. The expression of target genes was normalized to the internal reference *ubiquitin*. The primers used for qRT-PCR were, 5′- AACCAGCT GAGGCCCAAGA-3′/5′-ACGATTGATTTAACCAGTC CATGA-3′ (for *ubiquitin*), 5′- AAATCCCTGCCTCAG TTGCT-3′/5′-AACCCCGAGGTGATAATTCC-3′ (for *XA21*) and 5′- GGCTGCTGTAGACGCTAG-3′/5′-AC CTTCTCCTTGGCTTCAG-3′ (for *XB21*).

RNA Sequencing and Analysis

Three replicates, with each replicate of three leaf tissues, were collected from five-week-old Kitaake, transgenic plants carrying *XA21* (XA21, 23A-1-14) and transgenic plants overexpressing *XB21* in the *XA21* genetic background (XB21ox/XA21, progeny of homozygous 3A-11 line), with three replicates. RNA was isolated using TRIzol reagent, DNase-treated and purified using Qiagen RNeasy Plant mini kit. A total of nine cDNA libraries were constructed using Illumina reagents according to manufacturer's instructions. About 20 million reads were sequenced from each library by U.S. Department of Energy (DOE) Joint Genome Institute (JGI) using Illumina paired-end sequencing (2X76 bp). The high quality reads were mapped to rice reference sequence (MSU 6.1) using Tophat 3.1 (Trapnell et al. 2009), with default parameters. The mapped RNA sequencing (RNAseq) reads were assembled into transcripts by cufflinks (Trapnell et al. 2010). The data was analyzed in RobinA (http://mapman.gabipd.org/web/guest/robin) for differential expression (Lohse et al. 2012). The genes exhibiting ≥2-fold change in expression with FDR ≤0.01 in three biological replicates were defined as *'differentially expressed'* and used for the Gene Ontology (GO) enrichment analysis. The *'differentially expressed'* genes were categorized into pathways using Virtual Plant 1.3 (Katari et al. 2010) and mapped onto functional categories in the MapMan (Thimm et al. 2004).

Additional files

Additional file 1: Figure S1. Amino acids sequence alignment of ALPs from rice (XB21-Os12g36180 and Os11g43950) and *Arabidopsis* (At4g12770 and At4g12780). Dashes indicate gaps. **Figure S2.** Overexpression of *XB21* in Kitaake wild type plants exhibit enhanced resistance to *Xoo*. **Figure S3.** Kitaake plants overexpressing XB21 (XB21ox/Kit) display cell death lesions. **Figure S4.** T_1 generation of XB21ox/Kit carrying *XB21ox* construct display enhanced resistance to *Xoo*. **Figure S5.** Transgenic Kitaake plants carrying *XB21ox* construct (XB21ox/Kit) overexpress *XB21*. **Figure S6.** Overexpression of *XB21* in Kitaake wild type plants (XB21ox/Kit, T_2) displays resistant to a virulent strain of *Xoo*. **Figure S7.** Overexpression of *XB21* in XA21 plants (XB21ox/XA21) displays cell death lesions. **Figure S8.** T1 progeny from XB21 overexpressing XA21 transgenic plants (XB21ox/XA21) display cell death and enhanced resistance to *Xoo*. **Figure S9.** XA21 plants carrying *XB21ox* construct (XB21ox/XA21) overexpress *XB21*. **Figure S10.** Comparison of partial nucleotide sequence used for *XB21* RNA silencing and one of its closely related putative ALP Os11g43950. **Figure S11.** Silencing of *XB21* in Kitaake plants does not exhibit significant difference in lesion lengths after *Xoo* inoculation. **Figure S12.** T_1 generation of XB21RNAi/Kit do not display alteration of resistance to *Xoo*.

Additional file 2: Table S1. RNA sequencing-mediated transcriptome analysis of the Kitaake wild type, transgenic plants carrying *XA21* (XA21) (23A-1-14), and transgenic plants overexpressing *XB21* in the *XA21* genetic background (XB21ox/XA21, progeny of 3A-11). The genes, exhibiting ≥2-fold change in expression with FDR ≤0.01, were considered as *'differentially expressed'* and listed. (A) Up-regulated genes in XA21 plants compared to Kitaake plants. (B) down-regulated genes in XA21 plants compared to Kitaake plants. (C) Up-regulated genes in XB21ox/XA21 plants compared to XA21 plants. (D) Down-regulated genes in XB21ox/XA21 plants compared to XA21 plants.

Additional file 3: Table S2. List of up-regulated genes associated with GO terms, 'cell death' (A) and 'vesicle-mediated transport' (B) in XB21ox/XA21 plants compared to XA21 plants. The genes, exhibiting ≥2-fold change in expression with *p* value ≤0.01, were considered as *'differentially expressed'* and listed.

Abbreviations

Hsc: Heat shock cognate protein; Hsp: Heat shock protein; XB21: XA21 Binding Protein 21; *Xoo* PR6: *Xanthomonas oryzae* pv. *oryzae* Philippine Race 6

Acknowledgments

We thank Daniel F. Caddell for critical reading of the manuscript, Patrick E. Canlas for generation of transgenic rice plants and Randy Ruan for lab and green house management. This work was supported by grants from the National Science Foundation (NSF, IOS-0817738) and NIH (GM55962) to P.C.R., Korea IPET through Agri-Bio industry Technology Development Program by MAFRA (316087-4 to C.J.P.), and Ramalingaswami fellowship by Department of Biotechnology, Government of India (R.S.). The work conducted by the Joint BioEnergy Institute was supported by the Office of Science, Office of Biological and Environmental Research, of the US Department of Energy under Contract No. DE-AC02-05CH11231.

Author Contributions

CJP, TW, RS, and PCR conceived and designed the experiments. CJP, RS, and TW performed the experiments and analyzed the data. CJP, TW, RS, and PCR wrote the manuscript. All authors read and approved the final manuscript.

Competing Interests

The authors declare that they have no competing interests.

Accession Numbers

The rice cDNA locus identification numbers (MSU-RGAP; http://rice.plantbiology.msu.edu/) for *XB21* is Os12g36180. The GenBank accession number (http://www.ncbi.nlm.nih.gov/) for *XB21* cDNA sequence is AK072993.1.

Author details

[1]Department of Plant Pathology and the Genome Center, University of California Davis, Davis, CA 95616, USA. [2]Department of Bioresources Engineering and the Plant Engineering Research Institute, Sejong University, Seoul 05006, Republic of Korea. [3]School of Computational & Integrative Sciences, Jawaharlal Nehru University, New Delhi 110067, India. [4]Feedstocks Division, Joint BioEnergy Institute, Lawrence Berkeley National Laboratory, Berkeley, CA 94720, USA. [5]Environmental Genomics and Systems Biology, Lawrence Berkeley National Laboratory, Berkeley, CA 94720, USA.

References

Axtell MJ, Staskawicz BJ (2003) Initiation of RPS2-Specified Disease Resistance in Arabidopsis is Coupled to the AvrRpt2-Directed Elimination of RIN4. Cell 112(3):369–377

Bai T, Seebald JL, Kim KE, Ding HM, Szeto DP, Chang HC (2010) Disruption of Zebrafish Cyclin G-Associated Kinase (GAK) Function Impairs the Expression of Notch-Dependent Genes During Neurogenesis and Causes Defects in Neuronal Development. BMC Dev Biol 10:7. doi:10.1186/1471-213X-10-7

Bao YM, Sun SJ, Li M, Li L, Cao WL, Luo J, Tang HJ, Huang J, Wang ZF, Wang JF, Zhang HS (2012) Overexpression of the Qc-SNARE Gene OsSYP71 Enhances Tolerance to Oxidative Stress and Resistance to Rice Blast in Rice (Oryza Sativa L.). Gene 504(2):238–244. doi:10.1016/j.gene.2012.05.011

Bao YM, Wang JF, Huang J, Zhang HS (2008) Cloning and Characterization of Three Genes Encoding Qb-SNARE Proteins in Rice. Mol Genet Genomics 279(3):291–301. doi:10.1007/s00438-007-0313-2

Brett TJ, Traub LM, Fremont DH (2002) Accessory Protein Recruitment Motifs in Clathrin-Mediated Endocytosis. Structure 10(6):797–809

Cheetham ME, Caplan AJ (1998) Structure, Function and Evolution of DnaJ: Conservation and Adaptation of Chaperone Function. Cell Stress Chaperones 3(1):28–36

Chen F, Gao MJ, Miao YS, Yuan YX, Wang MY, Li Q, Mao BZ, Jiang LW, He ZH (2010a) Plasma Membrane Localization and Potential Endocytosis of Constitutively Expressed XA21 Proteins in Transgenic Rice. Mol Plant 3(5): 917–926. doi:10.1093/mp/ssq038

Chen X, Chern M, Canlas PE, Ruan D, Jiang C, Ronald PC (2010b) An ATPase Promotes Autophosphorylation of the Pattern Recognition Receptor XA21 and Inhibits XA21-Mediated Immunity. Proc Natl Acad Sci U S A 107(17): 8029–8034. doi:10.1073/pnas.0912311107

Chern M, Canlas PE, Fitzgerald HA, Ronald PC (2005) Rice NRR, a Negative Regulator of Disease Resistance, Interacts With Arabidopsis NPR1 and Rice NH1. Plant J 43(5):623–635. doi:10.1111/j.1365-313X.2005.02485.x

Danon A, Rotari VI, Gordon A, Mailhac N, Gallois P (2004) Ultraviolet-C Overexposure Induces Programmed Cell Death in Arabidopsis, Which is Mediated by Caspase-Like Activities and Which can be Suppressed by Caspase Inhibitors, p35 and Defender Against Apoptotic Death. J Biol Chem 279(1):779–787. doi:10.1074/jbc.M304468200

Dardick C, Ronald P (2006) Plant and Animal Pathogen Recognition Receptors Signal Through non-RD Kinases. PLoS Pathog 2(1):e2

Hagedorn EJ, Bayraktar JL, Kandachar VR, Bai T, Englert DM, Chang HC (2006) Drosophila Melanogaster Auxilin Regulates the Internalization of Delta to Control Activity of the Notch Signaling Pathway. J Cell Biol 173(3):443–452. doi:10.1083/jcb.200602054

Hatsugai N, Hillmer R, Yamaoka S, Hara-Nishimura I, Katagiri F (2016) The mu Subunit of Arabidopsis Adaptor Protein-2 Is Involved in Effector-Triggered Immunity Mediated by Membrane-Localized Resistance Proteins. Mol Plant Microbe Interact 29(5):345–351. doi:10.1094/MPMI-10-15-0228-R

Heese M, Gansel X, Sticher L, Wick P, Grebe M, Granier F, Jurgens G (2001) Functional Characterization of the KNOLLE-Interacting t-SNARE AtSNAP33 and its Role in Plant Cytokinesis. J Cell Biol 155(2):239–249. doi:10.1083/jcb.200107126

Hirst J, Sahlender DA, Li S, Lubben NB, Borner GH, Robinson MS (2008) Auxilin Depletion Causes Self-Assembly of Clathrin into Membraneless Cages in Vivo. Traffic 9(8):1354–1371. doi:10.1111/j.1600-0854.2008.00764.x

Jwa NS, Agrawal GK, Tamogami S, Yonekura M, Han O, Iwahashi H, Rakwal R (2006) Role of Defense/Stress-Related Marker Genes, Proteins and Secondary Metabolites in Defining Rice Self-Defense Mechanisms. Plant Physiol Biochem 44(5-6):261–273. doi:10.1016/j.plaphy.2006.06.010

Kabbage M, Dickman MB (2008) The BAG Proteins: A Ubiquitous Family of Chaperone Regulators. Cell Mol Life Sci 65(9):1390–1402. doi:10.1007/s00018-008-7535-2

Kampinga HH, Craig EA (2010) The HSP70 Chaperone Machinery: J Proteins as Drivers of Functional Specificity. Nat Rev 11(8):579–592. doi:10.1038/nrm2941

Kandachar V, Bai T, Chang HC (2008) The Clathrin-Binding Motif and the J-Domain of Drosophila Auxilin are Essential for Facilitating Notch Ligand Endocytosis. BMC Dev Biol 8:50. doi:10.1186/1471-213X-8-50

Karnik R, Grefen C, Bayne R, Honsbein A, Kohler T, Kioumourtzoglou D, Williams M, Bryant NJ, Blatt MR (2013) Arabidopsis Sec1/Munc18 Protein SEC11 is a Competitive and Dynamic Modulator of SNARE Binding and SYP121-Dependent Vesicle Traffic. Plant Cell 25(4):1368–1382. doi:10.1105/tpc.112.108506

Katari MS, Nowicki SD, Aceituno FF, Nero D, Kelfer J, Thompson LP, Cabello JM, Davidson RS, Goldberg AP, Shasha DE, Coruzzi GM, Gutierrez RA (2010) VirtualPlant: A Software Platform to Support Systems Biology Research. Plant Physiol 152(2):500–515. doi:10.1104/pp.109.147025

Kaur G, Subramanian S (2015) A Novel RING Finger in the C-Terminal Domain of the Coatomer Protein Alpha-COP. Biol Direct 10:70. doi:10.1186/s13062-015-0099-9

Koutras C, Braun JE (2014) J Protein Mutations and Resulting Proteostasis Collapse. Front Cell Neurosci 8:191. doi:10.3389/fncel.2014.00191

Lam BC, Sage TL, Bianchi F, Blumwald E (2001) Role of SH3 Domain-Containing Proteins in Clathrin-Mediated Vesicle Trafficking in Arabidopsis. Plant Cell 13(11):2499–2512

Lee DW, Zhao X, Yim YI, Eisenberg E, Greene LE (2008) Essential Role of Cyclin-G-Associated Kinase (Auxilin-2) in Developing and Mature Mice. Mol Biol Cell 19(7):2766–2776. doi:10.1091/mbc.E07-11-1115

Lee DW, Zhao X, Zhang F, Eisenberg E, Greene LE (2005) Depletion of GAK/Auxilin 2 Inhibits Receptor-Mediated Endocytosis and Recruitment of Both Clathrin and Clathrin Adaptors. J Cell Sci 118(Pt 18):4311–4321. doi:10.1242/jcs.02548

Lemmon SK (2001) Clathrin Uncoating: Auxilin Comes to Life. Curr Biol 11(2):R49–52

Lisak DA, Schacht T, Enders V, Habicht J, Kiviluoto S, Schneider J, Henke N, Bultynck G, Methner A (2015) The Transmembrane Bax Inhibitor Motif (TMBIM) Containing Protein Family: Tissue Expression, Intracellular Localization and Effects on the ER CA(2)(+)-Filling State. Biochim Biophys Acta 1853(9):2104–2114. doi:10.1016/j.bbamcr.2015.03.002

Liu JZ, Whitham SA (2013) Overexpression of a Soybean Nuclear Localized Type-III DnaJ Domain-Containing HSP40 Reveals its Roles in Cell Death and Disease Resistance. Plant J 74(1):110–121. doi:10.1111/tpj.12108

Lohse M, Bolger AM, Nagel A, Fernie AR, Lunn JE, Stitt M, Usadel B (2012) RobiNA: A User-Friendly, Integrated Software Solution for RNA-Seq-Based Transcriptomics. Nucleic Acids Res 40(Web Server issue):W622–627. doi:10.1093/nar/gks540

Mbengue M, Bourdais G, Gervasi F, Beck M, Zhou J, Spallek T, Bartels S, Boller T, Ueda T, Kuhn H, Robatzek S (2016) Clathrin-Dependent Endocytosis is Required for Immunity Mediated by Pattern Recognition Receptor Kinases. Proc Natl Acad Sci U S A 113(39):11034–11039. doi:10.1073/pnas.1606004113

Miki D, Shimamoto K (2004) Simple RNAi Vectors for Stable and Transient Suppression of Gene Function in Rice. Plant Cell Physiol 45(4):490–495

Park CJ, Bart R, Chern M, Canlas PE, Bai W, Ronald PC (2010) Overexpression of the Endoplasmic Reticulum Chaperone BiP3 Regulates XA21-Mediated Innate Immunity in Rice. PLoS One 5(2):e9262. doi:10.1371/journal.pone.0009262

Park CJ, Peng Y, Chen X, Dardick C, Ruan D, Bart R, Canlas PE, Ronald PC (2008) Rice XB15, a Protein Phosphatase 2C, Negatively Regulates Cell Death and XA21-Mediated Innate Immunity. PLoS Biol 6(9):e231. doi:10.1371/journal.pbio.0060231

Park M, Jurgens G (2011) Membrane Traffic and Fusion at Post-Golgi Compartments. Front Plant Sci 2:111. doi:10.3389/fpls.2011.00111

Peng Y, Bartley LE, Chen X, Dardick C, Chern M, Ruan R, Canlas PE, Ronald PC (2008) OsWRKY62 is a Negative Regulator of Basal and Xa21-Mediated Defense Against Xanthomonas Oryzae pv. Oryzae in Rice. Mol Plant 1(3):446–458, doi:ssn024

Piffanelli P, Zhou F, Casais C, Orme J, Jarosch B, Schaffrath U, Collins NC, Panstruga R, Schulze-Lefert P (2002) The Barley MLO Modulator of Defense and Cell Death is Responsive to Biotic and Abiotic Stress Stimuli. Plant Physiol 129(3):1076–1085. doi:10.1104/pp.010954

Pruitt RN, Schwessinger B, Joe A, Thomas N, Liu F, Albert M, Robinson MR, Chan LJ, Luu DD, Chen H, Bahar O, Daudi A, De Vleesschauwer D, Caddell D, Zhang W, Zhao X, Li X, Heazlewood JL, Ruan D, Majumder D, Chern M, Kalbacher H, Midha S, Patil PB, Sonti RV, Petzold CJ, Liu CC, Brodbelt JS, Felix G, Ronald PC (2015) The Rice Immune Receptor XA21 Recognizes a Tyrosine-Sulfated Protein from a Gram-Negative Bacterium. Sci Adv 1(6):e1500245. doi:10.1126/sciadv.1500245

Qiao Y, Jiang W, Lee J, Park B, Choi MS, Piao R, Woo MO, Roh JH, Han L, Paek NC, Seo HS, Koh HJ (2010) SPL28 Encodes a Clathrin-Associated Adaptor Protein Complex 1, Medium Subunit Micro 1 (AP1M1) and is Responsible for Spotted Leaf and Early Senescence in Rice (Oryza Sativa). New Phytol 185(1):258–274. doi:10.1111/j.1469-8137.2009.03047.x

Qiu XB, Shao YM, Miao S, Wang L (2006) The Diversity of the DnaJ/Hsp40 Family, the Crucial Partners for Hsp70 Chaperones. Cell Mol Life Sci 63(22):2560–2570. doi:10.1007/s00018-006-6192-6

Rajan VB, D'Silva P (2009) Arabidopsis Thaliana J-Class Heat Shock Proteins: Cellular Stress Sensors. Funct Integr Genomics 9(4):433–446. doi:10.1007/s10142-009-0132-0

Rehmany AP, Gordon A, Rose LE, Allen RL, Armstrong MR, Whisson SC, Kamoun S, Tyler BM, Birch PR, Beynon JL (2005) Differential Recognition of Highly Divergent Downy Mildew Avirulence Gene Alleles by RPP1 Resistance Genes from two Arabidopsis Lines. Plant Cell 17(6):1839–1850. doi:10.1105/tpc.105.031807

Rohila JS, Chen M, Chen S, Chen J, Ronald C, Dardick C, Canlas P, Xu X, Gribskov M, Kanrar S, Zhu J-K, Ronald P, Fromm ME (2006) Protein-protein interactions of tandem affinity purification-tagged protein kinases in rice. The Plant Journal 46 (1):1–13

Saibil H (2013) Chaperone Machines for Protein Folding, Unfolding and Disaggregation. Nature reviews 14(10):630–642. doi:10.1038/nrm3658

Sarkar NK, Thapar U, Kundnani P, Panwar P, Grover A (2013) Functional Relevance of J-Protein Family of Rice (Oryza Sativa). Cell Stress Chaperones 18(3):321–331. doi:10.1007/s12192-012-0384-9

Schwessinger B, Ronald PC (2012) Plant Innate Immunity: Perception of Conserved Microbial Signatures. Annu Rev Plant Biol 63:451–482. doi:10.1146/annurev-arplant-042811-105518

Sekhwal MK, Li P, Lam I, Wang X, Cloutier S, You FM (2015) Disease Resistance Gene Analogs (RGAs) in Plants. Int J Mol Sci 16(8):19248–19290. doi:10.3390/ijms160819248

Seo YS, Chern M, Bartley LE, Han M, Jung KH, Lee I, Walia H, Richter T, Xu X, Cao P, Bai W, Ramanan R, Amonpant F, Arul L, Canlas PE, Ruan R, Park CJ, Chen X, Hwang S, Jeon JS, Ronald PC (2011) Towards Establishment of a Rice Stress Response Interactome. PLoS Genet 7(4):e1002020. doi:10.1371/journal.pgen.1002020

Song WY, Wang GL, Chen LL, Kim HS, Pi LY, Holsten T, Gardner J, Wang B, Zhai WX, Zhu LH, Fauquet C, Ronald P (1995) A Receptor Kinase-Like Protein Encoded by the Rice Disease Resistance Gene, Xa21. Science 270(5243):1804–1806

Spielmeyer W, Lagudah S (2003) Homoeologous set of NBS-LRR Genes Located at Leaf and Stripe Rust Resistance Loci on Short Arms of Chromosome 1 of Wheat. Funct Integr Genomics 3(1-2):86–90. doi:10.1007/s10142-002-0074-2

Thimm O, Blasing O, Gibon Y, Nagel A, Meyer S, Kruger P, Selbig J, Muller LA, Rhee SY, Stitt M (2004) MAPMAN: A User-Driven Tool to Display Genomics Data Sets Onto Diagrams of Metabolic Pathways and Other Biological Processes. Plant J 37(6):914–939

Trapnell C, Pachter L, Salzberg SL (2009) TopHat: Discovering Splice Junctions With RNA-Seq. Bioinformatics 25(9):1105–1111. doi:10.1093/bioinformatics/btp120

Trapnell C, Williams BA, Pertea G, Mortazavi A, Kwan G, van Baren MJ, Salzberg SL, Wold BJ, Pachter L (2010) Transcript Assembly and Quantification by RNA-Seq Reveals Unannotated Transcripts and Isoform Switching During Cell Differentiation. Nat Biotechnol 28(5):511–515. doi:10.1038/nbt.1621

Ungewickell E, Ungewickell H, Holstein SE, Lindner R, Prasad K, Barouch W, Martin B, Greene LE, Eisenberg E (1995) Role of Auxilin in Uncoating Clathrin-Coated Vesicles. Nature 378(6557):632–635. doi:10.1038/378632a0

Wang YS, Pi LY, Chen X, Chakrabarty PK, Jiang J, De Leon AL, Liu GZ, Li L, Benny U, Oard J, Ronald PC, Song WY (2006) Rice XA21 Binding Protein 3 is a Ubiquitin Ligase Required for Full Xa21-Mediated Disease Resistance. Plant Cell 18(12):3635–3646. doi:10.1105/tpc.106.046730

Wick P, Gansel X, Oulevey C, Page V, Studer I, Durst M, Sticher L (2003) The Expression of the t-SNARE AtSNAP33 is Induced by Pathogens and Mechanical Stimulation. Plant Physiol 132(1):343–351. doi:10.1104/pp.102.012633

Wu B, Guo W (2015) The Exocyst at a Glance. J Cell Sci 128(16):2957–2964. doi:10.1242/jcs.156398

Wu G, Liu S, Zhao Y, Wang W, Kong Z, Tang D (2015) ENHANCED DISEASE RESISTANCE4 Associates With CLATHRIN HEAVY CHAIN2 and Modulates Plant Immunity by Regulating Relocation of EDR1 in Arabidopsis. Plant Cell 27(3):857–873. doi:10.1105/tpc.114.134668

Yun HS, Kwaaitaal M, Kato N, Yi C, Park S, Sato MH, Schulze-Lefert P, Kwon C (2013) Requirement of Vesicle-Associated Membrane Protein 721 and 722 for Sustained Growth During Immune Responses in Arabidopsis. Mol Cells 35(6):481–488. doi:10.1007/s10059-013-2130-2

Zhang L, Gjoerup O, Roberts TM (2004) The Serine/Threonine Kinase Cyclin G-Associated Kinase Regulates Epidermal Growth Factor Receptor Signaling. Proc Natl Acad Sci U S A 101(28):10296–10301. doi:10.1073/pnas.0403175101

8

Physiological and molecular responses of seedlings of an upland rice ('Tung Lu 3') to total submergence compared to those of a submergence-tolerant lowland rice ('FR13A')

Si-Yu Yang[1], Yu-Sian Wu[1], Chung-Tse Chen[2], Ming-Hsin Lai[3], Hsing-Mu Yen[3] and Chin-Ying Yang[1*]

Abstract

Background: Understanding the responses of rice to environmental stresses such as unscheduled submergence is of pressing important owing to increasing severity of weather thought to arise from global climate change. When rice is completely submerged, different types adopt either a quiescence survival strategy (i.e., minimal shoot elongation) or an escape strategy (i.e., enhanced shoot elongation). Each strategy can prolong survival depending on the circumstances. While submergence responses have been studied in rice typical of lowland and flood-prone areas, few studies have explored the physiological and molecular properties of upland rice under submergence. Here, we use seedlings of the upland rice 'Tung Lu 3' ('TL3') to analyze physiological and molecular responses to submergence. We compare them with those of 'FR13A', a lowland rice that tolerates submergence by adopting the quiescence strategy.

Results: Plant height and distance between leaf sheaths, increased rapidly in 'TL3' under submergence. Although this indicated a strong escape strategy the seedlings remained totally underwater for the duration of the experiments. In contrast, 'FR13A' elongated much less. Consequently, after 4 days complete submergence followed by drainage, 'TL3' lodged much more severely than 'FR13A'. After 10 d complete submergence, 55% of 'TL3' seedlings survived compared to 100% in 'FR13A'. Chlorophyll a, b and total chlorophyll concentrations of the 2nd oldest leaves of 'TL3' were also significantly above those of 'FR13A' (but were lower than 'FR13A' in the 3rd oldest leaves) and less hydrogen peroxide accumulated in 'TL3'. Peroxidase activity in submerged 'TL3' was also greater than in 'FR13A' 1 day after submergence. Quantitative RT–PCR showed increased expression of *sucrose synthase 1* and *alcohol dehydrogenases 1* after 2 days complete submergence with significantly higher levels in 'TL3' compared to 'FR13A'. Expression was also higher in 'TL3' under non-submerged conditions.

(Continued on next page)

* Correspondence: emiyang@dragon.nchu.edu.tw
[1]Department of Agronomy, National Chung Hsing University, Taichung 40227, Taiwan
Full list of author information is available at the end of the article

(Continued from previous page)

Conclusions: The upland rice line 'TL3' gave a stronger elongation response than 'FR13A' to complete submergence. This escape strategy is widely considered to prejudice survival when the plant remains totally submerged. However, contrary to expectations, 'TL3' survival rates were substantial although below those for 'FR13A' while physiological, biochemical and molecular parameters linked to adaptation differed in detail but appeared to be broadly comparable. These findings highlight that submergence tolerance is determine not only by the adoption of quiescence or escape strategies but maybe by metabolic and physiological properties unrelated to the underwater elongation rate.

Keywords: Upland rice, Submergence, ROS, Antioxidant enzyme activity, SUS1,

Background

The four major ecosystems in rice farming differ in terms of the amount of water required to provide a successful growing environment and can be characterized as follows: (I) the irrigated rice ecosystem. This comprises approximately 55% of the global rice land and produces 75% of the world's rice; (II) the rainfed lowland rice ecosystem. This comprises approximately 25% of the global rice land and produces 17% of global rice; (III) the upland rice ecosystem. This is typified by dry and infertile soil and makes-up approximately 13% of the global rice land area and produces 4% of global rice; and (IV) the flood-prone rice ecosystem. This is prevalent in South and Southeast Asia especially China, Cambodia, Bangladesh, and Thailand. Flood-prone ecosystems comprise approximately 7% of the global rice land area and produce 4% of global rice. Flood-prone rice is found mainly in river delta areas where deep flooding to 50–400 m is common during the growing season. In this ecosystem, rice is planted before the rainy season and stems later elongate with the rising river water with sufficient vigor to keep the upper shoot above water level (IRRI 1997; Ito et al. 1999).

Increased emissions of carbon dioxide and other 'greenhouse gases' such as methane are widely believed to result in global warming and accelerate global climate change (Karl and Trenberth 2003). Increases in global temperatures are causing sea levels to rise, rainfall and snowfall patterns to change and the frequency of extreme weather events to grow. One outcome is the greater prevalence of submergence stress in farming systems that severely depresses crop yields. This applies even to rice because of the severe restrictions submergence imposes for the inward movement of oxygen and carbon dioxide needed for respiration and photosynthesis. The gas transmission rate in water is much slower than in air, resulting in hypoxia and in photosynthetic deficiency (Gibbs and Greenway 2003). Various studies have revealed numerous changes in rice primary metabolism in the absence of oxygen. In particular, several reactions associated with sucrose metabolism and fermentation have been reported (Lakshmanan et al. 2014). Leaf photosynthetic capacity is also constrained under submergence stress because of the shading effect of water. The combined effects hinder leaf formation, reduce the total leaf area, promote leaf senescence, inhibit root growth, reduced tillering in rice and can be fatal (Kato et al. 2014).

'Escape' or 'quiescence' are two major types of adaptive response to submergence, seen in a wide range of species and based on the vigor of upward shoot elongation initiated by submergence (Ram et al. 2002). In most rice lines, the escape strategy takes the form of accelerated underwater elongation and is most marked in deep water lowland rice where vigorous stem extension prevents total submergence as floodwater gradually rises. However, leaves, stems and coleoptiles of almost all rice types elongate faster in response to submergence provided some oxygen is present. When fast shoot extension is insufficient to allow the plant to re-surface, the expenditure in energy and respirable substrates needed to support the faster underwater growth may threaten survival. The rice line 'FR13A' is an exception to this general rule. In contrast to most other rice lines 'FR13A' does not elongate faster underwater and thus conserves accumulated respirable biomass thereby adopting the so-called quiescence strategy (Singh et al. 2014). This has been linked to its ability to survive complete submergence for much longer than normal, mediated by energy and substrate conservation. This quiescence strategy is an outcome of insensitivity to the hormone ethylene, which normally promotes shoot elongation in rice (Jackson et al. 1987; Nagai et al. 2010; Perata and Voesenek 2007). The inheritable insensitivity to ethylene is located in the *Sub1* locus of chromosome 9 of 'FR13A'. It contains genes for Group VII apatala2/ethylene response factors. One of these (*Sub1A-1*) is a mutated form and responsible for suppressing ethylene responsiveness (Xu et al. 2006).

Upland rice possesses several characterstics that contribute to its tolerance of dry conditions (Bernier et al., 2007). For example, the roots of upland rice are thicker, larger, grow deeper and are more laterally spread. Roots of upland rice also have a higher osmolarity than roots of lowland rice and can thus retain moisture more effectively in dry soil (Asch et al. 2005; Ahmadi et al. 2014). Upland rice typically has a larger leaf area, fewer stomata, smaller thick-walled cells and larger vascular bundles compared to lowland rice. Although the term 'upland rice' is used widely, however episodes of waterlogging are also common in upland soils. There are many characteristics that vary continuously, with considerable overlap, between lowland and upland cultivars of rice (Chang et al. 1972; Colmer 2003). In contrast, little is known of the physiological and molecular properties of upland rice in relation to submergence stress. To rectify this shortcoming, we describe submergence responses of a typical upland rice line ('TL3') and compared them with those of a submergence-tolerant lowland rice ('FR13A'). We measured elongation, survival, activity of superoxide dismutase (SOD) and other enzymes associated with free radical scavaging, concentrations of total chlorophyll and expression levels of two genes associated with anaerobic metabolism *sucrose synthase 1* (*SUS1*) and *alcohol dehydrogenases 1* (*ADH1*). In this study, our results suggest that the upland rice line 'TL3' may exhibit some submergence tolerance, but that it is inferior to 'FR13A' in total submergence condition.

Results

'TL3' upland rice elongates its seedling shoot more than 'FR13A' when submerged but retains a substantial but inferior survival rate

Ten-day-old seedlings of 'TL3' and 'FR13A' were completely submerged for 2, 4, 6, 8, and 10 d and the phenotypic responses measured in terms of elongation, lodging and etiolation (Fig. 1a). After 4 days completely submergence, 'TL3' had lodging much more than 'FR13A' (Fig. 1a). The average shoot heights of 'TL3' were 19.7, 29.2, 37.8, 41.8, and 43.9 cm and those of the 'FR13A' were 17.8, 20.4, 27.5, 31.4, and 32.7 cm after complete submergence for 2, 4, 6, 8, and 10 d, respectively (Fig. 1b). Although extension growth by the leaf lamina is a major contributor to overall shoot height (Fig.1). Enhanced elongation by the leaf sheath also contributed to final shoot height of submerged 'TL3' (Fig. 2). In 'TL3', the length of the leaf sheath distance I (distance between 1st oldest leaf to the 2nd oldest leaf) (Fig.2a) was increased by up to 0.9 cm (Fig. 2b) and that from 2nd oldest leaf to the 3rd oldest leaf (distance II) was increased by up to 10.1 cm (Fig. 2c). Increases in leaf sheath lengths of 'FR13A' were very much less than this.

When seedlings were placed under complete submergence for 6, 8, and 10 d and then allowed to recover for 10 d, survival rate was assessed on the basis of an ability to form one or more new leaves. The survival rates of upland rice variety 'TL3' after 6, 8, or 10 d submergence were 100%, 84%, and 55%, respectively. Survival rates of the submergence tolerant 'FR13A', were, as expected, ~100% for all three submergence periods (Fig. 1c and d). The strongly increasing extension growth by 'TL3' during complete submergence is indicative of the escape strategy. But, in the absence of actual escape (the seedlings remained totally submerged during treatment) 55 % of 'TL3' seedlings survived even the longest period underwater (10d).

Submergence decreases leaf chlorophyll concentrations in both 'TL3' and 'FR13A' seedlings

Leaf color and chlorophyll concentration of plants under biotic stress (e.g., pathogens, insect pests) or abiotic stress (e.g., high temperature, low temperature, submergence and drought) are similar to those of plants undergoing natural senescence (Matile et al. 1996). To investigate changes in chlorophyll during submergence, 10-d-old 'TL3' and 'FR13A' seedlings were submerged for 6 d. The 2nd oldest leaves of both 'FR13A' and 'TL3' became yellow–green but visibly less so in 'TL3'. The 3rd oldest leaves of both lines yellowed to a similar extent although leaf apices remained green (Fig. 3a). Chlorophyll concentrations in 'TL3' and 'FR13A' were measured after submergence for 0, 2, 4, 6, 8, and 10 d (Fig. 3b). Chlorophyll a & b and total chlorophyll of the second oldest leaves decreased by about 90% in both lines after 10 d inundation. Concentrations were lower in 'FR13A' compared to 'TL3'. The reverse was true for the third oldest leaves. Here, chlorophyll loss was much stronger in 'TL3' throughout the 10d submergence treatment.

'TL3' accumulates less hydrogen peroxide (H_2O_2) during submergence than 'FR13A'

Nicotinamide adenine dinucleotide phosphate (NADPH) and H^+ derived from the light reactions of photosynthesis can accumulate in cell membranes generating superoxide anion radicals (O_2^-) in the chloroplast thylakoid. These are readily converted by superoxide dismutase (SOD) to H_2O_2 and stressed leaves are known to accumulate more H_2O_2 than unstressed leaves (Sagi and Fluhr 2006). To determine the amount of H_2O_2 accumulated under hypoxic stress, 10-d-old 'FR13A' and 'TL3' seedlings were submerged for 0, 6, and 10 d and H_2O_2 measured by 3,3′-diaminobenzidine staining. Staining of the second oldest leaf 2 was less in 'TL3'compared to 'FR13A' suggesting less H_2O_2 in 'TL3' (Fig. 4a). No reddish-brown staining was observed in the 3rd leaves of the 'FR13A' or 'TL3' rice. Cellular scavenging capacity of the antioxidative enzymes, such as catalase (CAT), ascorbate peroxidase (APX), SOD, and total peroxidase (POX) in 'FR13A' and 'TL3' rice seedlings were then determined after 2d submergence and again after 1 d recovery. Submergence decreased CAT activity and

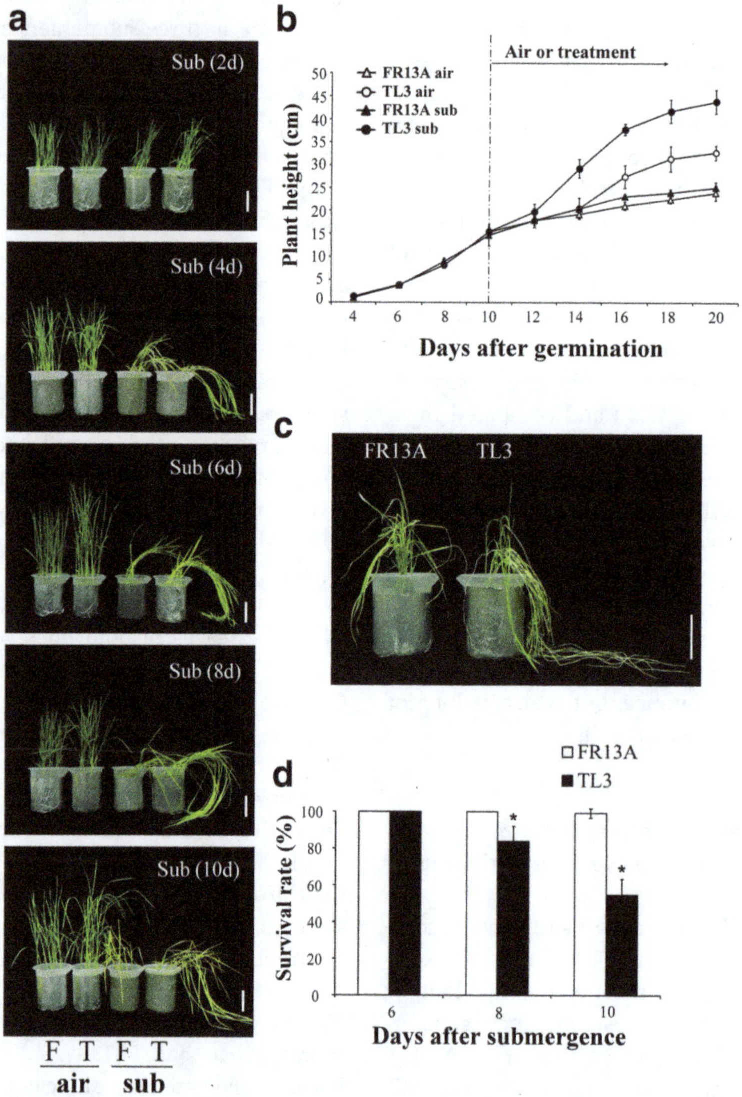

Fig. 1 Characterization of 'TL3' upland rice and 'FR13A' lowland rice seedlings grown under submergence conditions. **a** Photographs of 10-d-old upland and lowland rice seedlings grown in air or submerged for 2, 4, 6, 8, and 10 d. **b** Plant heights of 10-d-old upland and lowland rice seedlings grown in air or submerged for 2, 4, 6, 8, and 10 d. **c** Phenotypes of 10-d-old upland and lowland rice seedlings submerged for 10 d and then allowed to recover for 10 d. **d** Survival rates of seedlings after 6, 8, or 10 d submergence followed by 10 d recovery. Air, non-submerged controls; Sub, full submergence; 'TL3', upland rice; and 'FR13A', lowland rice. Bar = 7 cm. Data points are means ± standard deviation of three separate experiments each with at least 30 seedlings of each genotype. Asterisks indicate significant differences ($P < 0.05$; Student's *t*-test)

increased POX similarly in both 'FR13A'and 'TL3' (Fig. 4b). There was no effect of submergence on APX. However, SOD activity was significantly depressed by submergence but only in 'TL3'. After 1 d recovery, POX activity of 'TL3' was greater than in that of 'FR13A'.

Hypoxia-inducible gene expression levels in submerged 'TL3' and 'FR13A' seedlings

Previously published studies in rice show hypoxia can increase transcript levels of certain genes such as *SUS1* and *ADH1* that are involved in carbohydrate metabolism (Yang et al. 2011; Yang 2014). To assess the expression of these genes under submergence stress, 10-d-old 'FR13A' and 'TL3' seedlings were completely submerged for 2 d and transcript levels of *SUS1* and *ADH1* determined by quantitative RT–PCR. Levels of *SUS1* and *ADH1* transcripts increased under complete submergence stress in both 'FR13A' and 'TL3'. Stimulation was equally strong for *ADH1*. *SUS1* expression was approximately 3-fold higher in non-submerged 'TL3' compared to 'FR13A' but submergence raised transcripts more in 'FR13A' by 84% in 'TL3' and 228% in 'FR13A'. Thus, stimulation of levels of *SUS1* and *ADH1* expression by submergence was substantial in both 'TL3' and 'FR13A'.

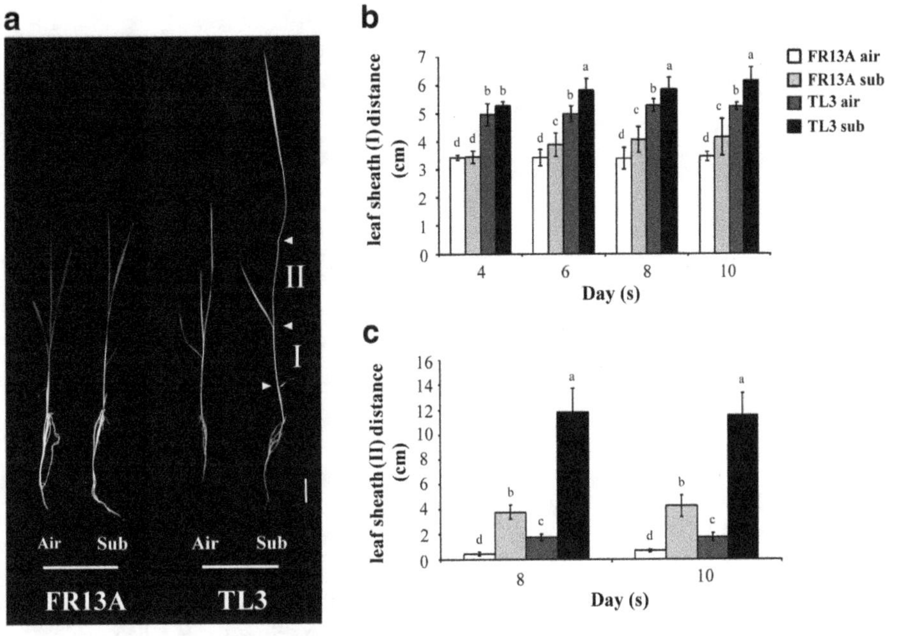

Fig. 2 Comparison of leaf sheath lengths in 'TL3' and 'FR13A' rice seedlings under submergence stress. **a** Photographs of 10-d-old upland and lowland rice seedlings after being exposed to air or submerged for 8 d. The distance from the 1st oldest leaf to the 2nd oldest leaf is leaf sheath distance I, and that from the 2nd oldest leaf to the 3rd oldest leaf is leaf sheath distance II. **b** and **c** Leaf sheath distances I and II of 10-d-old rice seedlings after submergence for 4, 6, 8, and 10 d. Air, non-submerged controls; Sub, full submergence; 'TL3', upland rice; and 'FR13A', lowland rice. Bar = 3 cm. Data points are means ± standard deviation of three separate experiments each with at least 30 seedlings of each genotype. Values with different letters are significantly different at $P < 0.05$, according to a post-hoc LSD test

Discussion

Most rice types elongates their shoot more rapidly when submerged and thus adopts the so-called escape strategy for survival. This reaction is seen in the leaves of small seedlings and is successful in promoting survival if, as a result, the shoot regains contact with the aerial environment. Exceptions to the rule are rice lines related to 'FR13A'. This is a mutated rice that originated in lowland India and was found to be unusually tolerant of complete submergence. The tolerance of FR13A is linked to its ability to conserve respirable reserves by not elongating underwater. This has been termed the quiescence strategy (Karin et al. 1982; Nagai et al. 2010; Perata and Voesenek 2007; Bailey-Serres and Voesenek 2008). However, the position regarding upland rice has not been explored before and we have therefore studied submergence responses of an upland rice cultivar ('TL3') and compared it to those of 'FR13A'. We found that 'TL3'elongated markedly compared to 'FR13A' and lodged severely after desubmergence (Fig. 1a and b). Elongation by the leaf sheath especially that between the second and third oldest leaves (distance II in Fig. 2) contributed to shoot lengthening. We further confirmed the submergence tolerance of 'FR13' and showed that although survival of 'TL3' is lower than that of 'FR13A' after 8–10 d of complete submergence, its survival ability remains substantial (Fig. 1c and d) despite substantial underwater shoot elongation. Therefore, for upland 'TL3' rice, the concepts of escape or quiescence strategies do not help understand the considerable resilience of 'TL3' to submergence stress (e.g., 84% survival after 8 d complete submergence). Therefore, 'TL3' may possess inherent metabolic tolerance to submergence that is probably linked to adaptive changes in its energy metabolism.

It is well-known that chloroplasts disintegrate and photosynthetic capacity is lost when rice is submerged (Mustroph et al. 2010). QTLs have been mapped to increased submergence tolerance through their effects on decreased underwater shoot elongation, to increased levels of chlorophyll and to loci of unknown function (Toojinda et al. 2003). One likely cause is membrane lipid peroxidation by $O_2^-$and H_2O_2. The latter is known to accumulate in low oxygen conditions (Santosa et al. 2007; Wu and Yang 2016). Accordingly, leaves of both the upland 'TL3' and lowland 'FR13A' yellowed when the seedlings were submerged. (Figure 1a). The rate of chlorophyll decline in the 2nd oldest leaves of 'TL3' was slower than in 'FR13A' but faster in the 3rd oldest leaf (Fig. 3). At the same time, accumulation of H_2O_2 in the 2nd oldest leaves of 'TL3' was less than in the 'FR13A' rice under submergence but greater in the 3rd oldest leaf (Fig. 4) thus linking higher H_2O_2 with greater loss of chlorophyll. Although chlorophyll in the 3rd oldest leaves of 'TL3' was significant lower than in 'FR13A', 80% of 'TL3' seedlings

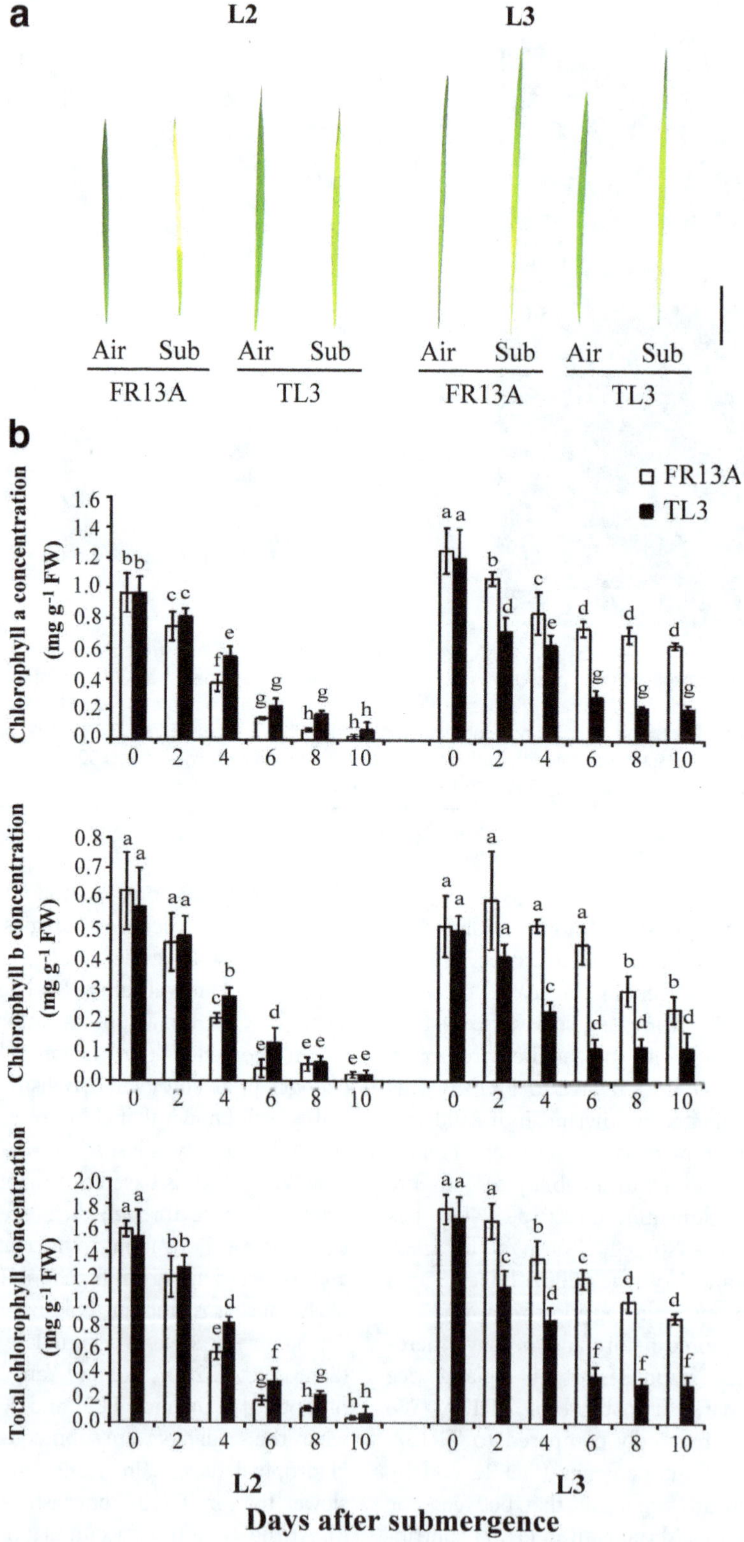

Fig. 3 Chlorophyll concentrations in 'TL3' and 'FR13A' rice seedlings under submerged conditions. **a** Photographs of 10-d-old upland and lowland rice seedlings after being exposed to air (control) or fully submerged (Sub) for 6 d. **b** The a, b and total chlorophyll concentration of 10-d-old seedlings after 0, 2, 4, 6, 8, and 10 d in air (control) or full submergence (Sub). L2, 2nd oldest leaf and L3, 3rd oldest leaf. Bar = 3 cm. Data points are means ± standard deviation of three separate experiments each with at least 30 seedlings of each genotype. Values with the different letters are significantly different ($P < 0.05$; post-hoc LSD test)

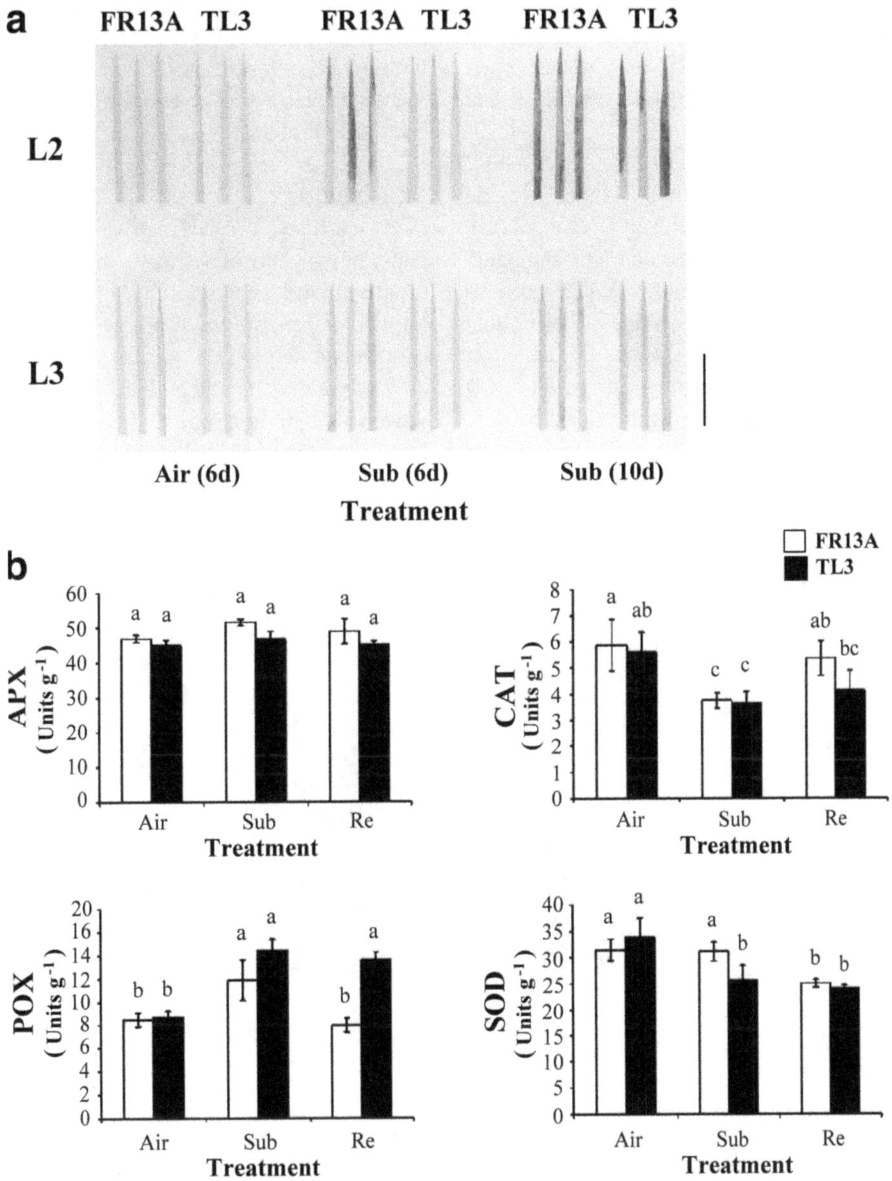

Fig. 4 Activities of antioxidative enzymes in 'TL3' and 'FR13A' rice seedlings under submerged conditions. **a** 3,3'-diaminobenzidine (DAB) staining of H_2O_2 in detached 2nd oldest leaves (L2) and 3rd oldest leaves (L3) of 10-d-old rice seedlings after 6 or 10 d in air (control) or full submergence (Sub). Similar results were obtained in three independent experiments. Photographs show the results of three independent leaves. Bar = 3 cm. **b** Enzyme activity of detached 2nd oldest leaves (L2) and 3rd oldest leaves (L3) of 10-d-old rice seedlings after 2d in air (control) or full submergence (Sub) and then allowed to recover for 1 d (Re). Ascorbate peroxidase (APX), catalase (CAT), total peroxidase (POX) and superoxide dismutase (SOD) activities are average values ± standard deviation of four biologically independent experiments each with duplicate samples. Values with the different letters are significantly different at $P < 0.05$, according to a post-hoc LSD test

survived 8 d submergence (Fig. 1d), thus demonstrating considerable resilience within the limitations of this study in not including a "submergence sensitive" control.

Potentially damaging reactive oxygen species (ROS) are generated more freely in plants under abiotic stress but elevated activities of antioxidant enzymes can reduce the injury and increase stress tolerance (Gill and Tuteja 2010). In our plants, total peroxidase activity (POX) increased in both 'FR13A'and 'TL3' during submergence and this may have lowered H_2O_2 sufficiently to contribute to their tolerance of submergence and especially so after submergence when activity in 'TL3' was maintained. However, the activity of catalase (CAT), an enzyme that degrades H_2O_2 to H_2O, declined in both 'TL3' and 'FR13A' during submergence (Fig. 4b) suggesting a minor role in regulating H_2O_2 levels. Activity of SOD, the enzyme responsible for converting highly active O_2^- was substantial in both rice lines and was little changed by submersion. Overall, higher H_2O_2 is a likely promoter of leaf senescence in submerged 'TL3' and 'FR13A' but assays of scavanging enzymes did not offer a clear explanation for this or for the different patterns of accumulation between the two lines and between older and younger leaves.

In non-submerged conditions, plants generate ATP through aerobic respiration of assimilates. Under hypoxic stress, the sucrose synthase (SUS) pathway is upregulated, and invertase activity and gene expression are inhibited. These changes can reduce consumption of ATP used to process sucrose into glycolysis and thus enhance plant survival possibly in conjunction with an enhanced fermentation pathway (Bieniawska et al. 2007). Our quantitative RT–PCR results revealed that *SUS1* expression was increased by submergence in both rice lines, with a proportionately much larger increase in 'FR13A'. However, in absolute terms, expression in 'TL3' was significantly greater than that in 'FR13A' regardless of submergence treatment (Fig. 5). Expression of *ADH1* was strongly increased by submergence indicating that fermentation was probably active during submergence. However, the levels of activity were similar in both 'TL3' and 'FR13A'.

Conclusions

We quantified survival rates and associated changes in plant height and leaf sheath length during submergence of an upland rice ('TL3') in comparison with a well-known submergence-tolerant lowland rice ('FR13A'). The tolerance of 'FR13A' was verified and linked to slow rates of underwater leaf and leaf sheath elongation (the quiesence strategy). In contrast, the upland 'TL3' elongated rapidly when submerged (the escape strategy). However, even though this extension growth was insufficient to regain contact with the air, survival rates exceeded 50% even after 10 d,

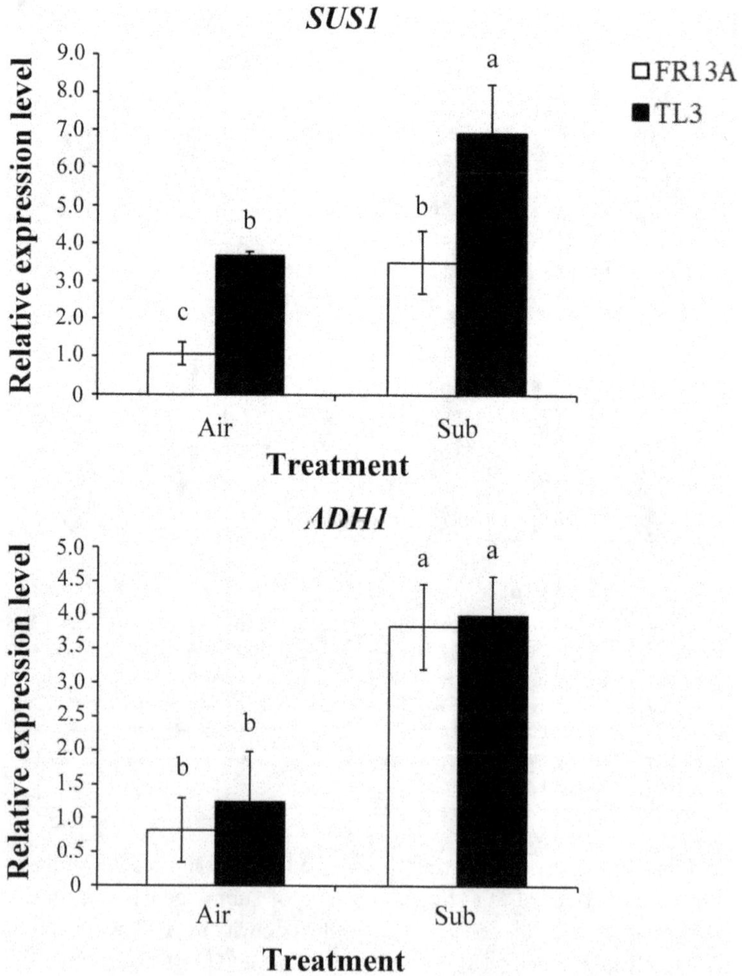

Fig. 5 Transcript levels of hypoxia-inducible genes in 'TL3' and 'FR13A' rice seedlings under submerged conditions. Total RNAs were isolated from shoots of 10-d-old 'TL3' upland rice and 'FR13A' lowland rice seedlings after 2d in air (control) or full submergence (Sub). Transcript levels of hypoxia-inducible *sucrose synthase 1* (*SUS1*) and *alcohol dehydrogenases 1* (*ADH1*) were determined using quantitative RT–PCR. Relative transcript levels were calculated and normalized to ubiquitin mRNA. Values are means ± standard deviation of three biologically independent experiments each with duplicate samples. Values with the different letters are significantly different at $P < 0.05$, according to a post-hoc LSD test

suggest a metabolic rather than morphological basis for the tolerance. Submergence promoted chlorosis in both rice lines although older leaves lost more chlorophyll in 'FR13A' and younger leaves lost more in 'TL3'. This pattern was reflected by H_2O_2 levels which were higher in the more chlorotic leaves. Additionally, POX activity was raised by submergence but to a similar extent in both lines although this fell-away more in 'FR13A' during post-submergence recovery. The hypoxia-related genes *SUS1* and *ADH1* were expressed more strongly under submergence in both rice lines. Overall, the regulatory mechanisms may explain the partial tolerance to submergence, whereby 'TL3' elongates in total submergence and thus expending considerable amounts of energy remain unclear and merit further study.

Methods

Plant materials and growth conditions

The submergence-tolerant rice variety 'FR13A' and upland rice variety 'Tung Lu 3' ('TL3') were used. Seeds were obtained from Dr. M-H Lai (Taiwan Agricultural Research Institute, Taiwan). Seeds were surface sterilized with 3% sodium hypochlorite for 30 min, washed several times with sterile deionized water and placed on wet filter paper for 3 d for germination in a growth chamber set at 28 °C with 6-h light (236 μmol m^{-2} s^{-1})/ 8-h dark cycle. Germinated seeds were transplanted onto a metal grid over a 500 mL beaker containing Kimura B solution (Yoshida et al. 1976). A total of 40 germinated seeds were placed on the grid of each beaker and the water culture solution changed every 2 d for 10 d (2–3 leaf stage). A total of 30 evenly-sized seedlings from each beaker were selected for subsequent experiments.

Seedling submergence treatments and survival rate determination

For submergence treatments, 10-d-old seedlings were placed in a water tank (40 cm × 40 cm × 60 cm) filled with 55 cm of water. At this depth, plants could not elongate sufficiently to reach the water surface within 10 d. Seedlings were subjected to complete submergence for 0, 2, 4, 6, 8, and 10 d under a 16-h light/8-h dark cycle. After each treatment, sampled tissues were immediately frozen in liquid nitrogen and stored at −80 °C. The ability to grow new leaves within 7 d of being submerged for 6, 8, and 10 d was taken as a measure of survival. Experiments were repeated three times and at least 30 seedlings were measured independently each time.

Plant height, leaf sheath length and chlorophyll concentration measurements

After treatment, lengths of each shoot and leaf sheaths of at least 30 plants were recorded at intervals indicated. Chlorophyll a and b were extracted from 50 mg of leaf tissue in 2 mL of sodium phosphate buffer (50 mM pH 6.8), 40 μL of which was added to 960 μL of 99% ethanol and incubated for 30 min at room temperature in the dark with gentle shaking. After centrifugation at 4 °C for 15 min at 1000 g, the absorbance values of the supernatant were measured at 665 and 649 nm with a spectrophotometer (Metertec SP8001).

Histochemical staining and antioxidative enzyme activity

Accumulation of H_2O_2 in cells was visualized by 3,3′-diaminobenzidine staining as previously described (Yang and Hong 2015). The experiments were repeated three times. For the antioxidative enzyme assays, seedlings were first subjected complete submergence for 2 d and then allowed to recover for additional 1 d. Control plants remained unsubmerged. Shoot tissue (50 mg) was excised and immediately used for enzyme extractions. Activity levels of CAT, APX, POX and SOD were analyzed as previously described (Wu and Yang 2016). Each experiment was repeated four times.

Quantitative RT–PCR analyses

Shoot samples were collected from 10-d-old seedlings and frozen in liquid nitrogen and stored in −80 °C until use. Total RNA was extracted using TRIzol (Invitrogen, Carlsbad, CA, USA) and then subjected to DNase treatment using the TURBO DNA-free Kit (Ambion, Austin, TX, USA). RNA concentrations were determined spectrophotometrically and then reverse transcribed into cDNA using Moloney murine leukemia virus reverse transcriptase (Invitrogen). Quantitative RT–PCR was performed using a Rotor-Gene 3000 instrument (Corbett Research, Sydney, Australia) with Power SYBR Green PCR Master Mix (GeneMark, Taipei, Taiwan) in accordance with the manufacturer's recommendations. The ubiquitin (*Os03g13170*) gene was used as an internal control to normalize cDNA levels. Relative expression levels were analyzed with Rotor-Gene 6 software (Corbett). Experiments were repeated three times independently with duplicate samples. The sequences of primers used for quantitative RT–PCR are in Table 1.

Table 1 Primers used for quantitative RT-PCR experiments

Gene name	Primer sequence
OsSUS1- forward	5′-catctcaggctgagactctga −3′
OsSUS1- reverse	5′- caaattcaatcgaccttactt −3′
OsADH1- forward	5′-gcaaatttctggctttgtcaatcagta −3′
OsADH1- reverse	5′-cgccaaaagatcactgattcttaacaa −3′
Osubiquitin - forward	5′-aaccagctgaggcccaaga-3′
Osubiquitin - reverse	5′-acgattgatttaaccagtccatga-3′

Abbreviations

APX: Ascorbate peroxidase; CAT: Catalase; H_2O_2: Hydrogen peroxide; POX: Total peroxidase; ROS: Reactive oxygen species; RT–PCR: Reverse transcriptase-polymerase chain reaction; SOD: Superoxide dismutase

Acknowledgments

This work was supported by a National Science Council grant (NSC 101-2311-B-005-001) to Chin-Ying Yang.

Authors' contributions

S-Y Yang, Y-S Wu and C-T Chen carried out the experiments and analyzed the data. M-H Lai and H-M Yen contributed to preparation of plant materials. C-Y Yang defined the research theme, designed experiments, and wrote the manuscript. All authors read and approved the final manuscript.

Competing interests

The authors declare that they have no competing interests.

Author details

[1]Department of Agronomy, National Chung Hsing University, Taichung 40227, Taiwan. [2]Graduate Institute of Biotechnology, National Chung Hsing University, Taichung 40227, Taiwan. [3]Crop Science Division, Taiwan Agricultural Research Institute, Taichung 41362, Taiwan.

References

Ahmadi N, Audebert A, Bennett MJ, Bishopp A, de Oliveira AC, Courtois B, Diedhiou A, Dievart A, Gantet P, Ghesquiere A, Guiderdoni E, Henry A, Inukai Y, Kochian L, Laplaze L, Lucas M, Luu DT, Manneh B, Mo XR, Muthurajan R, Perin C, Price A, Robin S, Sentenac H, Sine B, Uga Y, Very AA, Wissuwa M, Wu P, Xu J (2014) The roots of future rice harvests. Rice 7:29

Asch F, Dingkuhn M, Sow A, Audebert A (2005) Drought-induced changes in rooting patterns and assimilate partitioning between root and shoot in upland rice. Field Crop Res 93(2–3):223–236

Bailey-Serres J, Voesenek LACJ (2008) Flooding stress: acclimations and genetic diversity. Annu Rev Plant Biol 59:313–339

Bernier J, Kumar A, Ramaiah V, Spaner D, Atlin G (2007) A large-effect QTL for grain yield under reproductive-stage drought stress in upland rice. Crop Sci 47(2):507–518

Bieniawska Z, Barratt DHP, Garlick AP, Thole V, Kruger NJ, Martin C, Zrenner R, Smith AM (2007) Analysis of the sucrose synthase gene family in Arabidopsis. Plant J 49(5):810–828

Chang TT, Loresto GC, Tagumpay O (1972) Agronomic and growth characteristics of upland and lowland rice varieties. In: Rice breeding. International Rice Research Institute, Los Banos, Philippines, pp 645–661

Colmer TD (2003) Aerenchyma and an inducible barrier to radial oxygen loss facilitate root aeration in upland, paddy and deep-water rice (Oryza sativa L.) Ann Bot 91(2):301-309

Gibbs J, Greenway H (2003) Mechanisms of anoxia tolerance in plants. I. Growth, survival and anaerobic catabolism (vol 30, pg 1, 1993). Funct Plant Biol 30(3):353–U356

Gill SS, Tuteja N (2010) Reactive oxygen species and antioxidant machinery in abiotic stress tolerance in crop plants. Plant Physiol Biochem 48(12):909–930

IRRI (International Rice Research Institute) (1997) Rice Almanac, 2nd edn. International Rice Research Institute, Los Banos, Philippines, p 181

Ito O, Ella E, Kawano N (1999) Physiological basis of submergence tolerance in rainfed lowland rice ecosystem. Field Crop Res 64(1–2):75–90

Jackson MB, Waters I, Setter T (1987) Injury to rice plants caused by complete submergence: a contribution by ethylene (ethene). J Exp Bot 38:1826–1838

Karin S, Vergara BS, Mazaredo AM (1982) Anatomical and morphological studies of rice varieties tolerant of and susceptible to complete submergence at seedling stage. In: proceedings of the 1981 Intemational Deepwater Rice workshop. Temational Rice research institute, los Bafios, the Philippines, pp 287–291

Karl TR, Trenberth KE (2003) Modern global climate change. Science 302(5651): 1719–1723. https://doi.org/10.1126/science.1090228

Kato Y, Collard BCY, Septiningsih EM, Ismail AM (2014) Physiological analyses of traits associated with tolerance of long-term partial submergence in rice. AoB PLANTS 6:plu058

Lakshmanan M, Mohanty B, Lim SH, Ha SH, Lee DY (2014) Metabolic and transcriptional regulatory mechanisms underlying the anoxic adaptation of rice coleoptiles. AoB PLANTS 6:plu026

Matile P, Hortensteiner S, Thomas H, Krautler B (1996) Chlorophyll breakdown in senescent leaves. Plant Physiol 112(4):1403–1409

Mustroph A, Lee SC, Oosumi T, Zanetti ME, Yang HJ, Ma K, Yaghoubi-Masihi A, Fukao T, Bailey-Serres J (2010) Cross-kingdom comparison of Transcriptomic adjustments to low-oxygen stress highlights conserved and plant-specific responses. Plant Physiol 152(3):1484–1500

Nagai K, Hattori Y, Ashikari M (2010) Stunt or elongate? Two opposite strategies by which rice adapts to floods. J Plant Res 123(3):303–309

Perata P, Voesenek LACJ (2007) Submergence tolerance in rice requires Sub1A, an ethylene-response-factor-like gene. Trends Plant Sci 12(2):43–46

Ram PC, Singh BB, Singh AK, Ram P, Singh PN, Singh HP, Boamfa I, Harren F, Santosa E, Jackson MB, Setter TL, Reuss J, Wade LJ, Singh VP, Singh RK (2002) Submergence tolerance in rainfed lowland rice: physiological basis and prospects for cultivar improvement through marker-aided breeding. Field Crop Res 76:131–152

Sagi M, Fluhr R (2006) Production of reactive oxygen species by plant NADPH oxidases. Plant Physiol 141(2):336–340

Santosa IE, Ram PC, Boamfa EI, Laarhoven LJ, Reuss J, Jackson MB, Harren FJ (2007) Patterns of peroxidative ethane emission from submerged rice seedlings indicate that damage from reactive oxygen species takes place during submergence and is not necessarily a post-anoxic phenomenon. Planta 226(1):193–202

Singh S, Mackill DJ, Ismail AM (2014) Physiological basis of tolerance to complete submergence in rice involves genetic factors in addition to the SUB1 gene. AoB PLANTS 6: plu060

Toojinda T, Siangliw M, Tragoonrung S, Vanavichit A (2003) Molecular genetics of submergence tolerance in rice: QTL analysis of key traits. Ann Bot 91(2):243–253

Wu YS, Yang CY (2016) Physiological responses and expression profile of NADPH oxidase in Rice (Oryza Sativa) seedlings under different levels of submergence. Rice 9(1):2

Xu K, Xu X, Fukao T, Canlas P, Maghirang-Rodriguez R, Heuer S, Ismail AM, Bailey-Serres J, Ronald PC, Mackill DJ (2006) Sub1A is an ethylene-response-factor-like gene that confers submergence tolerance to rice. Nature 442(7103):705–708

Yang CY (2014) Hydrogen peroxide controls transcriptional responses of ERF73/HRE1 and ADH1 via modulation of ethylene signaling during hypoxic stress. Planta 239(4):877–885

Yang CY, Hong CP (2015) The NADPH oxidase Rboh D is involved in primary hypoxia signalling and modulates expression of hypoxia-inducible genes under hypoxic stress. Environ Exp Bot 115:63–72. https://doi.org/10.1016/j.envexpbot.2015.02.008

Yang CY, Hsu FC, Li JP, Wang NN, Shih MC (2011) The AP2/ERF transcription factor AtERF73/HRE1 modulates ethylene responses during hypoxia in Arabidopsis. Plant Physiol 156(1):202–212

Yoshida S, Forno DA, Cock JA, Gomez KA (1976) Laboratory manual for plant physiological studies of Rice, Ed 3. International Rice Research Institute, Manila, Philippines

9

Delineation of Inheritance Pattern of Aleurone Layer Colour Through Chemical Tests in Rice

Chandu Singh[1,2*], Sripathy K.V.[1], Jeevan Kumar S.P.[1], Bhojaraja Naik K.[1], Govind Pal[1], Udaya Bhaskar K.[1], Ramesh K.V.[1] and Somasundaram G.[1]

Abstract

Background: Rice aleurone layer develops different colours with various chemical tests that may help to develop some rapid tests for identification/grouping of rice varieties. Understanding the colour inheritance pattern could enable to develop chemical clues that may help for genetic purity analysis along with grow-out-test.

Results: In this study, inheritance pattern of aleurone layer colour was studied in parents, F_1 and F_2 progenies derived from the crosses IR 36 × Acc. No. 2693 and IR 64 × Acc. No. 2693. The parent IR 36 showed light yellow (NaOH/KOH) and brown (phenol/modified phenol test) colour; whereas, Acc. No. 2693 revealed wine red/dark wine red (NaOH/KOH) and light brown colour/no reaction (phenol/modified phenol test). In contrary, another parent IR 64 exhibited light yellow (KOH/NaOH) and dark brown (phenol, modified phenol) colour. Both the F_1 showed an intermediate light wine red colour (NaOH/KOH) and dark brown (phenol and modified phenol) colour, which is dominant over their one of the parents. The colour pattern with standard phenol/modified phenol, NaOH and KOH tests in F_2 progenies of both the crosses showed 9:7 (complementary gene interaction) and 11:5 ratios (reciprocal dominance modification of recessive alleles), respectively.

Conclusions: Our findings clearly elucidate the colour inheritance pattern in rice that may facilitate to develop rapid chemical tests to identify/ group the varieties for genetic purity analysis.

Keywords: Aleurone layer, Alleles, Complementary gene action, Duplicate gene action, Rice

Background

Chemical tests in rice, so far have been extrapolated for identification/characterization of cultivars on the basis of colour pattern in aleurone layer that might developed through enzyme mediated reactions. Aleurone layer is a living entity, which constitute outermost layer of endosperm, specialised in de novo synthesis of reserve mobilizing enzyme complex during seed germination process (Kumar et al. 2015). In addition, the aleurone layer is also involved in the synthesis of oxidase enzymes such as laccase, tyrosinase, polyphenol oxidase, monophenol oxidase and horse-radish peroxidase, which catalyzes to form a colour reaction (Cabaj et al. 2010; Fernandes et al. 2005). Among these oxidases, polyphenol oxidase (PPO) is one of the enzyme that is involved in oxidation of phenol colour reaction through formation of brown coloured melanin pigment (Steffens et al. 1994; Kumar et al. 2016). Polyphenol oxidases avail molecular oxygen, which undergoes hydroxylation and dehydrogenation of phenolic compounds to form reactive *o*-quinones. These *o*-quinones alkylate nucleophilic groups and self-polymerize to form melanin polymers (Fuerst et al. 2014). Role of oxidases is reported to be multifaceted, wherein they are involved in potential seed defense pathways and located in aleurone layer as indicated by their increased levels in the aleuronic fraction (Fraignier et al. 2000; Kumar et al. 2017a; Sinha et al. 2016). Proteomic analysis of aleurone layer in wheat suggested the presence of oxidases along with proteins involved in metabolism (Jerkovic et al. 2010; Kumar et al. 2017b).

* Correspondence: chandusinghrathod@gmail.com
[1]ICAR- Indian Institute of Seed Science, Mau, Uttar Pradesh 275103, India
[2]Seed Production Unit, ICAR-Indian Agricultural Research Institute, New Delhi 110012, India

In rice, genetic studies on inheritance of colour formation in aleurone layer have been limited. However, biochemical aspects of various enzymes secreted through aleurone layer especially PPO and other reserve mobilizing enzymes have been well studied. Many workers (Joshi and Banerjee 1970; Joshi and Banerjee 1969; Jimenez and Dubcovsky 1999) studied the colour formation of wheat polyploids that had developed colour when reacted with phenol and tyrosine substrates mediated by PPO. Further, reactions of whole-wheat seeds with phenol (Joshi et al. 1969: Maguire et al. 1975) and catechol (Milner and Gould 1951) have been used for cultivar purity testing. Miczynski (1938) reported the presence of one or two genes in bread wheat, which controlled the phenol colour reaction. Moreover, various chemical tests were used so far in identification and differentiation of crop varieties based on the colour differences generated in the aleurone layer when reacted with different chemical tests. Besides, genetic background of the varieties also plays an important role in the identification of varieties. The colour formation by enzyme system has been reported in pearlmillet (Varier et al. 1995), foxtail millet (Pallares et al. 2004), sorghum (Thangavel et al. 2005), rice (Mor et al. 2006; Dileepkumar et al., 2015; Chandu et al. 2017) and wheat (Joshi et al. 2007), respectively. Chemical tests such as $FeSO_4$ (Pallares et al. 2004), KOH (Mor et al. 2006), phenol and modified phenol tests (Joshi et al. 2007; Banerjee and Chandra 1977) etc. were studied for the development of seed keys. The purpose of the present study is to determine the inheritance pattern and segregation of colour formation trait in aleurone layer of rice using parents, F_1 and F_2 progenies derived from the crosses IR 36 × Acc. No. 2693 and IR 64 × Acc. No. 2693 by chemical tests.

Results and Discussion

Aleurone Layer Colouration in Parents IR 36 × Acc. No. 2693 and IR 64 × Acc. No. 2693

Studies on colour formation of aleurone layer revealed that IR 36 showed light yellow with NaOH and KOH tests, whereas, phenol and modified phenol tests recorded brown colour. The Acc. No. 2693 recorded wine red / dark wine red colour with NaOH, KOH and light brown colour/ no reaction colour with phenol and modified phenol tests, respectively. Similarly, IR 64 recorded light yellow colour with NaOH and KOH tests; while phenol and modified phenol tests were recorded dark brown colour as shown in Table 1. Colour formation with phenol test is depicted in Fig. 1, where the genotypes were grouped based on the biochemical tests and were in congruence with other studies (Thangavel et al. 2005; Nethra et al. 2007; Vijayalakshmi and Vijay 2009; Singh et al. 2011; Anitalakshmi et al. 2014; Kumar et al. 2015).

Table 1 Aleurone layer color reaction of parents and F_1 to different chemical tests

Chemical tests	IR 36	IR 64	Acc. No. 2693	F_1
NaOH	LY	LY	WR	LWR
KOH	LY	LY	DWR	LWR
Phenol	B	DB	LB/NC	DB
Modified Phenol	B	DB	LB/NC	DB

LY Light Yellow, *WR* Wine Red, *DWR* Dark Wine Red, *LWR* light wine red, *LB* Light Brown, *B* Brown, *DB* Dark Brown, *NC* No reaction

Aleurone Layer Colour Inheritance in F_1 Plant of IR 36 × Acc. No. 2693 and IR 64 × Acc. No. 2693

The F_1s were derived from the crosses IR 36 × Acc. No. 2693 and IR64 × Acc. No. 2693, respectively (Table 1). The freshly harvested F_1 seeds (derived from cross IR 36 × Acc. No. 2693, as depicted in Fig. 2) showed light wine red colour with NaOH and KOH tests; while brown colour was recorded with phenol and modified phenol tests, respectively.

Further, the freshly harvested F_1 seeds derived from the cross IR 64 × Acc. No. 2693 (Fig. 3) showed light wine red colour with NaOH and KOH tests, respectively. In contrary, reaction with phenol and modified phenol tests showed dark brown colour, respectively. NaOH and KOH tests showed intermediate colour segregation i.e. light wine red colour compared with parents. In the same way, in case of phenol and modified phenol tests showed dark brown colour from cross IR 64 × Acc. No. 2693 and the similar colour pattern was recorded in case of IR 36 × Acc. No. 2693 that implies that the observed colour is dominant over light yellow.

Aleurone Layer Colour Inheritance in F_2 Progenies Derived from IR 36 × Acc. No. 2693 and IR 64 × Acc. No. 2693 with Standard Phenol and Modified Phenol (copper sulphate) Tests

Out of 484 F_2 progenies of cross IR 36 × Acc. No. 2693; 273 and 255 F_2 progenies were brown/dark brown in colour, whereas 211 and 229 F_2 progenies showed light brown/no colour with standard phenol and modified phenol tests, respectively. Further, 420 F_2 progenies derived from cross IR 64 × Acc. No. 2693 were evaluated. It is observed that 251 and 246 F_2 progenies were brown/dark brown in colour; 169 and 174 F_2 progenies showed light brown/no reaction with standard phenol and modified phenol tests, respectively. Therefore, the investigation revealed that the colour trait in the aleurone layer of F_2 progenies were segregated with complementary gene interaction with a ratio of 9:7 indicates a goodness of fit with observed ratio (Table 2). The F_2 colour segregation was consistent with the complementary gene interaction (9:7) for all F_2 progenies of both the crosses. Therefore, two major genes and their alleles

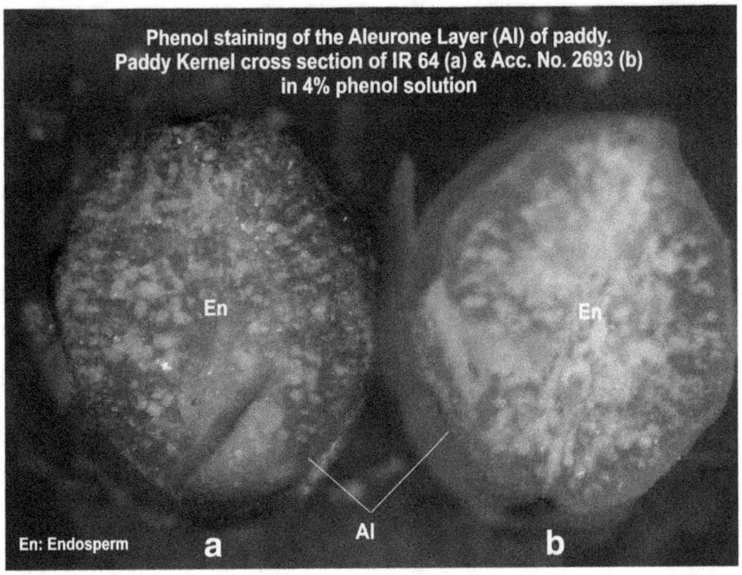

Fig. 1 Cross section of rice kernel derived from IR 64 (coloured) and Acc. No. 2693 after subjection to phenol test. Phenol staining of the Aleurone Layer (AL) of paddy. Paddy Kernel cross section of IR 64 (**a**) & Acc. No. 2693 (**b**) in 4% phenol solution

with complementary gene action controls the colour formation in aleurone layer.

The development of colour in aleurone layer (brown/ dark brown) with standard phenol and modified phenol tests (copper sulphate) in rice requires the presence of two dominant genes, B_1 and D_2 e.g. B_1B_1 D_2D_2 (Fig. 4). When either B_1 (e.g. b_1b_1 D_2D_2) or D_2 (B_1B_1 d_2d_2) or both the genes (e.g. b_1b_1 d_2d_2) are present in homozygous recessive condition, brown/ dark brown colour cannot be produced; as a result, light brown / no reactions were obtained. The brown/dark brown colour variety (IR 36 and IR 64) of rice (B_1B_1 D_2D_2) was crossed to a light brown/no reaction variety (Acc. No 2693) with the genotype b_1b_1 d_2d_2 showed dark brown colour in the F_1 (B_1b_1 D_2d_2) progeny. In the F_2 progenies, on an average nine progenies had one dominant allele of both the genes B_1 and D_2. These plants, therefore has brown/ dark brown colour. Three, out of sixteen F_2 progenies, shall have dominant B_1 but had homozygous recessive genotype b_1b_1; three others shall have dominant D_2 but had homozygous recessive genotype d_2d_2, while one plant had both the genes in homozygous recessive genotype b_1b_1 d_2d_2. All these (seven progenies) had light brown/no reaction. In this type of gene interaction, the production of one of the two phenotypes of colour trait requires the presence of dominant alleles of both the

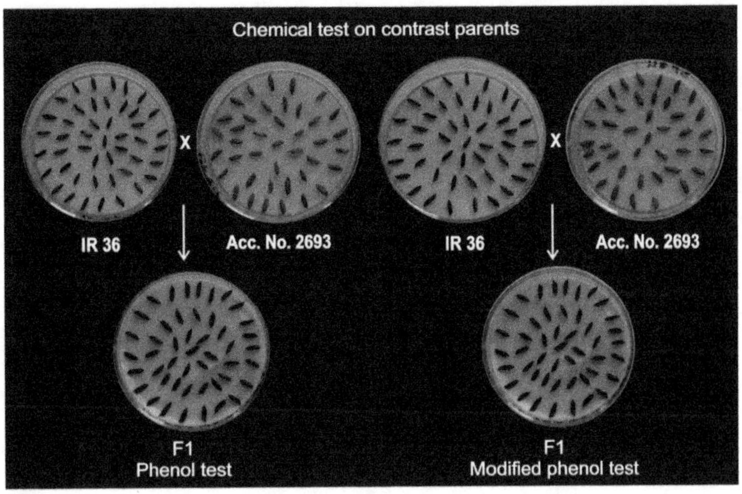

Fig. 2 Colour formation in aleurone layer of parents IR 36 (brown colour), Acc. No. 2693 (light brown/no reaction) and their F_1 progeny (brown colour) with phenol and modified phenol tests

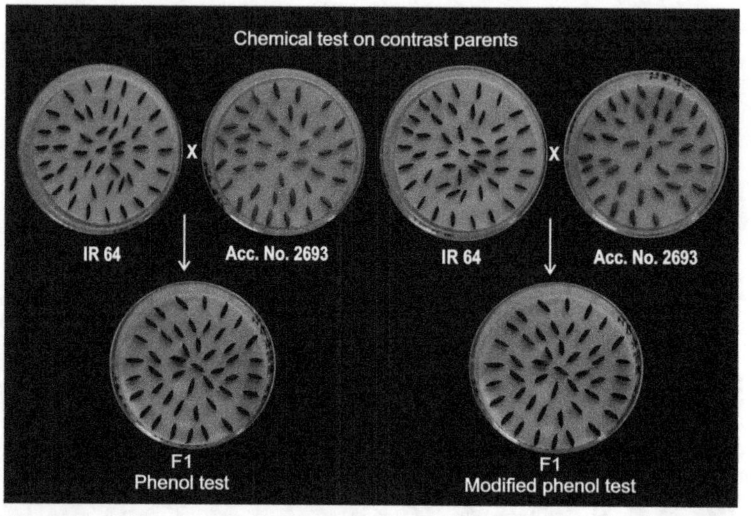

Fig. 3 Colour formation in aleurone layer of parents IR 64 (dark brown colour), Acc. No. 2693 (light brown/No reaction) and their F_1 progeny (dark brown colour) with phenol and modified phenol tests

genes controlling the concerned trait. When any one of the two or both the genes are present in the homozygous recessive state, the contrasting phenotype is produced, which leads to various modifications of the typical dihybrid, trihybrid etc. F_2 ratio.

The Mechanism of Colour Formation in Aleurone Layer

Phenol test, which is an index of polyphenol oxidase activity, has been reported to be associated with intra-varietal diversity that has been used in ascertaining varietal purity. Colour formation in aleurone layer occurs by phenol oxidation in two reactions (Fig. 5). In the first reaction, the aromatic ring of phenol can be hydroxylated to form catechols or quinols, respectively. In the second reaction, the quinols or catechols undergo oxidation to form quinones (Takahashi, 1984). Two major genes and their allelic interactions control this reaction, which is localized in seed aleurone layer in rice. The ability of genotype to form colour depends on the tyrosinase activity, which is located at aleuronic layer (Masuthi et al. 2015). The extent of colour intensity among germplasms varied because of differences in enzyme activity, temperature, light, aeration and genetic background, respectively (Sivasubramanian and Ramakrishnan 1974). On the other hand, the germplasms with no colour might have resulted because of inability to hydroxylate the aromatic ring of phenol either due to shortage of electron donor or hydroxylating enzyme (Takahashi and Hamza 1983).

The results of phenol test are usually distinct and easily interpreted. Walls (1965), reported that the phenol colour reaction depends on the quality and quantity of oxidases present in the seeds, whereas Takahashi and Hamza (1983), reported that monophenol oxidase was extremely localized in aleurone layer of grains even though it is present in all other plant parts of a variety. Presence of enzyme system has been utilized to distinguish the crop varieties in rice and tomato (Pieper 1922; Joshi and Banerjee 1970; Mansing 2010; Vijayalakshmi

Table 2 Aleurone layer color segregation in F_2 progenies of the crosses IR 36 × Acc.No. 2693 and IR 64 × Acc. No. 2693for standard phenol and modified phenol with copper sulphate tests

Cross	Chemical tests	Class	Brown/dark brown colour	Light brown/ No reaction	Total	Ratio	χ2	*P* value (at 1 degrees of freedom)
IR 36 × Acc. No. 2693	Phenol test	Observed	273.00	211.00	484	9:7	0.005	0.945
		Expected	272.25	211.75				
	Modified phenol test	Observed	255.00	229.00	484	9:7	2.498	0.114
		Expected	272.25	211.75				
IR 64 × Acc. No. 2693	Phenol test	Observed	251.00	169.00	420	9:7	2.104	0.147
		Expected	236.25	183.75				
	Modified phenol test	Observed	246.00	174.00	420	9:7	0.920	0.338
		Expected	236.25	183.75				

Parents Brown / Dark brown light brown/No reaction

B_1B_1 D_2D_2× b_1b_1 d_2d_2

Gametes (B_1D_2) ↓ (b_1d_2)

F_1Dark brown

$B_1b_1D_2d_2$

↓

F_2⚥	**B_1D_2**	**B_1d_2**	**b_1D_2**	**b_1d_2**
B_1D_2	B_1B_1 D_2D_2 Brown/Dark Brown	B_1B_1 D_2d_2 Brown/Dark Brown	B_1b_1 D_2D_2 Brown/Dark Brown	B_1b_1 D_2d_2 Brown/Dark Brown
B_1d_2	B_1B_1 d_2D_2 Brown/Dark Brrown	B_1B_1 d_2d_2 **Light brown/ No reaction**	B_1b_1 d_2D_2 Brown/Dark Brown	B_1b_1 d_2d_2 **Light brown/ No reaction**
b_1D_2	b_1B_1 D_2D_2 Brown/Dark Brown	b_1B_1 D_2d_2 Brown/Dark Brown	b_1b_1 D_2D_2 **Light brown/ No reaction**	b_1b_1 D_2d_2 **Light brown/No reaction**
b_1d_2	b_1B_2 d_2D_2 Brown/Dark Brown	b_1B_1 d_2d_2 **Light brown/No reaction**	b_1b_1 d_2D_2 **Light brown/No reaction**	b_1b_1 d_2d_2 **Light brown/No reaction**

Phenotypic ratio: 09 Brown/dark brown: **07** Light brown/No reaction

Fig. 4 Complementary gene interaction in the development of aleurone layer colour through standard phenol and modified phenol tests in rice seed giving rise to the phenotypic ratio of 9:7 (brown/dark brown: light brown/no reaction) in F_2 progenies

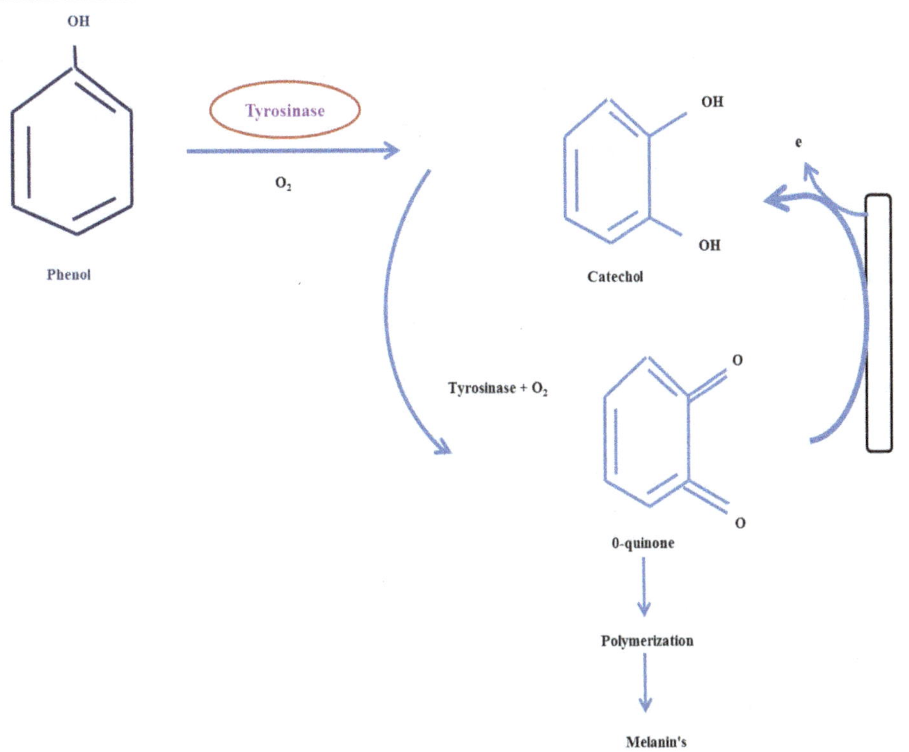

Fig. 5 Mechanism of melanin colour formation in seed aleurone layer using enzyme system upon reaction with phenol test

and Vijay 2009; Anitalakshmi et al. 2014; Sripunitha and Sivasubramaniam 2014; Vishwanath et al. 2013).

Qian et al. (2000) reported major QTL (qPH-4a) that is responsible for phenol colour has been located on chromosome 4, which explained the total phenotypic variation as high as 94.6%. Hence, this QTL is controlled by major gene. Moreover, two minor QTLs (qPH-1 and qPH-4b) located on chromosome 1 and 4 have account of total phenotypic variation of 14.9 and 29.5%, respectively. Hence, minor genes that code for these QTLs, which control phenol colour have showed positive additive effects. Phenol showed a bimodal distribution in the double haploid population and a major gene detected was close to the *Ph* gene located on chromosome 4 (Lin et al. 1994).

Aleurone Layer Colour Inheritance in F_2 Population Derived from IR 36 × Acc. No. 2693 and IR 64 × Acc. No. 2693 with Sodium Hydroxide (NaOH) and Potassium Hydroxide (KOH) Tests

The 484 F_2 progenies of the cross IR 36 × Acc. No. 2693 were evaluated. Among them, 319 and 335 F_2 progenies were wine red/dark wine red colour, whereas 165 and 149 F_2 progenies were light yellow coloured with NaOH and KOH tests, respectively. Further, 420 F_2 progenies derived from the cross IR 64 × Acc. No. 2693 were evaluated, of which 281 and 290 F_2 progenies were wine red/dark wine red colour and 139 and 130 F_2 progenies were light yellow coloured with NaOH and KOH tests, respectively.

The segregation of aleurone layer colour in 11:5 ratio for wine red/dark wine red and light yellow in F_2 progenies of both the crosses showed that the colour trait is governed by two genes, where a dominance of one gene is modified by homozygous recessive condition of another gene. The two genes that interact to produce a single character may also reciprocally modify the dominance relationship between the alleles at the other locus. Thus, the typical 15:1 ratio for duplicate gene action is modified as 11:5 due to the reciprocal dominance modification of recessive alleles. This indicates a goodness of fit with expected ratio of 11:5 for the reciprocal dominance of duplicate genes as given in Table 3.

Two genes, WR_1 and WR_2 showing duplicate interaction governs the development of aleurone layer colour (wine red / dark wine red/light) with NaOH and KOH in rice plants. However, the recessive homozygous condition of one gene, say wr_1 wr_1 reverses the dominance relationship at the other locus; hence, the genotype wr_1wr_1 WR_2 wr_2 produces the same phenotype as the homozygous double recessive wr_1 wr_1, wr_2 wr_2 of light yellow/no reaction. Similarly, homozygous recessive condition of the other gene, wr_2 wr_2 has the same effect on the dominance relationship at the wr_1 locus. As a consequence, the genotype WR_1 wr_1 wr_2 wr_2 produces the same phenotype as the double recessive homozygote wr_1 wr_1 wr_2 wr_2 of light yellow/no reaction. Therefore, the genotypes WR_1 wr_1 wr_2 wr_2; wr_1 wr_1 WR_2 wr_2 and wr_1 wr_1 wr_1 wr_2 all produce light brown/no reaction condition. Thus, the dominant genes WR_1 and WR_2 behave as if, they were recessive to their allele's wr_1 and wr_2, respectively; whenever they are present in association with the homozygous recessive state at the other locus that is with wr_2 wr_2 and wr_1 wr_1, respectively as depicted in Fig. 6.

The wine red/ dark wine red colour variety (Acc. No. 2693) of rice (WR_1WR_1 WR_2 WR_2) was crossed with a light yellow/no reaction variety (IR 36 and IR 64) with the genotype wr_1wr_1, wr_2wr_2, the derived F_1 (WR_1 wr_1 WR_2 wr_2) has produced light wine red colour (intermediate) even in the presence of dominant alleles of both the genes. In the F_2 generation, on an average nine plants out of 16, have at least one dominant allele of both the genes WR_1 and WR_2; these plants develop wine red and dark wine red colour.

One plant has the genotype WR_1 WR_1 wr_2 wr_2, while another has the genotype wr_1 wr_1 WR_2 WR_2. These two plants also develop wine red and dark wine red colour; since, they have either WR_1 / WR_2 in the homozygous state, which is able to produce wine red and dark wine

Table 3 Aleurone layer color segregation in F_2 progenies of the crosses IR 36 × Acc.No. 2693 and IR 64 × Acc. No. 2693 for NaOH and KOH tests

Cross	Chemical tests	Class	Wine red/dark wine red	Light yellow/ No reaction	Total	Ratio	χ2	P value(at 1 degrees of freedom)
IR 36 × Acc. No. 2693	NaOH	Observed	319.00	165.00	484	11:5	1.818	0.178
		Expected	332.75	151.25				
	KOH	Observed	335.00	149.00	484	11:5	0.048	0.825
		Expected	332.75	151.25				
IR 64 × Acc. No. 2693	NaOH	Observed	281.00	139.00	420	11:5	0.665	0.415
		Expected	288.75	131.25				
	KOH	Observed	290.00	130.00	420	11:5	0.017	0.895
		Expected	288.75	131.25				

Parents **Wine red / dark wine red** **Light yellow/ No reaction**

$WR_1WR_1\ WR_2WR_2 \times wr_1wr_1\ wr_2wr_2$

Gametes **$WR_1\ WR_2$** ↓ **$wr_1\ wr_2$**

F_1
Light wine red
$WR_1wr_1WR_2\ wr_2$

↓

F_2 ♂/♀	**$WR_1\ WR_2$**	**$WR_1\ wr_2$**	**$wr_1\ WR_2$**	**$wr_1\ wr_2$**
WR_1 WR_2	$WR_1WR_1\ WR_2WR_2$ Wine red / Dark wine red	$WR_1WR_1\ WR_2wr_2$ Wine red / Dark wine red	$WR_1wr_1\ WR_2WR_2$ Wine red / Dark wine red	$WR_1wr_1\ WR_2wr_2$ Wine red / Dark wine red
WR_1 wr_2	$WR_1WR_1\ wr_2WR_2$ Wine red / Dark wine red	$WR_1WR_1\ wr_2wr_2$ Wine red / Dark wine red	$WR_1wr_1\ wr_2WR_2$ Wine red / Dark wine red	$WR_1wr_1\ wr_2wr_2$ **Light yellow/No reaction**
wr_1 WR_2	$wr_1WR_1\ WR_2wr_2$ Wine red / Dark wine red	$wr_1WR_1\ WR_2wr_2$ Wine red / Dark wine red	$wr_1wr_1\ WR_2WR_2$ Wine red / Dark wine red	$wr_1wr_1\ WR_2wr_2$ **Light yellow/No reaction**
wr_1 wr_2	$wr_1WR_1\ wr_2WR_2$ Wine red / Dark wine red	$wr_1WR_1\ wr_2wr_2$ **Light yellow/ No reaction**	$wr_1wr_1\ wr_2WR_2$ **Light yellow / No reaction**	$wr_1wr_1\ wr_2wr_2$ **Light yellow/No reaction**

Phenotypic ratio: 11 Wine red/dark wine red: **05** Light yellow/No reaction

Fig. 6 Dominance modification of duplicate genes leading to a 11:5 phenotypic ratio in the F_2 progenies for the presence of wine red / dark wine red and light yellow/no reaction colouration of aleurone layer of rice seed with NaOH and KOH tests

red colour. Two plants out of 16 are heterozygous for WR_2 and homozygous for wr_1 and vice versa. These four plants do not develop wine red and dark wine red colour; since, the homozygous recessive state of wr_2 and wr_1, reverses the dominance relationship between WR_1/ wr_1, WR_2 / wr_2, respectively. The remaining one plant is also light yellow/no reaction because it is homozygous recessive for both the genes $wr_1\ wr_1\ wr_2\ wr_2$. Thus, the typical dihybrid 15:1 ratio for duplicate gene action is modified as 11:5 due to the reciprocal dominance modification of wr_1 and wr_2.

The present findings were reported in case of cotton (Fuchs et al. 1972). Two genes, G_1 and G_2 showing duplicate gene interaction governs the presence of pigment glands on cotton plants. However, the recessive homozygous condition of one gene, say $g_1\ g_1$, reverses the dominance relationship at the other locus so that genotype $g_1\ g_1\ G_2\ g_2$ produces the same phenotype as the homozygous double recessive $g_1\ g_1\ g_2\ g_2$. Similarly, homozygous recessive condition of the other gene, $g_2\ g_2$ has the same effect on the dominance relationship at the g_1 locus. As a consequence, the genotype $G_1\ g_1\ g_2\ g_2$ produces the same phenotype as the double recessive homozygote $g_1\ g_1\ g_2\ g_2$ produces glandless plants. Therefore, the genotypes $G_1\ g_1\ g_2\ g_2$, $g_1\ g_1\ G_2\ g_2$ and $g_1\ g_1\ g_2\ g_2$ all produce glandless condition; hence, the F_2 phenotypic ratio 11:5 was observed. Similar classification was noticed in the present study with NaOH, which is in congruence with the report in wheat (Mansing 2010), rice (Vanangamudi et al. 1988; Sripunitha and Sivasubramaniam 2014), *urdbean* (Chakrabarthy and Agrawal 1990); cotton (Ponnuswamy et al. 2003; Reddy et al. 2008), safflower (Biradar Patil et al. 2006). In addition, the same findings corroborate with the crops such as sesamum (Suhasini 2006) soybean (Chavan 2010), sunflower (Sathisha et al. 2012; Kallihal et al. 2013) and tomato (Qian et al. 2000), respectively. The reasons for various colour attributed when reacted with sodium hydroxide and potassium hydroxide might be due to inherent chemical difference, stability of genetic characters and secondary metabolites present in the seeds (Masuthi et al. 2015; Vanderburg and Vanzwol 1991; Chakrabarthy and Agrawal 1990).

Therefore, these studies are particularly useful, where non-availability of distinct stable morphological markers for identification of increased varieties. These chemical tests along with other parameters like 1000 seed weight, seed size, response to GA_3, 2,4-D and soluble proteins acts as a descriptors for identification of the rice varieties. Further, these tests could help to develop a rapid varietal identification that may help the breeders and seed inspectors to monitor the quality seed production

(Nethra et al. 2007). Thus, chemical tests are one of the important characters that help in easy identification of varieties for genetic purity.

Conclusions

Based on the response of biochemical tests with 904 F_2 progenies derived from crosses, IR 36 × Acc. No. 2693 and IR 64 × Acc. No. 2693 were utilised for delineation of inheritance pattern of aleurone layer colour in rice. In this investigation, it was found that the colour trait in aleurone layer of the F_2 progenies derived from both the crosses were segregated with complementary gene interaction of 9:7 ratio (brown/dark brown: light brown/no reaction), indicating a goodness of fit with observed ratio for standard phenol and modified phenol tests ($CuSO_4$), respectively. Further, the colour trait in F_2 progenies of both the crosses with NaOH and KOH tests were observed to segregate in 11:5 ratio (wine red/dark wine red: light yellow/ no reaction), wherein typical 15:1 ratio for duplicate gene action is modified as 11:5 due to the reciprocal dominance modification of recessive alleles. Therefore, it is deduced that the colour trait in aleurone layer was found to be controlled by two major genes and their allelic interactions. These findings could be utilised for easy identification of varieties in rice breeding programme, gene expression analysis, cloning and tagging of gene and also to develop the seed keys to precisely define cultivars that would serve an alternative for Grow-out-test.

Methods

Plant Materials

The present work was carried out using IR 36, IR 64 and Acc. No. 2693 (as parents), F_1 and F_2 progenies. The F_1s were derived from cross between IR 36 × Acc. No. 2693 and IR 64 × Acc. No. 2693, respectively during *kharif* 2014 at ICAR-Directorate of Seed Research (ICAR-DSR), Mau, Uttar Pradesh, India. Both the crosses (F_1 seeds) were raised during off-season 2014–15 at regional station, ICAR-DSR, GKVK campus, Bengaluru. Further, F_2 progenies of both the crosses were raised during *kharif* – 2015 at ICAR-DSR, Mau; 484 and 420 F_2 seeds of both crosses IR 36 × Acc. No. 2693 and IR 64 × Acc. No. 2693 were used for present investigation to delineate the inheritance pattern of colour trait in aleurone layer using chemical tests, respectively.

Chemical Tests

To know the segregation pattern of colour trait in aleuronic layer, study has been performed using different chemical tests such as standard phenol, modified phenol ($CuSO_4$), potasium hydroxide and sodium hydroxide tests. These chemical tests are insensitive to environment and serves not only as basis for grouping of varieties, but also used for genotype identification (Naga Padma et al. 1996) with consistent results.

Phenol Test

For phenol test, seeds were pre-soaked in distilled water for 24 h. Thereafter, they were transferred on two layers of Whatman No.1 filter paper saturated with 4 % phenol solution (Merck, Cat. No. AL8AF58565, Merck Specialities Private Ltd. Mumbai, India). The Petri-dishes were covered and incubated at 25 ± 1 °C and the change in colour of aleurone layer in response to phenol reaction was evaluated after 24 h. The parents, F_1 and F_2 progenies were categorized into five categories as no reaction, light brown, brown, dark brown and black colour (Jaiswal and Agrawal 1995).

Modified Phenol Test-A ($CuSO_4$)

Modified phenol test was conducted, which is alike to standard phenol test except that seeds were pre-soaked in 0.5% (*w*/*v*) copper sulphate (Helix Bio-Science, Cat. No.HBC043212, New Delhi, India) solution for 24 h. Colour reaction was noted after 48 h of incubation and the parents; F_1 and F_2 progenies were categorized into five categories as no reaction, light brown, brown, dark brown and black colour (Jaiswal and Agrawal 1995).

Sodium Hydroxide (NaOH) Test

Parents, F_1 and F_2 seeds were subjected to sodium hydroxide test where, seeds were pre-soaked in 5 % sodium hydroxide solution (Merck, Cat. No. MJ8D580230, Merck Specialities Private Ltd. Mumbai, India) and kept at room temperature for one hour and change in colour of the solution was observed. Chakrabarty et al. 1989, categorized the reaction into light yellow and wine red based on the intensity of change in colour solution.

Potassium Hydroxide (KOH) Test

Seeds of parents, F_1 and F_2 progenies were pre-soaked in 5 % potassium hydroxide solution (Helix Bio-Science, Cat. No. A3641, New Delhi, India) and kept at room temperature for 4 h and a change in colour of the solution was observed. Based on the intensity of reaction, the populations were categorized into various groups viz., light yellow, dark yellow, light wine red, dark wine red (Vanangamudi et al. 1988).

Chi-square (χ2) Goodness of Fit Test

Chi-square (χ2) goodness of fit test was performed to analyse phenotypic segregation between observed values (O) to the expected values (E) for F_2 population data, using $\chi^2 = \Sigma(O - E)^2 / E$.

Abbreviations

Acc. No.: Accession number; B: Brown; b_1b_1 d_2d_2: Light brown/no reaction; B_1B_1: Brown; D_2D_2: Dark brown; DB: Dark brown; DWR: Dark wine red; ICAR: Indian Council of Agricultural Research; KOH: Potassium hydroxide;

LB: Light brown; LWR: Light wine red; LY: Light yellow; NaOH: Sodium hydroxide; NR: No reaction; WR: Wine red; wr_1 wr_1 wr_2 wr_2: Light yellow/ no reaction; WR_1 WR_1: Wine red; WR_2 WR_2: Dark wine red

Acknowledgements
The authors are grateful to Mr. Sunil Kumar Kannaujiya and Mr. Satish Chandra Kumar for their assistance in the experimental work.

Authors' Contributions
Conceived, Designed & Performed the Experiments: CS. Formal analysis: CS; KBN. Contributing reagents: KVS; KUB; GS. Writing-original draft: CS; KVS. Writing-review & editing: CS; GP; SPJK. All authors read and approved the final manuscript.

Competing Interests
The authors declare that they have no competing interests.

References
Anitalakshmi V, Gowda R, Sathisha CS, Prasad R (2014) Varietal response to various chemical tests for their characterization in rice (*Oryza sativa* L.) Indian J Plant Sci 3(2):177–179
Banerjee SK, Chandra R (1977) Modified phenol test for the varietal identification of wheat seed. Seed Sci Technol 5:53–60
Biradar Patil NK, Sangeeta M, Motagi BN, Vijaykumar AG, Hanchinal RR (2006) Characterization of safflower varieties through chemical tests. Abst.XII Nation. Seed Sem, 24–26 February, ANGRAU, Hyderabad, pp.168
Cabaj J, Sołoducho J, Nowakowska-Oleksy A (2010) Langmuir-Blodgett film based biosensor for estimation of phenol derivatives. Sensors Actuators B 143(2): 508–515. 10.1016/j.snb.2009.09.047
Chakrabarthy SK, Agrawal RL (1990) Identification of black gram varieties-III: utilization of seedling growth response to added chemicals. Seed Res. 18(1):34–39
Chakrabarty SK, Agrawal RL, Bhowal (1989) Identification of blackgram varieties II. Utilization of morphological characteristics of seedlings. Seed Res 17(2):139–142
Chandu S, Jeevan Kumar SP, Sripathy KV, Somasundaram G, Udaya Bhaskar K, Ramesh KV, Madan K, Rajendra Prasad S (2017) Characterization and identification of rice germplasm accessions using chemical tests. Seed Res 45(1):75–83
Chavan NG (2010) Characterization of soybean genotypes [*Glycine max* (L.)Merrill] through morphological, chemical, molecular markers and image analyzer. M. Sc. Thesis submitted to University of Agricultural Sciences, Dharwad. http://krishikosh.egranth.ac.in/bitstream/1/84304/1/th10085.pdf
Dileepkumar M, Vyakarnahal BS, Deshpande VK, Jagadeesha RC, Mukesh L, Chavan, Khaja RS (2015) Characterization of traditional aromatic rice cultivars by chemical markers. Int J Adv Res 3(5):103–107
Fernandes KF, Lima CS, Lopes FM, Collins CH (2005) Hydrogen peroxide detection system consisting of chemically immobilised Peroxidase and spectrometer. Process Biochem 40(11):3441–3445
Fraignier MP, Michaux-Ferriere N, Kobrehel K (2000) Distribution of peroxidases in durum wheat (*Triticum durum*). Cereal Chem 77:11–17. 10.1094/CCHEM.
Fuchs JA, Smith JD, Birds LS (1972) Genetic basis for an 11:5 dihybrid ratio observed in *Gossypium hirsutum*. J Hered 63:300–303
Fuerst EP, Okubara PA, Anderson JV, Morris CF (2014) Polyphenol oxidase as a biochemical seed defense mechanism. Front Plant Sci 5:689
Jaiswal JP, Agrawal RL (1995) Varietal purity determination in rice: modification of the phenol test. Seed Sci Technol 23:33–42
Jerkovic A, Kriegel AM, Bradner JR, Atwell BJ, Roberts TH, Willows RD (2010) Strategic distribution of protective proteins within bran layers of wheat protects the nutrient-rich endosperm. Plant Physiol 152(3):1459–1470
Jimenez M, Dubcovsky J (1999) Chromosome location of genes affecting polyphenol oxidase activity in seeds of common and durum wheat. Plant Breed 118:395–398
Namrata-Gupta, Joshi MA, Sarao NK, Sharma RC (2007) Utility of phenol test in varietal characterization. Crop Improv 34(1):77–81
Joshi MG, Banerjee SK (1969) Multiple alleles for phenol colour reaction in emmer wheat. Indian J. Genet. Plant Breed 29(2):280–284
Joshi MG, Banerjee SK (1970) Genetics of phenol colour reaction in emmer wheat. Proc Int Seed Assoc 35:207
Joshi MG, Banerjee SK, Naik MS (1969) Tyrosinase activity as a quantitative indicator of genetic control of phenol colour reaction in tetraploid triticums. Indian J Exp Biol 7:189–190
Kallihal PK, Rajendra Prasad S, Shwetha KS (2013) Characterization of sunflower (*Helianthus annuus* L.) hybrid and its parental lines based on biochemical tests at seed and seedling stages. Annals Biol Res 4(4):96–99
Kumar SPJ, Kumar VG, Dash A, Scholz P, Banerjee R (2017) Sustainable green solvents and techniques for lipid extraction from microalgae: a review. Algal Res 21:138–147
Kumar SPJ, Prasad SR, Banerjee R, Agarwal DK, Kulkarni KS, Ramesh KV (2017) Green solvents and technologies for oil extraction from oilseeds. Chem Central J 11:1–9
Kumar SPJ, Rajendra Prasad S, Banerjee R, Thammineni C (2015) Seed birth to death: dual functions of reactive oxygen species in seed physiology. Ann Bot 116(4):663–668
Kumar SPJ, Rajendra Prasad S, Kumar M, Singh C, Sinha AK, Pathak A (2016) Seed quality markers: a review. Res Rev S3:13–17
Lin SY, Nagamura N, Kurata M (1994) DNA markers tightly linked to genes. Ph. Alk and Rc. Rice Genet Newslett 11:108
Maguire JD, Steen KM, Grzelak K (1975) Classification of pacific northwest winter and spring wheat cultivars by phenol reactions. Proc. off. Assn. Seed Analysts 65:143–146
Mansing JS (2010) Characterization of wheat (*Triticum* spp.) genotypes trough morphological, chemical and molecular markers. *M.Sc. (Agri.)* thesis submitted to University of Agricultural Sciences, Dharwad. http://krishikosh.egranth.ac.in/handle/1/84530
Masuthi D, Vyakarnahal BS, Deshpande VK, Jagadeesha RC, Mukesh LC, Khaja Rubina S (2015) Characterization of traditional aromatic rice cultivars by chemical markers. Int J Adv Res 3(5):103–107
Miczynski K (1938) Genetische Studien fiber die Phenol Farben Reaktion beim Weizen. Z, zucht 22:564–587
Milner M, Gould MR (1951) The quantitative determination phenol oxidase activity in wheat varieties. Cereal Chem 28:473–478
Mor RS, Battan KR, Mehla BS (2006) Identification of rice varieties by laboratory techniques. National J Pl Improvement 8(1):28–31
Naga Padma K, Muralimohan RB, Ankaiah R, Sarada P (1996) Characterisation of twenty-three maize inbred lines through seed morphology, leaf stomatal studies and response of seedlings to added chemicals. Seed Res 24(1):15–19
Nethra N, Rajendra Prasad S, Vishwanath K, Dhanraj KN, Gowda R (2007) Identification of rice hybrids and their parental lines based on seed, seedling characters, chemical tests and gel electrophoresis of Total soluble proteins. Seed Sci Technol 35:176–186
Pallares IN, Ferrari L, Postulka EB, Ritta MN (2004) Identification of cultivars of foxtail millet (*Setaria italica* (L.) Beauv.) from morphological and biochemical traits of their seed. Seed Sci Technol 32(1):187–196
Pieper H (1922) Ein Mittel zur Unterscheidung von Weizensorten am Korn. Deutsche Lantwritschaftlinche Presse 49(67):438–439
Ponnuswamy AS, Bhaskaran M, Sastri G (2003) Variety characterization in cotton by physical, chemical and bio-chemical methods. In: Training Mannual, variety characterization by image analysis and electrophoresis, pp 106–120
Qian Q, He P, Zheng X, Ying C, Zhu L (2000) Genetic analysis of morphological index and its related taxonomic traits for classification of indica/japonica rice. Sci China (series C) 43(2)
Reddy M, Hunje R, Nadaf HL, Biradar DP, Vyakarnahal BS (2008) Identification of cotton hybrids and parents through chemical tests. Agril Sci Digest 28(1):51–53
Sathisha CS, Rajendra Prasad S, Gowda R, Thimmegowda MN (2012) Comparison of various chemical tests for varietal characterization in sunflower (*Helianthus annuus* L.). Indian J. Plant Sci 1(1):39–43
Singh D, Kumar A, Kumar N, Kumar V, Kumar S, Bhardwaj A, Sirohi P, Chand S (2011) Comparison of traditional physico-chemical methods and molecular marker assays for characterization of basmati rice (*Oryza sativa* L.). African J. Of. Biotechnology 10(62):13390–13398
Sinha AK, Agarwal DK, Kumar SPJ, Chaturvedi A, Tiwari TN (2016) Novel technique for precluding hybrid necrosis in bread wheat. Int J Tro Agr 34(3): 761–765
Sivasubramanian S, Ramakrishnan V (1974) Use of phenol colour reaction for identification of rice varieties. Exp Agric 10:299–303
Sripunitha A, Sivasubramaniam K (2014) Varietal characterization rice varieties based on chemical methods. Trends Biosci 7(20):3139–3146
Steffens JC, Harel E, Hunt MD (1994) Polyphenol oxidase. In: Ellis BE et al (eds) Genetic engineering of plant secondary metabolism. Plenum Press, New York, pp 275–313
Suhasini KS (2006) Characterization of sesame genotypes through morphological,

chemical and RAPD markers. M.Sc (Agri.) Thesis submitted to University of Agricultural Sciences, Dharwad. http://krishikosh.egranth.ac.in/handle/1/81949

Takahashi N (1984) Differentiation of ecotypes in *Oryza sativa* L. In: Takahashi N, Tsunoda S (eds) Biology of Rice. Japan Scientific Societies Press, Tokyo, pp 31–67

Takahashi N, Hamza AHA (1983) Differentiation of ecotypesin *Oryza sativa* L. I: reexamination for colour reaction with phenol. Japanese J Breed 33(3):243–250

Thangavel P, Bharathi A, Natarajan N, Evera T (2005) Varietal grouping in sorghum by seed and seedling morphology and response to chemical testing. *Karnataka* JAgric Sci. 18 (3): 664-672.

Vanangamudi K, Palanisamy V, Natesan P (1988) Variety determination in rice phenol and potassium hydroxide tests. Seed Sci Technol 16:465–470

Vanderburg NJ, Vanzwol RA (1991) Rapid identification techniques used in laboratories of the international seed testing association: a survey. Seed Sci Technol 19:687–700

Varier A, Dadlani M, Sharma SP (1995) Phenol test and electrophoresis of seed esterases for testing genetic purity in seed lots of pearl millet (*Pennisetum glaucum*) inbreds. Indian J Agri Sci 65(11):789–792

Vijayalakshmi B, Vijay D (2009) Development of seed keys for varietal identification in Rice (*Oryza sativa* L.) using chemical tests. Seed Res 37(1&2): 56–61

Vishwanath K, Pallavi HM, Nethra N, Rajendra Prasad S (2013) Chemical tests for identification and characterization of tomato cultivars. Plant Breed Seed Sci 68:3–12

Walls WE (1965) A standardized phenol method for testing wheat seed for varietal purity. Contribution no. 28 to the Handbook on Seed Testing. The Assoc. of Official Seed Analysts, Lincoln, NE.

The rice *zebra3* (*z3*) mutation disrupts citrate distribution and produces transverse dark-green/green variegation in mature leaves

Suk-Hwan Kim[1†], Choon-Tak Kwon[1,4†], Giha Song[1], Hee-Jong Koh[1], Gynheung An[2] and Nam-Chon Paek[1,3*]

Abstract

Background: Rice *zebra* mutants are leaf variegation mutants that exhibit transverse sectors of green/yellow or green/white in developing or mature leaves. In most cases, leaf variegation is caused by defects in chloroplast biogenesis pathways, leading to an accumulation of reactive oxygen species in a transverse pattern in the leaves. Here, we examine a new type of leaf variegation mutant in rice, *zebra3* (*z3*), which exhibits transverse dark-green/green sectors in mature leaves and lacks the typical yellow or white sectors.

Results: Map-based cloning revealed that the *Z3* locus encodes a putative citrate transporter that belongs to the citrate-metal hydrogen symport (CitMHS) family. CitMHS family members have been extensively studied in bacteria and function as secondary transporters that can transport metal-citrate complexes, but whether CitMHS family transporters exist in eukaryotes remains unknown. To investigate whether Z3 acts as a citrate transporter in rice, we measured citrate levels in wild-type leaves and in the dark-green and green sectors of the leaves of *z3* mutants. The results showed that citrates accumulated to high levels in the dark-green sectors of *z3* mutant leaves, but not in the green sectors as compared with the wild-type leaves.

Conclusions: These results suggest that leaf variegation in the *z3* mutant is caused by an unbalanced accumulation of citrate in a transverse pattern in the leaves. Taking these results together, we propose that Z3 plays an important role in citrate transport and distribution during leaf development and is a possible candidate for a CitMHS family member in plants.

Keywords: Rice, *ZEBRA3* (*Z3*), Leaf variegation, CitMHS family, Citrate transporter

Background

Citrate (citric acid) is a weak organic acid found in nature, and ubiquitously acts as an energy intermediate, solubilizing agent, carbon source, and chelator (Sarantinopoulos et al. 2003). The roles of citrate are mostly related to cellular metabolism. For instance, citrate acts as a metabolite in the Krebs cycle under aerobic conditions (Krebs and Johnson 1937), and as a source of acetyl-CoA for fatty acid synthesis (Fritsch and Beevers 1979; Matto and Modi 1970; Nelson and Rinne 1975). In plant tissues, citrate also supplies carbon skeletons for NH_4^+ assimilation via the glutamine synthetase (GS) and glutamate synthase (GOGAT) pathway (Chen and Gadal 1990; Agata 2008). In plants, citrate can enter into the glyoxylate cycle, allowing the plant to use two-carbon acetate as a carbon source via gluconeogenesis (Tanner and Beevers 1965).

Citrate performs other roles when transported outside the cell. Because citrate is a tricarboxylic acid with good chelating ability for cations, such as manganese, magnesium, iron, and calcium (Glusker 1980; Milewska 1988), it can chelate these metal ions and thus change their

* Correspondence: ncpaek@snu.ac.kr
Suk-Hwan Kim and Choon-Tak Kwon contributed equally to this work.
†Equal contributors
[1]Department of Plant Science, Plant Genomics and Breeding Institute, Research Institute of Agriculture and Life Sciences, Seoul National University, Seoul 08826, Republic of Korea
[3]Crop Biotechnology Institute, Institutes of Green Bio Science and Technology, Seoul National University, Pyeongchang 25354, Republic of Korea
Full list of author information is available at the end of the article

physicochemical properties. The roles of citrate in iron translocation and aluminum detoxification have been extensively studied in plants. For example, citrate is needed to solubilize inorganic iron for long-distance transport via the xylem because free ionic forms of iron are toxic and subject to precipitation in the slightly acidic xylem sap (Rellán-Álvarez et al. 2010; Green and Rogers 2004; Tiffin 1970). In many plant species, citrate is excreted into the rhizosphere, where it chelates aluminum, thereby excluding it from entering the plant (Fuente et al. 1997; Miyasaka et al. 1991; Magalhaes et al. 2007). In addition, under phosphorus-limiting conditions in *Lupinus albus*, *Brassica napus*, and the Proteaceae family of plants, the increased excretion of organic acids, particularly citrate, may enhance phosphorus uptake by displacing the phosphorus from insoluble complexes in the soil (Dinkelaker et al. 1989; Dinkelaker et al. 1995; Zhang et al. 1997; Hoffland et al. 1989).

Citrate transporters that are known to transport free citrate as a chelating agent have been extensively studied in plants (Liu et al. 2009; Zhou et al. 2013; Yokosho et al. 2009; Yokosho et al. 2011). One particular family of citrate transporter that transports free citrate is the multidrug and toxic compound extrusion (MATE) protein family. Many of these family members, such as *OsFRDL1* in rice (Yokosho et al. 2009), *FRD3* in Arabidopsis (Durrett et al. 2007; Roschzttardtz et al. 2011), and *GmFRD3* and *GmFRD3b* in soybean (Rogers et al. 2009), function in the efficient translocation of iron by excreting citrate into the xylem. Other MATE family genes, such as *OsFRDL4* in rice (Yokosho et al. 2011), *AtMATE* in Arabidopsis (Liu et al. 2009), and *HvAACT1* in barley (Furukawa et al. 2007), play a role in secreting citrate into the rhizosphere to detoxify aluminums by chelating them.

Citrate transporters that recognize and/or transport metal-citrate complexes, such as iron-citrate complexes, have not been well studied in plants, whereas some have been identified in bacteria (Boorsma et al. 1996; Korithoski et al. 2005; Blancato et al. 2008; Lensbouer et al. 2008). One example is the citrate-metal hydrogen symport (CitMHS) family, which transports metal-citrate complexes in symport with a proton. This family of citrate transporters has evolved in some bacteria, such as *Bacillus*, *Streptomyces*, *Klebsiella*, and *Neisseria*. It is believed that these bacteria use metal–citrate complexes as a source of both carbon and metal, because metal–citrate complexes are predominantly accessible in their native environment, such as the rhizosphere (Dessureault-Rompré et al. 2008). It remains unknown if these CitMHS family transporters exist in eukaryotes.

Leaf variegation mutants have been reported in many species of higher plants (Li et al. 2010; Sakuraba et al. 2013; Martínez-Zapater 1993; Rédei 1967; Campitelli et al. 2008; Tsukaya et al. 2004). The leaves of these mutants have green and yellow (or white) sectors, and in most cases, this abnormal phenotype is caused by impaired chloroplast biogenesis in the off-colored cells (Aluru et al. 2006). While the green sectors of the leaves have morphologically normal plastids with the same levels of chlorophyll and carotenoids compared with wild-type leaves, the yellow (or white) sectors contain chloroplasts with defective thylakoid stacking and less photosynthetic pigments (Li et al. 2010; Han et al. 2012; Carol et al. 1999). In addition, leaf variegation can arise from excessive accumulation of reactive oxygen species (ROS), as is the case with *zebra-necrosis* (*zn*) mutants (Li et al. 2010) and *Staygreen* (*SGR*)-overexpressing plants in rice (Jiang et al. 2011), and *yellow variegated2* (*var2*) mutants in Arabidopsis (Kato et al. 2009).

Here, we investigated a new type of rice leaf variegation mutant, *zebra3* (*z3*), which has transverse dark-green/green stripes in the leaves with no yellow or white sectors. The dark green sectors of the *z3* mutant leaves had significantly higher levels of photosynthetic pigments and proteins compared with wild-type leaves, and the green sectors had levels similar to the wild type. Positional cloning revealed that the *Z3* gene encodes a protein predicted to be a CitMHS family citrate transporter, which have only been reported in bacteria thus far. Indeed, we found that the dark-green sectors accumulated more citrates than the green sectors, consistent with a putative role for Z3 in citrate transport. We discuss the possible function of Z3 as a candidate citrate transporter in eukaryotes.

Results

Phenotypic characterization of the *z3* mutant

The *z3* mutant produced slightly lighter green leaf blades and had a slower growth rate compared to its parental *japonica* cultivar 'Kinmaze' during early vegetative growth in a natural paddy. These phenotypes of the *z3* mutant disappeared gradually and no distinct differences in phenotype were observed at 80 days after sowing, although the height of the *z3* mutant was slightly shorter than that of the wild type, possibly due to its slower growth at the early stages (data not shown). Thereafter, transverse dark-green and green sectors began to appear in the mature leaves of the *z3* mutant (Fig. 1a and b). The leaf variegation became more obvious as the plants approached maturity and was most notable in the fully mature leaves during grain filling.

In addition to the leaf variegation, the *z3* mutant showed a late-flowering phenotype in the paddy field under natural long-day (NLD) conditions (~14 h light/day in Suwon, Korea, 37°N latitude) (Additional file 1: figure S1a-b). To investigate whether a reduction in vegetative growth rate or prolonged plastochron (i.e. the

Fig. 1 Phenotypic characterization of the *z3* mutant. **a** Phenotypes of the wild-type (WT) and *z3* mutant plants at 140 days after sowing (DAS) grown under natural long day conditions in the paddy field. **b** Leaf phenotypes of the 140-day-old WT and *z3* mutant plants grown in the paddy field. **c–d** Panicle (**c**) and grain (**d**) phenotypes of the WT and *z3* mutant plants at 140 DAS grown under natural long day conditions. WT, wild type; *z3*, *zebra3*; *z3* (du), dull-brown colored grain of the *z3* mutant; *z3* (iv), ivory colored grain of the *z3* mutant

time between one leaf initiation and the next) causes the late-flowering phenotype in the *z3* mutant, we compared leaf emergence rates between the wild type and the *z3* mutant under long-day conditions (14.5 h-light, 30 °C/9.5 h-dark, 24 °C) in the growth chambers. The wild type flowered around 110 days after sowing (DAS), whereas the *z3* mutant flowered around 119 DAS. At heading, the total number of leaves on the main culm of the *z3* mutant was less than the wild type (Additional file 1: figure S1c). This observation demonstrates that the leaf emergence rate of the *z3* mutant is significantly lower than the wild type, indicating that the *z3* mutation retards the vegetative growth rate or prolongs the plastochron.

Moreover, the *z3* mutant simultaneously produced two distinct defective panicles on the same plant; one was a dull brown color and the other was an ivory color instead of the normal golden-yellow color in the wild type (Fig. 1c and d). Seed coat discoloration is an early indicator of poor quality seed, resulting in dull brown-colored grains (Bodalkar and Awadhiya 2014), which negatively affects grain yield, especially by decreasing the number of filled grains per panicle (or seed-setting rate) and 1000-grain weight (Phat et al. 2005). Furthermore, ivory-colored spikelets, which are generally caused by excessive transpiration during panicle growth, often result in low fertility in rice (Tsuda et al. 2000). To examine the relationship between the defective seed phenotypes and yield components in the *z3* mutant, we evaluated several agronomic traits under NLD conditions: main panicle length, number of panicles per plant, number of spikelets per main panicle, seed setting rate, 500-grain weight, and grain yield per plant (Additional file 2: figure S2). All of the examined traits were inferior in the *z3* mutant (Additional file 2: figure S2a-e), and the grain yield of the *z3* mutant was significantly lower compared with the wild type (Additional file 2: figure S2f). These results indicated that seed discoloration in the *z3* mutant negatively affects important yield traits in rice.

The levels of photosynthetic pigments and proteins are altered in the leaves of the *z3* mutant

In typical leaf variegation mutants in rice, the relatively normal green sectors of the leaves are far lighter in color than the wild-type leaves. Consistent with the lighter color, typical leaf variegation mutants in rice also show substantial decreases in the concentrations of photosynthetic pigments and proteins, even in the green sectors of the leaves (Li et al. 2010; Han et al. 2012; Sakuraba et al. 2013). However, the relatively normal green colored sectors of the *z3* mutant were very similar in color to the wild-type leaves (Fig. 1b). Moreover, the levels of chlorophyll and carotenoids were higher than the wild type in the dark-green sectors of the *z3* mutant leaves, by approximately 19.6% and 20.4%, respectively, whereas their levels in the green sectors of the *z3* mutant leaves were statistically indistinguishable from the wild-type leaves (Fig. 2a and b). The levels of photosynthetic proteins (Lhca2, Lhcb2, Lhcb4, D1, and RbcL) showed similar patterns to those of the photosynthetic pigments (Fig. 2c) in that the quantities of all tested proteins were increased in the dark-green sectors of the *z3* mutant leaves, and the green sectors contained almost the same amount of photosynthetic proteins as the wild-type leaves (Fig. 2c). These results suggest that the *z3* mutation affects the levels of photosynthetic pigments and proteins in mature leaves in a transverse pattern.

Map-based cloning of the *z3* locus

The single recessive *z3* mutation was previously mapped onto the classical genetic map of the short arm of chromosome 3 (Iwata et al. 1979). To identify the *Z3* gene, we generated a mapping population of 1547 F_2 plants from a cross between the *z3* mutant (*japonica*) and Milyang23 (an *indica/japonica* hybrid). A total of 291 plants among the F_2 progenies clearly demonstrated the *z3* phenotype in the mature leaves and were used as the homozygous *z3* mutant genotype for map-based cloning. The *z3* locus was initially mapped to a 3.91-Mb region on the short arm of chromosome 3 using one sequence-tagged site (STS) and nine simple sequence repeat (SSR) markers (Fig. 3a). By fine mapping, the *z3* locus was further mapped to a 78-kb interval (Fig. 3b). Twelve predicted genes were identified in this region from the Rice Functional Genomic Express (RiceGE) database (Fig. 3c). To identify the *Z3* gene, we compared the sequence of the 78-kb candidate genomic region between the wild-type *japonica* cultivar 'Nipponbare' (NB) and the *z3* mutant. Of the twelve candidate genes, eleven open reading frames (ORFs) were exactly the same between NB and the *z3* mutant, and only one expressed a gene of unknown function (LOC_Os03g05390). This candidate gene contained a single base substitution from T to C in the third exon, leading to a missense mutation (serine to proline) at the 542th amino acid (Fig. 3d). This single-base substitution was further confirmed by dCAPS analysis (Fig. 3e).

To investigate the importance of a serine at the 542th amino acid (S542) for its role in protein function, we obtained a few protein sequences of predicted orthologs from higher plants using the NCBI-BLASTP program (https://blast.ncbi.nlm.nih.gov/Blast.cgi?PROGRAM=blastp&PAGE_TYPE=BlastSearch&LINK_LOC=blasthome) for amino acid alignment (Additional file 3: figure S3). This analysis revealed that the serine^{542} (S542) residue was highly conserved in higher plants including *Arabidopsis thaliana*, *Zea mays*, *Hordeum vulgare*, *Glycine max*, *Sorghum bicolor*, *Solanum tuberosum*, and *Oryza brachyantha*. This suggested that S542 is a critical residue in terms of the structure or function of the Z3 protein.

The *Z3* gene consists of a 1674-bp open reading frame encoding 557 amino acids with a molecular weight of 60.25 kDa. Z3 has a predicted N-terminal secretory pathway signal peptide, based on analysis with TargetP (Emanuelsson et al. 2000) (Fig. 3f, open rectangle). In addition, Z3 contains nine transmembrane domains (TMDs) and is predicted to be a CitMHS family citrate transporter by the Rice Genome Annotation Project database (http:// rice.plantbiology.msu.edu/), as previously described (Kawahara et al. 2013) (Fig. 3f). These data suggest that Z3 is an integral membrane protein.

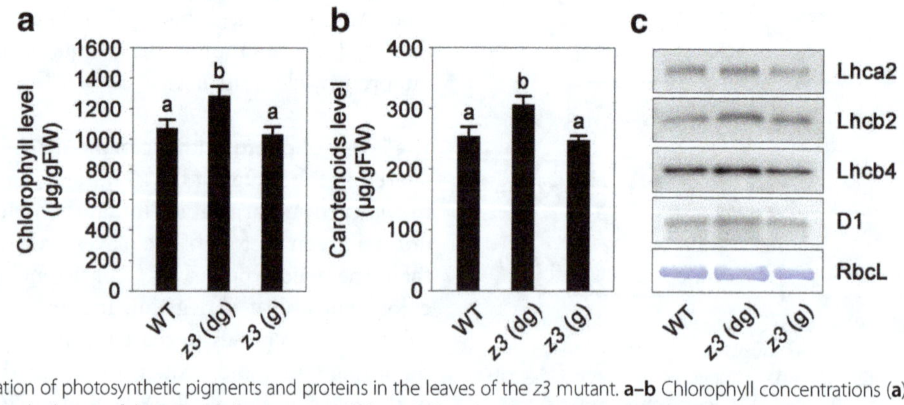

Fig. 2 Characterization of photosynthetic pigments and proteins in the leaves of the *z3* mutant. **a–b** Chlorophyll concentrations (**a**) and carotenoids concentrations (**b**) in the leaves of the WT and the *z3* mutant. Chlorophyll and carotenoids were extracted from the second leaves of plants at 110 DAS grown under natural long day conditions. Means and SD were obtained from five biological replicates. The letters a and b on the bars denote that they are significantly different at the 1% level according to the Duncan's multiple range test. **c** Immunoblot analysis of chloroplast proteins. Total protein was extracted from the second leaves of 110-day-old plants, and extracts were subjected to SDS-PAGE. Antibodies against Lhca2, Lhcb2, Lhcb4, and D1 were used for detection, and the Rubisco large subunit (RbcL) was detected by Coomassie brilliant blue staining. FW, fresh weight; WT, wild type; *z3* (dg), dark-green sectors of the *z3* mutant leaves; *z3* (g), green sectors of the *z3* mutant leaves

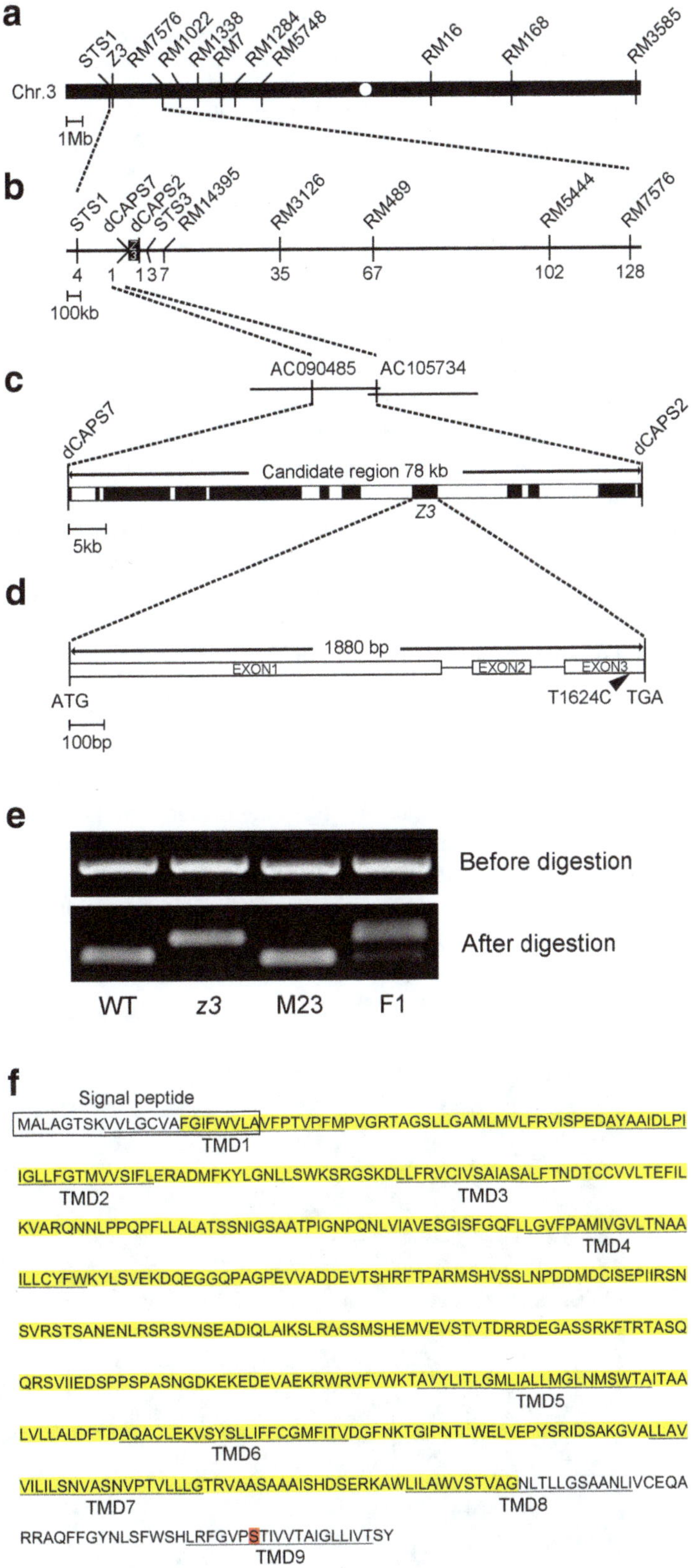

Fig. 3 (See legend on next page.)

(See figure on previous page.)

Fig. 3 Map-based cloning of the *z3* locus. **a** Physical mapping of the *z3* locus. The *z3* locus was initially mapped to a 3.91-Mb region between two markers, STS1 and RM7576, on the short arm of chromosome 3. The marker information is listed in Additional file 8: Table S1. **b** Fine mapping of the *z3* locus. The locus was further mapped within a 78-kb region between two markers, dCAPS7 and dCAPS2. Numbers below the line indicate F_2 recombinants at the marker regions. **c** Candidate genes (black boxes) in the 78-kb region. **d** T to C substitution of *Z3* in the *z3* mutant. Three exons and two introns are indicated as rectangles and lines, respectively. The position of the point mutation in the *z3* mutant is at the third exon and represented by a black arrowhead. **e** Derived cleaved amplified polymorphic sequence (dCAPS) analysis of the point mutation in the *z3* mutant. *Aat*II was able to digest the genomic PCR products amplified from the WT, but not from the *z3* mutant because of its T to C substitution in the *Aat*II restriction enzyme site. M23, a mapping parent 'Milyang23'; F1, F_1 hybrid (*z3*/Milyang23). **f** Z3 protein sequence. The predicted N-terminal signal peptide is marked with an open rectangle and the nine transmembrane domains (TMDs) are underlined. The yellow highlight represents regions predicted to be a CitMHS family. The position of the point mutation in the *z3* mutant is highlighted in red

To determine the intracellular localization of Z3, a reporter gene encoding yellow fluorescent protein (YFP) was fused to the C-terminus of Z3, and the construct was introduced into onion epidermal cells using particle bombardment. The fluorescent signal of YFP alone, used as a control, was observed throughout the cells (Fig. 4a–c), whereas the signal of the Z3-YFP fusion protein was observed only at the plasma membrane (Fig. 4d–f). These results suggest that Z3 localizes in the plasma membrane.

Complementation of the *z3* mutation by constitutive expression of *Z3* in Rice

To confirm that the missense mutation in Z3 (S542) is responsible for pleiotropic phenotype of the *z3* mutant, including leaf variegation, seed discoloration, and late flowering, we transformed the *z3* mutants with the full-length cDNA of *Z3* driven by the cauliflower mosaic virus 35S promoter (*35S::Z3*). We obtained six independent T_0 transgenic plants and observed the phenotypes of T_1 transgenic lines. To confirm the rice transformation, we performed PCR analysis using primers designed to amplify the *Z3* transgene and confirmed all six independent T_1 transgenic plants (Additional file 4: figure S4a). Overexpression of *Z3* by the CaMV 35S promoter in the transgenic lines was validated by semi-quantitative RT-PCR (Additional file 4: figure S4b). The six complementation lines (*35S:Z3*/*z3*) grew normally at the early vegetative stage uren contrast to the inferior growth of the *z3* mutant (data not shown), and flowered as early as the wild type (Fig. 5a), indicating that constitutive expression of *Z3* complemented the late-flowering phenotype of the *z3* mutants under NLD conditions in the field. Moreover, the leaf and grain phenotypes of these six complementation lines were indistinguishable from those of the wild-type plants (Fig. 5b–d), confirming that the leaf variegation and seed discoloration phenotypes of the *z3* mutant results from the missense mutation in Z3 (S542).

To further validate the effects of the z3 mutation on the defective leaf and seed phenotypes, we obtained two independent T-DNA insertional alleles for the *z3* mutation from the Salk Institute Genomic Analysis Laboratory (http://signal.salk.edu/cgi-bin/RiceGE) (Jeon et al. 2000).

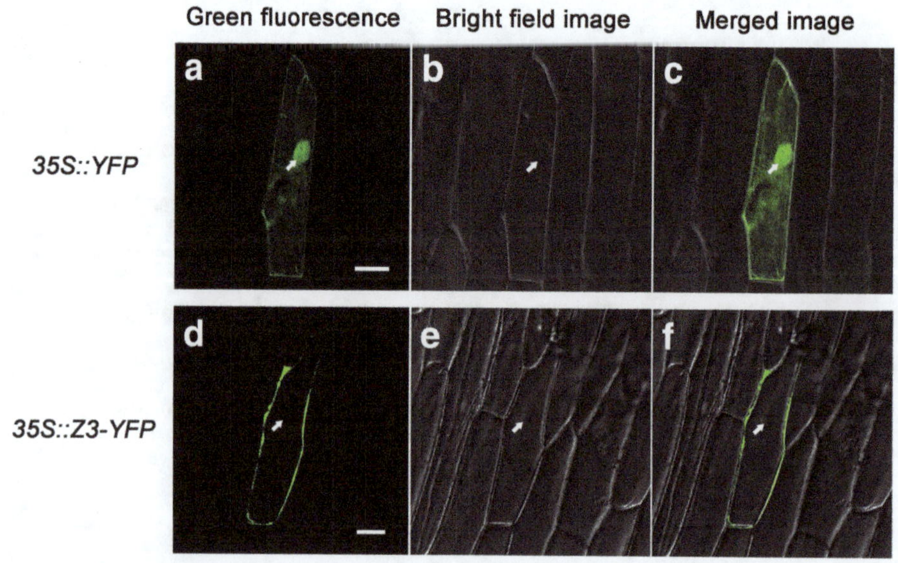

Fig. 4 Z3 localizes in the plasma membrane. **a–c** Onion epidermal cells transformed with the control construct (*35S::YFP*). **d–f** Onion epidermal cells transformed with the construct *35S::Z3-YFP*. **a, d** Green fluorescence; **b, e** bright field image; **c, f** merged image. Cells were analyzed by confocal microscopy 18 h after the particle bombardment. Arrows indicate the nucleus. Scale bars = 50 μm

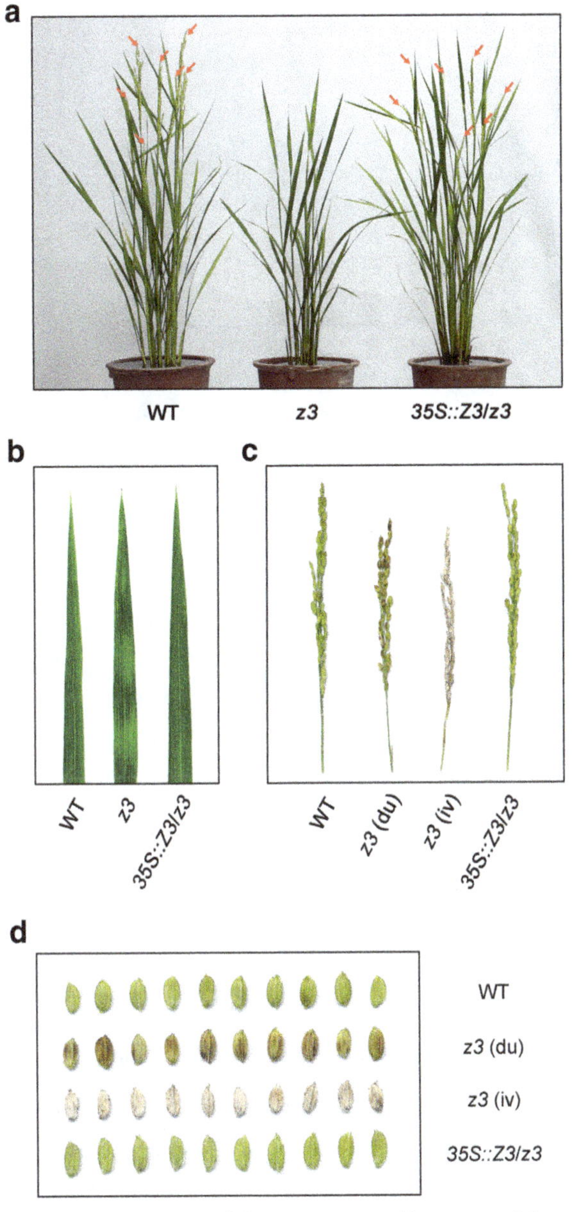

Fig. 5 Complementation of the *z3* mutation. **a** Phenotypes of the WT, the *z3* mutant, and the *35S::Z3/z3* transgenic plant at 80 DAS grown under natural long day conditions. Leaf variegation and delayed flowering (Additional file 1: figure S1) of the *z3* mutation were rescued by the transformation of *35S::Z3* construct. WT flowered around one month earlier than the days to heading data in Additional file 1: figure S1 because the seeds of the WT, the *z3* mutant, and the *35S::Z3/z3* transgenic plant in Fig. 5a were sown on June, not in April (Suwon, Korea, 37°N latitude). Arrows indicate the panicle. **b** Leaf phenotypes of the WT, the *z3* mutant, and the *35S::Z3/z3* transgenic plant at 80 DAS grown in the paddy field. **c–d** Panicle and grain phenotypes of 80-day-old WT, *z3* mutant, and *35S::Z3/z3* transgenic plants grown under natural long day conditions. *z3* (du), dull-brown colored grain of the *z3* mutant; *z3* (iv), ivory colored grain of the *z3* mutant

We designated these two independent alleles as *z3–2* and *z3–3*, and the *z3* allele identified by map-based cloning in this study as *z3–1*. For the *z3–2* and *z3–3* alleles, the T-DNA was inserted in the first exon of *Z3*, but in different regions (Additional file 5: figure S5a). We performed genotyping to confirm the T-DNA insertion in the *Z3* gene, and obtained homozygous mutants for each allele (Additional file 5: figure S5f). The absence of the full-length transcript of *Z3* in these two T-DNA lines was validated by semi-quantitative RT-PCR (Additional file 5: figure S5 g). The two independent *z3–2* and *z3–3* mutants began to produce leaf variegation at about 95 days after sowing, and transverse variegation of the leaf blades gradually became clearer thereafter, as in the *z3–1* mutant (Additional file 5: figure S5b). Furthermore, they also flowered late (Additional file 5: figure S5e) and exhibited two types of panicles on the same plant; an ivory colored head and a panicle with dull brown-colored grain (Additional file 5: figure S5c-d). These observations further demonstrated that the loss-of-function *z3* mutation is responsible for the pleiotropic phenotypes including leaf variegation, late flowering, and seed discoloration.

The *Z3* gene is mainly expressed in the leaf blade and panicle branches at the heading stage

To examine the potential tissue specificity of Z3, we used qRT-PCR to characterize the spatial patterns of *Z3* expression at the seedling and heading stages. In the 2-week-old seedling plants, the *Z3* gene appeared to be expressed evenly in all tissues, including leaf blade, leaf sheath, and root (Fig. 6a). However, at the heading stage, *Z3* was mainly expressed in leaf blades and panicle branches, although we could detect *Z3* transcripts in all tissues, including flag leaf, second leaf blade, leaf sheath, culm, root, and spikelet (Fig. 6b).

Leaf variegation of the *z3* mutant appears around three months after sowing, approximately at the panicle differentiation stage, in natural long-day conditions. This suggests that the Z3 protein has a functional role in the mature leaf mainly at the late vegetative stage. To test whether *Z3* expression changes in the wild type when leaf variegation appears in the *z3* mutant, we inspected the transcript levels of the *Z3* gene in the leaf blade with qRT-PCR at different growth stages of the wild type grown in the field under natural long-day conditions at 20-day intervals. This showed that *Z3* expression in the wild-type leaves was upregulated at three months after sowing, and continued to increase thereafter (Fig. 6c).

The leaf variegation in the *z3* mutant may be caused by different levels of citrate in the dark-green and green sectors

Excessive accumulation of ROS in leaves can lead to leaf variegation and/or necrosis in some rice mutants such

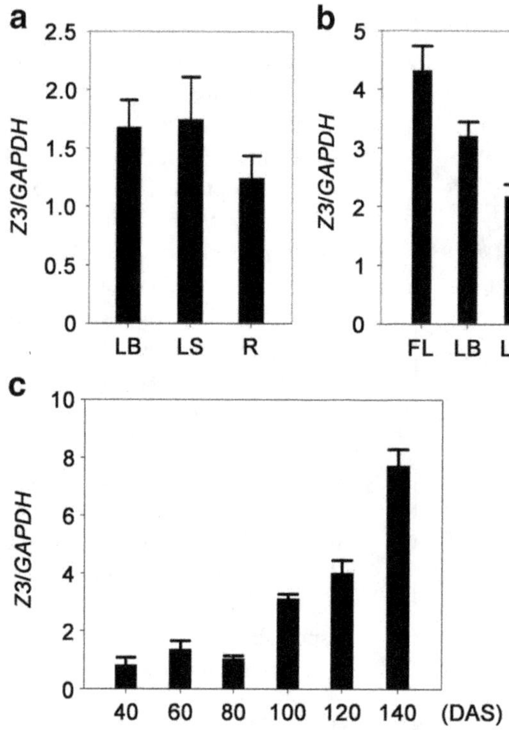

Fig. 6 Spatial and temporal expression of the *Z3* gene. **a** Expression of the *Z3* gene in different organs at the seedling stage. Different organs were sampled from 2-week-old seedlings (cv Kinmaze) grown under long-day conditions (14.5 h-light/9.5 h-dark) in the growth chamber. LB, leaf blade; LS, leaf sheath; R, root. **b** Spatial expression of the *Z3* gene in different rice tissues at heading stage. Different tissues were sampled from the 115-day-old wild type grown under natural long-day conditions in the paddy field. FL, flag leaf; LB, leaf blade of the second leaf; LS, leaf sheath; C, culm; R, root; S, spikelet; PB, panicle branch. **c** Temporal expression of the *Z3* gene. Samples were harvested from fully expanded young leaves of the wild type at different stages grown under natural long -day conditions. DAS, days after sowing. *GAPDH* was used as internal control. All data in Fig. 6 are means ± SD ($n = 3$)

as in the *preharvesting sprouting3* (*phs3*), *faded green leaf* (*fgl*), and *zebra2* (*z2*), *zebra-necrosis* (*zn*) mutants, and the *Staygreen* (*SGR*)-overexpressing plants (Han et al. 2012; Li et al. 2010; Sakuraba et al. 2013; Jiang et al. 2011; Fang et al. 2008). Therefore, we examined the levels of two types of ROS molecules, hydrogen peroxide and superoxide anion radicals, in the leaves of the *z3* mutant. Hydrogen peroxide and superoxide anion radicals were not detectable in the *z3* mutant leaves (Additional file 6: figure S6). These results suggested that ROS are not involved in the formation of variegated leaves in the *z3* mutant, unlike other leaf variegation mutants in rice.

Many leaf variegation mutants have defects in chloroplast development. For example, chloroplasts in the *fgl* mutant display defective thylakoid stacking and have several plastoglobules (Sakuraba et al. 2013), and chloroplasts in the *yellow-green leaf 2* (*ygl2*) and the naturally occurring Baihuaidio 7 mutants exhibit fewer lamellar structures (Chen et al. 2013; Li et al. 2012). To determine whether the leaves in the *z3* mutant show defects in chloroplast structure, we performed ultrastructural analysis of the chloroplasts in the leaves of the wild type and the *z3* mutant. The chloroplasts in the *z3* mutant leaves were indistinguishable from those of the wild type, with both exhibiting well-stacked grana thylakoids with no plastoglobules (Additional file 7: figure S7). Moreover, the green sectors of the *z3* mutant leaves contain almost the same levels of photosynthetic pigments compared with the wild type. These results indicated that the *z3* mutation has little effect on chloroplast development.

In the NCBI BLAST (https://blast.ncbi.nlm.nih.gov/Blast.cgi) and the Rice Genome Annotation Project databases (http:// rice.plantbiology.msu.edu/), the Z3 protein is predicted to be a CitMHS family citrate transporter (Fig. 3f). This suggests that Z3 might be involved in the distribution and/or translocation of citrate in rice. Therefore, we measured citrate concentrations in the leaves of the wild type and the *z3* mutant using HPLC. The concentration of citrate in 90-day-old *z3* mutant leaves (prior to visible leaf variegation) was similar to that of the wild type (Fig. 7a–b). However, citrates accumulated to high levels in the dark-green sectors of 117-day-old *z3* mutant leaves, but not in the green sectors (Fig. 7c–d). Therefore, the citrate concentration

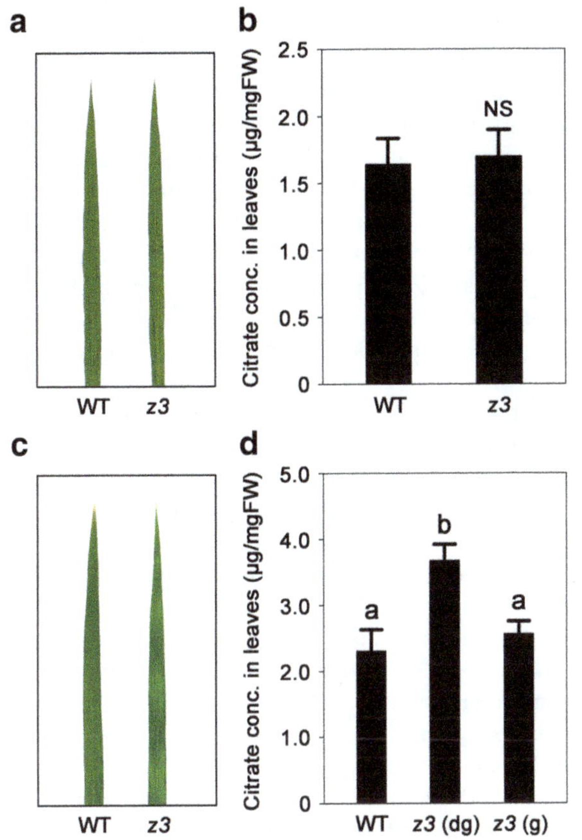

Fig. 7 Citrate concentrations in the leaves of the *z3* mutant. **a, c** The second leaves of 90-day-old (**a**) and 117-day-old (**c**) WT and the *z3* mutant grown in a natural paddy field. **b, d** Citrate concentrations in the second leaves of the WT and the *z3* mutant at 90 DAS (**b**) and 117 DAS (**d**) grown under natural long day conditions. Citrate concentrations were determined with HPLC. Data are means ± SD ($n = 4$). Different letters indicate a significant difference at $P <$ 0.01 (Duncan's test). Abbreviations: WT, wild type; *z3* (dg), dark-green sectors of the *z3* mutant leaves; *z3* (g), green sectors of the *z3* mutant leaves; FW, fresh weight; NS, not significant; conc., concentration

in the leaves was not affected by the *z3* mutation before leaf variegation was visible. These results suggest that Z3 might play a role in the distribution and/or translocation of citrate in leaves after the late vegetative stage, and thus the *z3* mutant produces variegated leaves after 95 days after sowing, possibly by accumulating citrate to different extents in the dark-green and green parts of the leaves.

Discussion

Here, we report a new type of leaf variegation mutant, *zebra3* (*z3*), which has transverse dark-green/green sectors in mature leaves without yellow or white sectors. In most cases, the leaves of variegated mutants have green/yellow (or white) sectors, and this abnormal phenotype is caused by impaired chloroplast biogenesis in the yellow or white sectors. Thus, the yellow (or white) sectors of these mutant's leaves often contain chloroplasts with defective thylakoid stacking, less photosynthetic pigments, and also accumulate excessive levels of ROS (Li et al. 2010; Han et al. 2012; Carol et al. 1999). However, the *z3* mutant leaves did not show any of these defects. Even the slightly lighter green colored sectors in the *z3* mutant leaves contained similar levels of photosynthetic pigments and proteins to the wild type (Fig. 2), and showed normal chloroplast structure with well-stacked grana thylakoids (Additional file 7: figure S7) when compared with the wild type. Moreover, the *z3* mutant leaves did not accumulate excessive ROS as is common in other leaf variegation mutants (Additional file 6: Figure S6). These results suggest that leaf variegation of the *z3* mutant is not associated with a defect in chloroplast biogenesis or an increase of ROS.

The *z3* mutant is further different from other *zebra* mutants in that leaf variegation occurs at the late vegetative stage, around the panicle differentiation stage, strongly suggesting that Z3 has an important role after the late vegetative stage. Consistent with this, *Z3* expression increased in mature leaves at this stage, and its transcript level increased further thereafter (Fig. 6c). These results may partially explain why leaf variegation of the *z3* mutants is visible only at the late growth stages.

In addition, the *z3* mutant produced pale-green leaves (data not shown) and had a slower growth rate at the seedling stage (Additional file 1: figure S1c), although the overall phenotype of the *z3* mutant is normal compared with the wild type during vegetative growth. Even though Z3 mainly functions after the late vegetative stage, it also has a basal function in the early vegetative stages, since *Z3* is expressed in all tested tissues, including leaf blade, leaf sheath, and root (Fig. 6a). Taken together, these results indicate that Z3 protein may be involved in the maintenance of chlorophyll biosynthesis and overall growth rate at the early vegetative stage.

Map-based cloning revealed that *Z3* encodes a putative citrate transporter that belongs to the citrate-metal hydrogen symport (CitMHS) family. CitMHS family members transport metal–citrate complexes in symport with a proton, and have only been identified in bacteria thus far. Dessureault-Rompré et al. (2008) reported that bacteria acquire citrate and metal ions by importing metal-citrate complexes against the cellular electrochemical potential gradient by coupling import to the transport of a proton. In this scenario, citrate functions not only as an energy source but also as a chelating agent. In plants, citrate is mostly involved in cellular metabolism, such as the Krebs cycle, fatty acid synthesis, NH_4^+ assimilation, and the glyoxylate cycle (Sarantinopoulos et al. 2003; Krebs and Johnson 1937; Fritsch and Beevers 1979; Mattoo and Modi 1970; Nelson and Rinne 1975; Chen and Gadal 1990; Sienkiewicz-Porzucek et al. 2008; Tanner and Beevers 1965). Because the Z3 protein localizes to the plasma membrane (Fig. 4) and is predicted to be a CitMHS family member (Fig. 3), it may transport citrate among leaf cells. Furthermore, as the leaf blades of the *z3* mutant accumulated citrate to different extents in the dark-green and green sectors (Fig. 7), it appears that Z3 can act as a citrate transporter between leaf cells, and thus may play an important role in the balanced distribution of citrates. However, when citrate is released and transported outside the cell, it can function as a chelator. Therefore, the citrates potentially transported by Z3 may act as chelators, rather than as intermediates, in cellular metabolism. In this respect, it can be speculated that the unusual leaf variegation phenotype in the *z3* mutant is associated with metal ions which play a key role in chlorophyll biosynthesis. However, we cannot rule out the possibility that the leaf variegation in the *z3* mutants is caused by the citrates by themselves.

Leaf variegation mutants can be classified into two types depending on the genotypes of the green sectors and yellow (or white) sectors. First, in some mutants, the green and the yellow (or white) sectors in the leaves have different genotypes. That is, the green sectors contain the wild-type alleles, while the yellow (or white) sectors contain only mutant alleles. In many cases, this type of leaf variegation results from transposable element activity. For example, *dcl* (*defective chloroplasts and leaves*) in tomato (Keddie et al. 1996), *dag* (*differentiation and greening*) in *Antirrhinum* (Chatterjee et al. 1996), and *pyl-v* (*pale-yellow-leaf variegated*) in rice (Tsugane et al. 2006) are well-studied mutable alleles of genes for chloroplast biogenesis. In the second type of leaf variegation, all cells in the mutant leaves have the same genotype. That is, both green and yellow (or white) sectors contain the same mutant alleles, but the phenotype is differentially expressed between subsets of the cells, leading to the generation of green and yellow/white sectors. Because the leaves of the *z3* mutant have a uniform mutant genotype (data not shown), the leaf variegation of the *z3* mutant belongs to the second type. However, the leaf variegation mechanism of this type remains poorly understood.

One outstanding question about the *z3* phenotype is how mutation of a putative citrate transporter produces variegated leaves, although the all cells in the *z3* mutant have the same mutant genotype. The *z3* mutant leaves accumulated different levels of citrates in the dark-green and the green sectors (Fig. 7c-d), and this might be associated with the production of variegated leaves. As the Z3 protein is thought to transport citrates and the *z3* mutant accumulates the citrates to high levels in the dark-green sectors of leaves, it is plausible that the Z3 is involved in dispersion of citrates in the leaves. In this scenario, there might be other citrate transporters which are involved in polar transport of citrates in leaves. Polar transport is a directional transport of functional molecules between cells, and plasma membrane transporters that localize in a polar manner are responsible for polar transport in most cases (Naramoto 2017). For example, AtPIN1, an auxin efflux carrier protein in *Arabidopsis*, localizes only at one side of the plasma membrane in cells, and transports auxin to convergence points in the epidermal tissues where the major veins are produced (Scarpella et al. 2006). Similarly, there might be other citrate transporters, possibly such as MATE family citrate transporters (Yokosho et al. 2009; Yokosho et al. 2011) or other CitMHS family citrate transporters, which transport citrates to convergence points in leaves in transverse patterns after the late vegetative stage. In this case, citrate transported transversely by these citrate transporters could be dispersed by Z3 to maintain a uniform distribution of citrate in the leaves of the wild type. Therefore, it could be speculated that due to the functional deficiency of Z3 protein, the uniform citrate distribution in leaves is disrupted, resulting in accumulation of citrate in transverse patterns. The formation of dark green sectors and different citrate concentrations in the sectors of *z3* mutant leaves remains a key question for future study.

The *z3* mutant exhibits a pleiotropic phenotype in addition to leaf variegation; the *z3* mutants flowered late compared with the wild type (Additional file 1: figure S1a-b). *Z3* was expressed throughout the tested tissues, even in early vegetative stages (Fig. 6a) and *Z3* expression occurs throughout development (Fig. 6c), indicating that the Z3 protein has basal roles in all tissues. As the Z3 protein may transport metal-citrate complexes, we speculate that the *z3* mutation might disrupt metal ion homeostasis, thus affecting overall growth rate. Consistent with our speculation, the *z3* mutant produced slightly lighter green leaves at the seedling stage (data not shown), and showed inferior growth rates during vegetative growth in comparison with the wild type (Additional file 1: figure S1c). These results suggest that the late flowering of the *z3* mutant is partially due to the retardation of growth resulting from unbalanced distribution of metal ions. However, based on our current data, we cannot rule out the possibility that the *z3* mutation also causes a delayed floral transition.

The *z3* mutant also showed a grain discoloration phenotype at the ripening stage under natural long-day conditions (Fig. 1c-d). Grain discoloration in rice is characterized by dull brown or black spots on the husk, and reduces grain quality (Navasero and Winslow 1987). In most cases, fungal or bacterial infections are responsible for grain discoloration (Ngala 1983), and deficiency of mineral nutrients, such as silicon, is involved in this symptom (Datnoff et al. 1991; Winslow 1992). In addition,

Dobermann and Fairhurst (2000) reported that the lack of metal ions, such as potassium, could cause irregular brown necrotic spots on developing panicles in rice. As *Z3* is highly expressed in panicle branches, similar to the expression levels in leaf blade (Fig. 6b), it is plausible that the Z3 protein also functions in translocating metal ions into grains. In this case, a functional deficiency of Z3 protein may disrupt the translocation of metal ions complexed with citrates into developing grains, and the resulting deficiency of metal ions, or possibly citrate itself, causes grain discoloration.

Conclusions

The *zebra* mutants in rice produce transverse sectors of green/yellow or green/white in developing or mature leaves. In most cases, this abnormal phenotype is caused by impaired chloroplast biogenesis or excessive accumulation of ROS in the yellow or white sectors. In contrast to these mutants, the *zebra3* (*z3*) mutant has transverse dark-green/green stripes in the leaves with no yellow or white sectors. In addition, the *z3* mutant flowers about 10 days later than the WT, and produces defective panicles. We found that he *Z3* locus encodes a putative citrate transporter by map-based cloning and complementation test. Though the *z3* mutation has little effect on chloroplast development, citrate levels were significantly higher in the dark-green sectors of the *z3* mutant leaves than in the green sectors or the normal leaves of the wild type. These results indicate that leaf variegation in the *z3* mutant is caused by an unbalanced distribution of citrate in a transverse pattern in leaf tissues. Taking all the results together, we propose that Z3 plays a role in the transport and distribution of citrates in leaves, and is a possible candidate for CitMHS family members in higher plants.

Methods

Plant materials and growth conditions

The *zebra3–1* (*z3–1*) mutant was isolated from a mutant pool produced by applying *N*-methyl-*N*-nitrosourea, to a *japonica* rice cultivar 'Kinmaze', as previously described (Iwata et al. 1979). T-DNA knockout mutants of *Z3* (LOC_Os03g05390; PFG_1E-02030.R, designated as *z3–2*; PFG_2D-00576.R, designated as *z3–3*) were derived from the Korean *japonica* rice cultivar 'Hwayoung' (HY) and obtained from the Salk Institute Genomic Analysis Laboratory (http://signal.salk.edu/cgi-bin/RiceGE) (Jeon et al. 2000). All rice plants were grown under natural long day conditions (approximately 14 h light/day) in a paddy field (Suwon, Korea, 37°N latitude). For spatial expression analysis of the *Z3* gene at the seedling stage, the rice cultivar Kinmaze was grown under long-day conditions (14.5 h-light, 30 °C/9.5 h-dark, 24 °C) under cool-white light (300 μmol m^{-2} s^{-1}) in the growth chamber.

Leaf emergence rate and heading date

To compare the leaf emergence rate between the *z3* mutant and the wild-type, the plants were grown in long-day conditions (14.5 h-light, 30 °C/9.5 h-dark, 24 °C) with 60% relative humidity. The light source was light-emitting diodes producing mixed red, green, and blue light, and photon flux density was about 300 μmol m^{-2} s $^{-1}$. Leaf emergence rate was estimated as previously described (Itoh et al. 1998). Briefly, when each new leaf blade had completely emerged from the sheath of the previous leaf, the number of leaves on the main culm was counted for each plant. Heading date (or flowering time) was recorded from sowing to emergence of the first panicle in the main culm.

Map-based cloning

A mapping population of 1547 F_2 individuals was generated by crossing a *japonica*-type *z3* mutant and a tongil-type cultivar, Milyang23 (M23). To confirm the chromosomal localization of the *Z3* locus, we initially mapped the *z3* mutation using 291 F_2 plants with the *z3* mutant phenotype, 9 simple sequence repeat (SSR) markers, and one sequence-tagged site (STS) marker distributed on chromosome 3. SSR marker information is available in GRAMENE (http://www.gramene.org). For fine mapping, one STS and two derived cleaved amplified polymorphic sequence (dCAPS) markers were designed by comparison of the genomic DNA sequences between M23 and the *z3* mutant using whole-genome re-sequencing data (Additional file 8: Table S1). Genomic DNA was extracted from the leaf blades of two-week-old M23 and *z3* mutant plants using the NucleoSpin Plant II kit (Macherey-Nagel, Germany) according to the manufacturer's instructions. Whole-genome re-sequencing was carried out using SolexaQA package v.1.13 (Cox et al. 2010) and BWA 0.6.1-r104 (Li and Durbin 2009). For dCAPS analysis, a mismatch forward primer (5′-CCAC CTCCGGTTCGG GGTGAC-3′) and the reverse primer (5′- CAAAGGGTTCGTTCATCTATCTC -3′) were designed using DCAPS FINDER2.0 (http://helix.wustl.edu/dcaps/dcaps.html). Each PCR product was digested with a restriction enzyme, *Aat*II, and separated on an agarose gel.

Subcellular localization of Z3

YFP cDNA was fused to the C-terminus of *Z3* cDNA in the pEarleyGate 101 (pEG101) vector through LR recombination (Lambda integrase/excisionase; Elpis-Biotech), resulting in the *35S:Z3-YFP* plasmid. The fusion construct, as well as the control (empty pEG101 vector; *35S:YFP*), were transformed into onion (*Allium cepa*) epidermal cells using a DNA particle delivery system (Biolistic PDS-1000/He; Bio-Rad, Hercules, CA, USA). The transformed onion epidermal layers were incubated at 25 °C on Murashige and

Skoog medium (pH 5.7) plates under the dark for 20 h. Then, the onion cell layers were examined with a confocal laser scanning microscope (SP8 X, Leica, Germany).

Vector construction and plant transformation

For complementation of the *z3* mutation, a full-length cDNA of *Z3* was cloned into the pMDC32 Gateway binary vector containing the cauliflower mosaic virus (CaMV) 35S promoter (Curtis and Grossniklaus 2003). The *35S:Z3* construct in the pMDC32 vector was introduced into calli generated from mature seed embryos of *z3* mutants through the *Agrobacterium* (strain EHA105)-mediated method (Jeon et al. 1999; Lee et al. 1999). The transgenic rice plants were selected on 2 N6 media containing hygromycin (50 mg L^{-1}) and confirmed by PCR using the specific primer set, *Z3* (Transgene-2) F/R (Additional file 8: Table S1).

Identification of the T-DNA insertional alleles *z3–2* and *z3–3*

To identify the homozygous *z3–2* and *z3–3* mutants, we extracted genomic DNA from the segregating T_1 populations using a cetyl trimethyl ammonium bromide (CTAB) method (Murray and Thompson 1980) and performed PCR analysis. PCR was conducted with a T-DNA plasmid pGA2707 right border primer (*z3–2* RB) in combination with a *Z3* left genomic primer (*z3–2* LP) for the *z3–2* mutant, and with a T-DNA plasmid pGA2772 right border primer (*z3–3* RB) in combination with a *Z3* left genomic primer (*z3–3* LP) for the *z3–3* mutant (Additional file 8: Table S1). PCR was performed with 30 cycles of 98 °C for 30 s, 55 °C for 30 s, and 72 °C for 45 s. PCR was also conducted with a *Z3* left genomic primer (*z3–2* LP) in combination with a *Z3* right genomic primer (*z3–2* RP) for the *z3–2* mutant, and with a *Z3* left genomic primer (*z3–3* LP) in combination with a *Z3* right genomic primer (*z3–3* RP) for the *z3–3* mutant (Additional file 8: Table S1). PCR was performed with 30 cycles of 98 °C for 30 s, 55 °C for 30 s, and 72 °C for 1 min.

Reverse transcription and semi-quantitative and quantitative RT-PCR

Total RNA was extracted from leaves using the MG Total RNA Extraction Kit (Macrogen, Seoul, Korea) according to the manufacturer's instructions. First-strand cDNAs were synthesized from 2 μg of total RNA using oligo$(dT)_{15}$ primers and M-MLV reverse transcriptase (Promega, Madison, WI, USA) and diluted with water to 100 μL. The transcript level of the *Z3* gene was examined by semi-quantitative RT-PCR using specific primers, and *Ubiquitin5* (*UBQ5*) (GenBank accession number: AK061988) or *Glyceraldehyde-3-phosphate dehydrogenase* (*GAPDH*) (GenBank accession number: AK064960) was used for a loading control as previously described (Jain et al. 2006).

To determine the transcript levels of the *Z3* gene by quantitative RT-PCR analysis, 20 μl of qRT-PCR mixture containing 2 μl of the first-strand cDNA mixture, 0.25 μM of the forward and reverse primers for each gene, and 10 μl of 2X QuantiTect LightCycler 480 SYBR Green I Master (Roche) were used. qRT-PCR analysis was performed on the Light Cycler 2.0 instrument (Roche Diagnostics, Germany). The mRNA levels of the *Z3* gene were normalized to those of *Glyceraldehyde-3-phosphate dehydrogenase* (*GAPDH*). Primer information is listed in Additional file 8: Table S1.

Measurement of photosynthetic pigments

To evaluate the concentration of total chlorophyll and carotenoids, pigments were extracted from equal fresh weights of leaves with 80% ice-cold acetone. Chlorophyll and carotenoid levels were measured with a UV/VIS spectrophotometer (BioTek) as previously described (Lichtenthaler 1987).

SDS-PAGE and immunoblot analysis

Photosynthesis-related proteins were detected as previously described, with some modifications (Han et al. 2012; Kwon et al. 2015; Kwon et al. 2017). Leaf tissue (10 mg) was homogenized with 100 μL of SDS-PAGE sample buffer [50 mM Tris pH 6.8, 2 mM EDTA, 10% (*w*/*v*) glycerol, 2% SDS, and 6% 2-mercaptoethanol] and denatured at 100 °C for 5 min, then samples were subjected to SDS-PAGE. For immunoblot analysis, 5 μL of each protein sample was used. The resolved proteins were electroblotted onto the Immobilon-P Transfer Membrane (Millipore). Antibodies against photosystem proteins (Lhca2, Lhcb2, Lhcb4, and D1) were obtained from Agrisera (Sweden). To detect the horseradish peroxidase activity of secondary antibodies (Sigma), an ECL detection kit, WESTSAVE (AbFRONTIER, Korea), and a chemiluminescence system (FUSION-FX7, France) were used according to the manufacturers' protocols. The Rubisco large subunit (RbcL) was visualized by staining with Coomassie brilliant blue reagent (Sigma-Aldrich) after immunoblot analysis.

Detection of reactive oxygen species

Detection of hydrogen peroxide (H_2O_2) and superoxide anion radicals (O_2^-) was carried out as previously described (Li et al. 2010; Han et al. 2012; Wi et al. 2010) with minor modifications. Hydrogen peroxide and superoxide anion radicals were detected by 3,3-diaminobenzidine (DAB) and nitroblue tetrazolium chloride (NBT), respectively. Flag leaves of 160-day-old plants grown in the field were sampled and incubated in 0.1% DAB (Sigma) in distilled water or 0.05% NBT (Duchefa) in 50 mM sodium phosphate buffer (pH 7.5) at room temperature overnight with gentle shaking. Chlorophyll was then removed by incubation in 90% ethanol at 80 °C. Hydrogen peroxide and superoxide anion radicals were visualized as reddish brown and dark blue stains, respectively.

Transmission electron microscopy

Leaf samples for transmission electron microscopy were harvested from five-month-old plants grown under natural long day conditions. Fixation and polymerization of leaf samples were carried out as described previously (Park et al. 2007). Segments of leaf tissues were fixed in modified Karnovsky's fixative (2% paraformaldehyde, 2% glutaraldehyde, and 50 mM sodium cacodylate buffer, pH 7.2) and washed three times with 50 mM sodium cacodylate buffer (pH 7.2) at 4 °C for 10 min. The samples were post-fixed with 1% osmium tetroxide in 50 mM sodium cacodylate buffer, pH 7.2, at 4 °C for 2 h and briefly washed twice with distilled water at 25 °C. The samples were then bloc stained in 0.5% uranyl acetate at 4 °C for a minimum of 30 min, dehydrated in a gradient series of ethanol and propylene oxide, and embedded in Spurr's resin. After polymerization at 70 °C for 24 h, the sections were sliced to 60 nm with an ultramicrotome (MT-X; RMC, http://www.rmcboeckeler.com/) and stained with 2% uranyl acetate for 5 min and Reynold's lead citrate for 2 min at 25 °C. The processed samples were finally examined using a JEM-1010 EX electron microscope (JEOL, https://www.jeol.co.jp/en/).

Estimation of citrates by high performance liquid chromatography (HPLC)

Organic acids, including citrate, were extracted from leaf samples as previously described (Phillips and Jennings 1976), with some modifications. Leaf tissues (100 mg) of 117-day-old plants grown under NLD conditions were sampled and homogenized with 1 mL of cold acidified 80% ethanol. After centrifugation at 5000 g for 15 min, the supernatant was decanted and kept. All operations up to this point were carried out at 4 °C. The samples were filtered with a Millex-GV syringe filter (0.22 μm), and citrate concentrations in the samples were analyzed with HPLC using an Aminex 87H column. The mobile phase was a 0.01 N H_2SO_4 run at 40 °C, and peaks were detected by a refractive index detector (ERC, Refracto-MAX520, Japan) at a wavelength of 210 nm.

Additional files

Additional file 1: Figure S1. Late flowering phenotypes of the *z3* mutant. a Flowering phenotypes of the 117-day-old WT and *z3* mutant grown under natural long day conditions (14 h light/day, 37o N latitude) in the paddy field. b Days to heading of the WT and the *z3* mutant in natural long day conditions. Means and SD were obtained from 15 plants of each genotype. Error bars indicate SD. Differences between means were compared using Student's *t*-test (*** $P < 0.001$). c Comparison of leaf emergence rates between the wild type and the *z3* mutants grown under long-day conditions (14.5 h-light/9.5 h-dark) in the growth chamber. Mean and standard deviation values are shown ($n = 10$). Leaf emergence rate was calculated according to the methods described by Itoh et al. (1998). The average heading dates of the wild type and the *z3* mutants are shown by closed and open arrows, respectively.

Additional file 2: Figure S2. Statistical analysis of agronomic traits in the *z3* mutant. a–f Agronomic traits were investigated between the WT and the *z3* mutant plants grown under natural conditions. Investigated traits are as follows: a main panicle length, b number of panicles per plant, c number of spikelets per main panicle, d seed setting rate, e 500-grain weight, and f yield per plant. Ten plants were used to measure each trait. Values are shown as means. Error bars indicate SD. Student's *t*-test was performed for statistical analysis (* $P < 0.05$, ** $P < 0.01$, *** $P < 0.001$).

Additional file 3: Figure S3. Amino acid alignment of Z3 homologs in higher plants. The amino acid sequences of Z3 homologs in higher plants were acquired from NCBI (http://www.ncbi.nlm.nih.gov/), and the amino acid alignment was obtained using the Clustal Omega EMBL-EBI (https://www.ebi.ac.uk/Tools/msa/clustalo/) and BoxShade 3.21 Server (https://www.ch.embnet.org/software/BOX_form.html). *O. brachyantha*, *Z. mays_1*, *Z. mays_2*, *H. vulgare*, *S. bicolor*, *S. tuberosum*, *G. max*, and *A. thaliana* have 99, 96, 95, 87, 72, 65, 64, and 62% sequence similarity to Z3, respectively. The mutated residue (S542P in *z3*) is indicated by a red arrowhead. *O. sativa_ZEBRA3* (*Oryza sativa* ZEBRA3, LOC_Os03g05390, NP_001048962.1); *O. brachyantha* (*Oryza brachyantha*, XP_006649400.1); *Z. mays_1* (*Zea mays*, NP_001151517.1); *Z. mays_2* (*Zea mays*, ACG43196.1); *H. vulgare* (*Hordeum vulgare*, BAK05230.1); *S. bicolor* (*Sorghum bicolor*, XP_002467148.1); *S. tuberosum* (*Solanum tuberosum*, XP_006363328.1); *G. max* (*Glycine max*, XP_003533988.1); *A. thaliana* (*Arabidopsis thaliana*, NP_171728.2).

Additional file 4: Figure S4. Confirmation of the overexpression of *Z3* by the CaMV 35S promoter in six independent complementation lines. a Genomic PCR analysis for confirmation of rice transformation. The genomic region between the vector and the *Z3* transgene was amplified in the transformed lines. All six independent lines were confirmed to be transformed. Primer information is listed in Additional file 8: Table S1. EV, empty vector; Control, vector used for transformation without the *Z3* gene. b The overexpression of *Z3* by the CaMV 35S promoter in six independent complementation lines. Primers were designed to amplify the transgene mRNA, and are listed in Additional file 8: Table S1. The *GAPDH* mRNA level was measured as a loading control.

Additional file 5: Figure S5. Phenotypes of the homozygous T-DNA null mutants of *Z3*. a Schematic diagrams of the T-DNA insertion. The *z3–2* and *z3–3* mutants have a T-DNA insertion in the first exon, but in different locations. These T-DNA knockout mutants of *Z3* were derived from the *japonica* rice cv 'Hwayoung'. Filled boxes and thick lines indicate exons and introns, respectively. Primers LP1 (left primer 1), RP1 (right primer 1), LP2 (left primer 2), RP2 (right primer 2), and RB (right border primer) used for genotyping are represented as arrows and listed in Additional file 8: Table S1. b Phenotypes of leaf blades from four-month-old WT (Hwayoung), and *z3–2* and *z3–3* mutants grown under natural long day (NLD) conditions in the paddy field. c–d Panicle and grain phenotypes of the WT (Hwayoung) and two *z3*-knockout mutants at 150 DAS grown in a paddy field. e Days to heading data of the WT, and the *z3–2*, and *z3–3* mutants in natural long-day conditions. Means and standard deviations were obtained from 5 plants of each genotype. Error bars indicate SD. Differences between means were compared using Student's *t*-test (*** $P < 0.001$). f Genotyping of T-DNA insertion lines. Each LP and RP primer should have amplified about 1.1 kb of genomic DNA if no T-DNA insertion was made. If T-DNA was inserted, the length between the two primers was too large to be amplified. The RB and each RP primer should have amplified an approximately 0.6-kb band if T-DNA was inserted in the first exon. WT (Hwayoung) was used for the control. g Expression of *Z3* in the WT and T-DNA knockout mutants by semi-quantitative RT-PCR. *Z3* transcripts were not detected in the T-DNA knockout mutants. *Ubiquitin 5* (*UBQ5*) mRNA was measured as a loading control. db, dull-brown colored grain; iv, ivory colored grain.

Additional file 6: Figure S6. Analysis of reactive oxygen species (ROS) in the *z3* mutant leaves. a-b Hydrogen peroxide (H2O2) and superoxide anion radicals (O2-) in flag leaves of the WT and *z3* mutants at 160 DAS grown under natural long day conditions were visualized by staining with DAB (a) and NBT (b), respectively. Leaves before (left) and after (right) staining are shown.

Additional file 7: Figure S7. Transmission electron microscopy analysis of chloroplasts in the *z3* mutant leaves. a-c Chloroplasts in the green leaves of the WT (a) and in the dark-green (b) and green (c) sectors of the *z3* mutant leaves. Flag leaves of the 150-day-old WT and *z3* mutant grown under natural long day conditions were sampled for analysis. G, grana thylakoid. Scale bars = 0.5 μm.

Additional file 8: Table S1. Primers used in this study.

Abbreviations

CaMV: Cauliflower mosaic virus; CitMHS: Citrate-metal hydrogen symport; CTAB: Cetyl trimethyl ammonium bromide; DAB: 3,3-Diaminobendizine; DAS: Days after sowing; dCAPS: Derived cleaved amplified polymorphic sequence; DTH: Days to heading; FW: Fresh weight; G: Grana thylakoid; GAPDH: Glyceraldehyde-3-phosphate dehydrogenase; GOGAT: Glutamate synthase; GS: Glutamine synthetase; HPLC: High performance liquid chromatography; HY: Hwayoung; KO: Knock-out; Lhca: Light-harvesting chlorophyll *a/b*-binding protein of photosystem I; Lhcb: Light-harvesting chlorophyll *a/b*-binding protein of photosystem II; LP: Left genomic primer; M23: Milyang23; MNU: *N*-methyl-*N*-nitrosourea; NB: Nipponbare; NBT: Nitroblue tetrazolium chloride; NLD: Natural long day; ORF: Open reading frame; PCR: Polymerase chain reaction; RB: Right border primer; ROS: Reactive oxygen species; RP: Right genomic primer; SD: Standard deviation; SDS-PAGE: Sodium dodecyl sulfate-polyacrylamide gel electrophoresis; SSR: Simple sequence repeat; STS: Sequence-tagged site; TEM: Transmission electron microscopy;; TMD: Transmembrane domain; *UBQ5*: *Ubiquitin5*; WT: Wild type; YFP: Yellow fluorescent protein; *Z3*: *ZEBRA3*

Acknowledgments

This work was carried out with the support of the Cooperative Research Program for Agriculture & Technology Development (Project No. PJ011063), Rural Development Administration, Republic of Korea.

Authors' contributions

SHK and CTK designed and performed all experiments. NCP designed and supervised the project. SHK, CTK, and NCP wrote and edited the manuscript. GS assisted to analyze the phenotypic characterization. HJK provided whole genome sequencing data for Milyang23. GA donated the *z3-2* and *z3-3* T-DNA insertional rice mutants. The authors read and approved the final manuscript.

Competing interests

The authors declare that they have no competing interests.

Author details

[1]Department of Plant Science, Plant Genomics and Breeding Institute, Research Institute of Agriculture and Life Sciences, Seoul National University, Seoul 08826, Republic of Korea. [2]Crop Biotech Institute and Graduate School of Biotechnology, Kyung Hee University, Yongin 17104, Republic of Korea. [3]Crop Biotechnology Institute, Institutes of Green Bio Science and Technology, Seoul National University, Pyeongchang 25354, Republic of Korea. [4]Present address: Cold Spring Harbor Laboratory, Cold Spring Harbor, NY 11724, USA.

References

Aluru MR, Yu F, Fu A, Rodermel S (2006) *Arabidopsis* variegation mutants: new insights into chloroplast biogenesis. J Exp Bot 57(9):1871–1881

Blancato VS, Repizo GD, Suárez CA, Magni C (2008) Transcriptional regulation of the citrate gene cluster of *Enterococcus faecalis* involves the GntR family transcriptional activator CitO. J Bacteriol 190(22):7419–7430

Bodalkar C, Awadhiya GK (2014) Assessment of percent grain discoloration in important rice varieties. Int J Curr Res Biosci Plant Biol 1(4):61–64

Boorsma A, Rest ME, Lolkema JS, Konings WN (1996) Secondary transporters for citrate and the mg(2+)-citrate complex in *Bacillus subtilis* are homologous proteins. J Bacteriol 178(21):6216–6222

Campitelli BE, Stehlik I, Stinchcombe JR (2008) Leaf variegation is associated with reduced herbivore damage in *Hydrophyllum virginianum*. Botany 86(3):306–313

Carol P, Stevenson D, Bisanz C, Breitenbach J, Sandmann G, Mache R, Coupland G, Kuntz M (1999) Mutations in the Arabidopsis gene *IMMUTANS* cause a variegated phenotype by inactivating a chloroplast terminal oxidase associated with phytoene desaturation. Plant Cell 11(1):57–68

Chatterjee M, Sparvoli S, Edmunds C, Garosi P, Findlay K, Martin C (1996) DAG, a gene required for chloroplast differentiation and palisade development in Antirrhinum Majus. EMBO J 15(16):4194

Chen H, Cheng Z, Ma X, Wu H, Liu Y, Zhou K, Chen Y, Ma W, Bi J, Zhang X, Guo X, Wang J, Lei C, Wu F, Lin Q, Liu Y, Liu L, Jiang K (2013) A knockdown mutation of *YELLOW-GREEN LEAF2* blocks chlorophyll biosynthesis in rice. Plant Cell Rep 32(12):1855–1867

Chen RD, Gadal P (1990) Structure, functions and regulation of NAD and NADP dependent isocitrate dehydrogenases in higher plants and in other organisms. Plant Physiol Biochem (Paris) 28(3):411–427

Cox MP, Peterson DA, Biggs PJ (2010) SolexaQA: at-a-glance quality assessment of Illumina second-generation sequencing data. BMC Bioinforma 11(1):485

Curtis MD, Grossniklaus U (2003) A gateway cloning vector set for high-throughput functional analysis of genes in planta. Plant Physiol 133(2):462–469

Datnoff LE, Raid RN, Snyder GH, Jones DB (1991) Effect of calcium silicate on blast and brown spot intensities and yields of rice. Plant Dis 75:729–732

Dessureault-Rompré J, Nowack B, Schulin R, Tercier-Waeber ML, Luster J (2008) Metal solubility and speciation in the rhizosphere of *Lupinus albus* cluster roots. Environ Sci Technol 42(19):7146–7151

Dinkelaker B, Hengeler C, Marschner H (1995) Distribution and function of proteoid roots and other root clusters. Plant Biol 108(3):183–200

Dinkelaker B, Römheld V, Marschner H (1989) Citric acid excretion and precipitation of calcium citrate in the rhizosphere of white lupin (*Lupinus albus* L.) Plant Cell Environ 12(3):285–292

Dobermann A, Fairhurst T (2000) Rice: Nutrient disorders & nutrient management. Handbook Series, Potash & Phosphate Institute (PPI), Potash & Phosphate Institute of Canada (PPIC) and International Rice Research Institute, 191 p. (http://books.irri.org/9810427425_content.pdf)

Durrett TP, Gassmann W, Rogers EE (2007) The FRD3-mediated efflux of citrate into the root vasculature is necessary for efficient iron translocation. Plant Physiol 144(1):197–205

Emanuelsson O, Nielsen H, Brunak S, Heijne G (2000) Predicting subcellular localization of proteins based on their N-terminal amino acid sequence. J Mol Biol 300(4):1005–1016

Fang J, Chai C, Qian Q, Li C, Tang J, Sun L, Huang Z, Guo X, Sun C, Liu M, Zhang Y, Lu Q, Wang Y, Lu C, Han B, Chen F, Cheng Z, Chu C (2008) Mutations of genes in synthesis of the carotenoid precursors of ABA lead to pre-harvest sprouting and photo-oxidation in rice. Plant J 54(2):177–189

Fritsch H, Beevers H (1979) ATP citrate lyase from germinating castor bean endosperm. Plant Physiol 63(4):687–691

Fuente JM, Ramírez-Rodríguez V, Cabrera-Ponce JL, Herrera-Estrella L (1997) Aluminum tolerance in transgenic plants by alteration of citrate synthesis. Science 276(5318):1566–1568

Furukawa J, Yamaji N, Wang H, Mitani N, Murata Y, Sato K, Katsuhara M, Takeda K, Ma JF (2007) An aluminum-activated citrate transporter in barley. Plant Cell Physiol 48(8):1081–1091

Glusker JP (1980) Citrate conformation and chelation: enzymic implications. Acc Chem Res 13(10):345–352

Green LS, Rogers EE (2004) *FRD3* controls iron localization in Arabidopsis. Plant Physiol 136(1):2523–2531

Han SH, Sakuraba Y, Koh HJ, Paek NC (2012) Leaf variegation in the rice *zebra2* mutant is caused by photoperiodic accumulation of tetra-*Cis*-lycopene and singlet oxygen. Mol Cells 33(1):87–97

Hoffland E, Findenegg GR, Nelemans JA (1989) Solubilization of rock phosphate by rape. Plant Soil 113(2):155–160

Itoh JI, Hasegawa A, Kitano H, Nagato Y (1998) A recessive heterochronic mutation, plastochron1, shortens the plastochron and elongates the vegetative phase in rice. Plant Cell 10(9):1511–1522

Jain M, Nijhawan A, Tyagi AK, Khurana JP (2006) Validation of housekeeping genes as internal control for studying gene expression in rice by quantitative real-time PCR. Biochem Biophys Res Commun 345(2):646–651

Jeon JS, Chung YY, Lee S, Yi GH, BG O, An G (1999) Isolation and characterization of an anther-specific gene, RA8, from rice (Oryza Sativa L.) Plant Mol Biol 39(1):35–44

Jeon JS, Lee S, Jung KH, Jun SH, Jeong DH, Lee J, Kim C, Jang S, Lee S, Yang K (2000) T-DNA insertional mutagenesis for functional genomics in rice. Plant J 22(6):561–570

Jiang H, Chen Y, Li M, Xu X, Wu G (2011) Overexpression of *SGR* results in oxidative stress and lesion-mimic cell death in rice seedlings. J Integr Plant Biol 53(5):375–387

Kato Y, Miura E, Ido K, Ifuku K, Sakamoto W (2009) The variegated mutants lacking chloroplastic FtsHs are defective in D1 degradation and accumulate reactive oxygen species. Plant Physiol 151(4):1790–1801

Kawahara Y, Bastide M, Hamilton JP, Kanamori H, McCombie WR, Ouyang S, Schwartz DC, Tanaka T, Wu J, Zhou S, Childs KL, Davidson RM, Lin H, Quesada-Ocampo L, Vaillancourt B, Sakai H, Lee SS, Kim J, Numa H, Itoh T, Buell CR, Matsumoto T (2013) Improvement of the *Oryza sativa* Nipponbare reference genome using next generation sequence and optical map data. Rice 6(1):4

Keddie JS, Carroll B, Jones JD, Gruissem W (1996) The DCL gene of tomato is required for chloroplast development and palisade cell morphogenesis in leaves. EMBO J 15(16):4208

Korithoski B, Krastel K, Cvitkovitch DG (2005) Transport and metabolism of citrate by *Streptococcus mutans*. J Bacteriol 187(13):4451–4456

Krebs HA, Johnson WA (1937) The role of citric acid in intermediate metabolism in animal tissues. Enzymologia 4:148–156

Kwon CT, Kim SH, Song G, Kim D, Paek NC (2017) Two NADPH: protochlorophyllide oxidoreductase (POR) isoforms play distinct roles in environmental adaptation in rice. Rice 10(1):1

Kwon CT, Koo BH, Kim D, Yoo SC, Paek NC (2015) Casein kinases I and 2α phosphorylate Oryza Sativa pseudo-response regulator 37 (OsPRR37) in photoperiodic flowering in rice. Mol Cells 38(1):81–88

Lee S, Jeon JS, Jung KH, An G (1999) Binary vectors for efficient transformation of rice. J Plant Biol 42(4):310

Lensbouer JJ, Patel A, Sirianni JP, Doyle RP (2008) Functional characterization and metal ion specificity of the metal-citrate complex transporter from *Streptomyces coelicolor*. J Bacteriol 190(16):5616–5623

Li H, Durbin R (2009) Fast and accurate short read alignment with burrows–wheeler transform. Bioinformatics 25(14):1754–1760

Li J, Pandeya D, Nath K, Zulfugarov IS, Yoo SC, Zhang H, Yoo JH, Cho SH, Koh HJ, Kim DS, Seo HS, Kang BC, Lee CH, Paek NC (2010) ZEBRA-NECROSIS, a thylakoid-bound protein, is critical for the photoprotection of developing chloroplasts during early leaf development. Plant J 62(4):713–725

Li Y, Wang B, Dai Z, Li A, Liu G, Zuo S, Zhang H, Pan X (2012) Morphological structure and genetic mapping of new leaf-color mutant gene in rice (*Oryza sativa*). Rice Sci 19(2):79–85

Lichtenthaler HK (1987) Chlorophyls and carotenoids: pigments of photosynthetic biomembranes. Methods Enzymol 148(34):350–382

Liu J, Magalhaes JV, Shaff J, Kochian LV (2009) Aluminum-activated citrate and malate transporters from the MATE and ALMT families function independently to confer Arabidopsis aluminum tolerance. Plant J 57(3):389–399

Magalhaes JV, Liu J, Guimarães CT, Lana UGP, Alves VMC, Wang YH, Schaffert RE, Hoekenga OA, Piñerps MA, Shaff JE, Klein PE, Carneiro NP, Coelho CM, Trick HN, Kochian LV (2007) A gene in the multidrug and toxic compound extrusion (MATE) family confers aluminum tolerance in sorghum. Nat Genet 39(9):1156–1161

Martínez-Zapater JM (1993) Genetic analysis of variegated mutants in Arabidopsis. J Hered 84(2):138–140

Mattoo AK, Modi VV (1970) Citrate cleavage enzyme in mango fruit. Biochem Biophys Res Commun 39(5):895–904

Milewska MJ (1988) Citric acid — its natural and synthetic derivatives. Z Chem 28(6):204–211

Miyasaka SC, Buta JG, Howell RK, Foy CD (1991) Mechanism of aluminum tolerance in snapbeans. Plant Physiol 96(3):737–743

Murray MG, Thompson WF (1980) Rapid isolation of high molecular weight plant DNA. Nucleic Acids Res 8(19):4321–4326

Naramoto S (2017) Polar transport in plants mediated by membrane transporters: focus on mechanisms of polar auxin transport. Curr Opin Plant Biol 40:8–14

Navasero EP, Winslow MD (1987) Physicochemical properties of discolored grains. Int Rice Res Newl 12:13–14

Nelson DR, Rinne RW (1975) Citrate cleavage enzymes from developing soybean cotyledons. Plant Physiol 55(1):69–72

Ngala GN (1983) *Sarocladium attenuatum* as one of the causes of rice grain spotting in Nigeria. Plant Pathol 32(3):289–293

Park SY, JW Y, Park JS, Li J, Yoo SC, Lee NY, Lee SK, Jeong SW, Seo HS, Koh HJ, Jeon JS, Park YI, Paek NC (2007) The senescence-induced staygreen protein regulates chlorophyll degradation. Plant Cell 19(5):1649–1664

Phat CT, Duong NT, Du LT (2005) Influence of grain discoloration to seed quality. Omonrice 13:139–144

Phillips RD, Jennings DH (1976) The estimation of plant organic acids by gas-liquid chromatography. New Phytol 77(2):333–339

Rédei GP (1967) Biochemical aspects of a genetically determined variegation in Arabidopsis. Genetics 56(3):431–443

Rellán-Álvarez R, Giner-Martínez-Sierra J, Orduna J, Orera I, Rodríguez-Castrillón JÁ, García-Alonso JI, Abadía J, Álvarez-Fernández A (2010) Identification of a tri-iron(III), tri-citrate complex in the xylem sap of iron-deficient tomato resupplied with iron: new insights into plant iron long-distance transport. Plant Cell Physiol 51(1):91–102

Rogers EE, Wu X, Stacey G, Nguyen HT (2009) Two. MATE proteins play a role in iron efficiency in soybean 166(13):1453–1459

Roschzttardtz H, Séguéla-Arnaud M, Briat JF, Vert G, Curie C (2011) The FRD3 citrate effluxer promotes iron nutrition between symplastically disconnected tissues throughout *Arabidopsis* development. Plant Cell 23(7):2725–2737

Sakuraba Y, Rahman ML, Cho SH, Kim YS, Koh HJ, Yoo SC, Paek NC (2013) The rice *faded green leaf* locus encodes protochlorophyllide oxidoreductase B and is essential for chlorophyll synthesis under high light conditions. Plant J 74(1):122–133

Sarantinopoulos P, Makras L, Vaningelgem F, Kalantzopoulos G, Vuyst LD, Tsakalidou E (2003) Growth and energy generation by *Enterococcus faecium* FAIR-E 198 during citrate metabolism. Int J Food Microbiol 84(2):197–206

Scarpella E, Marcos D, Friml J, Berleth T (2006) Control of leaf vascular patterning by polar auxin transport. Genes Dev 20(8):1015–1027

Sienkiewicz-Porzucek A, Nunes-Nesi A, Sulpice R, Lisec J, Centeno DC, Carillo P, Leisse A, Urbanczyk-Wochniak E, Fernie AR (2008) Mild reductions in mitochondrial citrate synthase activity result in a compromised nitrate assimilation and reduced leaf pigmentation but have no effect on photosynthetic performance or growth. Plant Physiol 147(1):115–127

Tanner W, Beevers H (1965) The competition between the glyoxylate cycle and the oxidative breakdown of acetate in Ricinus endosperm. I. The metabolism of isocitrate in endosperm extracts. Z Pflanzenphysiol 53:72–85

Tiffin LO (1970) Translocation of iron citrate and phosphorus in xylem exudate of soybean. Plant Physiol 45(3):280–283

Tsuda M, Morita M, Makihara D, Hirai Y (2000) The involvement of silicon deposition in salinity-induced white head in rice (*Oryza sativa* L.) Plant Prod Sci 3(3):328–334

Tsugane K, Maekawa M, Takagi K, Takahara H, Qian Q, Eun C-H, Iida S (2006) An active DNA transposon *nDart* causing leaf variegation and mutable dwarfism and its related elements in rice. Plant J 45(1):46–57

Tsukaya H, Okada H, Mohamed M (2004) A novel feature of structural variegation in leaves of the tropical plant *Schismatoglottis calyptrata*. J Plant Res 117(6):477–480

Wi SJ, Jang SJ, Park KY (2010) Inhibition of biphasic ethylene production enhances tolerance to abiotic stress by reducing the accumulation of reactive oxygen species in *Nicotiana tabacum*. Mol Cells 30(1):37–49

Winslow MD (1992) Silicon, disease resistance and yield of rice genotypes under upland cultural conditions. Crop Sci 32:1208–1213

Yokosho K, Yamaji N, Ma JF (2011) An al-inducible MATE gene is involved in external detoxification of al in rice. Plant J 68(6):1061–1069

Yokosho K, Yamaji N, Ueno D, Mitani N, Ma JF (2009) OsFRDL1 is a citrate transporter required for efficient translocation of iron in rice. Plant Physiol 149(1):297–305

Zhang FS, Ma J, Cao YP (1997) Phosphorus deficiency enhances root exudation of low-molecular weight organic acids and utilization of sparingly soluble inorganic phosphates by radish (*Raghanus satiuvs* L.) and rape (*Brassica napus* L.) plants. Plant Soil 196(2):261–264

Zhou G, Delhaize E, Zhou M, Ryan PR (2013) The barley *MATE* gene, *HvAACT1*, increases citrate efflux and Al^{3+} tolerance when expressed in wheat and barley. Ann Bot 112(3):603–612

11

Identification of a cold-tolerant locus in rice (*Oryza sativa* L.) using bulked segregant analysis with a next-generation sequencing strategy

Jian Sun[1†], Luomiao Yang[1†], Jingguo Wang[1], Hualong Liu[1], Hongliang Zheng[1], Dongwei Xie[2], Minghui Zhang[3], Mingfang Feng[3], Yan Jia[1], Hongwei Zhao[1] and Detang Zou[1*]

Abstract

Background: Cold stress can cause serious abiotic damage that limits the growth, development and yield of rice. Cold tolerance during the booting stage of rice is a key factor that can guarantee a high and stable yield under cold stress. The cold tolerance of rice is controlled by quantitative trait loci (QTLs). Based on the complex genetic basis of cold tolerance in rice, additional efforts are needed to detect reliable QTLs and identify candidate genes. In this study, recombinant inbred lines (RILs) derived from a cross between a cold sensitive variety, Dongnong422, and strongly cold-tolerant variety, Kongyu131, were used to screen for cold-tolerant loci at the booting stage of rice.

Results: A novel major QTL, *qPSST6*, controlling the percent seed set under cold water treatment (PSST) under the field conditions of 17 °C cold water irrigation was located on the 28.4 cM interval on chromosome 6. Using the combination of bulked-segregant analysis (BSA) and next-generation sequencing (NGS) technology (Seq-BSA), a 1. 81 Mb region that contains 269 predicted genes on chromosome 6 was identified as the candidate region of *qPSST6*. Two genes, *LOC_Os06g39740* and *LOC_Os06g39750*, were annotated as "response to cold" by gene ontology (GO) analysis. qRT-PCR analysis revealed that *LOC_Os06g39750* was strongly induced by cold stress. Haplotype analysis also demonstrate a key role of *LOC_Os06g39750* in regulating the PSST of rice, suggesting that it was the candidate gene of *qPSST6*.

Conclusions: The information obtained in this study is useful for gene cloning of *qPSST6* and for breeding cold-tolerant varieties of rice using marker assisted selection (MAS).

Keywords: *Oryza sativa* L, Cold tolerance, QTL, Seq-BSA, Candidate gene

Background

Rice, a staple food crop cultivated worldwide, feeds more than half of the world's population. Rice originates from tropical and subtropical regions and is more sensitive to low temperature than other crops originating in temperate zones (Wang et al. 2016). In high-latitude or high-altitude regions of Asia, Europe, America and other rice cultivation areas of the world, the temperature is not consistently high enough for the growth of rice. Remarkably, low temperatures that occur frequently during the reproductive stage of rice can cause a fatal yield loss (Jena et al. 2012; Shirasawa et al. 2012). Male sterility, arising from low temperature (lower than 19 °C) during the period of microspore development at the booting stage, is the key reason for the reduction of percent seed set and the resulting loss in yield (Satake and Hayase 1970). Therefore, the breeding of cold-tolerant varieties at the booting stage is an effective method to maintain high and stable yields in rice cultivation regions.

The cold tolerance of rice is a complex trait, and many methods have been established to evaluate and select

* Correspondence: zoudt@sina.cn
Jian Sun and Luomiao Yang contributed equally to this work.
†Equal contributors
[1]College of Agriculture, Northeast Agricultural University, Harbin 150030, China
Full list of author information is available at the end of the article

cold-tolerant varieties of rice (Zhang et al. 2014). A cold-water irrigation system has been developed as a reliable identified method of determining cold tolerance at the booting stage of rice. Rice plants are maintained in a cold deep-water irrigated pool during the entire booting stage, and the spikelet fertility was used to examine the cold tolerance of rice varieties (Shirasawa et al. 2012). This method exposes rice plants in field growth conditions, and its accurate evaluation results are still widely used for selecting cold-tolerant lines and developing cold-tolerant rice varieties (Matsunaga 2005; Jia et al. 2015).

Many studies agree that the cold tolerance of rice is controlled by quantitative trait loci (QTLs) (Andaya and Mackill 2003a; Zhang et al. 2005; Suh et al. 2010; Zhang et al. 2014). QTL mapping is the main approach to excavate and clone cold-tolerant related genes in rice. At the booting stage of rice, a number of QTLs about cold tolerance have been reported, including *qCTB3* (Andaya and Mackill 2003b), *qPSST-3* (Suh et al. 2010) and *qLTB3* (Shirasawa et al. 2012) on chromosome 3, *Ctb1* (Saito et al. 2004), *Ctb2* (Saito et al. 2001), *qCTB-4-1*, *qCTB-4-2* (Xu et al. 2008), and *CTB4a* (Zhang et al. 2017) on chromosome 4, and *qPSST-7* (Suh et al. 2010) and *qCTB7* (Zhou et al. 2010) on chromosome 7. For the different rice materials and research backgrounds, QTLs for cold tolerance at the booting stage were mapped at the various locations on different chromosomes. However, the above QTLs identified by the biparental cross linkage mapping method are labor and time intensive to map the genotypes a large number of individuals in the segregated population and to finely map the target QTL (Salvi and Tuberosa 2005).

A bulked segregant analysis (BSA) is a simple and rapid approach to identify molecular markers tightly linked to the target genes or QTLs (Michelmore et al. 1991). With the rapid development of next-generation sequencing (NGS) technology, the combination of the BSA with NGS strategy (Seq-BSA) is becoming a widely used approach in the mapping of major QTLs and gene identification (Wenger et al. 2010; Takagi et al. 2013). This strategy has been demonstrated in many plants, such as Arabidopsis (Schneeberger et al. 2009), rice (Abe et al. 2012; Wambugu et al. 2017), soybean (Song et al. 2017), wheat (Trick et al. 2012), pigeon pea (Singh et al. 2016), sunflower (Livaja et al. 2013) and sorghum (Han et al. 2015), and it has identified some QTLs and genes for important traits. In the present study, two strategies, traditional QTL mapping and Seq-BSA were employed to identify the genes for cold tolerance at the booting stage in recombinant inbred lines (RILs) derived from a cross between a cold sensitive variety, Dongnong422, and a strongly cold-tolerant variety, Kongyu131. The results contribute to the understanding of the genetic bases for cold tolerance at the booting stage, and future cloning of the candidate gene will facilitate the molecular breeding of cold tolerance in rice.

Methods

Plant materials

Two *japonica* varieties, "Dongnong422" (DN422) and "Kongyu131 (KY131)" were used as parental lines to develop the RIL population. DN422 is a cold-sensitive variety that was obtained from Northeast Agriculture University. KY131, a strongly cold-tolerant variety, is widely cultivated in the northeast region of China. The mapping populations of 190 F_7, F_8 and F_9 RILs were produced by single seed descent (SSD) from an F_2 population of a cross between DN422 and KY131.

Cold tolerance evaluation

An evaluation of cold tolerance was performed on the experimental farm of Northeast Agricultural University, Harbin, China (47°98′N, 128°08′E) in 2014, 2015 and 2016. The parents and RIL populations were grown in a randomized block design with two replications of double row plots, a 2-m row length, a 30-cm row spacing and a 10-cm hill spacing. The 190 RILs were divided into three groups, early maturing, middle-maturing and late-maturing groups, according to their heading dates. DN422 belongs to the late-maturing group, and KY131 belongs to the early maturing group. For the evaluation of cold tolerance, each group was irrigated with 17 °C water in an irrigated pool (25 m × 5 m) independently from the panicle initiation stage (approximately 35 days after transplanting for the early maturing group, approximately 40 days after transplanting for the middle-maturing group, and approximately 45 days after transplanting for the late-maturing group) to the full heading stage. When the auricle of the flag leaf is approximately 5 cm below the auricle of the penultimate leaf on each plant, the pollen should have undergone meiosis, which was indicative of panicle initiation stage (Satake and Hayase 1970). Water at 17 °C was prepared in a storage pool with a cool and warm water mixture measured by the temperature sensors. Flood irrigation was performed from the inlet to the outlet (5 m) of the irrigated pool. The depth of the irrigated water was 18–20 cm, and the irrigated time was from 8 a.m. to 16 p.m. every day.

The percent seed set under the cold-water treatment (PSST), the ratio of the number of fertile seeds in the number of total seeds, was used as the index of cold tolerance. A basic statistical analysis was implemented by the SPSS16.0 software (SPSS Inc., Chicago, IL, USA). The mean data of PSST in RILs over the two replications was used for QTL analysis, and the extreme cold-tolerant and sensitive plants in the RIL population used for Seq-BSA were selected according to the mean data of F_7, F_8 and F_9 RILs.

QTL analysis

One hundred eighty-five polymorphic SSR markers between DN422 and KY131 covering the rice genome were used for genotyping the RIL population. A PCR was performed according to the procedure of Chen et al. (1997), and the PCR products were then separated on a 6% polyacrylamide gel, followed by silver staining.

The MAP function of QTL IciMapping 3.2 (Wang et al. 2012) was used to construct the genetic linkage map of the RIL population, and the Kosambi's mapping function was used to calculate the genetic distances. QTLs were detected using the inclusive composite interval mapping (ICIM) module of QTL IciMapping 3.2. The threshold of the LOD score for declaring the presence of a significant QTL was determined by the permutation test with 1000 repetitions at $P < 0.05$. The QTL was named according to the trait and its chromosome location.

Construction of segregating pools and sequencing

For Seq-BSA, two DNA pools were developed by selecting the extreme cold-tolerant and extreme cold-sensitive individuals according to the PSST of the RIL population in the range from 0.50–0.95. The tolerant pool (T-pool) was made by mixing equal amounts of DNA from 20 extreme cold-tolerant RILs with a PSST above 0.90, and the sensitive pool (S-pool) was made by mixing equal amounts of DNA from 20 extreme cold-sensitive RILs with a PSST below 0.63 (Additional file 2: Table S1). The DNA isolated from DN422 and KY131 and the two DNA pools were prepared for sequencing.

Libraries for all the DNA pools were prepared according to the Illumina TruSeq DNA sample Preparation v2 Guide. The DNA libraries were sequenced on Illumina MiSeq platform using MiSeq Reagent Kit v2 (500 cycles) (Illumina Inc., San Diego, CA, USA). The short reads from both parents and the two DNA pools were aligned to *Nipponbare* reference genome (IRGSP 2005) using the BWA software (Li and Durbin 2009). Reads of the T-pool and S-pool were separately aligned to KY131 and DN422 consensus sequence reads to call SNPs with the SAM tools software (Li and Durbin 2009).

Analysis of the Seq-BSA data

According to the locating results of clean reads among the reference genome, duplicate reads were removed using the Picard tool (http://sourceforge.net/projects/picard/), and GATK software (McKenna et al. 2010) was used to perform the local realignment and base recalibration to insure the accuracy of the SNP detecting. The SNP loci between the test samples and reference genome were obtained using the GATK software according to the best practices on the GATK website (https://www.broadinstitute.org/gatk/guide/best-practices.php). All the SNP loci between the test samples were summarized according to the alignment results of test samples and the reference genome.

The Euclidean distance (ED) and SNP-index were calculated to identify the candidate regions of the genome associated with PSST. The ED algorithm is a method of searching markers with significant differences between the pools according to the sequencing data and evaluating the associated regions between markers and traits (Hill et al. 2013).

The calculation formula of the ED algorithm was as follows:

$$ED = \sqrt{(Aaa-Aab)^2 + (Caa-Cab)^2 + (Gaa-Gab)^2 + (Taa-Tab)^2}$$

where A_{aa}, C_{aa}, G_{aa} and T_{aa} represent the frequency of bases A, C, G and T in the T-pool, respectively. A_{ab}, C_{ab}, G_{ab} and T_{ab} represent the frequency of bases A, C, G and T in the S-pool, respectively. The depth of each base in different pools and the ED value of each SNP loci were calculated. To eliminate the background noise, the ED value was powered, and the ED^5 was used as the associated value (Hill et al. 2013).

The SNP-index association analysis is a method used to calculate genotype frequency differences between two pools (Takagi et al. 2013; Fekih et al. 2013). A SNP-index is the proportion of reads harboring the SNP that are different from the reference sequence. The Δ(SNP-index) of each locus was calculated by subtraction of the SNP-index of the T-pool from that of the S-pool. SNP-index = 0 if the entire short reads contain genomic fragments from DN422; SNP-index = 1 if all the short reads were from KY131. The average of the SNP-index was calculated in a 1 Mb interval using a sliding window analysis with 10 kb. The SNP-index graphs of the T-pool, S-pool and corresponding Δ(SNP-index) graphs were plotted. The statistical confidence intervals of Δ(SNP-index) were calculated under the null hypothesis of no QTLs following the description of Takagi et al. (2013).

qRT-PCR analysis of the candidate genes

Two parents, DN422 and KY131, were used to identify the expression patterns of the putative candidate genes. Samples were collected when they reached the panicle initiation stage after cold water irrigation in 2017. Plants with normal temperature water irrigation were used as the control. The leaves of the two parents were collected at 0, 1, 3, 5, 7, 9, 11 and 13 d, for three repetitions. The samples were placed in liquid nitrogen immediately and then stored at − 80 °C for total RNA isolation.

Total RNA was isolated using the Trizol reagent (Invitrogen) according to the manufacturer's instructions and was purified using the DNA-free RNA kit. The first-strand cDNA was synthesized using the

Fermentas RevertAid First Strand cDNA Synthesis Kit. A pair of primers was designed for each candidate gene using the Primer Premier 5.0 software. The housekeeping gene *Actin1* (*Os05g36290*) was used as the internal control (Siahpoosh et al. 2011). The information of the specific primers is provided in Additional file 2: Table S2. The qRT-PCR reactions were performed using a Roche LightCycler 2.10 with a 2 × SYBR Green I PCR Master Mix. The PCR reaction procedure was as follows: 95 °C for 5 min; 45 cycles of 95 °C for 10 s, 60 °C for 20 s, and 72 °C for 20 s. The mean Ct values of all the biological replicates were normalized with the Ct values of *Actin1*. The relative expression level was calculated using the $2^{-\Delta\Delta Ct}$ method (Livak and Schmittgen 2001), where $\Delta\Delta Ct = (Ct,_{target} - Ct,_{actin})_{time\ x} - (Ct,_{target} - Ct,_{actin})_{time\ 0}$.

Sequencing of the nsSNP loci in RILs

Genomic DNA of the 190 RILs was extracted using the CTAB method (Murray and Thompson 1980).The eight SNP loci in *LOC_Os06g39750* were sequenced to identify the haplotypes of RIL population. Primers for amplifying the SNP were designed using the Primer Premier 5.0 software, and the sequences of specific primers are provided in Additional file 2: Table S3.

Total 20 μL PCR reaction mixture contained 2 μL of genomic DNA (50 ng/μL), 1.5 μL of forward primer (10 μM), 1.5 μL of reverse primer (10 μM), 5 μL ddH_2O, and 10 μL of Pfu Master Mix (CWBio, Inc., Beijing, China) including Taq DNA Polymerase, PCR Buffer, Mg^{2+} and dNTPs. The PCR reaction was carried out in Eppendorf Mastercycler 5333 thermal cycler, and the procedure was same as the above qRT-PCR. The amplified products were checked by electrophoresis in 1% agarose gel. PCR products were directly sequenced by the BGI Life Tech Co., Ltd.

Results

Phenotypic characterization of the cold tolerance in DN422/KY131 RILs

In the present study, two *japonica* varieties, DN422 and KY131, were crossed to develop the RIL populations for the QTL analysis and the Seq-BSA of cold tolerance. The phenotypic data were collected using the F_7, F_8 and F_9 RILs in 2014, 2015 and 2016, continuously, and the basic statistical analysis of the tested materials is provided in Table 1. Extensive phenotyping data under cold stress and control conditions for heading time, plant height, panicle number, and grain yield per plant were provide in Additional file 2: Table S4 to insights into the cold tolerance of the test materials. The mean PSST values of DN422, KY131 and the RIL population in the 3 years were 0.59, 0.92 and 0.76, respectively, and their mean percent seed set values under normal water irrigation (PSSN) were 0.90, 0.97 and 0.89, respectively. The correlation analysis showed a significantly positive correlation among the PSST in 2014, 2015 and 2016, respectively, and the PSSN had the same pattern (Additional file 2: Table S5). This indicates that the cold-water irrigation was moderate and could differentiate the plants between cold stress and control effectively. The PSST values of the cold-tolerant parent KY131 were extremely significant higher than those of the cold-sensitive DN422 in all 3 years (Fig. 1, Table 1), indicating the stronger cold tolerance of KY131 compared to DN422. Among the RIL populations, the absolute values of skewness and kurtosis for PSST in 3 years were all close to 1 which indicate that the data of PSST were suitable for the QTL analysis (Table 1) (Li et al. 2012).

SSR-based QTL mapping

One thousand SSR markers with a uniform distribution throughout the 12 rice chromosomes were selected to screen the polymorphism between DN422 and KY131.

Table 1 PSST and PSSN of parents and RIL population in 3 years

Traits	Year	DN422[a]	KY131	RIL populations[b]				
				Mean	Range	SD	Skewness	Kurtosis
PSST	2014	0.57 ± 0.09	0.92** ± 0.11	0.78	0.42~ 0.94	0.14	− 1.08	1.16
	2015	0.61 ± 0.13	0.93** ± 0.08	0.79	0.50~ 0.98	0.12	−0.42	− 0.87
	2016	0.58 ± 0.07	0.91** ± 0.14	0.70	0.31~ 0.99	0.15	0.02	−0.65
	Mean	0.59 ± 0.10	0.92** ± 0.09	0.76	0.50~ 0.95	0.10	−0.10	−0.53
PSSN	2014	0.91 ± 0.05	0.97 ± 0.09	0.95	0.87~ 1.00	0.07	0.25	0.43
	2015	0.88 ± 0.06	0.98 ± 0.11	0.85	0.66~ 0.99	0.03	−0.27	−0.99
	2016	0.90 ± 0.14	0.96 ± 0.06	0.87	0.67~ 0.99	0.08	−0.36	−0.49
	Mean	0.90 ± 0.11	0.97 ± 0.08	0.89	0.73~ 0.99	0.06	−0.07	−0.93

[a] Means ± SD (standard deviation)
[b] Population sample size $n = 190$, replications $r = 2$
*,** significance at the level of 5 and 1%, respectively, according to Student's t-test

Fig. 1 Panicles of DN422 and KY131 planted under cold water irrigation condition. Scale bars, 3 cm

Among them, 158 SSR markers were polymorphic between the two parents and were used to construct the genetic linkage map of the RIL population. The map included 12 chromosomes and covered 2355.3 cM of the rice genome with an average distance of 14.91 cM between markers.

In total, three QTLs for the PSST located on chromosomes 5, 6 and 7 were detected by using the ICIM module of QTL IciMapping 3.2 in 3 years (Table 2). The total phenotypic variation explained (PVE) for all identified QTLs ranged from 9.01 to 47.94%. *qPSST5* and *qPSST7* were detected in only 1 year, and their PVE were 9.01% and 14.95%, respectively. Their facticity in controlling the PSST of rice is suspicious. Remarkably, *qPSST6* in the RM20261-RM20356 interval of chromosome 6 was detected in all 3 years, and the PVE values of the 3 years were 24.30%, 47.94% and 19.11%, respectively, with a mean of 30.45%. In addition, their additive effect values were all negative, indicating that KY131 had a positive effect on increasing the PSST. Thus, *qPSST6* is a major QTL for a high PVE and existed stably in the 3 years, suggesting a key role in controlling the PSST of rice. In contrast, six QTLs with small PVE for the PSSN were detected on chromosomes 2, 4, 6, 7, 8, and major QTL was not found (Additional file 2: Table S6).

Sequencing of the parents and extreme pools

Twenty RILs with extreme cold tolerance (PSST ranged from 0.90–0.95) and 20 RILs with extreme cold sensitivity (PSST ranged from 0.50–0.63) were selected to prepare the T-pool and S-pool (Additional file 2: Table S1). Illumina high-throughput sequencing generated 233.62 million raw reads, and 232.30 million clean reads (99.43%) were obtained after filtering. The mean value of Q30 was 91.92%, indicating that most of the bases were high quality. The sequencing depths were 38-fold in DN422, 42-fold in KY131, 31-fold in the T-pool, and 45-fold in the S-pool, which could guarantee the accuracy of the BSA analysis (Additional file 1: Figure S1, Additional file 2: Table S7). There were 336,632 SNPs between DN422 and KY131, which included 36,448 non-synonymous SNPs (nsSNP); there were 184,917 SNPs between the T-pool and the S-pool, which included 18,013 nsSNP. The raw sequencing data were deposited in NCBI (https://www.ncbi.nlm.nih.gov/), and the accession number were SRR6327815, SRR6327816, SRR6327817, SRR6327818 for KY131, DN422, T-pool and S-pool, respectively.

ED and SNP-index analysis

The two analysis methods of Seq-BSA and the ED and SNP-index were used to identify the candidate genome

Table 2 QTLs for PSST of rice detected in 3 years

QTL	Years	Chr.[a]	Marker interval	LOD[b]	A[c]	PVE%[d]
qPSST5	2015	5	RM146-RM18332	4.69	0.04	9.01
qPSST6	2014	6	RM20261-RM20356	5.43	−0.05	24.30
	2015	6	RM20261-RM20356	12.29	−0.08	47.94
	2016	6	RM20261-RM20356	5.52	−0.07	19.11
qPSST7	2016	7	RM1335-RM182	3.88	0.02	14.95

[a] Chromosome
[b] Logarithm of odd score for each QTL
[c] Additive effect of the QTL. A positive value indicates that DN422 contributes the positive allele; a negative value indicates that KY131 contributes the positive allele
[d] Phenotypic variation explained by A effects

regions associated with PSST. The 184,917 SNPs between the T-pool and the S-pool were used for an association analysis through the two methods. The association threshold of the ED method was 3.82, and six genome regions were significantly correlated with the PSST. The result of the ED association analysis is shown in Fig. 2 and Table 3. The candidate regions had a total size of 23.48 Mb, distributed on chromosomes 4, 5 and 6, and contained 4338 genes with 792 nsSNP loci among them.

The SNP-index of the T-pool and S-pool were calculated for each identified SNP in the genome, and an average SNP-index was computed in a 1 Mb interval using a 10-kb sliding window. By combining the information of the SNP-index of the T-pool and S-pool, the Δ(SNP-index) was calculated, and the Δ(SNP-index) trends were visualized by means of a sliding window. Using the association threshold of 0.9532, six genome regions distributed on chromosomes 5 and 6 that were significantly correlated with the PSST were identified. The total length of these regions was 3.68 Mb, and they contained 649 genes with 93 nsSNP loci (Fig. 3, Table 4). By comparing the association results of ED and the SNP-index methods, their intersections of the genome regions were consistent to the association regions of the SNP-index method (Table 4). The 3.68 Mb regions of chromosome 6 by Seq-BSA was the hot region for the PSST of rice.

To narrow the genome region of PSST further, mapping interval of the major QTL *qPSST6* was compared to the hot region of Seq-BSA. *qPSST6* was detected in the interval between RM20261-RM20356 with a genetic distance of 28.4 cM by using the ICIM module of QTL IciMapping 3.2. RM20261 is located at 21,304,727–21,304,750 bp, and RM20356 is located at 23,654,188–23,654,207 bp on chromosome 6 of the *Nipponbare*

Table 3 Information of the association region by the ED association analysis method

Chromosomes	Chromosome locations (bp)		Size (Mb)	Gene number
	Start	End		
4	5,830,000	7,910,000	2.08	324
4	20,560,000	24,640,000	4.08	770
5	26,010,000	29,950,000	3.94	873
6	2,530,000	6,440,000	3.91	756
6	19,480,000	28,910,000	9.43	1601
6	28,930,000	28,970,000	0.04	14
Total	–	–	23.48	4338

genome. *qPSST6* spanned a physical distance of 2.35 Mb and intersected with the hot region of Seq-BSA between 21,840,000 bp and 23,654,188 bp (Fig. 4). Thus, the region containing the PSST locus was narrowed to a region of 1.81 Mb, which contains 269 predicted genes.

Identification of the candidate genes for PSST

To identify the candidate genes for PSST, the 269 predicted genes were searched in gene ontology (GO) (Ashburner et al. 2000), non-redundant protein (NR) (Deng et al. 2006), Swiss-Prot (Apweiler et al. 2004), Kyoto Encyclopedia of Genes and Genomes (KEGG) (Kanehisa et al. 2004) and Cluster of Orthologous Groups of proteins (COG) (Tatusov et al. 2000) databases by the BLAST software (Altschul et al. 1997). The results revealed that among the 269 predicted genes, 212 were successfully annotated. The most enriched terms of biological process ontology were metabolic processes, such as carbohydrate metabolic processes (GO: 0005975), DNA metabolic processes (GO:0006259), cellular macromolecule metabolic processes (GO:0044260),

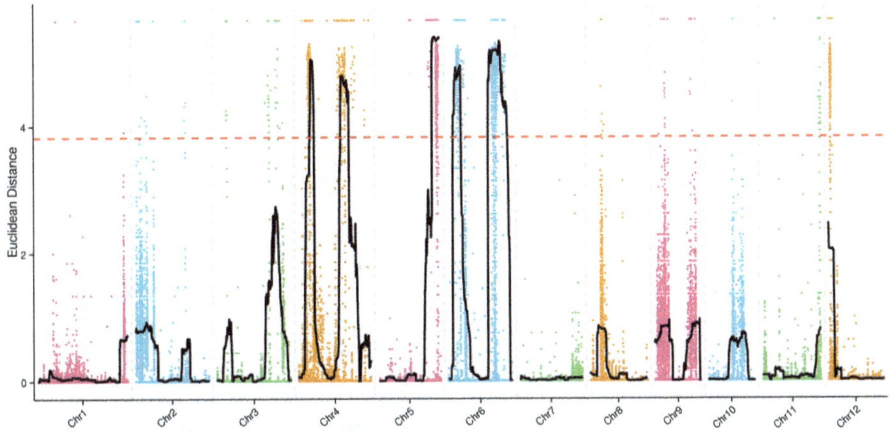

Fig. 2 Identification of the hot-region for PSST through the ED association analysis method. X-axis represents the position of 12 chromosomes of rice and Y-axis represents the ED. The color dots represent the ED value of every SNP locus. The black lines show the ED value of fitting results, and the red imaginary lines show the association threshold of ED

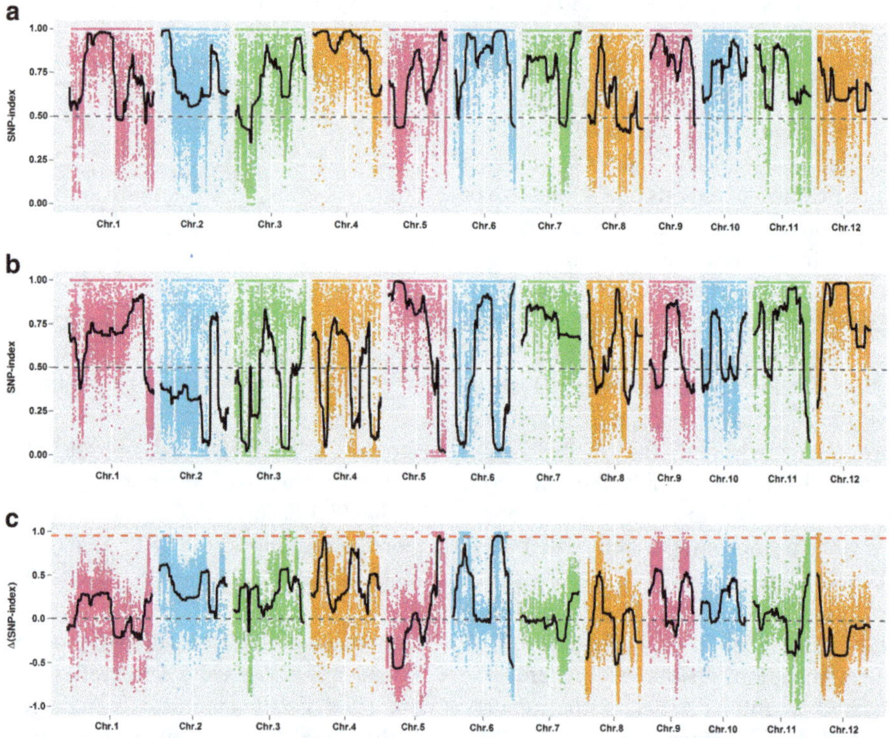

Fig. 3 Identification of the hot-region for PSST through the SNP-index association analysis method. X-axis represents the position of 12 chromosomes of rice and Y-axis represents the SNP-index or Δ (SNP-index). The color dots represent the SNP-index or Δ (SNP-index) value of every SNP locus. The black lines show the SNP-index or Δ (SNP-index) value of fitting results. a: the SNP-index graph of T-pool. b: the SNP-index graph of S-pool. c: the Δ (SNP-index) graph. The red dotted line shows the association threshold value (0.9532)

and nucleic acid metabolic processes (GO:0090304). The most enriched terms of molecular function were binding, such as nucleic acid binding (GO:0003676), iron ion binding (GO:0005506), and zinc ion binding, (GO: 0008270). The most enriched terms of cellular component were membranes, such as chloroplast outer membranes (GO:0009707), cytoplasmic membrane-bounded vesicles (GO:0016023), and plasma membranes (GO:0005886). All the annotated information is listed in Additional file 2: Table S8. Out of the 212 annotated genes, only two, *LOC_Os06g39740* and *LOC_Os06g39750*, were annotated as the function of "response to cold (GO: 0009409)", suggesting their key roles in regulating cold tolerance in rice. Moreover, 13 predicted genes (4.83%) contained 52 nsSNP loci including three nsSNP in *LOC_Os06g39750*, but no non-synonymous substitution occurred in *LOC_Os06g39740* (Additional file 2: Table S9).

Table 4 Information of the association region by the SNP-index association analysis method

Chromosomes	Chromosome locations (bp)		Size (Mb)	Gene number
	Start	End		
Chr5	26,700,000	27,370,000	0.67	158
Chr6	21,840,000	22,070,000	0.23	33
Chr6	22,140,000	24,450,000	2.31	388
Chr6	24,480,000	24,700,000	0.22	27
Chr6	24,750,000	24,830,000	0.08	16
Chr6	24,980,000	25,150,000	0.17	27
Total	–	–	3.68	649

To determine the expression patterns of the two genes annotated as "response to cold" in response to cold stress, a qRT-PCR was performed using the total RNA of the two parents DN422 and KY131 subjected to cold water conditions. Two primers pairs for the qRT-PCR analysis were designed based on the CDS sequences of the two genes (Additional file 2: Table S2). According to the qRT-PCR results (Fig. 5), *LOC_Os06g39740* was slightly induced by the cold stress after 3 days in DN422 and after 1 day in KY131. The largest relative expression quantity was nearly 2-fold at day 5 in DN422 and 2.5-fold at day 7 in KY131. The expression patterns of *LOC_Os06g39740* have no obvious up or down-regulation for all days at the control conditions in DN422 and KY131. *LOC_Os06g39750* exhibits a strong inducement by cold stress at days 5, 7, 9, 11, and 13 after treatment in KY131, and the highest relative expression quantity was above 9-fold at day 9. Compared to KY131, the up-regulated expression of

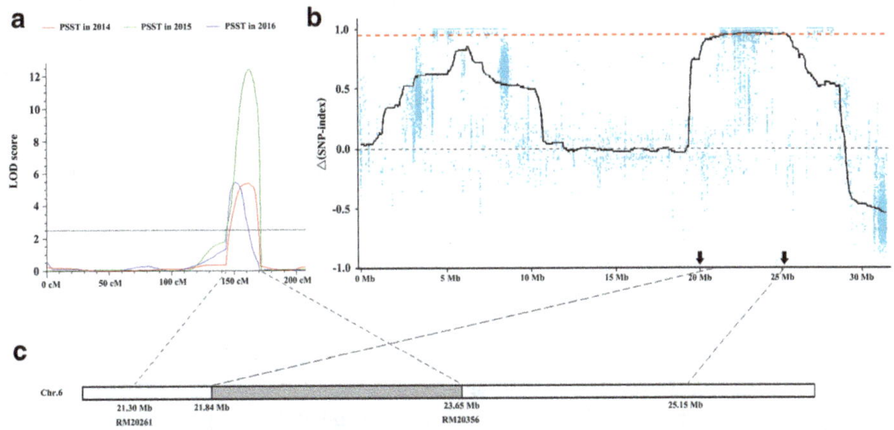

Fig. 4 Mapping of *qPSST6* on chromosome 6 using the methods of QTL analysis and Seq-BSA. a: *qPSST6* detected in the RIL population by using the ICIM module of QTL IciMapping 3.2. The LOD scores are shown with three distinct peaks corresponding to *qPSST6* in 2014, 2015 and 2016, respectively. b: hot-region for PSST identified by SNP-index method on chromosome 6. The black line shows the Δ (SNP-index) value of fitting results. The red dotted line shows the association threshold value (0.9532), and the black arrows on the x-axis represent the location of hot-region for PSST. c: the intersection of *qPSST6* interval identified by QTL analysis and Seq-BSA. The gray region on the chromosome shows the 1.81 Mb intersection of *qPSST6*

LOC_Os06g39750 under cold stress in DN422 were weak, and the highest relative expression quantity was approximately 2.6-fold at day 7. *LOC_Os06g39750* exhibited no obvious regularity in the control condition in both DN422 and KY131. The result of the qRT-PCR suggested that *LOC_Os06g39750* was the candidate gene controlling the PSST in rice.

Among the *LOC_Os06g39750*, eight SNP loci were identified, and they were all in the CDS region of *LOC_Os06g39750* (Additional file 2: Table S10). Sequencing of the eight SNP loci in *LOC_Os06g39750* revealed that four haplotypes (HapI, HapII, HapIII, and HapIV) existed among the RIL population (Fig. 6, Additional file 2: Table S11). HapI, DN422 genotype, which contains 20 RILs

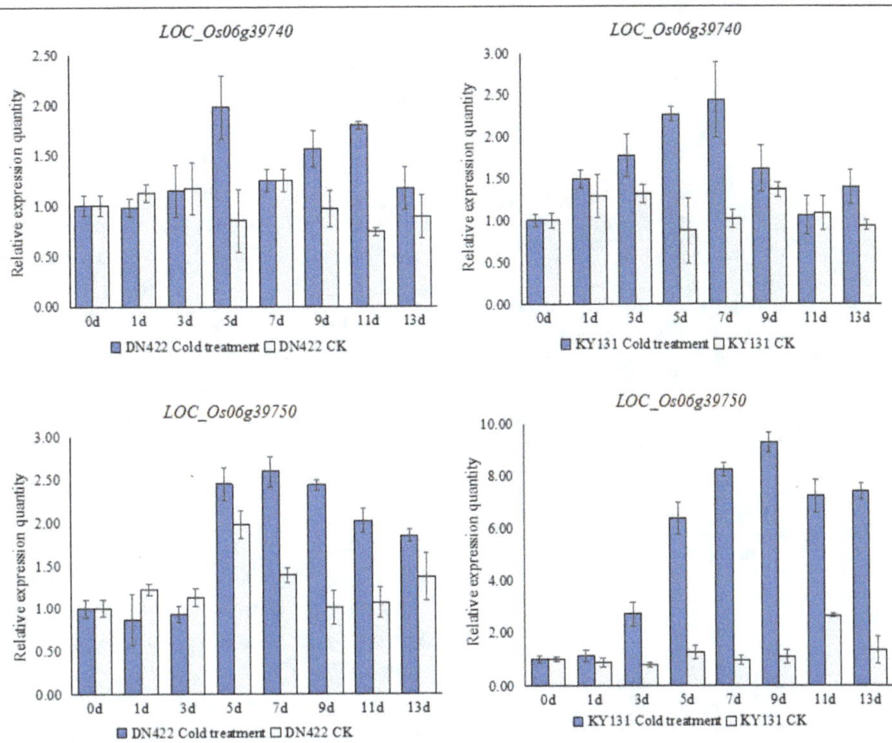

Fig. 5 Expression patterns of *LOC_Os06g39740* and *LOC_Os06g39750* under the conditions of cold water irrigation and CK. Each line with different patterns on the X-axes from left to right indicate the sampling times of 0, 1, 3, 5, 7, 9, 11 and 13 day after cold water irrigating in KY131 and DN422. The Y-axes indicate the relative expression quantity. The error bars indicate SD for the data obtained in three biological replicates

of S-pool and other 24 RILs. HapII, KY131 genotype, which contains 20 RILs of T-pool and other 9 RILs. HapIII contains 92 RILs, which has seven same SNPs to DN422, and only one same SNP to KY131. HapIV contains 25 RILs, which has six same SNPs to DN422, and two same SNPs to KY131. The mean PSST of the 44 HapI and 29 HapII RILs was 0.63 and 0.91, respectively, indicating that RILs with HapII of KY131 genotype usually have a higher mean PSST. Haplotype analysis demonstrate a key role of *LOC_Os06g39750* in regulating the PSST of rice once again.

Discussion

Cold stress is one of the major abiotic environmental stresses that significantly affects rice yield. Improving the cold tolerance of rice varieties using cold-tolerant genes is a fast and efficient approach to reduce the yield loss in rice cultivation regions. KY131, a *japonica* rice, has been widely cultivated in the northeast region of China for nearly 20 years. A prominent characteristic of KY131 is the strong cold tolerance to cope with the frequent low temperatures that occur during the booting stage in this high-latitude region (Yao et al. 2012; Wang et al. 2016). In this study, a RIL population derived from a cross between DN422, a cold sensitive *japonica* rice, and KY131 was exposed in a cold-water environment over 3 years to acquire the PSST to map cold-tolerant genes. The identification of cold tolerance at the booting stage is more difficult compared to other agronomic traits in rice. Therefore, we divided the RILs into early maturing, middle-maturing and late-maturing groups, and irrigated them independently to evaluate the PSST of the parents and each RIL line. The critical temperature for the cold stress treatment at the booting stage of rice is 17–20 °C (Zhou et al. 2010). The cold-water temperature of 17 °C selected in this study could differentiate the PSST of the two parents significantly, which made the RILs vary widely (Table 1, Fig. 1). All the measures adopted in this study were effective at obtaining accurate PSST phenotypic data.

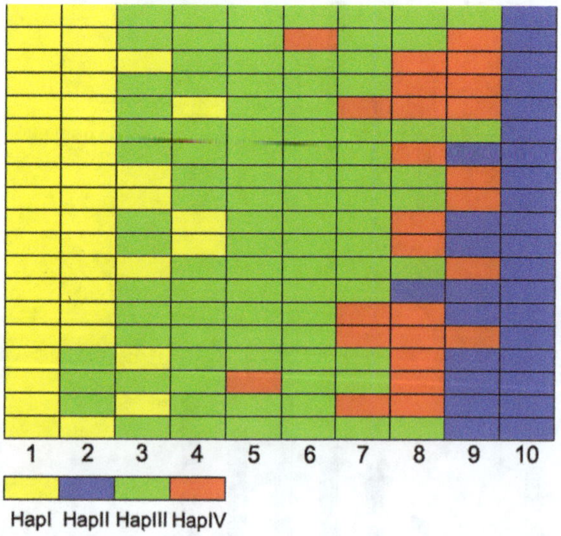

Fig. 6 Haplotype distribution of *LOC_Os06g39750* in RIL population. Every block represents a RIL, and all RILs are arranged according to their PSST value. 1, 2, 3, 4, 5, 6, 7, 8, 9, 10 column represent the RILs with PSST of 0.56~ 0.63, 0.63~ 0.67, 0.68~ 0.70, 0.70~ 0.72, 0.73~ 0.75, 0.75~ 0.77, 0.77~ 0.82, 0.82~ 0.85, 0.85~ 0.90, 0.90~ 0.95 from top block to bottom block, respectively. Yellow, blue, green, and red block represent HapI, HapII, HapIII, and HapIV, respectively

Conventional methods of fine mapping and map-based cloning of QTLs were very difficult for the development of high-density genetic map and a series of near-isogenic lines, especially the genetic population derived from *japonica* × *japonica* or *indica* × *indica* crosses. For example, a long time was spent from the preliminary mapping to the cloning of the cold-tolerance gene *Ctb1* (Saito et al. 1995; Saito et al. 2001; Saito et al. 2010). Seq-BSA technology provides a powerful, time-saving method to narrow the chromosome interval harboring the target genes/QTLs. Using the strategy of traditional QTL mapping combined with Seq-BSA, Zheng et al. (2016) rapidly mapped and identified a novel broad-spectrum resistance gene to rice blast (*Pi65(t)*). The *Pi65(t)* region has been narrowed to 60 Kb, which contains 4 predicted *R* genes. In the present study, a major QTL *qPSST6* that explained 30.45% of the phenotypic variation was initially mapped using the traditional QTL mapping method. Despite the relatively low density of the genetic map containing only 158 SSR markers, the 3 years of phenotypic data of PSST demonstrate a major QTL existing within the 28.4 cM interval between SSR markers RM20261-RM20356. By employing the Seq-BSA method, *qPSST6* was quickly delimited to a 1.81 Mb physical interval on chromosome 6 (Fig. 4), and two genes, *LOC_Os06g39740* and *LOC_Os06g39750*, were locked according to their annotated biological function as "responds to cold". Thus, our study provided a fast and cost-effective strategy to identify the quantitative locus of complex trait variation. Some QTLs about cold tolerance at the booting stage have been reported in recent years. These QTLs were mainly distributed on chromosome 3 (Andaya and Mackill 2003b; Suh et al. 2010; Shirasawa et al. 2012; Zhu et al. 2015; Ulziibat et al. 2016), chromosome 4 (Saito et al. 2004; Saito et al. 2001; Xu et al. 2008; Endo et al. 2016; Zhang et al. 2017), and chromosome 7 (Suh et al. 2010; Zhou et al. 2010). Andaya and Mackill (2003b) and Oh et al. (2004) identified a QTL controlling spikelet fertility (*qCTB6*) and the days to heading (*dth6*), respectively, located on chromosome 6, but the two QTLs were approximately 16.4 Mb and 11.9 Mb distant to *qPSST6* in this study. Beyond those, we did not find any QTL loci distributed on chromosome 6. Thus, *qPSST6* was a new locus for cold tolerance at the booting stage in rice.

Among the 269 predicted genes identified by QTL mapping and Seq-BSA, *LOC_Os06g39740* and *LOC_Os06g39750* were taken into full consideration as the candidate genes of *qPSST6* according to the results of the GO annotation generated by Seq-BSA. The results of the qRT-PCR showed that *LOC_Os06g39750* was induced strongly by cold stress in the cold-tolerant parent KY131. Haplotype analysis also demonstrated the RILs with HapII of KY131 genotype usually have a higher mean PSST. These all suggesting that *LOC_Os06g39750* was the candidate gene of *qPSST6*. So far, there has not been any report about *LOC_Os06g39750* in rice. By searching the Arabidopsis Information Resource (TAIR, http://www.arabidopsis.org/) database using the gene sequence of *LOC_Os06g39750*, ten Arabidopsis homologous genes were identified (Additional file 2: Table S12). Eight of the ten genes were the members of the 3-ketoacyl-CoA synthase family involved in the biosynthesis of VLCFA (very long chain fatty acids). Among the ten Arabidopsis homologous genes, *At5g43760*, *At2g26640*, *At2g16280*, *At1g25450*, *At1g01120*, and *At2g26250* were annotated as functions of "response to cold" in TAIR, indicating that they may participate the regulation of reactions to cold in Arabidopsis. An expression profiling analysis revealed that most of the 3-ketoacyl-CoA synthase (KCS) family genes in Arabidopsis, including the above six genes, responded to cold stress (Joubes et al. 2008). Therefore, it is reasonable to predict that *LOC_Os06g39750* is the candidate gene for *qPSST6*. However, gene clone, genetic transformation, and further studies are needed to functionally validate this conclusion.

Conclusions

The present study using the Dongnong422/Kongyu131 RIL population and a genetic linkage map containing 158 SSR markers identified a novel major QTL *qPSST6* for cold tolerance at the booting stage of rice in the 28.4 cM region on chromosome 6. By using the combined Seq-BSA strategy, *qPSST6* was narrowed to a region of 1.81 Mb, which contains 269 predicted genes. According to the results of the gene ontology (GO) analysis, *LOC_Os06g39740* and *LOC_Os06g39750* were annotated as "response to cold". *LOC_Os06g39750* exhibited strong up-regulated expressions at the 5, 7, 9, 11, and 13 day after the cold-water treatment in KY131. Haplotype analysis also demonstrate a key role of *LOC_Os06g39750* in regulating the PSST of rice, suggesting that it was the candidate gene of *qPSST6*. This study provided a fast and cost-effective strategy to identify cold-tolerant genes at the booting stage in rice.

Additional files

Additional file 1: Figure S1. Distributions of coverage depth on chromosomes of the sequencing samples. X-axe indicates the 12 chromosomes of rice. Y-axe indicates the log2 value of coverage depth corresponding to chromosome location.

Additional file 2: Table S1. The PSST of T-pool and S-pool. **Table S2.** Sequences of specific primers for PSST candidate genes. **Table S3.** Sequences of specific primers for SNP loci in LOC_Os06g39750. **Table S4.** Heading time, plant height, panicle number, and grain yield per plant of parents and RIL population in 2014, 2015 and 2016. **Table S5.** Coefficients of correlation (r) among PSST and PSSN in 2014, 2015 and 2016. **Table S6.** QTLs for PSSN of rice detected in 3 years. **Table S7.** Summary of Seq-BSA data for each sample. **Table S8.** Annotation information of predicted genes for PSST. **Table S9.** nsSNPs between DN422 and KY131 along with their S-pool and T-pool. **Table S10.** Information of SNP/InDel in LOC_Os06g39740 and LOC_Os06g39750. **Table S11.** Haplotype of LOC_Os06g39750 in RIL population and mean PSST of each RIL. **Table S12.** The Arabidopsis homologous genes of LOC_Os06g39750.

Abbreviations

BSA: Bulked-segregant analysis; COG: Cluster of orthologous groups of proteins; ED: Euclidean distance; GO: Gene ontology; KEGG: Kyoto Encyclopedia of Genes and Genomes; MAS: Marker assisted selection; NGS: Next-generation sequencing; NR: Non-redundant protein; PSSN: Percent seed set under normal water irrigation; PSST: Percent seed set under cold water treatment; PVE: Phenotypic variation explained; QTL: Quantitative trait loci; RIL: Recombinant inbred line; S-pool: Sensitive pool; SSD: Single seed descent; T-pool: Tolerant pool

Acknowledgments

We would like to thank the "Young Talents" Project of Northeast Agricultural University for its research funding support, and the National Innovation Center of Regional Technology in Rice for the laboratory and field support.

Funding

This work was supported by the "Young Talents" Project of Northeast Agricultural University (16QC03), National Natural Science Foundation (31701507), Youth Science Foundation of Heilongjiang Province (QC2017015), Major Scientific and Technological Bidding Projects of Heilongjiang Province (GA14B102-02).

Authors' contributions

SJ and ZDT designed research; YLM performed QTL mapping; WJG analyzed BSA sequence data; LHL, ZHL, and JY irrigated the cold water and got the phenotyping data; XDW and ZMH performed qRT-PCR analysis, FMF analyzed candidate genes; ZHW prepared the T-pool and S-pool; SJ wrote manuscript; ZDT corrected manuscript. All authors have read and approved the manuscript.

Competing interests

The authors declare that they have no competing interests.

Author details

[1]College of Agriculture, Northeast Agricultural University, Harbin 150030, China. [2]The Institute of Industrial Crops of Heilongjiang Academy of Agricultural Sciences, Harbin 150086, China. [3]College of Life Science, Northeast Agricultural University, Harbin 150030, China.

References

Abe A, Kosugi S, Yoshida K, natsume S, Takagi H, Kanzaki H, Matsumura H, Yoshida K, Mitsuoka C, Tamiru M (2012) Genome sequencing reveals agronomically important loci in rice using MutMap. Nat Biotech 30:174–178

Altschul SF, Madden TL, Schäffer AA, Zhang J, Zhang Z, Miller W, Lipman DJ (1997) Gapped BLAST and PSI-BLAST: a new generation of protein database search programs. Nucleic Acids Res 25(17):3389–3402

Andaya V, Mackill D (2003b) QTLs conferring cold tolerance at the booting stage of rice using recombinant inbred lines from a *japonica* × *indica* cross. Theor Appl Genet 106(6):1084–1090

Andaya VC, Mackill DJ (2003a) Mapping of QTLs associated with cold tolerance during the vegetative stage in rice. J Exp Bot 54(392):2579

Apweiler R, Bairoch A, Wu CH, Barker WC, Boeckmann B, Ferro S, Gasteiger E, Huang HZ, Lopez R, Magrane M, Martin MJ, Natale DA, O'Donovan C, Redaschi N, Yeh SL (2004) UniProt: the universal protein knowledgebase. Nucleic Acids Res 32:115–119

Ashburner M, Ball CA, Blake JA, Botstein D, Butler H, Cherry JM, Davis AP, Dolinski K, Dwight SS, Eppig JT, Harris MA, Hill DP, Issel-Tarver L, Kasarskis A, Lewis S, Matese JC, Richardson JE, Ringwald M, Rubin GM, Sherlock G (2000) Gene ontology: tool for the unification of biology. Nat Genet 25(1):25–29

Chen X, Temnykh S, Xu Y, Cho YG, McCouch SR (1997) Development of a microsatellite framework map providing genome-wide coverage in rice (*Oryza sativa* L.). Theor Appl Genet 95(4):553–567

Deng YY, Li JQ, Wu SF, Zhu YP, Chen YW, He FC (2006) Integrated nr database in protein annotation system and its localization. Comput Eng 32(5):71–74

Endo T, Chiba B, Wagatsuma K, Saeki K, Ando T, Shomura A, Mizubayashi T, Ueda T, Yamamoto T, Nishio T (2016) Detection of QTLs for cold tolerance of rice cultivar "Kuchum" and effect of QTL pyramiding. Theor Appl Genet 129(3):631

Fekih R, Takagi H, Tamiru M, Abe A, Natsume S, Yaegashi H, Sharma S, Sharma S, Kanzaki H, Matsumura H, Saitoh H, Mitsuoka C, Utsushi H, Saitoh H (2013) MutMap+: genetic mapping and mutant identification without crossing in rice. PLoS One 8(7):e68529

Han YC, Lv P, Hou SL, Li SY, Ji GS, Ma X, Du RH, Liu GQ (2015) Combining next generation sequencing with bulked segregant analysis to fine map a stem moisture locus in Sorghum (*Sorghum bicolor* L. Moench). PLoS One 10(5): e0127065

Hill JT, Demarest BL, Bisgrove BW, Gorsi B, Su YC, Yost HJ (2013) MMAPPR: mutation mapping analysis pipeline for pooled RNA-seq. Genome Res 23(4): 687–697

International Rice Genome Sequencing Project (IRGSP) (2005) The map-based sequence of the rice genome. Nature 436:793–800

Jena KK, Kim SM, Suh JP, Yang CI, Kim YJ (2012) Identification of cold-tolerant breeding lines by quantitative trait loci associated with cold tolerance in rice. Crop Sci 51(2):517–523

Jia Y, Zou DT, Wang JG, Liu HL, Inayat MA, Sha HJ, Zheng HL, Sun J, Zhao HW (2015) Effect of low water temperature at reproductive stage on yield and glutamate metabolism of rice (*Oryza sativa* L.) in China. Field Crops Res 175:16–25

Joubes J, Raffaele S, Bourdenx B, Garcia C, Laroche-Traineau J, Moreau P, Domergue F, Lessire R (2008) The VLCFA elongase gene family in Arabidopsis thaliana: phylogenetic analysis, 3D modelling and expression profiling. Plant Mol Biol 67:547–566

Kanehisa M, Goto S, Kawashima S, Okuno Y, Hattori AM (2004) The KEGG resource for deciphering the genome. Nucleic Acids Res 32:277–280

Li H, Durbin R (2009) Fast and accurate short read alignment with burrows-wheeler transform. Bioinformatics 25:1754–1760

Li SH, Wang CS, Chang XP, Jing RL (2012) Genetic dissection of developmental behavior of grain weight in wheat under diverse temperature and water regimes. Genetica 140:393–405

Livaja M, Wang Y, Wieckhorst S, Haseneyer G, Seidel M, Hahn V, Knapp SJ, Taudien S, Schön CC, Bauer E (2013) BSTA: a targeted approach combines bulked segregant analysis with next-generation sequencing and de novo transcriptome assembly for SNP discovery in sunflower. BMC Genomics 14:628

Livak KJ, Schmittgen TD (2001) Analysis of relative gene expression data using real-time quantitative pcr and the $2^{-\Delta\Delta Ct}$ method. Methods 25:402–408

Matsunaga K (2005) Establishment of an evaluation method for cold tolerance at the booting stage of rice using deep water irrigation system and development of highly cold-tolerant rice varieties by combining cold tolerance genes. Bull Miyagi Furukawa Agric Exp Stn 4:1–78

McKenna A, Hanna M, Banks E, Sivachenko A, Cibulskis K, Kernytsky A, Garimella K, Altshuler D, Gabriel S, Daly M, DePristo MA (2010) The genome analysis toolkit: a MapReduce framework for analyzing next-generation DNA sequencing data. Genome Res 20(9):1297–1303

Michelmore RW, Paran I, Kesseli RV (1991) Identification of markers linked to disease-resistance genes by bulked segregant analysis: a rapid method to detect markers in specific genomic regions by using segregating populations. Proc Natl Acad Sci U S A 88:9828–9832

Murray MG, Thompson WF (1980) Rapid isolation of high molecular weight plant DNA. Nucl Acids Res 8:4321–4325

Oh CS, Choi YH, Lee SJ, Yoon DB, Moon HP, Ahn SN (2004) Mapping of quantitative trait loci for cold tolerance in weedy rice. Breeding Sci 54(4):373–380

Saito K, Hayano Y, Kuroki M, Sato Y (2010) Map-based cloning of the rice cold tolerance gene *Ctb1*. Plant Sci 179:97–102

Saito K, Hayano-Saito Y, Maruyama-Funatsuki W, Sato Y, Kato A (2004) Physical mapping and putative candidate gene identification of a quantitative trait locus Ctb1 for cold tolerance at the booting stage of rice. Theor Appl Genet 109(3):515–522

Saito K, Miura K, Nagano K, Hayano-Saito Y, Araki H, Kato A (2001) Identification of two closely linked quantitative trait loci for cold tolerance on chromosome 4 of rice and their association with anther length. Theor Appl Genet 103(6):862–868

Saito K, Miura K, Nagano K, Hayano-Saito Y, Saito A, Araki H, Kato A (1995) Chromosomal location of quantitative trait loci for cool tolerance at the booting stage in rice variety "Norin-PL8". Japanese J Breeding 45(3):337–340

Salvi S, Tuberosa R (2005) To clone or not to clone plant QTLs: present and future challenges. Trends Plant Sci 10:297–304

Satake T, Hayase H (1970) Male sterility caused by cooling treatment at the young micro-spore stage in rice plants. V. Estimations of pollen developmental stage and the most sensitive stage to coolness. Proc Crop Sci Soc Jpn 39(4):468–473

Schneeberger K, Ossowski S, Lanz C, Juul T, Petersen AH, Nielsen KL, Jorgensen JE, Weigel D, Andersen SU (2009) SHOREmap: simultaneous mapping and mutation identification by deep sequencing. Nat Methods 6:550–551

Shirasawa S, Endo T, Nakagomi K, Yamaguchi M, Nishio T (2012) Delimitation of a QTL region controlling cold tolerance at booting stage of a cultivar, 'Lijiangxintuanheigu', in rice, *Oryza sativa* L. Theor Appl Genet 124(5):937–946

Siahpoosh MR, Sanchez DH, Schlereth A, Scofield GN, Furbank RT, Dongen JT, Kopka J (2011) Modification of *OsSUT1* gene expression modulates the salt response of rice *Oryza sativa* cv. Taipei 309. Plant Sci 182:101–111

Singh VK, Khan AW, Saxena RK, Kumar V, Kale SM, Sinha P, Chitikineni A, Pazhamala LT, Garg V, Sharma M, Kumar CVS, Parupalli S, Vechalapu S, Patil S, Muniswamy S, Ghanta A, Yamini KN, Dharmaraj PS, Varshney RK (2016) Next-generation sequencing for identification of candidate genes for fusarium wilt and sterility mosaic disease in pigeonpea (*Cajanus cajan*). Plant Biotech J 14(5):1183–1194

Song J, Li Z, Liu Z, Guo Y, Qiu LJ (2017) Next-generation sequencing from bulked-segregant analysis accelerates the simultaneous identification of two qualitative genes in soybean. Frontiers Plant Sci 8. https://doi.org/10.3389/fpls.2017.00919

Suh JP, Jeung JU, Lee JI, Choi YH, Yea JD, Virk PS, Mackill DJ, Jena KK (2010) Identification and analysis of QTLs controlling cold tolerance at the reproductive stage and validation of effective QTLs in cold-tolerant genotypes of rice (*Oryza sativa* L.). Theor Appl Genet 120:985–995

Takagi H, Abe A, Yoshida K, Kosugi S, Natsume S, Mitsuoka C, Uemura A, Utsushi H, Tamiru M, Takuno S, Innan H, Cano LM, Kamoun S, Terauchi R (2013) QTL-seq: rapid mapping of quantitative trait loci in rice by whole genome resequencing of DNA from two bulked populations. The. Plant J 74(1):174–183

Tatusov RL, Galperin MY, Natale DA, Koonin EV (2000) The COG database: a tool for genome scale analysis of protein functions and evolution. Nucleic Acids Res 28:33–36

Trick M, Adamski NM, Mugford SG, Jiang CC, Febrer M, Uauy C (2012) Combining SNP discovery from next-generation sequencing data with bulked segregant analysis (BSA) to fine-map genes in polyploid wheat. BMC Plant Biol 12:14

Ulziibat B, Ohta H, Fukushima A, Shirasawa S, Kitashiba H, Nishio T (2016) Examination of candidates for the gene of cold tolerance at the booting stage in a delimited QTL region in rice cultivar 'Lijiangxintuanheigu'. Euphytica 211(3):331–341

Wambugu P, Ndjiondjop MN, Furtado A, Henry R (2017) Sequencing of bulks of segregants allows dissection of genetic control of amylose content in rice. Plant Biotech J. https://doi.org/10.1111/pbi.12752

Wang D, Liu JL, Li CG, Kang HX, Wang Y, Tan XQ, Liu MH, Deng YF, Wang ZL, Liu Y, Zhang DY, Xiao YH, Wang GL (2016) Genome-wide association mapping of cold tolerance genes at the seedling stage in rice. Rice 9(1):61

Wang LZ, Wang LM, Xiang HT, Luo Y, Li R, Li ZJ, Wang CY, Meng Y (2016) Relationship of photosynthetic efficiency and seed-setting rate in two contrasting rice cultivars under chilling stress. Photosynthetica 54(4):581–588

Wenger WJ, Schwartz K, Sherlock G (2010) Bulk segregant analysis by high-throughput sequencing reveals a novel xylose utilization gene from Saccharomyces cerevisiae. PLoS Genet 6:e1000942

Xu LM, Zhou L, Zeng YW, Wang FM, Zhang HL, Shen SQ, Li ZC (2008) Identification and mapping of quantitative trait loci for cold tolerance at the booting stage in a *japonica* rice near-isogenic line. Plant Sci 174(3):340–347

Yao Y, Miao Y, Huang S, Gao L, Ma X, Zhao G, Gnyp ML (2012) Active canopy sensor-based precision N management strategy for rice. Agron Sustain Dev 32(4):925–933

Zhang Q, Chen QH, Wang SL, Hong YH, Wang ZL (2014) Rice and cold stress: methods for its evaluation and summary of cold tolerance-related quantitative trait loci. Rice 7:24

Zhang ZH, Su L, Li W, Chen W, Zhu YG (2005) A major QTL conferring cold tolerance at the early seedling stage using recombinant inbred lines of rice (*Oryza sativa* L.). Plant Sci 168(2):527–534

Zhang ZY, Li JJ, Pan YH, Li JL, Zhou L, Shi HL, Zeng YW, Guo HF, Yang SM, Zheng WW, Yu JP, Sun XM, Li GL, Ding YL, Ma L, Shen SQ, Dai LY, Zhang HL, Yang SH, Guo Y, Li ZC (2017) Natural variation in *CTB4a* enhances rice adaptation to cold habitats. Nat Commun 8:14788

Zheng W, Wang Y, Wang L, Ma ZB, Zhao JM, Wang P, Zhang LX, Liu ZH, Lu XC (2016) Genetic mapping and molecular marker development for *Pi65(t)*, a novel broad-spectrum resistance gene to rice blast using next-generation sequencing. Theor Appl Genet 129(5):1035

Zhou L, Zeng YW, Zheng WW, Tang B, Yang SM, Zhang HL, Li JJ, Li ZC (2010) Fine mapping a QTL *qCTB7* for cold tolerance at the booting stage on rice chromosome 7 using a near-isogenic line. Theor Appl Genet 121(5):895

Zhu Y, Chen K, Mi X, Chen T, Ali J, Ye G, Xu JL, Li ZK (2015) Identification and fine mapping of a stably expressed QTL for cold tolerance at the booting stage using an interconnected breeding population in rice. PLoS One 10(12):e0145704

12

Genome-wide analyses of late pollen-preferred genes conserved in various rice cultivars and functional identification of a gene involved in the key processes of late pollen development

Sunok Moon[1†], Moe Moe Oo[2†], Backki Kim[3], Hee-Jong Koh[3], Sung Aeong Oh[2], Gihwan Yi[4], Gynheung An[1], Soon Ki Park[2*] and Ki-Hong Jung[1*]

Abstract

Background: Understanding late pollen development, including the maturation and pollination process, is a key component in maintaining crop yields. Transcriptome data obtained through microarray or RNA-seq technologies can provide useful insight into those developmental processes. Six series of microarray data from a public transcriptome database, the Gene Expression Omnibus of the National Center for Biotechnology Information, are related to anther and pollen development.

Results: We performed a systematic and functional study across the rice genome of genes that are preferentially expressed in the late stages of pollen development, including maturation and germination. By comparing the transcriptomes of sporophytes and male gametes over time, we identified 627 late pollen-preferred genes that are conserved among japonica and indica rice cultivars. Functional classification analysis with a MapMan tool kit revealed a significant association between cell wall organization/metabolism and mature pollen grains. Comparative analysis of rice and *Arabidopsis* demonstrated that genes involved in cell wall modifications and the metabolism of major carbohydrates are unique to rice. We used the *GUS* reporter system to monitor the expression of eight of those genes. In addition, we evaluated the significance of our candidate genes, using T-DNA insertional mutant population and the CRISPR/Cas9 system. Mutants from T-DNA insertion and CRISPR/Cas9 systems of a rice gene encoding glycerophosphoryl diester phosphodiesterase are defective in their male gamete transfer.

(Continued on next page)

* Correspondence: psk@knu.ac.kr; khjung2010@khu.ac.kr
Sunok Moon and Moe Moe Oo contributed equally to this work.
†Equal contributors
[2]School of Applied Biosciences, Kyungpook National University, Daegu 702-701, South Korea
[1]Graduate School of Biotechnology and Crop Biotech Institute, Kyung Hee University, Yongin 446-701, South Korea
Full list of author information is available at the end of the article

(Continued from previous page)

Conclusion: Through the global analyses of the late pollen-preferred genes from rice, we found several biological features of these genes. First, biological process related to cell wall organization and modification is over-represented in these genes to support rapid tube growth. Second, comparative analysis of late pollen preferred genes between rice and *Arabidopsis* provide a significant insight on the evolutional disparateness in cell wall biogenesis and storage reserves of pollen. In addition, these candidates might be useful targets for future examinations of late pollen development, and will be a valuable resource for accelerating the understanding of molecular mechanisms for pollen maturation and germination processes in rice.

Keywords: GUS assay, Pollen maturation process, Pollen germination process, Rice, T-DNA insertional mutant, CRISPR/Cas9, Network analysis,

Background

The life cycles of all land plants alternate between diploid sporophyte and haploid gametophyte. By the time gametophyte development is finished, a floret of rice contains one carpel with a single egg cell, plus six stamens, each filled with approximately 1000 pollen grains (Prasad et al. 2006; Zhang et al. 2011). Pollen development covers several steps, from stamen meristem specification to pollen grain formation and pollination, and various cell and tissue types are involved. Not only dynamic changes in gene expression but also post-translational control of proteins are important for those processes (Guan et al. 2014). Development of the gametes is modulated by gametophytic and sporophytic gene expression. In the case of the pollen wall, the tapetum contributes to exine and tryphine formation while the gametophytes are responsible for intine formation (Shi et al. 2015). The tapetum, the innermost anther wall layer, serves as a nutritive tissue that secretes pollen wall components, nutrients, and metabolites (Xu et al. 2014). Many tapetum-expressed genes have been shown to regulate anther and pollen wall development (Xu et al. 2014; Shi et al. 2015). In addition, plant hormones such as auxin and gibberellin (GA) control pollen development in sporophytic tissues (Sakata et al. 2014; Cecchetti et al. 2017).

The male gametophyte originates from a pollen mother cell that undergoes meiosis to produce tetrads of haploid microspores (Owen and Makaroff 1995). After their release from those tetrads, the microspores enlarge and then undergo asymmetric mitosis to form bi-cellular pollen with different cell fates (pollen mitosis I). Afterward, the smaller generative cell divides symmetrically to produce two sperm cells (pollen mitosis II). Tri-cellular mature pollen is further processed to complete its development and is then ready to germinate (Bedinger 1992).

High-throughput microarray data provide a powerful tool for identifying genes at the genome scale that control pollen development. Genome-wide investigations have already been performed with the male gametophytic transcriptomes in *Arabidopsis* and rice (Becker et al. 2003; Honys and Twell 2003; Honys and Twell 2004; Suwabe et al. 2008; Fujita et al. 2010; Wei et al. 2010; Aya et al. 2011). Those studies have revealed that mature pollen grains have a unique transcriptome profile with a higher proportion of selectively expressed genes than in any other tissues (Becker et al. 2003; Wei et al. 2010). This unique transcriptome represents the biological processes required for pollen development, i.e., cell wall metabolism, signalling, and cytoskeleton dynamics. Honys and Twell (2004) and Wei et al. (2010) have reported dynamic changes in the transcriptomes of *Arabidopsis* and rice, based on their sequential stages of pollen development. Patterns of changes in their transcripts are conserved between those two species. The diversity of these transcripts is greatly decreased as the pollen progresses from uni-cellular microspores to mature grains (Wei et al. 2010). The transition from bi-cellular to tri-cellular pollen was also observed by an increase in the proportion of male gametophyte-specific transcripts (Wei et al. 2010). Competition among pollen grains for an egg cell to produce a diploid zygote is a common phenomenon. Because rice pollen begins to germinate within 2 min after pollination (Chen et al. 2008), most of the required components for this are already present in the mature grains. Analysis by Wei et al. (2010) has revealed that the gene expression profiles of mature grains and germinated grains are significantly and positively correlated ($r = 0.99$), suggesting that the former stores a set of transcripts for germination. Moreover, inhibition of RNA synthesis by actinomycin D does not interrupt germination, indicating that the RNAs necessary for germination are already present in mature grains (Mascarenhas 1993; Hao et al. 2005).

Success during the late stages of pollen development is critical if crop yields are to be sustained. Applying microarray technologies with in rice, a model system, several research groups have conducted genome-wide transcriptome analyses and identified genes that are preferentially expressed in the pollen (Fujita et al. 2010; Deveshwar et al. 2011; Russell et al. 2012). Although different numbers of genes – 453 (Fujita et al., 2010), 735 (Deveshwar et al., 2011), and 1376 (Russell et al., 2012) – have been uncovered in those independent studies, the consistency of expression patterns has not been assayed over a range of growing environments and genetic backgrounds. Using

Arabidopsis, Suzuki et al. (2008) and Dobritsa et al. (2011) have performed large-scale screening of pollen exine mutants from randomly selected lines and have identified 12 and 14 genes involved in forming the pollen exine. Although numerous gene expression data and genome-wide gene-indexed mutant populations are available for rice, no previous attempts have been described to understand the roles of genes in controlling late pollen development on a large scale. Thus, such systematic approaches might help accelerate our identification of rice genes that function in mature pollen development and germination.

A shift in gene expression exists during pollen development and bi-cellular pollen may be a key point for the regulation of shift (Wei et al. 2010). "Late pollen-preferred genes" was defined as preferentially expressed genes at pollen following the bi-cellular stage. In this study, we identified late pollen-preferred genes conserved in japonica and indica cultivars of rice. Here we provide detailed results from our genome-wide analysis of such genes conserved in two subspecies of rice. We also discuss mutants obtained through T-DNA and gene-editing that show defects in gene-transfer through the male gamete.

Results

Meta-expression analysis and genome-wide identification of late pollen-preferred genes conserved in various rice cultivars

To identify the late pollen-preferred genes that are conserved among japonica and indica cultivars of rice, we examined publicly available Affymetrix rice microarray data. From the National Center for Biotechnology Information Gene Expression Omnibus (NCBI GEO; http://www.ncbi.nlm.nih.gov/geo/), we downloaded and analyzed six series of data prepared from developing anthers and pollen grains (Additional file 1: Table S1) (Edgar et al. 2002). The samples were arranged in order according to the process of anther and pollen development in each cultivar. Intensity values for these six series were first normalized with the Affy package in the R program and then $\log_2$-transformed. As a control, all of the Affymetrix anatomical meta-expression profiles in the Rice Oligonucleotide Array Database (ROAD;), except for two anther samples, were used to check expression patterns in other tissues/organs (Cao et al. 2012). Afterward, we performed *K*-means clustering (KMC) analysis with the Euclidean Distance Metric and grouped 57,382 probes into 36 clusters based on their expression patterns. Clusters 2 and 35 exhibited the highest expression at the later stages, after bi-cellular pollen (Additional file 2: Figure S1). As a result, we found that 627 genes, represented by 750 probe sets in the rice affymetrix array, showed late pollen-preferred expression patterns in both subspecies (Fig. 1). These gene clusters are listed in Table S2 (Additional file 1).

Expression patterns for late pollen-preferred genes confirmed through the *GUS* reporter system

To verify the patterns of late pollen-preferred expression in our candidate genes, we used two strategies: a promoter trap system and generation of transgenic plants under the control of selected promoters. We have previously described this promoter trap technique using T-DNA that

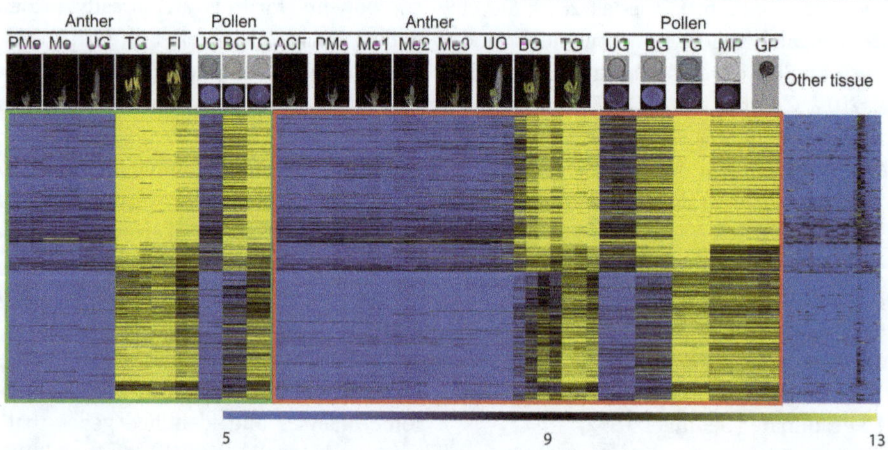

Fig. 1 Heatmap expression analysis of 627 late pollen-preferred genes. Expression patterns conserved among indica (green box) and japonica (red box) rice types are shown over several stages of development. Microarray data numbered 22 for indica rice (5 stages for anthers and 3 for pollen) and 42 for japonica rice (8 stages for anthers and 5 for pollen). Anatomical meta-expression data from ROAD were used to check expression patterns in other tissues. ACF, archesporial cell-forming stage; BG, bi-cellular gametophyte stage; FI, flowering stage; GP, germinating pollen; Me, meiotic stage; Me1, meiotic leptotene stage; Me2, meiotic zygotene-pachytene stage; Me3, meiotic diplotene-tetrad stage; MP, mature pollen stage; PMe, pre-meiosis; TG, tri-cellular pollen stage; UG, uni-cellular gametophyte stage. Yellow color in heatmap indicates high level of expression; dark-blue, low expression

carries the promoterless *GUS* reporter gene in japonica rice (Jeon et al. 2000). In all, 22 lines were chosen that had an in-frame fusion of the promoterless *GUS* in the genic region. Expression of that reporter gene was detected at three developmental stages. Subsequently, we identified two lines showing late pollen-preferred GUS-staining patterns (Fig. 2a and b), as had been expected from our meta-expression analysis depicted in Fig. 1. The efficiency of *GUS* expression was 9.09% (2/22), which was approximately 16-fold higher than the 0.58% we had previously found (Jeong et al. 2002). This was because, for our current investigation, we chose lines with T-DNA insertions within the late pollen-preferred genes whereas earlier studies used randomly selected samples (Jeong et al. 2002). The T-DNAs were inserted into LOC_Os11g20384, encoding SacI homology domain-containing protein (Line 1A-13,819), and into LOC_Os07g17310, encoding B12D protein (Line 2D-41,188). Schematic diagrams of those T-DNA insertions are presented in Figure S2 (Additional file 2). In a separate examination, we used transgenic plants harboring promoter and *GUS* fusion constructs. Six genes in Clusters 2 and 35 were chosen that exhibited expression values above 13 $\log_2$ in mature pollen; their locus numbers and promoter regions are shown in Table S3 (Additional file 1). Their promoter activities had already been tested in *Arabidopsis* (Oo et al. 2014). *LOC_Os11g45730* encodes pectinesterase; *LOC_Os02g50770,* plant peroxidase family protein; *LOC_Os01g69020,* cell division protein FtsZ family protein; *LOC_Os05g46530,* plant invertase/pectin methylesterase (PME) inhibitor domain-containing protein; *LOC_Os07g14340,* pectinesterase inhibitor domain-containing protein; and *LOC_Os04g25190,* pollen allergen Lol p2 family protein. The GUS expression patterns within transgenic lines for these genes were monitored in rice at three sequential stages of pollen development (Fig. 2c-g). Strong GUS activity was detected at the mature pollen stage, as indicated in the heatmap. The pattern of late

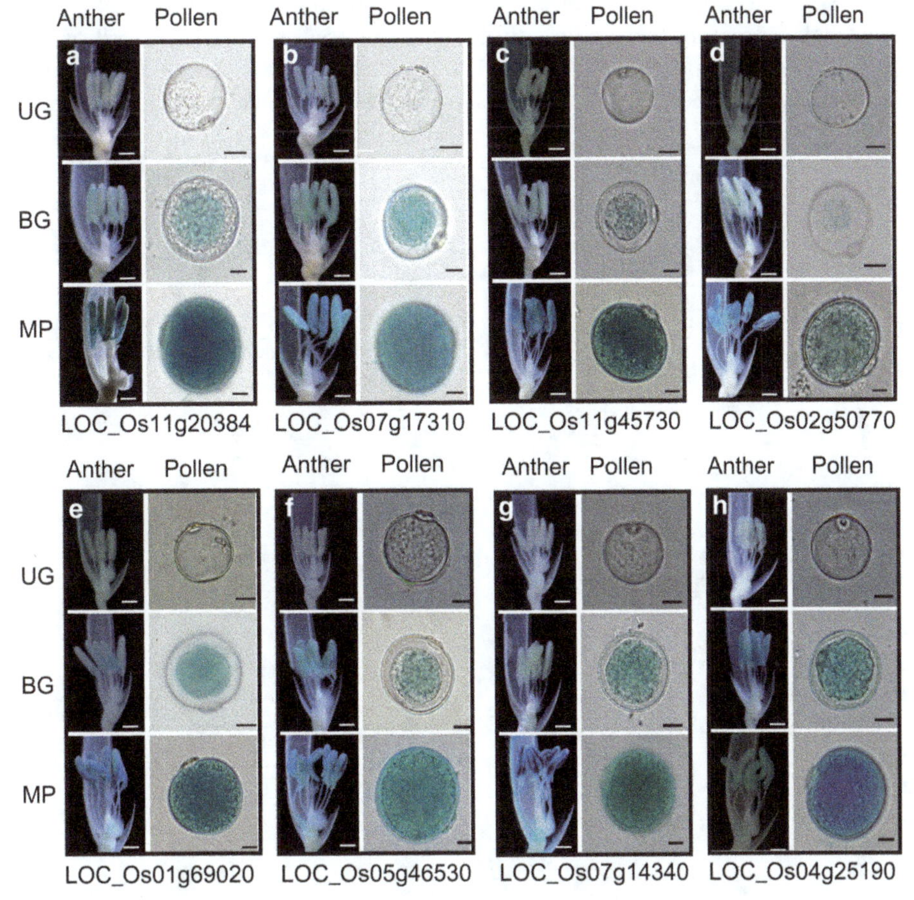

Fig. 2 Evaluation of late pollen-preferred genes using *GUS* reporter systems. Images of flower and pollen from promoter trap lines 1A-13,819 having T-DNA insertion within *LOC_Os11g20384* (**a**) and 2D-41,188 harboring T-DNA insertion within *LOC_Os07g17310* (**b**) at sequential stages of pollen development. Images of flower and pollen from transgenic plants harboring promoter::*GUS* reporter constructs. Promoters from *LOC_Os11g45730* (**c**), *LOC_Os02g50770* (**d**), *LOC_Os01g69020* (**e**), *LOC_Os05g46530* (**f**), *LOC_Os07g14340* (**g**), and *LOC_Os04g25190* (**h**) were used to control *GUS* expression. BG, bi-cellular gametophyte stage; MP, mature pollen stage; UG, uni-cellular gametophyte stage. *GUS* expression patterns were used from promoter trap lines and promoter-*GUS* transgenic lines. Bar in flower image = 0.5 mm; bar in pollen image = 10 μm

pollen-preferred expression by these genes, as shown in Fig. 2, was reconfirmed by real-time PCR analysis. For all eight genes, strong expression was detected at the tri-cellular mature pollen stage (Fig. 3). All of our results demonstrated that the meta-expression data for these genes are a valuable source of novel promoters for controlling traits associated with late pollen development or germination and for discovering the novel functions of such genes.

Biological processes specific to late pollen-preferred genes

We conducted a Gene Ontology (GO) enrichment analysis in ROAD (Jung et al. 2008) to query 627 late pollen-preferred genes belonging to Clusters 2 and 35. Our objective was to examine GO enrichment within the category of 'biological process' with the ROAD GO enrichment tool. As a result, we identified 547 GO terms assigned to 308 genes; the other 319 genes did not have GO annotations (Additional file 1: Table S4). Significant terms in that category were selected with hypergeometric *p*-values ≤0.05 and enrichment values of at least two-fold. Of these, 16 GO terms were over-represented in late pollen-preferred genes (Table 1). Significantly enriched terms were found for biological processes corresponding to phosphatidylcholine metabolism (22.3 GO fold-enrichment value), clathrin coat assembly (19.7), GA metabolism (18.1), cell wall modifications (17.3), phosphatidyinositol metabolism (12.1), glycogen biosynthesis (10.9), starch biosynthesis (9.7), polysaccharide catabolism (8.5), sexual reproduction (7.4), cytoskeleton organization (7.0), cellulose biosynthesis (6.4), potassium ion transport (3.2), and ATP biosynthesis (2.9). The most abundant terms were for 'protein amino acid phosphorylation', representing 56 genes. This was followed by 34 genes for transport and 23 for carbohydrate metabolism.

Late pollen-preferred genes functionally classified through MapMan analysis

MapMan allows one to group genes into different functional categories and visualize data through diagrams (Yoo et al. 2015). To classify these late pollen-preferred genes, we primarily used the metabolism and regulation overviews installed in the MapMan tool kit (Fig. 4a). In the metabolism overview, cell wall organization and modification (48 genes), lipid metabolism (10), and metabolism of major carbohydrates (9) were the most

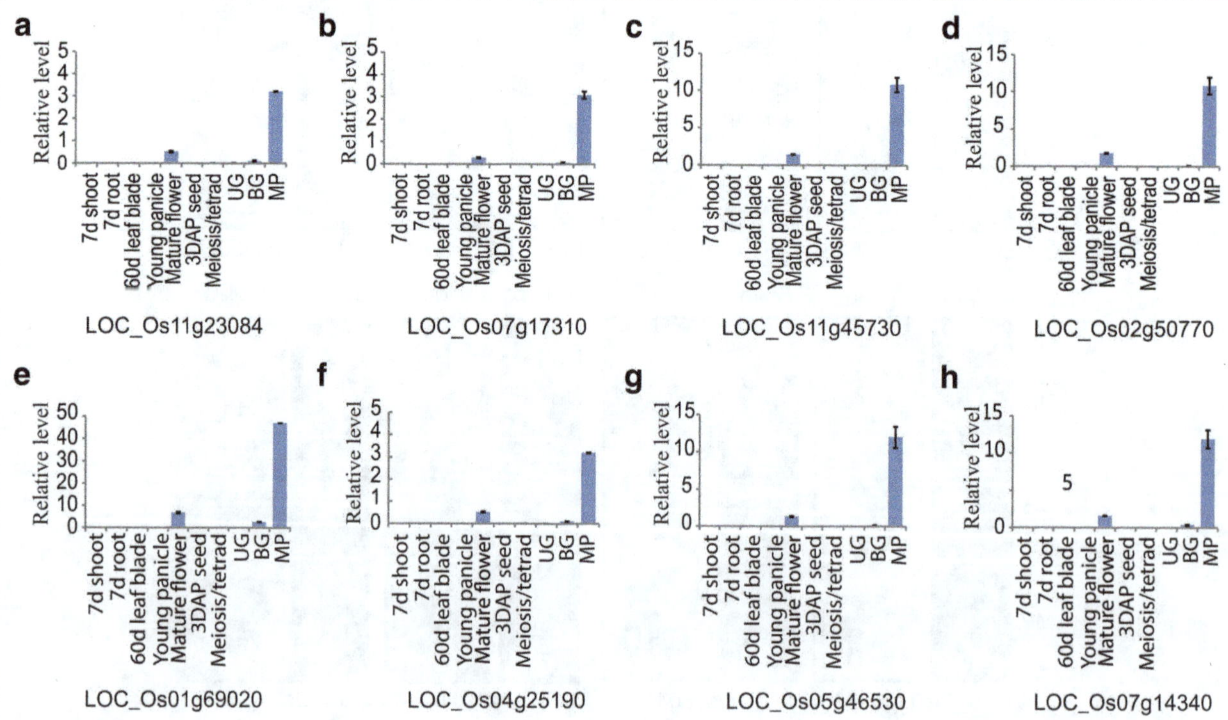

Fig. 3 Real-time RT-PCR analysis of 8 genes shown in Fig. 2. Expression patterns of *LOC_Os11g20384* (**a**), *LOC_Os07g17310* (**b**), *LOC_Os11g45730* (**c**), *LOC_Os02g50770* (**d**), *LOC_Os01g69020* (**e**), *LOC_Os05g46530* (**f**), *LOC_Os07g14340* (**g**), and *LOC_Os04g25190* (**h**) were analyzed by quantitative real-time PCR for various organs. Transcript levels were normalized to rice *Ubi5* and calculated by comparative cycle threshold method. Error bars indicate standard deviation (sd). y-axis, relative expression level to rice *Ubi5*; x-axis, sample names used for analyses. 3DAPS, seeds at 3 d after pollination; 7DS, 7-day-old shoots; 7DR, 7-day-old roots; ML, mature leaves; MF, mature flower; YP, young panicles (2 cm long); and anthers at bi-cellular gametophyte stage (BG); tri-cellular mature pollen stage (MP); meiosis and tetrad (M/T); and uni-cellular gametophyte stage (UG)

Table 1 Analysis of significantly enriched Gene Ontology terms for genes preferentially expressed in late from rice

GO category	No. of GO repeats[a]	No. of GO repeats in queried genes[b]	No. of expected GO repeats[c]	Fold-enrichment value[d]
Phosphatidylcholine metabolic process	13	4	0.2	22.3
Clathrin coat assembly	11	3	0.2	19.7
Gibberellin metabolic process	20	5	0.3	18.1
Cell wall modification	46	11	0.6	17.3
Phosphatidylinositol metabolic process	18	3	0.3	12.1
Glycogen biosynthetic process	20	3	0.3	10.9
Starch biosynthetic process	15	2	0.2	9.7
Polysaccharide catabolic process	34	4	0.5	8.5
Sexual reproduction	39	4	0.5	7.4
Cytoskeleton organization	31	3	0.4	7.0
Cellulose biosynthetic process	34	3	0.5	6.4
Potassium ion transport	203	9	2.8	3.2
ATP biosynthetic process	99	4	1.4	2.9
Carbohydrate metabolic process	600	23	8.3	2.8
Protein amino acid phosphorylation	1593	56	22.0	2.5
Transport	978	34	13.5	2.5

Total number of GO terms in rice genome is 39,571; total number of GO terms in queried pollen-preferred genes, 547
[a]number of selected GO Slim terms annotated in rice genome
[b]observed number of selected GO Slim terms in queried pollen-preferred genes
[c]expected number of selected GO Slim terms in queried pollen-preferred genes
[d]relative ratio of observed number to expected number for a selected GO Slim term

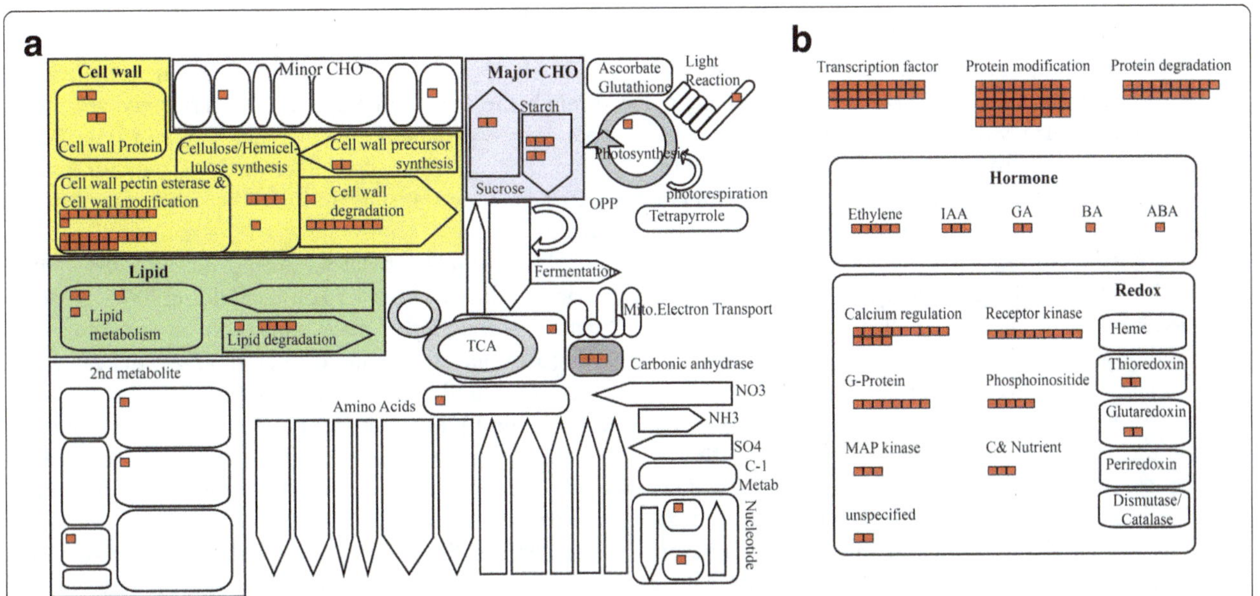

Fig. 4 MapMan analysis of late pollen-preferred genes. **a** Metabolic overview and (**b**) regulation overview were analyzed with 627 genes. Cell wall organization and modification, lipid metabolism, and carbohydrate metabolism were dominant MapMan terms in metabolic overview. Red boxes in these overviews indicate late pollen-preferential. Genes associated with transcription factors, protein modification, and protein degradation were identified from regulation overview. In addition, diverse hormone responses, signaling pathway components, and redox responses for late pollen development were visualized. Detailed information about overviews is shown in Table S5.

common (Additional file 1: Table S5). The predominance of the cell wall organization and modification terms was well-matched with our finding that GO terms related to those processes were enriched in the biological process category (Table 1). In particular, cell wall-modifying enzymes such as pectin methylesterases (PMEs) and cellulose synthases were functionally well-characterized during pollen tube growth because the cell walls of those tubes consist of callose in the inner layer plus pectin, cellulose, and hemicellulose in the outer layer (Krichevsky et al. 2007). Our list included 11 PMEs and 4 cellulose synthases, implying that those genes are involved in pollen maturation and germination in rice.

Our regulation overview showed that genes associated with transcriptional regulation, protein modification, and protein degradation were the most frequently coupled with late pollen-preferred genes (Additional file 1: Table S5). Of them, 26 genes were assigned to the transcription category and 47 genes were related to protein modification (Fig. 4b; Additional file 1: Table S5). For protein degradation, 19 genes were found. We also found 12 genes in the category of hormone metabolism – five for ethylene synthesis, three for auxin, and two for GA (Additional file 1: Table S6).

Comparative analysis of late pollen-preferred genes from rice and *Arabidopsis*

Late pollen-preferred genes in *Arabidopsis* were identified using a meta-anatomical expression database that we developed from the *Arabidopsis* Affymetrix array data deposited in the NCBI GEO database (See Materials and Methods). This produced 773 genes (Additional file 2: Figures S3 and S4; Additional file 1: Table S7). We then searched for orthologs between rice and *Arabidopsis* with a database of orthologous groups among rice, *Arabidopsis, Brachypodium*, maize (*Zea mays*), *Populus*, *Vitis vinifera*, and *Sorghum bicolor* from the Rice Genome Annotated Project (RGAP; http://rice.plantbiology.msu.edu/annotation_pseudo_ortho.shtml). Rice orthologs for 380 *Arabidopsis* genes were identified based on the sequence information. Among them, orthologs for 175 *Arabidopsis* genes exhibited patterns of late pollen-preferential expression (Additional file 1: Table S8). Another search based on 627 rice late pollen-preferred genes led to the identification of 220 *Arabidopsis* orthologs (Additional file 1: Table S9). Of them, 133 had *Arabidopsis* orthologs with late pollen-preferential expression. Therefore, all of these results indicated that functional conservancy was approximately 20%, with 175 (22.64%) of 773 *Arabidopsis* having rice orthologs and 133 (21.21%) of 627 rice genes having *Arabidopsis* orthologs. We believe that genes unique to each species, i.e., 407 in rice and 393 in *Arabidopsis*, will be good targets for uncovering the evolutionary developmental differences in mature and germinating pollens between those two plant systems.

Enriched biological processes in late pollen-preferred genes compared between rice and *Arabidopsis*

Through our comparative genome and expression analyses, we determined that approximately 20% of late pollen-preferred genes are conserved between rice and *Arabidopsis* in terms of their sequence homology and expression patterns. We performed MapMan analysis for rice divergent genes without ortholog(s) in *Arabidopsis* as well as common genes conserved between the two. Several MapMan terms did not overlap between the divergent and common genes. Moreover, MapMan terms for cell wall modification and major carbohydrate metabolism were found only in the rice divergent genes (Fig. 5).

Genetic study of T-DNA insertional mutants for a male-gene transfer defective (*MTD1*) gene involved in late pollen development

To identify the functional significance of our candidates, we analyzed T-DNA insertion lines for 28 genes (An et al. 2003; Ryu et al. 2004; Jeong et al. 2006). Until now, a mutant line was identified because of distortions in segregation ratios, which were close to 1:1:0 (wild type:heterozygote:homozygote) (Fig. 6a). The gene that causes those distortions in corresponding mutants was named *Male-gene Transfer Defective* (*MTD*). We then called the first gene *MTD1*. *MTD1/LOC_Os02g09450* encodes glycerophosphoryl diester phosphodiesterase (GPD), with 767 amino acids. It has nine exons and eight introns (Fig. 6b). We also identified a T-DNA insertional mutant of this gene (*mtd1–1*), which has a T-DNA insertion in the sixth exon (Fig. 6b).

To clarify how *MTD1* participates in gene-transfer through the male gametophyte, we performed reciprocal crosses between heterozygotes and the wild type (WT). From this, we determined that *mtd1–1* mutants have a defect in gene-transfer through the male gametophyte (Table 2). That is, a cross between the male (pollen) from an *MTD1/mtd1–1* heterozygous plant and the female (ovary) of the WT shows a complete defect (0/35) in gene-transfer through the male gametophyte. In contrast, a cross between the female from an *MTD1/mtd1* heterozygous plant and a WT male leads to successful gene-transfer through the female gametophyte. To examine any morphological defects, we observed mature grains under a bright field microscope but did not found any differences between WT and *mtd1–1* mutant pollen.

To confirm the phenotype of *mtd1* and understand the function of *MTD1*, we utilized the CRISPR/Cas9 system, selecting the third exon as a target region of the guide RNA (Fig. 6b). From sequence analysis of the primary transgenic lines, we obtained a homozygote with a 1-bp deletion in the *MTD1* coding sequence. That deletion mutant was named *mtd1–2*. Because we already knew that

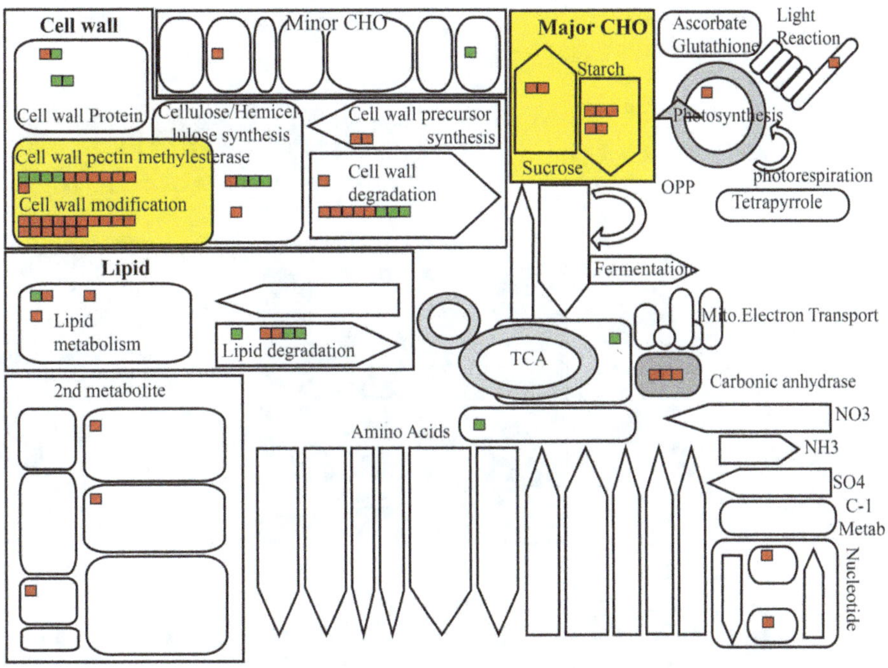

Fig. 5 Enriched metabolic pathways in rice late pollen-preferred genes compared to those of *Arabidopsis*. Red squares, genes unique to rice; green squares, conserved genes in terms of their sequence homology and expression patterns between 2 plant systems

mtd1–1 has a defect in gene-transfer through the male gametophyte, we checked the mature pollen status of *mtd1–2* by staining with auramine O, calcofluor white, and Lugol's iodine (Moon et al. 2013). No differences between WT and mutant pollen were observed with regard to the accumulation of exine (stained by auramine O), intine (calcofluor white), or starch (Lugol's iodine) (Fig. 6c). This meant that *MTD1* has no essential role in the production of mature pollen. Therefore, we speculated that this gene might function at the progamic phase, i.e., during pollen germination, pollen tube elongation, and commutation with female organs. We then tested in vitro pollen germination of *mtd1–2*/*mtd1–2*. Whereas the WT pollen had elongated tubes, the *mtd1–2* pollen exhibited only a small protrusion of those tubes (Fig. 6c). Because of this defect, homozygotic plants of *mtd1–2* were sterile.

We also found that there are 13 genes encoding GPD in rice, and nine of them are clustered more closely with MTD1 (Fig. 6d). Of these, *MTD1* was predominantly expressed in late pollen stage, which supported our supposition that it has a primary role in this developmental process, as revealed in the loss-of-function mutants (Fig. 6d).

Discussion

In silico validation of late pollen-preferred genes in various rice cultivars

We delved into the large collection of microarray data focused on anther and pollen development in rice. Having identified candidate genes with late pollen-preferred expression that are conserved among japonica and indica rice enabled us to investigate relevant developmental processes across a more diverse range of varieties than would be possible from examinations with only single variety. We have recently reported that at least 10% of genes showing light-responsive expression patterns differ among japonica ('Nipponbare', 'TP309', and 'LG') and indica ('IR24') cultivars (Jung et al. 2008). By analyzing their late pollen-preferred genes, we found that at least 17. 7% of the top 300 in each cultivar do not overlap. This result proved the relevance of our candidate genes for studying late pollen development in japonica and indica rice. In addition, we determined that late pollen-preferred genes that were not conserved among rice types would be good targets for investigating late pollen development within each cultivar.

Hormone metabolism and late pollen-preferred genes

Our MapMan analysis revealed 12 genes within the category of hormone metabolism – five for ethylene synthesis, three for auxin, and two for GA (Additional file 1: Table S6). All of these hormones are involved in late pollen development. For example, in *Petunia hybrida*, the level of 1-aminocyclopropane-1-carboxylic acid (ACC) in anthers is very low until the day before anthesis, when it then increases by up to 100-fold (Lindstrom et al. 1999), possibly as a result of activity by pPHACS2, an ACC synthase. Research with the ethylene-insensitive

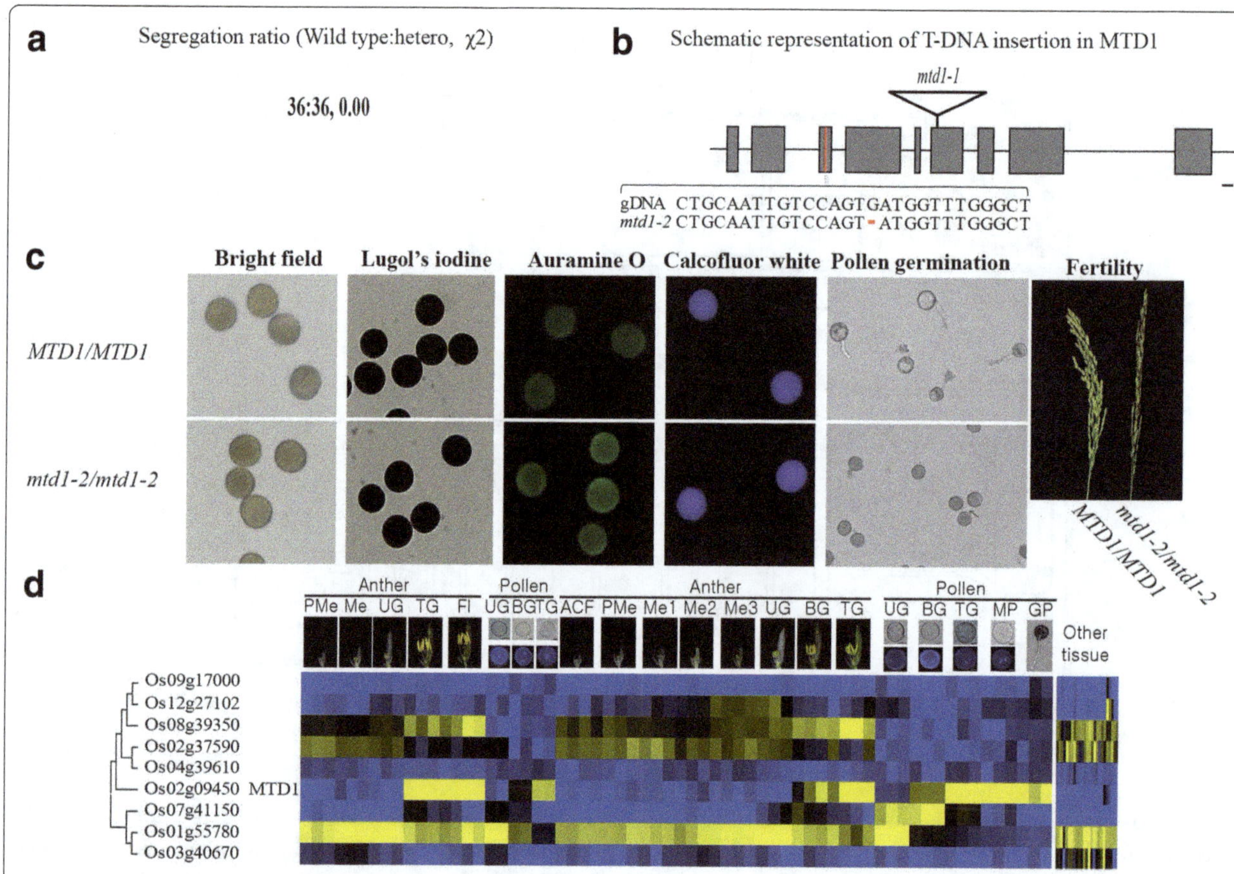

Fig. 6 Estimation of functional dominancy by MTD genes through phylogenomic analysis (**a**) segregation ratios of heterozygotes and wild-type segregants in mutants of *MTD1* by T-DNA insertion, and results of Chi-square test (χ^2). **b** Schematic view of *MTD1* structure and mutated position by T-DNA insertions and deletion generated by CRISPR/Cas9. Grey boxes indicate exon; lines, UTR or intron; reverse triangles, T-DNA; red box, target region of CRISPR/Cas9. Genomic DNA sequences from wild type and *gtd1–2* are shown. **c** Phenotypes of pollen grains from wild type and *mtd1–2*. Germinated pollen and grains stained with Lugol's iodine, auramine O, or calcofluor white were observed under bright field and fluorescence microscopes. Bar = 20 μm. Panicles at filling stage of wild type and *mtd1–2* were observed. **d** Expression patterns according to developmental stages of pollen were analyzed using 22 microarray data from indica rice (5 stages for anthers, 3 for pollen) and 42 microarray data from japonica rice (8 stages for anthers, 5 for pollen). Anatomical meta-expression data from ROAD were used to check expression patterns in other tissues. ACF, archesporial cell-forming stage; BG, bi-cellular gametophyte stage; FI, flowering stage; GP, germinating pollen; Me, meiotic stage; Me1, meiotic leptotene stage; Me2, meiotic zygotene-pachytene stage; Me3, meiotic diplotene tetrad stage; MP, mature pollen stage; PMe, pre-meiosis; TG, tri-cellular pollen stage; UG, uni-cellular gametophyte stage. Yellow color in heatmap indicates high level of expression; dark-blue, low expression

mutant *Never ripe* (*Nr*) has shown that ethylene plays a significant role in thermotolerance by regulating the level of sucrose in tomato (*Solanum lycopersicum*) (Firon et al. 2012). Auxin is also necessary for pollen maturation. For example, the auxin biosynthesis-defective *yuc2 yuc6* double mutant cannot produce grains (Cheng et al. 2006). The importance of auxin perception has also been demonstrated in a loss-of-function study with auxin receptor genes. There, the pollen grains from *tir1* and *afb1afb2afb3* multiple mutants age prematurely (Cecchetti et al. 2008). Gibberellic acid also has an essential role; GA-deficient mutants are mainly defective in pollen germination and elongation, while GA-insensitive mutants are mainly defective in pollen development (Chhun et al. 2007). *LOC_Os02g17780* encodes ent-copalyl diphosphate synthase1 (OsCPS1) while *LOC_Os06g02019* encodes ent-kaurenoic acid oxidase (OsKAO). Both enzymes are active in GA biosynthesis (Sakamoto et al. 2004), and mutants of both genes have significantly reduced frequencies of transmission through the male gametophytes, indicating that pollen germination and elongation essentially depend upon the de novo synthesis of GA in rice (Chhun et al. 2007). In addition,

Table 2 Genetic transmission analysis of *MTD1/mtd1* by reciprocal crossing

Genotype (female x male)	Wild type	Hetero	TE
MTD1/mtd1–1 X WT	33	28	84.8
WT X *MTD1/mtd1–1*	35	0	0.0

Segregation ratio was analyzed in F1 progeny of represented crosses. The transmission efficiency (TE) represents the percentage of *mtd1–1* mutant allele successfully transmitted through male or female gametes expression of *OsCPS1* and *OsKAO* peaks at the mature pollen stage (Chhun et al. 2007; Hirano et al. 2008). We also checked the expression patterns of other genes involved in GA biosynthesis and signalling pathways by using our meta-expression data with OsCPS1 and OsKAO (Additional file 2: Fig. S5). Among them, *GA3ox1*, *GA20ox3*, and *KO2* also showed late pollen-preferred expression patterns. Application of GA suppresses low temperature-induced male sterility (Sakata et al. 2014). Because pollen development is a complicated process accomplished by both microspores and the tapetum, various hormones, e.g., ethylene, auxin, and GA, might be involved in this process. Further experiments should focus on how hormone synthesis and signaling participate in the functioning of late pollen-preferred genes identified here.

Major metabolic processes of cell wall organization during late pollen development in rice and *Arabidopsis*

During the late stages of pollen development in rice, cell wall organization is the most common MapMan term. Similar results have been reported for *Arabidopsis* pollen-preferred genes. One of the most distinctive structural features of a pollen grain is its double wall layer – intine and exine (Vizcay-Barrena and Wilson, 2006). While the tapetum plays a pivotal role in exine formation, intine synthesis is largely dependent upon activity by the microspore (Nakamura et al. 2010; Yeung et al. 2011). The inner cell wall, intine, is simply composed of cellulose, pectin, and various other proteins (Noher de Halac et al., 2003). During pollen germination, the pollen tubes produce a new wall with two layers, the inner callosic layer and the outer layer, which again contains mainly cellulose, pectin, and other proteins (Donaldson and Knox, 2012). Among the late pollen-preferred genes found in our examination, 45 from *Arabidopsis* and 48 from rice are related with cell wall synthesis, proving their conserved roles in both species (Additional file 1: Table S10).

Our list shows that five genes from *Arabidopsis* and four from rice are related to cellulose synthesis. Mutations in any of three unique types of components within the cellulose synthase (CESA) complex – CESA1, CESA3, or CESA6 – result in the deformation of pollen grains owing to a deficiency in cellulose (Persson et al. 2007). Moreover, two *Arabidopsis* cellulose synthase-like D (CSLD) genes that are preferentially expressed in late pollen – *CSLD1* and *CSLD4* – have functions in pollen tube growth (Wang et al. 2011). Proteins from both are located in the Golgi apparatus and are transported to the plasma membrane in the tip region of the elongating tube, where cellulose is actively synthesized (Wang et al. 2011). In *csld1* and *csld4* mutants, the pollen tube wall is disorganized and exhibits less cellulose deposition. All of those observations suggest that the roles of rice homologs are conserved in their late pollen-preferred expression.

Among the genes investigated here, nine from *Arabidopsis* and five from rice are involved in the synthesis of cell wall proteins. In *Arabidopsis*, pollen-preferred *Arabinogalactan glycoprotein* (*AGP*)*6* and *AGP11* encode arabinogalactan glycoproteins associated with the cell wall. More than 50% of the grains from *agp6 agp11* are collapsed while the remainder display reduced germination and elongation (Coimbra et al. 2009; Coimbra et al. 2010). *Brassica campestris* male fertility 8, another pollen-specific AGP gene, functions in maintaining normal intine formation (Lin et al. 2014). One fasciclin-like AGP (FLA) gene, *FLA3*, is also involved in the development of intine walls (Li et al. 2010). All of this is evidence of conserved, pollen-preferred roles by rice homologs.

The grains of *Arabidopsis* and rice have 13 and 11 pollen-preferred PMEs, respectively. Among the 60 PME genes in the *Arabidopsis* ATH1 genome array, 18 are expressed in the pollen (Bosch et al. 2005). In many species, including tobacco (*Nicotiana tabacum*) and *Arabidopsis*, the apical end of the pollen tube is one layer composed mainly of pectin (Leroux et al. 2015). Pectins are synthesized in the Golgi apparatus and deposited as highly methyl-esterified polymers (Leroux et al. 2015). The PMEs de-esterify pectins and their activity can lead to cell-wall loosening or stiffening, depending on the apoplastic pH (Tian et al. 2006). Moreover, PME activity in pollen from a *vangard1* mutation is only 82% of that in the WT, and grains of the former have reduced tube growth (Jiang et al. 2005). Another pollen-specific PME, AtPPME1, is associated with slower pollen growth and irregularly shaped tubes (Tian et al. 2006). Rapid pollen tube growth after pollination might require strong activity by PME enzymes. In fact, such growth is completed within 50 min for rice while pollen tube discharge occurs in 4. 0 to 9.5 h for *Arabidopsis* (Faure et al. 2002; Chen et al. 2008). Therefore, for both species, numerous genes related to cell wall organization and modification must be expressed in the mature grains in order to support rapid tube growth.

The biochemical function of MTD1 in plant pollen development has not yet been determined. One such gene from Arabidopsis, SHAVEN3 (SHV3), is required for root hair elongation (Parker et al. 2000; Jones et al. 2006). Furthermore, SHV3 and its paralogs are important factors for primary cell wall organization (Hayashi et al. 2008). Both SHV3-like 4 (SVL4) and SVL5 are specifically expressed in Arabidopsis pollen (Lalanne et al. 2004), suggesting conserved roles in pollen development for MTD1 in rice and SVL4 and SVL5 in Arabidopsis. More detailed functions should be analyzed in future molecular and biochemical studies.

Cell wall modifications and metabolism of major carbohydrates for late pollen development are more important in rice than in *Arabidopsis*

Among the 16 genes related to cell wall modifications, six encode expansin proteins that are known to have cell-wall loosening activity. Although three expansin genes show a pattern of late pollen-preferred expression in *Arabidopsis* (Mollet et al. 2013), the six in rice are not homologous to those in *Arabidopsis*. Research with maize has indicated that pollen extracts have expansin-like activity that is unique to the cell walls of grass species but not eudicots (Cosgrove et al. 1997). Expansin facilitates pollen tube penetration through the stigma and style (Tabuchi et al. 2011). Maize pollen β-expansin preferentially binds to xylans and also solubilizes feruloylated arabinoxylan from cell walls in grasses (Tabuchi et al. 2011). Arabinoxylan is enriched in those grass cell walls (Sampedro et al. 2015). Thus, the arabinoxylan–cellulose interaction might be a target of grass pollen β-expansins (Sampedro et al. 2015). These distinct differences in structure between grass and dicot pollen cell walls can explain why MapMan terms related to cell wall modifications are more enriched in rice divergent genes than in those of *Arabidopsis* (Cho and Kende, 1997; Sampedro et al. 2015).

In *Arabidopsis* pollen, the development of starch grains begins at the vacuolated microspore stage, and grains are highly accumulated at the bi-cellular stage. At pollen maturity, only a few starch grains remain in the *Arabidopsis* plastid (Kuang and Musgrave, 1996; Tang et al. 2009). This contrasts with the features of rice pollen, where starch granules are abundant at the mature stage (Zhang et al. 2010). The higher starch content in mature rice pollen might explain why genes related to major carbohydrate metabolism are enriched in rice. Among the genes with MapMan terms related to carbohydrate metabolism that occur only in rice divergent genes, two are involved in the starch synthesis pathway while five genes are part of the pathway for starch degradation. We found one starch synthase and one ADP-glucose pyrophosphorylase (AGPase) in the starch synthesis pathway (Additional file 1: Table S5). Among them, *OsAGPL4* encodes AGPase and is involved in starch accumulation during pollen development. The *osagl4* mutant pollen exhibits a starch deficiency in pollen grains that causes defects in gene-transfer through the male gametophyte (Lee et al. 2016). Three amylase and two invertase genes participate in the starch degradation pathway. Because rapid growth by pollen tubes is a highly energy-consuming process, storage materials within pollen grains are mobilized for fuel (Goetz et al. 2017). Amylase catalyzes the hydrolysis of starch into sugar and invertase participates in the hydrolysis of sucrose into glucose and fructose. Regarding these genes, one starch degradation-defective mutant, tomato *alpha-Glucan water dikinase (LeGWD)/legwd*, interrupts gene-transfer through the male gametophyte (Nashilevitz et al. 2009). These differences between rice and *Arabidopsis* in their storage reserves for late pollen might explain why genes related to the metabolism of major carbohydrates are enriched in rice divergent genes, as coupled with late pollen development.

Conclusion

By comparing the publicly available transcriptomes of sporophytes and male gametes throughout their development, we identified 627 late pollen-preferred genes that are conserved among japonica and indica cultivars of rice. We then performed global analyses of these candidate genes to study the processes of late pollen development, including maturation and germination. Functional classification analysis with a MapMan tool kit revealed that cell wall organization and metabolism are significantly associated with that stage. Comparative analysis of late pollen-preferred genes from rice and *Arabidopsis* demonstrated that those involved in cell wall modifications and major carbohydrate metabolisms are more important in rice. To evaluate the significance of our candidate genes, we used T-DNA insertional mutant population and the CRISPR/Cas9 system to determine the function of a rice gene encoding GPD protein by identifying defects in transfer through the male gamete. These candidates might be useful targets for future examinations of late pollen development in rice. Our results provide new tools to understand late pollen development.

Methods

Collection of microarray data and identification of mature pollen-preferred genes

We used 64 publicly available rice Affymetrix microarray data prepared from anthers and pollen in NCBI GEO to identify late pollen-preferred genes. For the comparative transcriptome analyses between rice and *Arabidopsis*, we downloaded *Arabidopsis* Affymetrix microarray data series GSE5630, GSE5631, GSE5632, GSE5633, GSE6162, GSE6696, GSE12316, GSE17343, and GSE27281. To examine these data, we used the Affy package encoded by R language to normalize the signal intensity and then transformed them to $\log_2$ values. The normalized data with averaged Affymetrix anatomical meta-expression data were then used for further investigations, e.g., KMC analysis, heatmap construction, and identification of the late pollen-preferred genes (Chandran et al. 2016).

MapMan analysis

The MapMan program allows one to group genes into different functional categories and visualize data through various diagrams (Jung and An, 2012). To obtain their functional classifications, we uploaded RGAP locus IDs for 627 rice late pollen-preferred genes to the MapMan tool kit. We then investigated the metabolism and regulation overviews based on diverse overviews installed in that kit (Fig. 4). All data are detailed in Table S5 (Additional file 1).

Histochemical GUS assay and microscopic analyses

Histochemical GUS-staining was performed as described by Jefferson et al. (1987) and Dai et al. (1996). The assayed flowers and pollen grains were photographed with an Olympus BX61 microscope (Olympus, Tokyo, Japan).

RNA isolation and RT-PCR analysis

Total RNAs were extracted with Tri Reagent (MRC Inc., Cincinnati, OH, USA). Complementary DNA (cDNA) was synthesized as described previously (Lee and An, 2015). To evaluate the expression patterns of eight pollen-preferred genes that showed *GUS* activity, we prepared samples from the shoots and roots of 7-day-old seedlings, mature leaves, young panicles (2 cm), mature flowers, developing seeds at 3 d after pollination, and anthers at four different developmental stages. All primers for RT-PCR are listed in Table S11 (Additional file 1). The ubiquitin 5 gene (*Ubi5*) was used as an internal control.

Rice T-DNA mutant screening

We searched for lines from our T-DNA insertional mutant population that had insertions within our candidate late pollen-preferred genes (Jeon et al. 2000; An et al. 2003). For *MTD1*, one T-DNA insertional line (*mtd1–1*) was selected. Seedlings were grown in a half-strength Murashige and Skoog (MS) medium. Their DNA was prepared from 7-day-old plants via the hexadecylotrimethylammonium method. Genotypes were determined by PCR using gene-specific primers and T-DNA primers (Additional file 1: Table S11). The T-DNA insertional lines without homozygous progenies or with rare homologous progenies were further genotyped to check for segregation distortion.

Construction of the CRISPR/Cas9 vector to produce a homozygous mutant

To generate the single guide RNA (sgRNA) and Cas9 expression vector, we synthesized one set of oligomers targeting the third exon of MTD1 and inserted them into the Bsa I sites of the RGEB32 binary vector (Addgene plamid ID: 63142). Ligation products were transformed into *Escherichia coli*. The RGEB32 vector containing the sgRNA and Cas9 expression cassette was transformed into *Agrobacterium tumefaciens* LBA4404.

Cytochemical analysis

Pollen grains were immersed in various staining solutions. Grains stained for 15 min with 0.1% calcofluor white were then monitored for the presence of intine under a UV light with an Olympus BX61 microscope. Grains stained with 0.001% auramine O in 17% sucrose were examined for exine, using the FITC channel of the Olympus BX61 microscope. Staining with Lugol's iodine was used to detect the presence of starch.

Additional files

Additional file 1: Table S1. Six series of microarray data comprising 64 slides (GPL2025) associated with anthers/pollen in rice. **Table S2.** Locus IDs and putative functions of late pollen-preferred genes from rice. **Table S3.** Locus IDs and promoter regions of genes used for promoter analysis with *GUS* reporter. **Table S4.** Classification of GO terms for biological processes associated with late pollen-preferred genes. **Table S5.** MapMan classification of late pollen-preferred genes. **Table S6.** Genes related to hormone metabolism term in MapMan. **Table S7.** Late pollen-preferred genes in *Arabidopsis*. **Table S8.** Assignment of rice orthologs to *Arabidopsis* late pollen-preferred genes. Locus numbers are shown in by red. **Table S9.** Assignment of *Arabidopsis* orthologs to rice late pollen-preferred genes. Locus numbers are shown in red. **Table S10.** MapMan terms related to cell wall organization and modifications in rice and *Arabidopsis*. **Table S11.** Primer sequences used in genotyping and real-time PCR.

Additional file 2: Figure S1. a. Order of developmental stages for anatomical samples. ACF, formation of archesporial cells; BG, bi-cellular gametophyte; FI, flowering; Me, meiosis; Me1, meiotic leptotene; Me2, meiotic zygotene-pachytene; Me3, meiotic diplotene-tetrad; MP, mature pollen; PMe, pre-meiosis; GP, germinating pollen; TG, tri-cellular pollen; UG, uni-cellular gametophyte. Red bar, sample containing late pollen. b. Expression graph of 36 clusters after KMC analysis with 57,382 probes. Clusters 2 and 35 exhibited mature pollen-preferential patterns of expression and are marked with red boxes. **Figure S2.** Schematic representation of 3 promoter trap lines for T-DNA insertions. a. T-DNA was inserted into 17th intron of SacI homology domain-containing protein (*LOC_Os11g20384*) in Line 1A-13,819 (*mtd1–1*). b. Line 3A-05916 has T-DNA insertion in B12D protein (*LOC_Os07g17310*). BL, left T-DNA border; RB, right T-DNA border; Gray boxes, exons; lines, introns. **Figure S3.** Expression graph after KMC analysis of meta-expression data from *Arabidopsis*. Clusters marked with red box showed late pollen-preferred patterns. **Figure S4.** Heatmap for expression profiles of late pollen-preferred genes in *Arabidopsis*. **Figure S5.** Heatmap of genes involved in GA biosynthesis and signaling. *CPS*, *GA3ox1*, *KAO*, *GA20ox3*, and *KO2* showed late pollen-preferred expression patterns and are outlined with red boxes.

Abbreviations

ACC: 1-aminocyclopropane-1-carboxylic acid; AGPase: ADP-glucose pyrophosphorylase; CSLD: Cellulose synthase-like D; DAG: Diacylglycerol; DGK: Diacylglycerol kinase; GA: Gibberellic acid; GO: Gene Ontology; GPD: Glycerophosphoryl diester phosphodiesterase; KMC: K-means clustering; MTD: Male-gene Transfer Defective; NCBI GEO: National Center for Biotechnology Information Gene; PLD: Phospholipase; PME: Pectin methylesterase; RGAP: Rice Genome Annotated Project; ROAD: Expression Omnibus the Rice Oligonucleotide Array Database; TF: Transcription factor

Acknowledgements

We appreciate the support of Dr. Pamela Ronald and Dr. Peijian Cao in developing the meta-expression profiling database for rice.

Funding

This work was supported in part by grants from the Next-Generation BioGreen 21 Program (PJ01325901 to KHJ and PJ01369001 to SKP) and the National Research Foundation of Korea (20151973 to SM).

Authors' contributions

SM, MMO, GA, HJK, SKP, and KHJ designed the research. SM, MMO, BK, SAO, and GY performed experiments. SM, MMO, SAP, BK, SAO, and GY analyzed data. SM, MMO, SKP, and KHJ wrote the manuscript. All authors read and approved the final manuscript.

Competing interests

The authors declare that they have no competing interests.

Author details

[1]Graduate School of Biotechnology and Crop Biotech Institute, Kyung Hee University, Yongin 446-701, South Korea. [2]School of Applied Biosciences, Kyungpook National University, Daegu 702-701, South Korea. [3]Department of Plant Science, Research Institute of Agriculture and Life Sciences, and Plant Genomics and Breeding Institute, Seoul National University, Seoul 151-921, South Korea. [4]College of Agriculture and Life Science, Daegu 702-701, South Korea.

References

An S, Park S, Jeong DH, Lee DY, Kang HG, Yu JH et al (2003) Generation and analysis of end sequence database for T-DNA tagging lines in rice. Plant Physiol 133:2040–2047

Aya K, Suzuki G, Suwabe K, Hobo T, Takahashi H, Shiono K et al (2011) Comprehensive network analysis of anther-expressed genes in rice by the combination of 33 laser microdissection and 143 spatiotemporal microarrays. PLoS One 6:e26162

Becker JD, Boavida LC, Carneiro J, Haury M, Feijo JA (2003) Transcriptional profiling of Arabidopsis tissues reveals the unique characteristics of the pollen transcriptome. Plant Physiol 133:713–725

Bedinger P (1992) The remarkable biology of pollen. Plant Cell 4:879–887

Bosch M, Cheung AY, Hepler PK (2005) Pectin methylesterase, a regulator of pollen tube growth. Plant Physiol 138:1334–1346

Cao P, Jung KH, Choi D, Hwang D, Ronald PC (2012) The Rice oligonucleotide Array database: an atlas of rice gene expression. Rice 5:17

Cecchetti V, Altamura MM, Falasca G, Costantino P, Cardarelli M (2008) Auxin regulates Arabidopsis anther dehiscence, pollen maturation, and filament elongation. Plant Cell 20:1760–1774

Cecchetti V, Celebrin D, Napoli N, Ghelli R, Brunetti P, Costantino P et al (2017) An auxin maximum in the middle layer controls stamen development and pollen maturation in Arabidopsis. New Phytol 213:1194–1207

Chandran AKN, Jeong HY, Jung KH, Lee C (2016) Development of functional modules based on co-expression patterns for cell-wall biosynthesis related genes in rice. J Plant Biol 59:1–15

Chen SQ, Wang Z, Liu MX, Xie ZW, Wang HH (2008) Pollen grain germination and pollen tube growth in pistil of rice. Rice Sci 15:125–130

Cheng Y, Dai X, Zhao Y (2006) Auxin biosynthesis by the YUCCA flavin monooxygenases controls the formation of floral organs and vascular tissues in Arabidopsis. Genes Dev 20:1790–1799

Chhun T, Aya K, Asano K, Yamamoto E, Morinaka Y, Watanabe M et al (2007) Gibberellin regulates pollen viability and pollen tube growth in rice. Plant Cell 19:3876–3888

Cho HT, Kende H (1997) Expansins and internodal growth of Deepwater rice. Plant Physiol 113:1145–1151

Coimbra S, Costa M, Jones B, Mendes MA, Pereira LG (2009) Pollen grain development is compromised in Arabidopsis agp6 agp11 null mutants. J Exp Bot 60:3133–3142

Coimbra S, Costa M, Mendes MA, Pereira AM, Pinto J, Pereira LG (2010) Early germination of Arabidopsis pollen in a double null mutant for the arabinogalactan protein genes AGP6 and AGP11. Plant Reprod 23:199–205

Cosgrove DJ, Bedinger P, Durachko DM (1997) Group I allergens of grass pollen as cell wall-loosening agents. Proc Natl Acad Sci U S A 94:6559–6564

Dai Z, Gao J, An K, Lee JM, Edwards GE, An G (1996) Promoter elements controlling developmental and environmental regulation of a tobacco ribosomal protein gene L34. Plant Mol Biol 32:1055–1065

Deveshwar P, Bovill WD, Sharma R, Able JA, Kapoor S (2011) Analysis of anther transcriptomes to identify genes contributing to meiosis and male gametophyte development in rice. BMC Plant Biol 11:78

Dobritsa AA, Geanconteri A, Shrestha J, Carlson A, Kooyers N, Coerper D et al (2011) A large-scale genetic screen in Arabidopsis to identify genes involved in pollen exine production. Plant Physiol 157:947–970

Donaldson LA, Knox JP (2012) Localization of cell wall polysaccharides in normal and compression wood of radiata pine: relationships with lignification and microfibril orientation. Plant Physiol 158:642–653

Edgar R, Domrachev M, Lash AE (2002) Gene expression omnibus: NCBI gene expression and hybridization array data repository. Nucleic Acids Res 30:207–210

Faure JE, Rotman N, Fortune P, Dumas C (2002) Fertilization in *Arabidopsis thaliana* wild type: developmental stages and time course. Plant J 30:481–488

Firon N, Pressman E, Meir S, Khoury R, Altahan L (2012) Ethylene is involved in maintaining tomato (Solanum lycopersicum) pollen quality under heat-stress conditions. AoB Plants 2012:pls024

Fujita M, Horiuchi Y, Ueda Y, Mizuta Y, Kubo T, Yano K et al (2010) Rice expression atlas in reproductive development. Plant Cell Physiol. 51:2060–2081

Goetz M, Guivarch A, Hirsche J, Bauerfeind MA, Gonzalez MC, Hyun TK et al (2017) Metabolic control of tobacco pollination by sugars and invertases. Plant Physiol 173:984–997

Guan Y, Meng X, Khanna R, LaMontagne E, Liu Y, Zhang S (2014) Phosphorylation of a WRKY transcription factor by MAPKs is required for pollen development and function in Arabidopsis. PLoS Genet 10:e1004384

Hao H, Li Y, Hu Y, Lin J (2005) Inhibition of RNA and protein synthesis in pollen tube development of *Pinus bungeana* by actinomycin D and cycloheximide. New Phytol 165:721–729

Hayashi S, Ishii T, Matsunaga T, Tominaga R, Kuromori T, Wada T et al (2008) The glycerophosphoryl diester phosphodiesterase-like proteins SHV3 and its homologs play important roles in cell wall organization. Plant Cell Physiol. 49:1522–1535

Hirano K, Aya K, Hobo T, Sakakibara H, Kojima M, Shim RA et al (2008) Comprehensive transcriptome analysis of phytohormone biosynthesis and signaling genes in microspore/pollen and tapetum of rice. Plant Cell Physiol 49:1429–1450

Honys D, Twell D (2003) Comparative analysis of the Arabidopsis pollen transcriptome. Plant Physiol 132:640–652

Honys D, Twell D (2004) Transcriptome analysis of haploid male gametophyte development in Arabidopsis. Genome Biol 5:R85

Jefferson RA, Kavanagh TA, Bevan MW (1987) GUS fusions: beta-glucuronidase as a sensitive and versatile gene fusion marker in higher plants. EMBO J 6:3901–3907

Jeon JS, Lee S, Jung KH, Jun SH, Jeong DH, Lee J et al (2000) T-DNA insertional mutagenesis for functional genomics in rice. Plant J 22:561–570

Jeong DH, An S, Kang HG, Moon S, Han JJ, Park S et al (2002) T-DNA insertional mutagenesis for activation tagging in rice. Plant Physiol 130:1636–1644

Jeong DH, An S, Park S, Kang HG, Park GG, Kim SR et al (2006) Generation of a flanking sequence-tag database for activation-tagging lines in japonica rice. Plant J 45:123–132

Jiang, L., Yang, S.L, Xie L.F., Puah, C.S., Zhang, X.Q., Yang, W.C., *et al.* (2005) VANGUARD1 encodes a pectin methylesterase that enhances pollen tube growth in the Arabidopsis style and transmitting tract. Plant Cell 17, 584–596

Jones MA, Raymond MJ, Smirnoff N (2006) Analysis of the root-hair morphogenesis transcriptome reveals the molecular identity of six genes with roles in root-hair development in Arabidopsis. Plant J 45:83–100

Jung KH, An G (2012) Application of MapMan and RiceNet drives systematic analyses of the early heat stress transcriptome in rice seedlings. J Plant Biol 55:436–449

Jung KH, Dardick C, Bartley LE, Cao PJ, Phetsom J, Canlas P et al (2008) Refinement of light-responsive transcript lists using rice oligonucleotide arrays, evaluation of gene-redundancy. PLoS One 3:e3337

Krichevsky A, Kozlovsky SV, Tian GW, Chen MH, Zaltsman A, Citovsky V (2007) How pollen tubes grow. Dev Biol 303:405–420

Kuang A, Musgrave ME (1996) Dynamics of vegetative cytoplasm during generative cell formation and pollen maturation in *Arabidopsis thaliana*. Protoplasma 194:81–90

Lalanne E, Honys D, Johnson A, Borner GH, Lilley KS, Dupree P et al (2004) SETH1 and SETH2, two components of the glycosylphosphatidylinositol anchor biosynthetic pathway, are required for pollen germination and tube growth in Arabidopsis. Plant Cell 16:229–240

Lee SK, Eom JS, Hwang SK, Shin D, An G, Okita TW et al (2016) Plastidic phosphoglucomutase and ADP-glucose pyrophosphorylase mutants impair starch synthesis in rice pollen grains and cause male sterility. J Exp Bot 67:5557–5569

Lee YS, An G (2015) OsGI controls flowering time by modulating rhythmic flowering time regulators preferentially under short day in rice. J Plant Biol 58:137–145

Leroux C, Bouton S, Kiefer-Meyer MC, Fabrice TN, Mareck A, Guenin S et al (2015) PECTIN METHYLESTERASE48 is involved in Arabidopsis pollen grain germination. Plant Physiol 167:367–380

Li J, Yu M, Geng LL, Zhao J (2010) The fasciclin-like arabinogalactan protein gene, FLA3, is involved in microspore development of Arabidopsis. Plant J 64:482–497

Lin S, Dong H, Zhang F, Qiu L, Wang FZ, Cao JS et al (2014) BcMF8, a putative arabinogalactan protein-encoding gene, contributes to pollen wall development, aperture formation and pollen tube growth in *Brassica campestris*. Ann Bot 113:777–788

Lindstrom JT, Lei CH, Jones ML, Woodson WR (1999) Accumulation of 1-aminocyclopropane-1-carboxylic acid (ACC) in petunia pollen is associated with expression of a pollen-specific ACC synthase late in development. J Am Soc Hortic Sci 124:145–151

Mascarenhas JP (1993) Molecular mechanisms of pollen tube growth and differentiation. Plant Cell 5:1303–1314

Mollet JC, Leroux C, Dardelle F, Lehner A (2013) Cell wall composition, biosynthesis and remodeling during pollen tube growth. Plants 2:107–147

Moon S, Kim SR, Zhao G, Yi J, Yoo Y, Jin P et al (2013) Rice GLYCOSYLTRANSFERASE1 encodes a GLYCOSYLTRANSFERASE essential for pollen wall formation. Plant Physiol 164:663–675

Nakamura AT, Longhi-Wagner HM, Scatena VL (2010) Anther and pollen development in some species of Poaceae (Poales). Braz J Biol 70:351–360

Nashilevitz S, Melamed-Bessudo C, Aharoni A, Kossmann J, Wolf S, Levy A (2009) The legwd mutant uncovers the role of starch phosphorylation in pollen development and germination in tomato. Plant J 57:1–13

Noher de Halac I, Cismondi IA, Rodriguez-Garcia MI, Fama G (2003) Distribution of pectins in the pollen apertures of *Oenothera hookeri*.Velans ster/+ster. Biocell 27:11–18

Oo MM, Bae HK, Nguyen TD, Moon S, Oh SA, Kim JH et al (2014) Evaluation of rice promoters conferring pollen-specific expression in a heterologous system, Arabidopsis. Plant Reprod 27:47–58

Owen HA, Makaroff CA (1995) Ultrastructure of microsporogenesis and microgametogenesis in *Arabidopsis thaliana* (L.) Heynh. Ecotype Wassilewskija (Brassicaceae). Protoplasma 185:7–21

Parker JS, Cavell AC, Dolan L, Roberts K, Grierson CS (2000) Genetic interactions during root hair morphogenesis in Arabidopsis. Plant Cell 12:1961–1974

Persson S, Paredez A, Carroll A, Palsdottir H, Doblin M, Poindexter P, Khitrov N, Auer M, Somerville CR et al (2007) Genetic evidence for three unique components in primary cell-wall cellulose synthase complexes in Arabidopsis. Proc Natl Acad Sci U S A 104:15566–15571

Prasad PW, Boote KJ, Allen LH, Sheehy JE, Thomas JMG (2006) Species, ecotype and cultivar differences in spikelet fertility and harvest index of rice in response to high temperature stress. Field Crops Res 95:398–411

Russell SD, Gou X, Wong CE, Wang X, Yuan T, Wei X et al (2012) Genomic profiling of rice sperm cell transcripts reveals conserved and distinct elements in the flowering plant male germ lineage. New Phytol 195:560–573

Ryu CH, You JH, Kang HG, Hur J, Kim YH, Han MJ et al (2004) Generation of T-DNA tagging lines with a bidirectional gene trap vector and the establishment of an insertion-site database. Plant Mol Biol 54:489–502

Sakamoto T, Miyura K, Itoh H, Tatsumi T, Ueguchi-Tanaka M, Ishiyama K et al (2004) An overview of gibberellin metabolism enzyme genes and their related mutants in rice. Plant Physiol 134:1642–1653

Sakata T, Oda S, Tsunaga Y, Shomura H, Kawagishi-Kobayashi M, Aya K et al (2014) Reduction of gibberellin by low temperature disrupts pollen development in rice. Plant Physiol 164:2011–2019

Sampedro J, Guttman M, Li LC, Cosgrove DJ (2015) Evolutionary divergence of beta-expansin structure and function in grasses parallels emergence of distinctive primary cell wall traits. Plant J 81:108–120

Shi J, Cui M, Yang L, Kim YJ, Zhang D (2015) Genetic and biochemical mechanisms of pollen wall development. Trends Plant Sci 20:741–753

Suwabe K, Suzuki G, Takahashi H, Shiono K, Endo M, Yano K et al (2008) Separated transcriptomes of male gametophyte and tapetum in rice: validity of a laser microdissection (LM) microarray. Plant Cell Physiol. 49:1407–1416

Suzuki T, Masaoka K, Nishi M, Nakamura K, Ishiguro S (2008) Identification of Kaonashi mutants showing abnormal pollen exine structure in *Arabidopsis thaliana*. Plant Cell Physiol. 49:1465–1477

Tabuchi A, Li LC, Cosgrove DJ (2011) Matrix solubilization and cell wall weakening by β-expansin (group-1 allergen) from maize pollen. Plant J 68:546–559

Tang LY, Nagata N, Matsushima R, Chen Y, Yoshioka Y, Sakamoto W (2009) Visualization of plastids in pollen grains: involvement of FtsZ1 in pollen plastid division. Plant Cell Physiol. 50:904–908

Tian GW, Chen MH, Zaltsman A, Citovsky V (2006) Pollen-specific pectin methylesterase involved in pollen tube growth. Dev Biol 294:83–91

Vizcay-Barrena G, Wilson ZA (2006) Altered tapetal PCD and pollen wall development in the Arabidopsis ms1 mutant. J Exp Bot 57:2709–2717

Wang W, Wang L, Chen C, Xiong G, Tan XY, Yang KZ et al (2011) Arabidopsis CSLD1 and CSLD4 are required for cellulose deposition and normal growth of pollen tubes. J Exp Bot 62:5161–5177

Wei LQ, Xu WY, Deng ZY, Su Z, Xue Y, Wang T (2010) Genome-scale analysis and comparison of gene expression profiles in developing and germinated pollen in *Oryza sativa*. BMC Genomics 11:338

Xu J, Ding Z, Vizcay-Barrena G, Shi J, Liang W, Yuan Z et al (2014) ABORTED MICROSPORES acts as a master regulator of pollen wall formation in Arabidopsis. Plant Cell 26:1544–1556

Yeung EC, Oinam GS, Yeung SS, Harry I (2011) Anther, pollen and tapetum development in safflower, *Carthamus tinctorius* L. Plant Reprod 24:307–317

Yoo YH, Choi HK, Jung KH (2015) Genome-wide identification and analysis of genes associated with lysigenous aerenchyma formation in rice roots. J Plant Biol 58:117–127

Zhang DB, Luo X, Zhu L (2011) Cytological analysis and genetic control of rice anther development. J Genet Genom 38:379–390

Zhang H, Liang WQ, Yang XJ, Luo X, Jiang N, Ma H et al (2010) Carbon starved anther encodes a MYB domain protein that regulates sugar partitioning required for rice pollen development. Plant Cell 22:672–689

13

Outgrowth of Rice Tillers Requires Availability of Glutamine in the Basal Portions of Shoots

Miwa Ohashi[1,3*], Keiki Ishiyama[1], Soichi Kojima[1], Noriyuki Konishi[1,4], Kazuhiro Sasaki[2], Mitsue Miyao[1], Toshihiko Hayakawa[1] and Tomoyuki Yamaya[1,5]

Abstract

Background: Our previous studies concluded that metabolic disorder in the basal portions of rice shoots caused by a lack of cytosolic glutamine synthetase1;2 (GS1;2) resulted in a severe reduction in the outgrowth of tillers. Rice mutants lacking GS1;2 (*gs1;2* mutants) showed a remarkable reduction in the contents of both glutamine and asparagine in the basal portions of shoots. In the current study, we attempted to reveal the mechanisms for this decrease in asparagine content using rice mutants lacking either GS1;2 or asparagine synthetase 1 (AS1). The contributions of the availability of glutamine and asparagine to the outgrowth of rice tillers were investigated.

Results: Rice has two *AS* genes, and the enzymes catalyse asparagine synthesis from glutamine. In the basal portions of rice shoots, expression of *OsAS1*, the major species in this tissue, was reduced in *gs1;2* mutants, whereas *OsAS2* expression was relatively constant. *OsAS1* was expressed in phloem companion cells of the nodal vascular anastomoses connected to the axillary bud vasculatures in the basal portions of wild-type shoots, whereas cell-specific expression was markedly reduced in *gs1;2* mutants. *OsAS1* was up-regulated significantly by NH_4^+ supply in the wild type but not in *gs1;2* mutants. When GS reactions were inhibited by methionine sulfoximine, *OsAS1* was up-regulated by glutamine but not by NH_4^+. The rice mutants lacking AS1 (*as1* mutants) showed a decrease in asparagine content in the basal portions of shoots. However, glutamine content and tiller number were less affected by the lack of AS1.

Conclusion: These results indicate that in phloem companion cells of the nodal vascular anastomoses, asparagine synthesis is largely dependent on glutamine or its related metabolite-responsive AS1. Thus, the decrease in glutamine content caused by a lack of GS1;2 is suggested to result in low expression of *OsAS1*, decreasing asparagine content. However, the availability of asparagine generated from AS1 reactions is apparently less effective for the outgrowth of tillers. With respect to the tiller number and the contents of glutamine and asparagine in *gs1;2* and *as1* mutants, the availability of glutamine rather than asparagine in basal portions of rice shoots may be required for the outgrowth of rice tillers.

Keywords: Asparagine, Asparagine synthetase, Cytosolic glutamine synthetase, Glutamine, Tiller, Rice

* Correspondence: ohashi@agr.nagoya-u.ac.jp
[1]Graduate School of Agricultural Science, Tohoku University, 468-1 Aoba, Aramaki-Aza, Aoba-ku, Sendai 980-8572, Japan
[3]Present Address: Graduate School of Bioagricultural Sciences, Nagoya University, Furo-cho, Chikusa-ku, Nagoya 464-8601, Japan
Full list of author information is available at the end of the article

Background

Glutamine and asparagine play crucial roles in plant growth and development as the major nitrogen forms for transport via vasculatures from source to sink tissues within various plants (Urquhart and Joy 1981; Ireland and Lea 1999; Lea et al. 2007; Gaufichon et al. 2010). In rice plants, these amides are major nitrogen forms in both phloem (Hayashi and Chino 1990) and xylem sap (Fukumorita and Chino 1982; Funayama et al. 2013; Ohashi et al. 2015a). In higher plants, including rice, glutamine is synthesized by glutamine synthetase (GS) (Lea and Miflin 1974; Yamaya and Oaks 2004), and asparagine is synthesized from glutamine by asparagine synthetase (AS) (Lea et al. 2007; Gaufichon et al. 2010; Ohashi et al. 2015a). Asparagine is catabolized by asparaginase into aspartate as the common precursor of the essential amino acids (Azevedo et al. 2006; Lea et al. 2007; Yabuki et al. 2017).

Rice has three cytosolic GS isoenzymes (GS1, *OsGS1;1* – *OsGS1;3*) and one chloroplastic GS (GS2) (Tabuchi et al. 2007). The occurrence of two AS isoenzymes (*OsAS1* and *OsAS2*) was recently confirmed (Ohashi et al. 2015a). Both GS1;2 and AS1 are important in the primary assimilation of NH_4^+ taken up by rice roots (Funayama et al. 2013; Ohashi et al. 2015a). *OsGS1;2* and *OsAS1* were specifically accumulated in three cell layers of root surfaces (epidermis, exodermis and sclerenchyma) after NH_4^+ supply to rice roots (Ishiyama et al. 2004; Ohashi et al. 2015a). In both roots and xylem sap after NH_4^+ supply, rice mutants lacking GS1;2 showed a decrease in glutamine and asparagine content, while mutants lacking AS1 showed a decrease in asparagine content (Funayama et al. 2013; Ohashi et al. 2015a).

Tiller number is a critical agronomic trait defining grain yields in rice and is influenced by the availability of nitrogen (Mae 1997; Sakamoto and Matsuoka 2008; Liu et al. 2011). We found that the lack of GS1;2 suppressed the outgrowth of the tiller axillary bud and hence a substantial decrease in active tiller number and yields (Funayama et al. 2013; Ohashi et al. 2015b). The outgrowth of tiller axillary buds has been proposed to be related to metabolite use efficiency and hormone signalling networks (Domagalska and Leyser 2011; Evers et al. 2011). We recently showed that metabolic disorder in the basal portions of rice shoots lacking GS1;2 caused a severe reduction in the outgrowth of tiller axillary buds, and this reduction was independent of the content of strigolactone (Ohashi et al. 2015b), a phytohormone inhibiting tiller development (Umehara et al. 2008). The basal portions of rice shoots are important organs for the circulation of metabolites because of vascular networks (Hoshikawa 1989). These organs consist of axillary buds, internodes, and shoot apical meristems (SAMs) (Hoshikawa 1989; Ohashi et al. 2015b). In the basal portions of rice shoots, GS1;2 protein localizes in phloem companion cells of the nodal vascular anastomoses connecting to the axillary bud vasculatures (Ohashi et al. 2015b). The lack of GS1;2 caused a large decrease in both glutamine and asparagine contents in roots and the basal portions of shoots (Ohashi et al. 2017). This decrease in glutamine content in the basal portions of shoots caused a deficiency of active cytokinin, which is required for the outgrowth of rice tillers, via the down-regulation of glutamine or its related metabolite-dependent cytokinin synthesis (Ohashi et al. 2017). Thus, the availability of glutamine and/or asparagine is required for the outgrowth of rice tillers (Ohashi et al. 2015b, 2017).

In the present study, we focused on the reduction of asparagine content in the basal portions of rice mutants lacking GS1;2 (*gs1;2* mutants). Using rice mutants lacking either GS1;2 or AS1 (*as1* mutants), we investigated (1) the molecular mechanisms involved in the decrease in asparagine content in the basal portions of *gs1;2* shoots and (2) the relation between low availability of asparagine and the reduction in tiller number.

Results

Reduced Expression of *OsAS1* in Phloem Companion Cells of the Nodal Vascular Anastomoses in the Basal Portions of Shoots Lacking GS1;2

Expression of *OsAS1* in the basal portions of shoots of NH_4^+-fed *gs1;2* mutant seedlings at the fourth-leaf stage showed an approximately 60% reduction compared with the level in the wild-type rice (Fig. 1). However, the expression of *OsAS2* in the *gs1;2* mutants was less affected in the basal portions (Fig. 1). *OsAS1* was expressed at twice the level of *OsAS2* in the basal portions of wild-type shoots (Fig. 1).

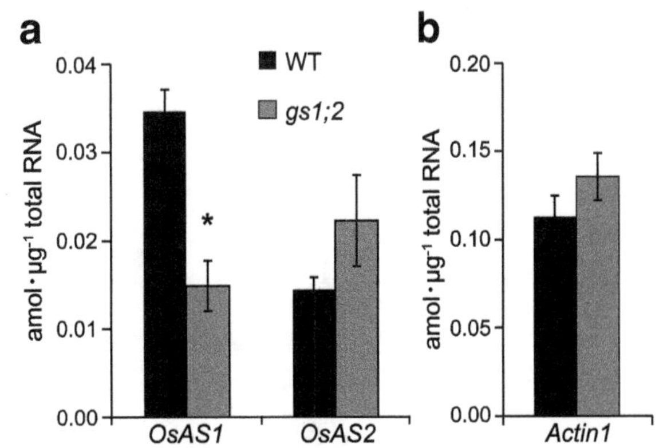

Fig. 1 qPCR analysis of *OsASs* in the basal portions of wild-type and *gs1;2* mutant shoots. qPCR analysis of *OsAS1* and *OsAS2* genes (**a**) and a control *Actin1* (**b**) in the basal portions of shoots of wild-type plants (WT: black column) and *gs1;2* mutants (*gs1;2*: grey column) were performed. Rice seedlings were grown hydroponically in the presence of 1 mM NH_4Cl until the fourth-leaf stage. Mean values with SE of four independent samples are shown. An asterisk denotes a statistically significant difference between wild type and *gs1;2* mutants (*, $P < 0.05$ by Student's *t*-test)

In situ hybridization revealed strong signals for the *OsAS1* transcript in phloem companion cells of the nodal vascular anastomoses in the basal portion of wild-type shoots (Fig. 2d, e). However, the signal intensity for *OsAS1* in phloem companion cells of the nodal vascular anastomoses in the *gs1;2* mutants was much lower than in the wild-type rice. Weak signals for *OsAS1* were also detected in the SAMs and immature leaves in both the *gs1;2* mutants and wild-type rice (Fig. 2a, b, g, h). Signals for the *OsAS2* transcript were also detected in phloem companion cells of the nodal vascular anastomoses, and weak signals were detected in the SAM and immature leaves in both the *gs1;2* mutants and wild-type rice. Unlike *OsAS1*, the signal intensity of the *OsAS2* transcript was not significantly different between *gs1;2* mutants and wild-type rice (Fig. 3a, b, d, e, g, h). These results fit well with the results obtained from quantitative real-time PCR (qPCR) analysis (Fig. 1). When the control sense probes were used, only background levels of the *OsAS1* and *OsAS2* transcript were detected (Figs. 2c, f, i, 3c, f, i).

Reduction in *OsAS1* Expression and Asparagine Content after NH_4^+ Supply to the *gs1;2* Mutants

In rice roots, up-regulation of *OsAS1* expression and accumulation of asparagine content were observed after NH_4^+ supply (Ohashi et al. 2015a). In the current study,

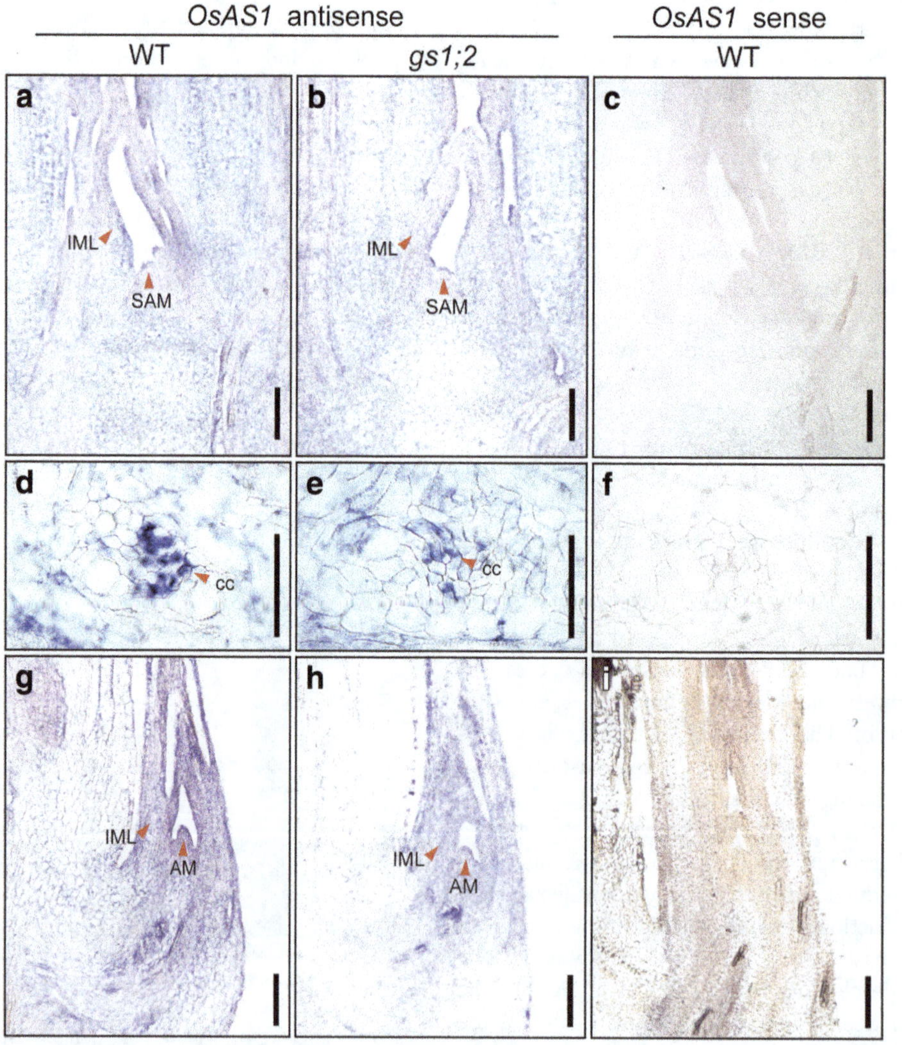

Fig. 2 In situ hybridization for *OsAS1* in the basal portions of wild-type and *gs1;2* mutant shoots. Longitudinal sections of basal portions of shoots were prepared from the wild-type rice (WT) (**a**, **c**, **d**, **f**, **g**, **i**) and from *gs1;2* mutants (*gs1;2*) (**b**, **e**, **h**) grown hydroponically in the presence of 1 mM NH_4Cl until the fourth-leaf stage. A sense probe was hybridized to WT sections (**c**, **f**, **i**) as a negative control. Red arrowheads indicate hybridization signals for *OsAS1* transcripts in the shoot apical meristem (SAM) (**a**, **b**), phloem companion cell of the nodal vascular anastomoses (cc) (**d**, **e**), an immature leaf (IML) area (**a**, **b**, **g**, **h**), and the tiller containing the axillary bud meristem (AM) (**g**, **h**). Note that the signal intensity in *gs1;2* (**e**) was far weaker than in the WT (**d**). Scale bars = 100 μm (**a** - **c**, **g** - **i**) and 50 μm (**d** - **f**)

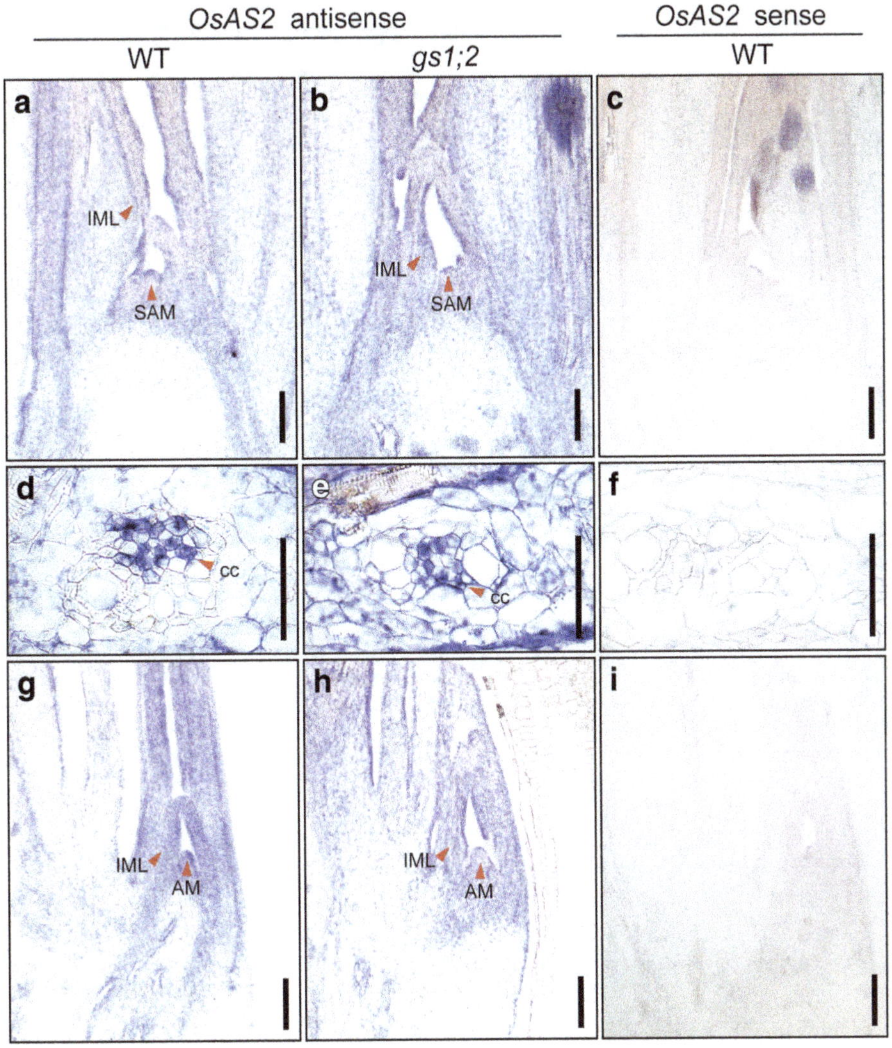

Fig. 3 In situ hybridization for *OsAS2* in the basal portions of wild-type and *gs1;2* mutant shoots. Longitudinal sections of basal portions of shoots were prepared from the wild-type rice (WT) (**a**, **c**, **d**, **f**, **g**, **i**) and the *gs1;2* mutants (*gs1;2*) (**b**, **e**, **h**) grown hydroponically in the presence of 1 mM NH_4Cl until the fourth-leaf stage. A sense probe was hybridized to WT sections (**c**, **f**, **i**) as a negative control. Red arrowheads indicate hybridization signals for *OsAS2* transcripts in the shoot apical meristem (SAM) (**a**, **b**), phloem companion cell of the nodal vascular anastomoses (cc) (**d**, **e**), an immature leaf (IML) area (**a**, **b**, **g**, **h**), and the tiller containing the axillary bud meristem (AM) (**g**, **h**). Scale bars = 100 μm (**a** - **c**, **g** - **i**) and 50 μm (**d** - **f**)

OsAS1 expression and asparagine content after NH_4^+ supply were determined in the basal portions of shoots using *gs1;2* mutant and wild-type plants. *OsAS1* expression was increased over twofold at 8 h after NH_4^+ supply in the shoot basal portions of the wild type, whereas it was less changed in the *gs1;2* mutants (Fig. 4a). In contrast, there were no significant differences in the expression of *OsAS2* and *Actin1* between the shoot basal portions of wild type and *gs1;2* mutants treated with or without NH_4^+ (Fig. 4b, c). Asparagine was accumulated remarkably at 24 h after NH_4^+ supply in the shoot basal portions of the wild type, whereas it was relatively constant in the *gs1;2* mutants (Fig. 4d). NH_4^+ was highly accumulated in these tissues in *gs1;2* mutants after NH_4^+ supply (Fig. 4e), indicating impairment of NH_4^+ assimilation. Under the same conditions, glutamine content was less affected 24 h after NH_4^+ supply in the shoot basal portions of *gs1;2* mutants (Ohashi et al. 2017), as in the case of asparagine content in the current study (Fig. 4d).

Further qPCR analysis was carried out in the presence or absence of 1 mM methionine sulfoximine (MSX), an inhibitor of GS reactions. As shown in Fig. 4a, a substantial increase in *OsAS1* expression was observed at 8 h after NH_4^+ supply in the basal portions of shoots, but this increase was not observed when MSX was treated prior to NH_4^+ supply (Fig. 5a). Over a three-fold increase in *OsAS1* expression was observed after glutamine supply with or without MSX pre-treatment (Fig. 5a). As in *OsAS1* expression, the significant increase in *OsGS1;2* expression was

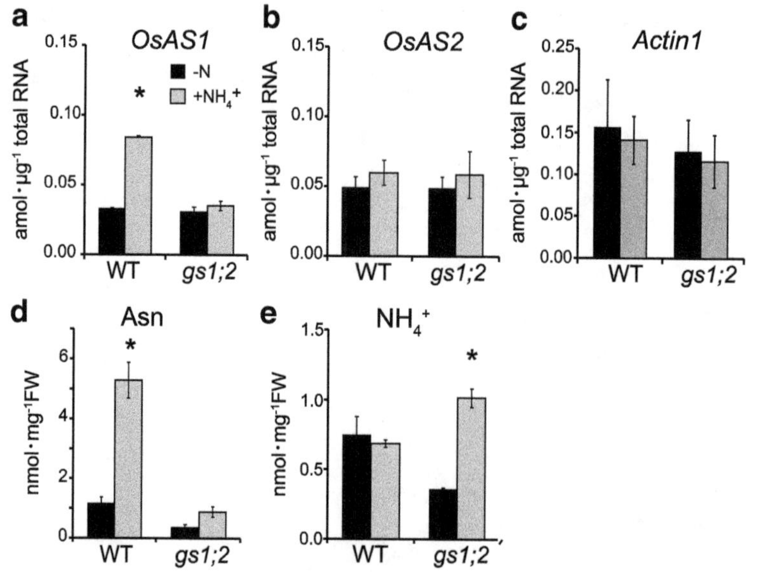

Fig. 4 Expression levels of *OsASs* and contents of asparagine and NH_4^+ in the basal portions. **a** - **c** Transcript contents of the *OsAS1* (**a**), *OsAS2* (**b**) and a control *Actin1* (**c**) in the basal portions of shoots of wild-type plants (WT: black column) and *gs1;2* mutants (*gs1;2*: grey column) at the seventh-leaf stage for seedlings grown in water for 3 d followed by treatments with (+NH_4^+) or without (-N) 1 mM NH_4Cl for 8 h. (**d**, **e**) Contents of asparagine (**d**) and NH_4^+ (**e**) in the basal portions of shoots of WT (black column) and *gs1;2* (grey column) at the fourth-leaf stage for seedlings grown in water for 3 d followed by treatments with or without 1 mM NH_4Cl for 24 h. Mean values with SE of four independent samples are shown. Asterisks denote statistically significant differences between samples treated with and without 1 mM NH_4Cl (*, $P < 0.05$ by Student's *t*-test)

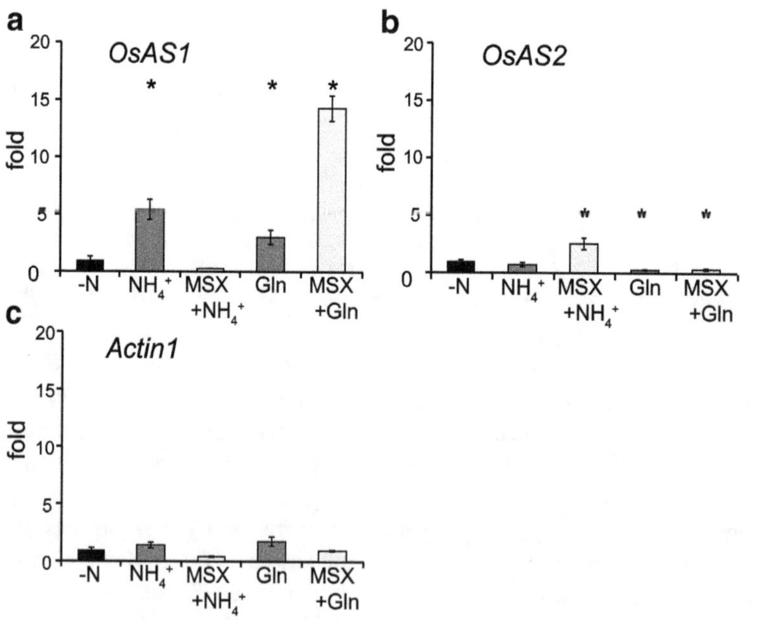

Fig. 5 qPCR analysis of *OsASs* in the basal portions of wild-type shoots using MSX. Wild-type rice seedlings were grown hydroponically in the presence of 1 mM NH_4Cl until the fourth-leaf stage and then transferred into water for 3 d. After pre-treatment with 1 mM MSX for 2 h, the seedlings were treated for 8 h with 1 mM NH_4^+ (MSX + NH_4^+) or 5 mM glutamine (MSX + Gln). The seedlings without MSX pre-treatment were also treated for 8 h with 1 mM NH_4^+ (+NH_4^+), 5 mM glutamine (Gln) or without nitrogen nutrients (-N). qPCR analyses of *OsAS1* (**a**), *OsAS2* (**b**), and *actin1* transcripts (**c**) were performed in the shoot basal parts. Fold changes of transcript levels in each treatment relative to those in –N were calculated, and mean values with SE of four independent samples are shown. Asterisks denote statistically significant differences between the samples treated without nitrogen nutrients (-N) and each treated sample (*, $P < 0.05$ by Student's *t*-test)

observed after NH_4^+ or glutamine supply, but an increase in *OsGS1;2* expression was not observed for the MSX treatment (Additional file 1: Figure S1b). The expression of *OsAS2* was stable after NH_4^+ supply, while glutamine supply caused a decrease in its expression (Fig. 5b). There were slight fluctuations in the expression of *OsGS1;1* and *Actin1* used as a control treated with or without NH_4^+ (Fig. 5c, Additional file 1: Figure S1a).

In the roots, *OsAS1* expression was increased significantly after NH_4^+ supply, but this increase was not observed in the MSX treatment (Additional file 2: Figure S2a). Increases in *OsAS1* expression were observed after glutamine supply with or without MSX treatment (Additional file 2: Figure S2a).

Lack of AS1 Caused a Reduction in Asparagine Content in the Basal Portions of Rice Shoots but not in Tiller Number

To examine the contribution of availability of asparagine generated from AS1 reactions in the basal portions of shoots to the outgrowth of rice tillers, reverse genetics studies were carried out using homozygous rice mutants lacking *OsAS1* (Ohashi et al. 2015a). Our previous study showed that the lack of AS1 did not affect the expression of other genes related to nitrogen metabolism in roots (Ohashi et al. 2015a). In the basal portions of shoots growing with NH_4^+, the *as1* mutants exhibited no significant difference in *OsAS2* expression, but an approximately 80% decrease in asparagine content was observed compared with the content in the wild type (Fig. 6a, c). In addition, transient NH_4^+ supply for 24 h to nitrogen-depleted *as1* mutants caused an approximately 60% decrease in asparagine content in the shoot basal portions (Additional file 3: Figure S3a). There were slight fluctuations in the content of glutamine and NH_4^+ between the basal portions of shoots of *as1* mutants and the wild type when plants were grown with NH_4^+ (Fig. 6d, c, Additional file 3: Figure S3b, d).

No significant differences between *as1* mutants and the wild type were observed with respect to tiller number and leaf stage under 1 mM and 2 mM NH_4^+ (Fig. 7). The outgrowth of primary tiller was little affected in *as1* mutants grown with 1 mM NH_4^+ until fifth leaf stage (Additional file 4: Figure S4). The same is true when those plants were grown in the paddy field under normal fertilization (Table 1). There were also no significant differences in shoot length, dry weight, or panicle number between *as1* mutants and wild-type plants grown in the paddy field (Table 1). In our experiments in the paddy field, the lodging phenotype was not observed in *as1* mutants (Additional file 5: Figure S5).

Wild-type plants and two lines of *as1* mutants (*as1-m1* and *as1-m2*) were grown in the paddy field under normal fertilization. Tiller number was measured at the panicle initiation stage. After panicle ripening, rice plants were harvested and shoot length, shoot dry weight without panicle and panicle number were measured. Mean values with SE of four independent plots (five average plants selected from each plot) are shown. One-way ANOVA followed by Bonferroni tests were used to identify significant differences between the wild type and *as1* mutants ($P < 0.05$).

Discussion

A lack of GS1;2 caused a reduction in asparagine content in the basal portions of shoots (Ohashi et al. 2017). The current study showed that the expression of *OsAS1*, the major species in the basal portions of shoots, was decreased by the lack of GS1;2, whereas *OsAS2*, which is dominant in rice leaves, was less affected (Fig. 1) (Ohashi et al. 2015a). In fact, the lack of AS1 severely decreased the asparagine content in the basal portions of shoots (Fig. 6c). These results indicate that the low expression of *OsAS1* caused a decrease in asparagine content in the shoot basal portions of *gs1;2* mutants. In the basal portions of shoots, GS1;2 and AS1 are suggested to be important in the biosynthesis of asparagine, whereas the contribution of AS2 to this process may be limited (Fig. 1).

Gene expression analyses clearly showed that the expression of *OsAS1*, but not *OsAS2*, in both roots and shoot basal portions was enhanced by NH_4^+ and glutamine (Figs. 4, 5, Additional file 2: Figure S2) (Ohashi et al. 2015a). The NH_4^+-induced *OsAS1* expression was inhibited when GS1;2 was missing or GS activity was inhibited by MSX treatment (Figs. 4, 5, Additional file 2: Figure S2). In addition, *OsAS1* expression was enhanced by glutamine when GS activity was inhibited by MSX treatment. These results are in close agreement with a previous study by Kawachi et al. (2002), which showed that the NH_4^+-induced accumulation of AS protein in rice roots was completely inhibited by MSX treatment. These results indicate that glutamine or its related metabolite, but not NH_4^+ itself, acts as a signal molecule for the stimulation of *OsAS1* expression in both roots and shoot basal portions, where AS1 is the major isoform (Fig. 5, Additional file 2: Figure S2) (Ohashi et al. 2015a). In addition, the expression of *OsAS1* in phloem companion cells of nodal vascular anastomoses, overlapping with *OsGS1;2* expression (Ohashi et al. 2015b), was reduced by the lack of GS1;2 (Fig. 2d, e). These results possibly allow glutamine, which is produced by GS1;2, to directly induce the expression of *OsAS1* and to be used for asparagine synthesis within these cells. We conclude that the low availability of glutamine caused reduced expression of *OsAS1*, decreasing asparagine synthesis, in phloem companion cells of the nodal vascular anastomoses in the basal portions of shoots in the *gs1;2* mutants.

There are also several genes reported to be up-regulated by glutamine or related metabolites in rice and other

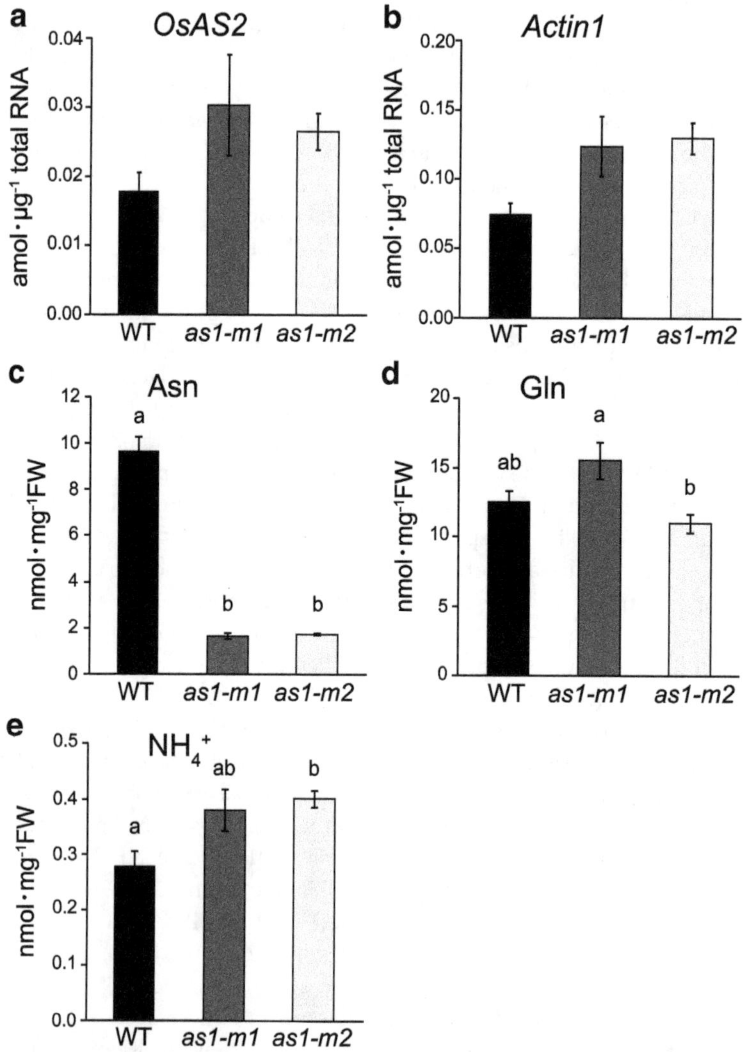

Fig. 6 qPCR and amino acid analysis in the basal portions of wild-type and *as1* mutant shoots. qPCR analysis of *OsAS2* (**a**) and *Actin1* (**b**) and analyses of asparagine (**c**), glutamine (**d**) and NH_4^+ contents (**e**) were performed in the basal portions of shoots of the wild type (WT: black column) and two lines of *as1* mutants (*as1-m1* and *as1-m2*: dark and light grey column, respectively). Seedlings were grown hydroponically in the presence of 1 mM NH_4Cl until the fourth-leaf stage. Mean values with SE of four independent samples are shown. One-way ANOVAs followed by Bonferroni tests were used to identify significant differences between the wild type and *as1* mutants ($P < 0.05$)

plants, i.e., in rice, *ammonium transporter1;1* (*OsAMT1;1*) and *OsAMT1;2* (Sonoda et al. 2003), *NADH-glutamate synthase1* (Hirose et al. 1997), and *adenosine phosphate-isopentenyltransferases4* (*IPT4*) and *IPT5* (Kamada-Nobusada et al. 2013); in Arabidopsis, *glutamine symtnetase1* (Oliveira and Coruzzi 1999), *AtAMT1;1* (Rawat et al. 1999), and *AtAS* (Lam et al. 1994); and nitrate reductase in tobacco leaves (Vincentz et al. 1993). The signalling mechanism in response to glutamine apparently contributes to modulating plant growth. However, few details are known regarding glutamine signalling, unlike the case of nitrate signalling (Sakakibara et al. 2006). It is worth examining the glutamine responsive induction of *OsAS1* to understand the mechanism of glutamine signalling in response to NH_4^+.

As discussed previously (Ohashi et al. 2015b), GS1;2 has a role in the re-assimilation of NH_4^+ released from the phenylalanine ammonia-lyase (PAL) reaction during lignification in the basal portions of shoots. In fact, the biosynthesis of lignin accounts for 10–23% in rice internodes (Ookawa and Ishihara 1993; Ookawa et al. 1993), and PAL activity in the basal portions of rice shoots was approximately 20-fold higher than in leaf blade at the seedling stage (Additional file 6: Figure S6). Thus, in the basal portions of rice shoots, NH_4^+ released during lignification must be re-assimilated immediately to avoid

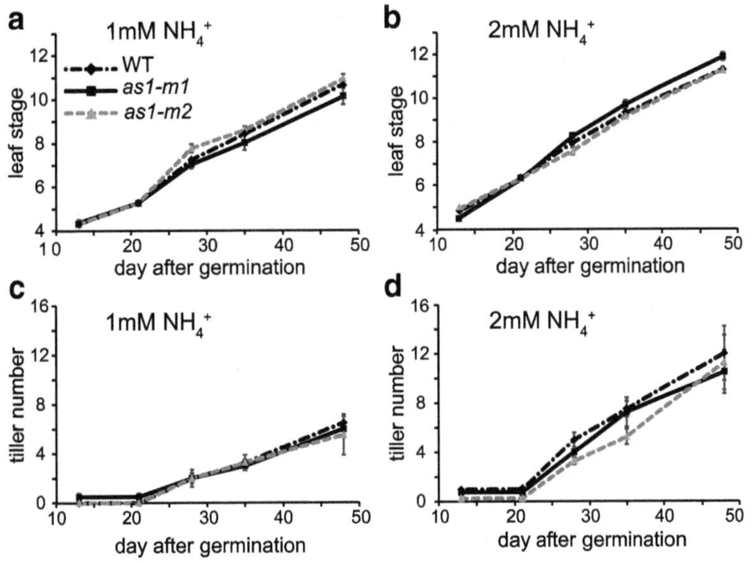

Fig. 7 Comparison of leaf stage and tiller number between the wild type and *as1* mutants. Wild-type (WT: black diamond) plants and two lines of *as1* mutants (*as1-m1* and *as1-m2*: black square and grey triangle, respectively) were grown hydroponically in the presence of 1 mM (**a**, **c**) or 2 mM NH_4Cl (**b**, **d**) until 50 days after germination. Time course studies on measurements of leaf stage and tiller number were conducted. Mean values with SE of four independent samples are shown. One-way ANOVAs followed by Bonferroni tests were used to identify significant differences between the wild type and *as1* mutants ($P < 0.05$)

the potential toxicity of NH_4^+ and the loss of nitrogen (Sakurai et al. 2001). In the basal portions of shoots, AS1 and GS1;2 could contribute to the re-assimilation of NH_4^+ during lignification. Considering its cellular localization in phloem companion cells of the nodal vascular anastomoses, the asparagine from AS1 reactions in these cells is apparently transported to sink organs, such as axillary buds, for its outgrowth and development (Fig. 2). Asparagine is catabolized by asparaginase into aspartate as the common precursor of the essential amino acids (Azevedo et al. 2006; Lea et al. 2007; Yabuki et al. 2017).

The lack of GS1;2 caused decreases in tiller number, rice yields and glutamine and asparagine content in the basal portions of shoots (Funayama et al. 2013; Ohashi et al. 2015b, 2017), while a lack of AS1 caused a decrease in only asparagine content and had less of an effect on tiller number and glutamine content (Figs. 6, 7, Table 1). These results indicate that the availability of asparagine generated from AS1 reactions apparently contributes less to the outgrowth of tillers in rice plants under conditions of sufficient glutamine. Thus, we conclude that the reduction in tiller number in *gs1;2* mutants is independent of the low availability of asparagine in the basal portions of shoots. Previous research has revealed that glutamine generated from GS1;2 reactions contributes to the outgrowth of tillers via glutamine-dependent cytokinin synthesis (Ohashi et al. 2017). Compared to that of glutamine, the contribution of asparagine to the outgrowth of rice tillers is suggested to be limited.

Table 1 Tiller number, panicle number and parameters of growth of plants growing in the paddy field

	wild type	*as1-m1*	*as1-m2*
Tiller number per plant	25.7 ± 1.83	25.35 ± 1.01	22.35 ± 0.49
	a	a	a
Panicle number per plant	20.32 ± 1.19	20.55 ± 0.71	18.66 ± 0.98
	a	a	a
Shoot length (cm)	100.9 ± 2.23	100.65 ± 1.62	94.1 ± 0.99
	a	a	a
Shoot dry weight per plant (g)	57.69 ± 4.47	65.35 ± 1.79	50.76 ± 3.13
	ab	a	b

Conclusion

Our results demonstrated that the expression of *OsAS1* could be induced by glutamine or its related metabolite in both roots and the basal portions of rice shoots. Especially in phloem companion cells of the nodal vascular anastomoses, asparagine synthesis is largely dependent on glutamine-responsive AS1. Thus, we conclude that the low availability of asparagine in *gs1;2* mutants was caused by the reduction in glutamine content, which is required for the up-regulation of *OsAS1* expression. Availability of glutamine is suggested to be more important than that of asparagine for the outgrowth of rice tillers.

Methods

Plant Materials

This study utilized the rice (*Oryza. sativa* L.) cultivar 'Nipponbare' as the wild-type plants as well as a retro-transposon *Tos17*-inserted line of the *gs1;2* mutant and two lines of *as1* mutants (Funayama et al. 2013; Ohashi et al. 2015a). The seeds of these rice plants were germinated and grown hydroponically until either the fourth- or seventh-leaf stage in an outdoor greenhouse. The temperature was controlled at 26 °C during the day with supplemental light for 13 h. The hydroponic culture solution was renewed once per week with 1 mM NH_4Cl as described in Ohashi et al. (2017). Longitudinal sections (5 mm) of basal portions of shoots including axillary buds, internodes and a SAMs were prepared by removing the primary and secondary leaves, seeds and roots as described in Ohashi et al. (2017). The basal portions of shoots at the fourth-leaf stage or seventh-leaf stage were used for the determination of the amino acid and NH_4^+ content, in situ hybridization, and qPCR analyses.

When the short-term effects of NH_4^+ supply to roots on gene expression and determination of the free amino acid and NH_4^+ content were tested, the seedlings at the fourth-leaf stage or seventh-leaf stage were further grown in water for 3 d to deplete nitrogen and were then treated with or without 1 mM NH_4Cl for 8 h or 24 h (Ohashi et al. 2017). In MSX-treatment experiments, the seedlings at the fourth-leaf stage grown in water for 3 d were pre-treated with or without 1 mM MSX (Sigma-Aldrich Japan, Tokyo, Japan) for 2 h and then transferred to media containing no nitrogen, 1 mM NH_4C1 or 5 mM glutamine for 8 h according to Kamada-Nobusada et al. (2013) with slight modifications.

For measuring the tiller number, panicle number, shoot length and shoot dry weight at a paddy field, rice plants were grown in the paddy field of the Institute for Sustainable Agro-ecosystem Services in University of Tokyo, Japan (35°44' N, 139°32′ E, 60 m altitude), during the summer (May to October) in 2015 under normal fertilizer, light irradiance and temperature. A germinated seed was sown into each cell of a cell tray and grown in a greenhouse under natural light conditions for four weeks. One seedling was then transplanted into each hill at 18 cm between hills and 30 cm between rows. Each plot was composed of four rows with ten hills with four replicates. Commercial compound fertilizer including N, P_2O_5 and K_2O was used to fertilize at 60, 90, and 80 kg ha^{-1} as a basal fertilizer before transplanting. At the panicle initiation stage, tiller numbers of five average plants selected from each plot were measured. After panicle ripening, ten plants in each plot were harvested and shoot length, shoot dry weight without panicle, and panicle number were measured.

qPCR Analysis

The qPCR analysis was performed according to Konishi et al. (2014). The nucleotide sequences for *OsAS1, OsAS2, OsGS1;1, OsGS1;2* and *Actin1* are registered in the Rice Annotation Project Database (RAP-DB; http://rapdb.dna.affrc.go.jp/) with accession numbers Os03g0291500, Os06g0265000, Os02g0735200, Os03g0223400 and Os03g0718100, respectively. Each gene-specific primer for qPCR analyses is shown in Tabuchi et al. (2007) and Ohashi et al. (2015a). The PCR products were amplified from the single-stranded cDNA as the template and quantified using a Light Cycler 480 (Roche Diagnostics K. K., Tokyo, Japan) according to the following program: 10 s at 95 °C, followed by 50 cycles of 95 °C for 5 s and 60 °C for 34 s. The transcript contents were quantitatively determined using each purified cDNA clone as a calibration standard.

In situ Hybridization of *OsAS* Genes

Preparation of RNA probes for *OsAS1* and *OsAS2* and in situ hybridization analysis for the basal portions of shoots were performed as described in Ohashi et al. (2015a, 2017).

Determination of Glutamine, Asparagine and NH_4^+ Contents

Free amino acids and NH_4^+ contents in the basal portions of shoots at the fourth-leaf stage were determined as described in Konishi et al. (2014). Derivatization of amino acids and NH_4^+ was carried out using the AccQ-Tag Ultra Derivatization Kit (Nihon Waters K.K., Tokyo, Japan). AccQ-Tag-labelled derivatives were separated and quantified using an ACQUITY UPLC H-Class with a tunable UV detector (Nihon Waters K.K.).

Analysis of Phenylalanine Ammonia-lyase Activity

The assay for PAL activity was conducted according to Olsen et al. (2008) with slight modifications. Shoot basal portion tissues (80–170 mg fresh weight) were pre-chilled in liquid nitrogen and then milled with a Tissue Lyser II (Qiagen, K. K., Tokyo, Japan) at 20 Hz for 2 min. Powdered samples were resolved in 5 volumes of extraction buffer (100 mM Tris–HCl (pH 8.8) with 12 mM β-mercaptoethanol) of sample weight, mixed in Tissue Lyser II at 20 Hz for 2 min, and centrifuged at 16,000 g for 15 min at 4 °C. The supernatant was passed through a MicroSpin S-200 HR Column (GE Healthcare Japan, Tokyo, Japan), and the elution as enzyme solution was used for the PAL activity assay. The PAL activity assay was performed at 37 °C for 60 min in an assay mixture containing 500 μl enzyme extract, 450 μl 100 mM Tris–HCl (pH 8.8) and 50 μl 100 mM L-phenylalanine. The reaction was terminated by adding 50 μl of 5 M HCl, and absorbance was recorded at 290 nm against blanks made in the same way as the assays but with 50 μl of 5 M HCl added before L-phenylalanine. The amount of product formed was calculated from the increase in absorbance using an extinction coefficient for cinnamate of 10,000 L/cm/mole (Zimmerman and Hahlbrock 1975).

Statistics

All data sets were analysed using Microsoft Excel add-in software (Social Survey Research Information Co., Ltd., Tokyo, Japan).

Additional files

Additional file 1: Figure S1. qPCR analysis of *OsGS1;1* and *OsGS1;2* in the basal portions of wild-type shoots using MSX. Wild-type rice seedlings were grown hydroponically in the presence of 1 mM NH_4Cl until the fourth-leaf stage and then in water for 3 d. After pre-treatment with 1 mM MSX for 2 h, the seedlings were treated for 8 h with 1 mM NH_4^+ (MSX + NH_4^+) or 5 mM glutamine (MSX + Gln). The seedlings without MSX pre-treatment were also treated for 8 h with 1 mM NH_4^+ (+NH_4^+), 5 mM glutamine (Gln) or without nitrogen nutrients (-N). qPCR analysis of *OsGS1;1* (a) and *OsGS1;2* (b) were performed in the shoot basal parts. Fold changes of transcript levels in each treatment relative to those in –N were calculated, and mean values with SE of four independent samples are shown. Asterisks denote statistically significant differences between samples treated without 1 mM NH_4^+ and each treated sample (*, $P < 0.05$ by Student's *t*-test).

Additional file 2: Figure S2. qPCR analysis of *OsAS1* and *OsAS2* in roots of wild-type rice using MSX. Wild-type rice seedlings were grown hydroponically in the presence of 1 mM NH_4Cl until the fourth-leaf stage and then transferred into water for 3 d. After pre-treatment with 1 mM MSX for 2 h, the seedlings were treated for 8 h with 1 mM NH_4^+ (MSX + NH_4^+) or 5 mM glutamine (MSX + Gln). The seedlings without MSX pre-treatment were also treated for 8 h with 1 mM NH_4^+ (+NH_4^+), 5 mM glutamine (Gln) or without nitrogen nutrients (-N). qPCR analyses of *OsAS1* (a), *OsAS2* (b), and *actin1* transcripts (c) were performed in roots. Fold changes of transcript levels in each treatment relative to those in –N were calculated, and mean values with SE of four independent samples are shown. Asterisks denote statistically significant differences between samples treated without 1 mM NH_4^+ and each treated sample (*, $P < 0.05$ by Student's *t*-test).

Additional file 3: Figure S3. Amino acid analysis in the basal portions of shoots supplied with transient NH_4^+. Seedlings of the wild type (WT: black column) and two lines of *as1* mutants (*as1-m1* and *as1-m2*: dark and light grey column, respectively) at the fourth-leaf stage were grown in water for 3 d, then treated with (+NH_4^+) or without (-N) 1 mM NH_4Cl for 24 h. Mean values with SE of four independent samples are shown. One-way ANOVAs followed by Bonferroni tests were used to identify significant differences between the WT and *as1* mutants ($P < 0.05$).

Additional file 4: Figure S4. Stereoscopic microscope observation of the primary tiller at fifth leaf stage. Seedlings of the wild type (WT) (a) and two lines of *as1* mutants (*as1-m1* and *as1-m2*) (b, c) were grown hydroponically in the presence of 1 mM NH_4^+ until the fifth leaf stage. The primary tiller was observed by microscope. Scale bars = 2 mm. (JPG 1217 kb)

Additional file 5: Figure S5. Photos of wild type and *as1* mutants at heading stage. Wild-type plants (WT) (a) and two lines of *as1* mutants (*as1-m1* and *as1-m2*) (b, c) were grown in the paddy field under normal fertilization. Each two plants were shown.

Additional file 6: Figure S6. Comparison of the PAL activity between basal portions of shoots and the third-leaf blade. Rice seedlings were grown hydroponically in the presence of 1 mM NH_4Cl until the fourth-leaf stage, and the basal portions of shoots (black column) and third-leaf blades (grey column) were harvested. Mean values with SE of five independent samples are shown. Asterisks denote statistically significant differences between each organ (*, $P < 0.05$ by Student's *t*-test).

Abbreviations

AMT: Ammonium transporter; AS: Asparagine synthetase; GS: Glutamine synthetase; IPT: Adenosine phosphate-isopentenyltransferase; MSX: Methionine sulfoximine; PAL: Phenylalanine ammonia-lyase; qPCR: Quantitative real-time PCR; SAM: Shoot apical meristem

Acknowledgements

We would like to express our deep gratitude to the technical support staff of the Institute for Sustainable Agro-ecosystem Services (ISAS) at the University of Tokyo for their help in the cultivation and management of rice. We thank Drs. Hirohiko Hirochika and Akio Miyao, The Project for Rice Genome Research, and The National Institute of Agrobiological Sciences, Tsukuba, Japan, for providing seeds of *Tos17*-inserted rice mutants. We would like to thank American Journal Experts (www.aje.com) for English language editing.

Funding

This work was supported by the Ministry of Education, Culture, Sports, Science and Technology of Japan [a Grant-in-Aid for Scientific Research on Innovative Area (22119003 to TY) and a GRENE NC-CARP project]; the Japan Society for the Promotion of Science (JSPS) KAKENHI [Grant No. 15 K14674 and 15H04468 to TH, 26450073 to SK, and 16 J01799 to MO]; and the Division for International Advanced Research and Education (DIARE), Tohoku University.

Authors' Contributions

MO and KI designed this study, performed most of the experiments and wrote the manuscript. NK operated UPLC for amino acid analysis. KS conducted the paddy field trials. SK, MM, TH and TY advised and assisted in the design of this study and in writing the manuscript. All authors read and approved the final manuscript.

Competing Interests

The authors declare that they have no competing interests.

Author details

[1]Graduate School of Agricultural Science, Tohoku University, 468-1 Aoba, Aramaki-Aza, Aoba-ku, Sendai 980-8572, Japan. [2]The University of Tokyo, Graduate School of Agricultural and Life Sciences, Institute of Sustainable Agro-ecosystem Services (ISAS), 1-1-1 Midori-cho, Nishitokyo, Tokyo 188-0002, Japan. [3]Present Address: Graduate School of Bioagricultural Sciences, Nagoya University, Furo-cho, Chikusa-ku, Nagoya 464-8601, Japan. [4]Present Address: Institute of Plant Science and Resources, Okayama University, 2-20-1 Chuo, Kurashiki 710-0046, Japan. [5]Present Address: Division for Interdisciplinary Advanced Research and Education, Tohoku University, 6-3 Aoba, Aramaki-Aza, Aoba-ku, Sendai 980-0845, Japan.

References

Azevedo RA, Lancien M, Lea PJ (2006) The aspartic acid metabolic pathway, an exciting and essential pathway in plants. Amino Acids 30:143–162

Domagalska MA, Leyser O (2011) Signal integration in the control of shoot branching. Nat Rev Mol Cell Biol 12:211–221

Evers JB, van der Krol AR, Vos J, Struik PC (2011) Understanding shoot branching by modelling form and function. Trends Plant Sci 16:464–467

Fukumorita T, Chino M (1982) Sugar, amino-acid, and inorganic contents in rice phloem sap. Plant Cell Physiol 23:273–283

Funayama K, Kojima S, Tabuchi-Kobayashi M, Sawa Y, Nakayama Y, Hayakawa T, Yamaya T (2013) Cytosolic glutamine synthetase1;2 is responsible for the primary assimilation of ammonium in rice roots. Plant Cell Physiol 54:934–943

Gaufichon L, Reisdorf-Cren M, Rothstein SJ, Chardon F, Suzuki A (2010) Biological functions of asparagine synthetase in plants. Plant Sci 179:141–153

Hayashi H, Chino M (1990) Chemical composition of phloem sap from the upper most internode of the rice plant. Plant Cell Physiol 31:247–251

Hirose N, Hayakawa T, Yamaya T (1997) Inducible accumulation of mRNA for NADH-dependent glutamate synthase in rice roots in response to ammonium ions. Plant Cell Physiol 38:1295–1297

Hoshikawa K (1989) The growing Rice Plant. Nosan Gyoson Bunka Kyokai, Tokyo

Ireland RJ, Lea PJ (1999) The enzymes of glutamine, glutamate, asparagine, and aspartate metabolism. In: Singh BK (ed) In plant amino acids: biochemistry and biotechnology. Marcel Dekker, Inc, New York, pp 49–109

Ishiyama K, Inoue E, Tabuchi M, Yamaya T, Takahashi H (2004) Biochemical background and compartmentalized functions of cytosolic glutamine synthetase for active ammonium assimilation in rice roots. Plant Cell Physiol 45:1640–1647

Kamada-Nobusada T, Makita N, Kojima M, Sakakibara H (2013) Nitrogen-dependent regulation of de novo cytokinin biosynthesis in rice: the role of glutamine metabolism as an additional signal. Plant Cell Physiol 54:1881–1893

Kawachi T, Sueyoshi K, Nakajima A, Yamagata H, Sugimoto T, Oji Y (2002) Expression of asparagine synthetase in rice (*Oryza sativa*) roots in response to nitrogen. Physiol Plant 114:41–46

Konishi N, Ishiyama K, Matsuoka K, Maru I, Hayakawa T, Yamaya T, Kojima S (2014) NADH-dependent glutamate synthase plays a crucial role in assimilating ammonium in the Arabidopsis root. Physiol Plant 152:138–151

Lam HM, Peng SS, Coruzzi GM (1994) Metabolic regulation of the gene encoding glutamine-dependent asparagine synthetase in *Arabidopsis thaliana*. Plant Physiol 106:1347–1357

Lea PJ, Miflin BJ (1974) Alternative route for nitrogen assimilation in higher plants. Nature 251:614–616

Lea PJ, Sodek L, Parry MAJ, Shewry PR, Halford NG (2007) Asparagine in plants. Ann Appl Biol 150:1–26

Liu Y, Gu D, Ding Y, Wang Q, Li G, Wang S (2011) The relationship between nitrogen, auxin and cytokinin in the growth regulation of rice (*Oryza sativa* L.) tiller buds. Aust J Crop Sci 5:1019–1026

Mae T (1997) Physiological nitrogen efficiency in rice: nitrogen utilization, photosynthesis, and yield potential. Plant Soil 196:201–210

Ohashi M, Ishiyama K, Kojima S, Kojima M, Sakakibara H, Yamaya T, Hayakawa T (2017) Lack of cytosolic glutamine synthetase1;2 activity reduces nitrogen-dependent biosynthesis of cytokinin required for axillary bud outgrowth in rice seedlings. Plant Cell Physiol 58:679–690

Ohashi M, Ishiyama K, Kojima S, Konishi N, Nakano K, Kanno K, Hayakawa T, Yamaya T (2015a) Asparagine synthetase1, but not asparagine synthetase2, is responsible for the biosynthesis of asparagine following the supply of ammonium to rice roots. Plant Cell Physiol 56:769–778

Ohashi M, Ishiyama K, Kusano M, Fukushima A, Kojima S, Hanada A, Kanno K, Hayakawa T, Seto Y, Kyozuka J, Yamaguchi S, Yamaya T (2015b) Lack of cytosolic glutamine synthetase1;2 in vascular tissues of axillary buds caused severe reduction in their outgrowth and disorder of metabolic balance in rice seedlings. Plant J 81:347–356

Oliveira IC, Coruzzi GM (1999) Carbon and amino acids reciprocally modulate the expression of glutamine synthetase in Arabidopsis. Plant Physiol 121:301–310

Olsen KM, Lea US, Slimestad R, Verheul M, Lillo C (2008) Differential expression of four Arabidopsis PAL genes; PAL1 and PAL2 have functional specialization in abiotic environmental-triggered flavonoid synthesis. J Plant Physiol 165:1491–1499

Ookawa T, Ishihara K (1993) Varietal difference of the cell wall components affecting the bending stress of the culm in relation to the loading resistance in paddy rice. Jpn J Crop Sci 62:378–384

Ookawa T, Todokoro Y, Ishihara K (1993) Changes in physical and chemical characteristics of culm associated with lodging resistance in paddy rice under different growth conditions and varietal difference of their changes. Jpn J Crop Sci 62:525–533

Rawat SR, Silim SN, Kronzucker HJ, Siddiqi MY, Glass AD (1999) *AtAMT1* gene expression and NH_4^+ uptake in roots of *Arabidopsis thaliana*: evidence for regulation by erroot glutamine levels. Plant J 19:143–152

Sakakibara H, Takei K, Hirose N (2006) Interactions between nitrogen and cytokinin in the regulation of metabolism and development. Trends Plant Sci 11:440–448

Sakamoto T, Matsuoka M (2008) Identifying and exploiting grain yield genes in rice. Curr Opin Plant Biol 11:209–214

Sakurai N, Katayama Y, Yamaya T (2001) Overlapping expression of cytosolic glutamine synthetase and phenylalanine ammonia-lyase in immature leaf blades of rice. Physiol Plant 113:400–408

Sonoda Y, Ikeda A, Saiki S, von Wirén N, Yamaya T, Yamaguchi J (2003) Distinct expression and function of three ammonium transporter genes (*OsAMT1;1–1; 3*) in rice. Plant Cell Physiol 44:726–734

Tabuchi M, Abiko T, Yamaya T (2007) Assimilation of ammonium ions and reutilization of nitrogen in rice (*Oryza sativa* L.). J Exp Bot 58:2319–2327

Umehara M, Hanada A, Yoshida S, Akiyama K, Arite T, Takeda-Kamiya N, Magome H, Kamiya Y, Shirasu K, Yoneyama K, Kyozuka J, Yamaguchi S (2008) Inhibition of shoot branching by new terpenoid plant hormomes. Nature 455:195–200

Urquhart AA, Joy KW (1981) Use of phloem exudate technique in the study of amino acid transport in pea plants. Plant Physiol 68:750–754

Vincentz M, Moureaux T, Leydecker MT, Vaucheret H, Caboche M (1993) Regulation of nitrate and nitrite reductase expression in *Nicotiana plumbaginifolia* leaves by nitrogen and carbon metabolites. Plant J 3:315–324

Yabuki Y, Ohashi M, Imagawa F, Ishiyama K, Beier MP, Konishi N, Umetsu-Ohashi T, Hayakawa T, Yamaya T, Kojima S (2017) A temporal and spatial contribution of asparaginase to asparagine catabolism during development of rice grains. Rice 10. https://doi.org/10.1186/s12284-017-0143-8

Yamaya T, Oaks A (2004) Metabolic regulation of ammonium uptake and assimilation. In: Amâncio S, Stulen I (eds) In nitrogen acquisition and assimilation in higher plants. Kluwer Academic Publishers, Dordrecht, pp 35–63

Zimmerman A, Hahlbrock K (1975) Light induced changes in enzyme activities in parsley cell suspension. Purification and some properties of phenylalanine ammonia lyase. Arch Biochem Biophys 166:54–62

14

DS1/OsEMF1 interacts with OsARF11 to control rice architecture by regulation of brassinosteroid signaling

X. Liu[1†], C. Y. Yang[1†], R. Miao[1], C. L. Zhou[1], P. H. Cao[1], J. Lan[1], X. J. Zhu[1], C. L. Mou[1], Y. S. Huang[1], S. J. Liu[1], Y. L. Tian[1], T. L. Nguyen[1], L. Jiang[1*] and J. M. Wan[1,2*]

Abstract

Background: Plant height and leaf angle are important determinants of yield in rice (*Oryza sativa* L.). Genes involved in regulating plant height and leaf angle were identified in previous studies; however, there are many remaining unknown factors that affect rice architecture.

Results: In this study, we characterized a dwarf mutant named *ds1* with small grain size and decreased leaf angle,selected from an irradiated population of ssp. *japonica* variety Nanjing35. The *ds1* mutant also showed abnormal floral organs. *ds1* plants were insensitive to BL treatment and expression of genes related to BR signaling was changed. An F_2 population from a cross between *ds1* and *indica* cultivar 93–11 was used to fine map *DS1* and to map-based clone the *DS1* allele, which encoded an EMF1-like protein that acted as a transcriptional regulator. *DS1* was constitutively expressed in various tissues, and especially highly expressed in young leaves, panicles and seeds. We showed that the DS1 protein interacted with auxin response factor 11 (OsARF11), a major transcriptional regulator of plant height and leaf angle, to co-regulate *D61/OsBRI1* expression. These findings provide novel insights into understanding the molecular mechanisms by which *DS1* integrates auxin and brassinosteroid signaling in rice.

Conclusion: The *DS1* gene encoded an EMF1-like protein in rice. The *ds1* mutation altered the expression of genes related to BR signaling, and *ds1* was insensitive to BL treatment. DS1 interacts with OsARF11 to co-regulate *OsBRI1* expression.

Keywords: Brassinosteroids (BRs), Plant architecture, *Oryza sativa*, *D61/OsBRI1*

Background

Rice is an important food crop that feeds more than half of the world population. With an increasing global population, food security and safety are becoming serious issues. Given the limited availability of arable land and the growing human population, a future increase in rice production will be a major challenge for rice breeders (Wang et al. 2008). Plant height and leaf angle are important agronomic traits directly affecting grain yield and crop architecture, which are major objectives of crop improvement (Ikeda et al. 2013).

* Correspondence: jiangling@njau.edu.cn; wanjm@njau.edu.cn
†X. Liu and C. Y. Yang contributed equally to this work.
[1]State Key Laboratory for Crop Genetics and Germplasm Enhancement, Jiangsu Plant Gene Engineering Research Center, Nanjing Agricultural University, Nanjing 210095, China
Full list of author information is available at the end of the article

Plant height, an important trait in crop improvement, is related to lodging resistance, grain yield, and biomass production. Reduced plant height and associated lodging resistance are predominant strategies in crop improvement (Ayano et al. 2014). Various factors cause reduced height in plants; gibberellin (GA) and brassinosteroids (BRs) are the most widely investigated factors affecting plant height in rice. Many rice GA- and BR-related mutants show dwarf or semi-dwarf phenotypes, such as *sd1, d1* and *d61* (Sasaki et al. 2002; Fujisawa et al. 1999; Zhang et al. 2016; Yamamuro et al. 2000). These dwarf mutants result from reduced cell numbers or cell length in stems, and analyzing more dwarf mutants may provide novel insights into the mechanisms controlling stem elongation.

Leaf angle, the degree of bending between the leaf blade and culm, is a key factor determining the plant architecture and grain yield (Zhao et al. 2010). Crops with erect leaves have increased photosynthetic effciency and nitrogen storage for grain flling and are suitable for dense planting (Sakamoto et al. 2006). Many genes or QTLs have been reported to control leaf angle, including *D61/OsBRI1, ILI1, LC2, ILA1, RAV6, OsARF19,* and *SLG* (Yamamuro et al. 2000; Zhang et al. 2009; Zhao et al. 2010; Ning et al. 2011; Zhang et al. 2015b; Zhang et al. 2015a; Feng et al. 2016). Most identified rice mutants with altered leaf inclination have abnormal cell division and or expansion and altered cell wall composition at the leaf:-stem joint (Zhang et al. 2009; Zhao et al. 2010). Nevertheless, BR and auxin synergistically control leaf inclination in rice (Wada et al. 1981; Hirano et al. 2017).

The synergetic effects of auxin and BRs on various physiological events have strongly suggested the interdependency of their signaling, and molecular evidence for this has been reported. For example, the auxin-inducible genes, *indole-5-acetic acid (IAA5), IAA19* and *SAUR-AC1*, are induced by brassinolide (BL) treatment, whereas the expression of *IAA5* and *IAA19* are down-regulated in the *Arabidopsis* BR-deficient mutant *de-etiolated 2* (*det2*) (Nakamura et al. 2003). In addition, expression of BR biosynthetic gene *DWARF4* and the BR receptor *BRI1* are induced by auxin (Park et al. 2011; Sakamoto et al. 2012). *BRASSINAZOLE RESISTANT 1* (*BZR1*) directly regulates many auxin-responsive genes (Sun et al. 2010). Nemhauser et al. (2004) found that 48 genes were co-regulated by auxin and BL (e.g. *SAUR, Aux/IAA* and *GH3*). They also found that a TGTCTC element in the auxin-responsive element (AuxRE) was enriched in genes up-regulated by both IAA and BL. These results strongly suggest that the two hormone pathways synergistically affect gene expression at the transcriptional level.

In this study, we identified a reduced BR-sensitivity mutant, *ds1*, showing dwarfness, smaller seed and decreased leaf angle compared to wild type. The small plant stature of the plants and decreased leaf angle resulted from reduced cell elongation. Our genetic mapping and molecular biology experiments revealed that the underlying *DS1*, a new allele of *OsEMF1*, encodes an EMF1-like protein involved in the development of plant height and leaf angle by interacting with OsARF11 to co-regulate *OsBRI1* expression.

Results

Characterization of the *ds1* mutant

To identify genetic factors regulating rice architecture, we screened our Nanjing 35 radiation mutagenesis population by isolating dwarf mutants. A mutant was identified with dwarfism and decreased leaf angles, *ds1* (Fig. 1a-d).

Differences between wild type and *ds1* mutant plants became apparent only after internode elongation when reduced height and darker leaf colour of the mutant were expressed. There was no difference between *ds1* and WT in young seedlings. Leaves of the mutant remained erect until maturity (Fig. 1a). Spikelet number per panicle, numbers of primary and secondary panicle branches of *ds1* were less than the WT (Additional file 1: Table S1). The panicles of *ds1* at maturity were almost erect whereas the panicles of WT clearly bend down. All internodes of *ds1* were shorter than those of WT (Fig. 1b-c). The dwarfism of *ds1* belonged to the dn type based on the classification of Yamamuro et al. (2000). Flag and second leaf angles of *ds1* were smaller than WT (Fig. 2a-b). These altered morphologies of *ds1* suggested that *DS1* participated in rice growth and development.

The *ds1* mutant is less sensitive to BRs

As *ds1* had dwarfism, erect and dark green leaves, characteristic phenotypes of BR-related mutants, we

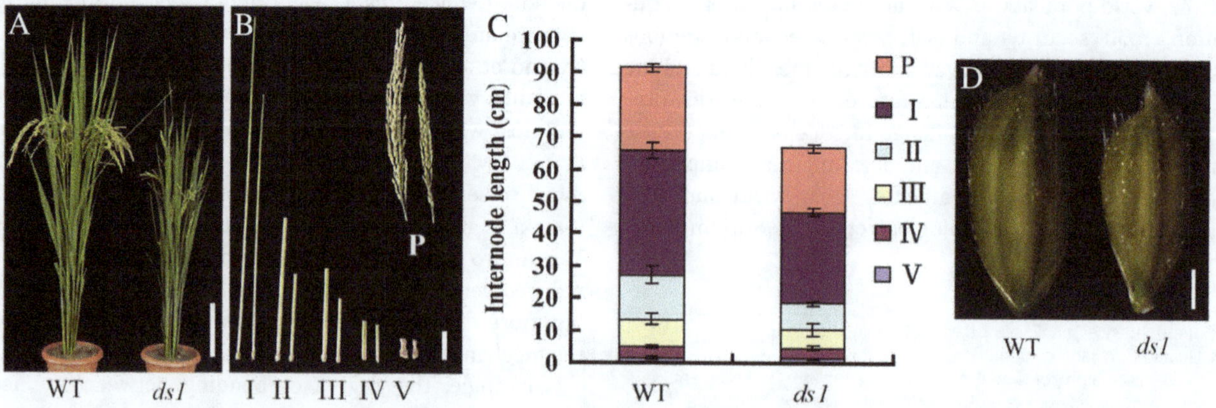

Fig. 1 Phenotypic comparison of wild type and *ds1* mutant. **a** Gross morphology of the wild type and *ds1* plants. Scale bar, 10 cm. **b** Comparison of internode and panicle lengths between WT and *ds1*. Scale bar, 2 cm. **c** Statistical data for internode lengths described in (**b**). Data are presented as means ± SD. **d** Young spikelet hulls of WT (left) and *ds1* (right) plants. Scale bar, 1 mm

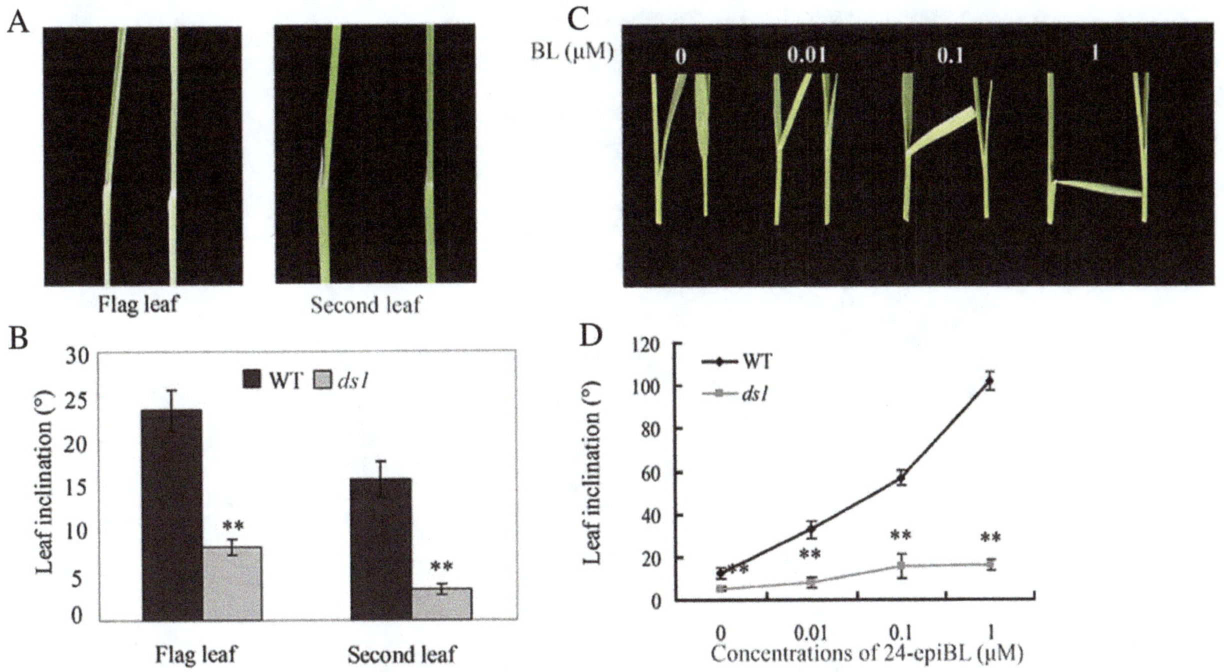

Fig. 2 *DS1* is a positive regulator of BR signaling in rice. **a** Flag and second leaf angles of WT (left) and *ds1* (right) at maturity. **b** Statistical data for flag and second leaf angles shown in (**a**). Data are presented as means ± SD. **c** Lamina joint bending response to various amounts of 24-epiBL in WT (left) and *ds1* mutant (right) seedlings. **d** Statistical data for lamina joint bending angle assay described in (**c**). Data are means ± SD.**, significant difference between WT and *ds1* at $P = 0.01$

hypothesized that *DS1* was involved in the BR pathway. Leaf angle in rice was known to be sensitive to the concentration of BL or related compounds (Wada et al. 1981). We tested the sensitivity of *ds1* to 24-epiBL in a lamina joint bending experiment. BL treatment caused a dose-dependent lamina joint inclination in WT, whereas *ds1* plants were insensitive. Treatment with BL showed an increased level leaf angle in WT plants, but little change in *ds1* plants (Fig. 2c-d).

DS1 regulates plant height and leaf angle by promoting cell elongation

To investigate the mutant phenotype of *ds1* at a cellular level we performed scanning electron microscopy (SEM) on *ds1* plants along with the WT (Fig. 3a-c). Dwarfism of *ds1* plants was mainly due to reduced cell length as measured in 2nd internodes (Fig. 3a-b). Observations on adaxial cells, which influence leaf angle, showed that those cells were also shorted in the *ds1* mutant (Fig. 3c). These results indicated that reduced cell length caused the dwarf phenotype of *ds1*.

As the cell wall is the major factor restricting cell elongation, many genes involved in cell wall synthesis are reported to regulate cell elongation (Ning et al. 2011). We investigated the expression levels of cell wall synthesis-related genes in WT and the *ds1* mutant. Expression levels of *IRX10L*, *CESA6*, and *UGA4e* were significantly down-regulated in *ds1* compared to WT (Additional file 1: Figure S1). Consistent with the morphological observations, this indicated that the dwarfism in *ds1* was caused by reduced cell length. To eliminate the possibility of decreased cell size in *ds1*, we analysed the expression levels of nine cell expansion regulators in WT and the *ds1* mutant (Additional file 1: Figure S1). Seven genes were significantly down-regulated in *ds1*, indicating that cell expansion was reduced. These results further confirmed that reduced cell elongation was the cause of the phenotype of *ds1*.

Map-based cloning and functional analysis of *DS1*

To determine the gene responsible for the *ds1* phenotype, we made a cross between *ds1* and the wild type. F_1 plants showed the WT phenotype and the F_2 segregation ratio was consistent with that expected a single locus ($\chi^2_{3:1} = 0.98$, $P > 0.05$). These results indicated that *ds1* phenotype was due to a single recessive gene.

To map the *DS1* locus, we next generated an F_2 population from a cross between *ds1* and *indica* cultivar 93–11. *DS1* was roughly mapped on the short arm of chromosome 1 between SSR markers N1–12 and N1–017. We then used 568 recessive individuals from the same F_2 population and narrowed down the *DS1* locus to a 80 kb region between markers L-3 and L-2 within BAC clones P0452F10 and P0485D09. A total of 11

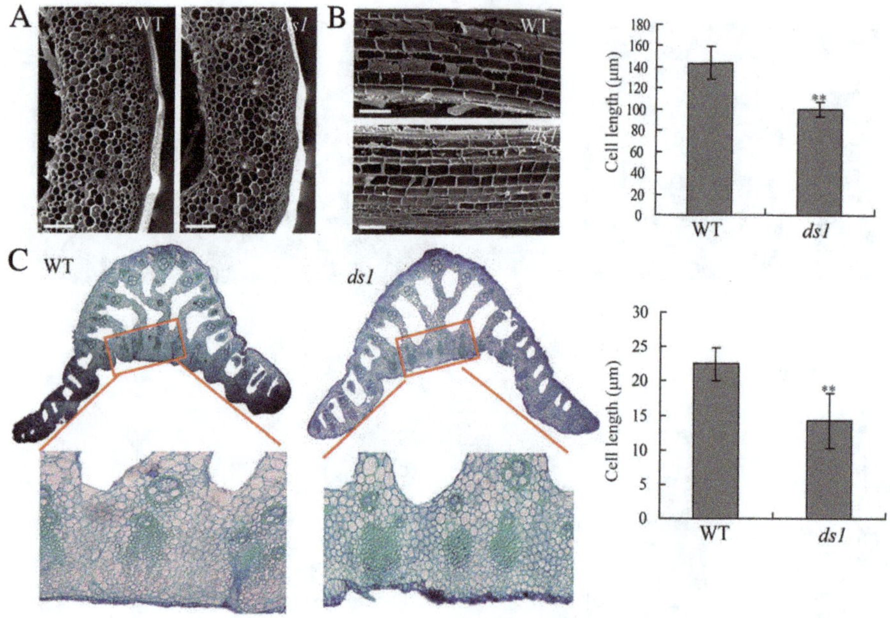

Fig. 3 Histological analysis of the 2nd internodes and lamina joints. **a** Transverse sections of WT and *ds1* mutant 2nd internodes. Bar, 100 μm. **b** Longitudinal sections of WT and *ds1* mutant 2nd internodes. Bar, 500 μm.** indicates significant difference at $P < 0.01$. **c** Cross section of lamina joints of flag leaves in WT (left) and *ds1* (right). ** indicates significant difference at $P < 0.01$

putative genes were localized in the 80 kb region according to the genome annotation of Nipponbare (http://www.gramene.org/). On comparison of the mapped genomic region between WT and *ds1*, we detected a single nucleotide deletion at the first exon and two nucleotide deletions in the second exon of *LOC_Os01g12890* (Fig. 4a). We identified three domains in the DS1 polypeptide, including a nuclear localization signal (NLS), an ATP/GTP binding motif, and an LXXLL motif (Fig. 4b). The LXXLL motif had been demonstrated to mediate interactions that can activate or repress transcription (Heery et al. 1997; Torchia et al. 1997; Calonje et al. 2008).

The *OsEMF1* gene, annotated as the *LOC_Os01g12890* gene model, encodes an EMF1-like protein, involved in H3K27me3-mediated epigenetic silencing. In *Arabidopsis*, *EMBRYONIC FLOWER 1* (*EMF1*), a plant-specific protein, participated in H3K27me3-mediated silencing of target genes by acting downstream of *EMF2* and likely interacted with both PRC1 RING-finger proteins and PRC2 component *MULTICOPY SUPRESSOR OFIRA1* (*MSI1*) (Aubert et al. 2001; Calonje et al. 2008). In rice, *OsEMF1* plays an important role in palea development through maintaining H3K27me3-mediated epigenetic repression on *OsMADS58* (Zheng et al. 2015; Yan et al. 2015). To confirm whether mutation of *DS1* was responsible for the *ds1* phenotype, a CDS fragment of 3174 bp containing the entire *DS1* coding sequence and the *DS1* promoter were cloned into the pCUbi1390 binary vector, and the complementation construct was then transformed into the *ds1* mutant. All 8 independent *DS1*-complementation lines rescued normal leaf angle and plant height of *ds1* plants (Fig. 4c). These genetic evidence and transformation results confirmed that *DS1* was a newly identified *OsEMF1* allele, which influenced leaf angle and plant height. Homology analysis found that amino acid sequence of DS1 had 37% similarity and 20% homology to *Arabidopsis thaliana* EMF1 (AtEMF1) (Additional file 1: Figure S2).

Expression pattern of *DS1*

To further explore the biological functions of *DS1*, we found that *DS1* was highly expressed in the seeds,young leaves and panicles according to the Rice eFP Browser (http://bar.utoronto.ca/efprice/, Fig. 5a). Then we tested the *DS1* expression pattern in various organs from WT and *ds1* by quantitative real-time PCR. The results showed that *DS1* was highly expressed in young leaves, panicles and seeds, with the lowest expression in the roots. In comparison with the WT, the expression level was not significantly different in all tested organs of the *ds1* mutant (Fig. 5b). To determine whether the expression of *DS1* was affected by BR, seeds were cultured on half-strength medium containing 1 μM 24-epiBL and expression of *DS1* was determined by quantitative real-time PCR. As shown in Fig. 5c, the *DS1* expression was induced by BL treatment in WT.

DS1 interacts with OsARF11

To gain a deeper insight into the molecular function of DS1, we performed a yeast two-hybrid (Y2H) screening using DS1 as bait. We obtained many candidate

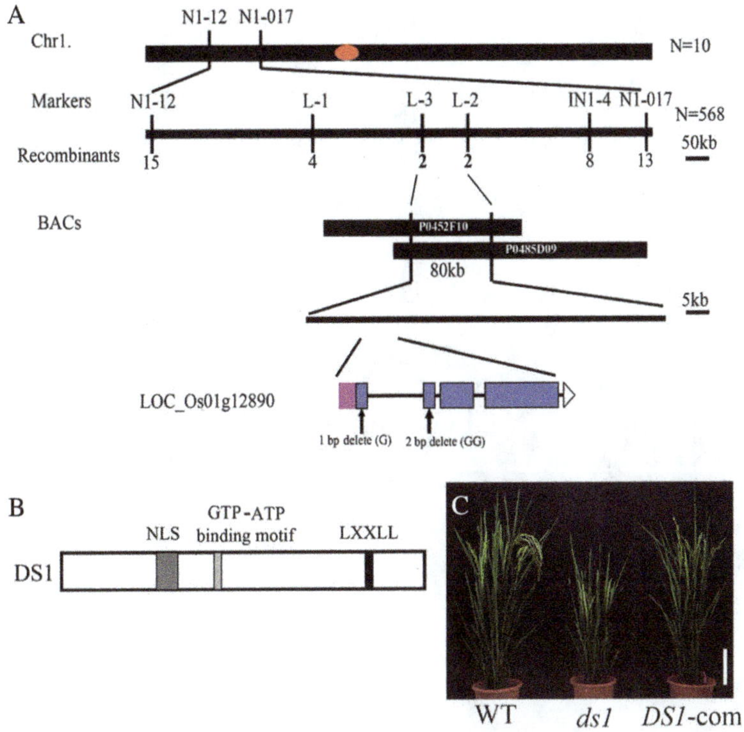

Fig. 4 Map-based cloning of *DS1*. **a** Fine mapping of the *DS1* gene on chromosome 1. The *DS1* locus was mapped to an 80 kb region that contained 11 predicted open reading frames (ORFs). **b** DS1 protein structure. Gray box, NLS; light gray box, GTP-ATP binding motif; black box, LXXLL motif. **c** Characterization of T_1 transgenic plants when *ds1* was transformed with the introduction of *DS1* fragment. Bar, 10 cm

interacting proteins, including auxin response factor 11 (OsARF11). The interaction between DS1 and OsARF11 was further confirmed using the full-length cDNA in yeast (Fig. 6a-b). Auxin response factors (ARFs), including *OsARF11* and *OsARF19*, were reported to regulating plant height and leaf angle. Loss-of-function *osarf* mutants showed decreased leaf angle and dwarfism, similar to the *ds1* mutant. Also, OsARFs directly regulate expression of *OsBRI1* by directly binding to the AuxRE element in the promoter of *OsBRI1* (Shen et al. 2010; Sakamoto et al. 2013; Zhang et al. 2015a).To validate this interaction, we performed a bimolecular fluorescence complementation (BiFC) study. The N-terminal half of YFP was fused to DS1, and the C-terminal half of YFP was fused to OsARF11. Both constructs were then transiently expressed in tobacco (*Nicotiana bethamiana*). YFP fluorescence was observed in the nucleus, confirming that DS1 interacted with OsARF11 in planta (Fig. 6c). Taken together, these results indicated that DS1 might regulate plant height and leaf angle through interaction with OsARF11.

Expression pattern and subcellular localization of *OsARF11*

To verify whether the expression pattern of *OsARF11* is similar to that of *DS1*, we found that *OsARF11* was highly expressed in the shoot apical meristem and young panicles according to the Rice eFP Browser (http://bar.utoronto.ca/efprice/, Fig. 7a). We employed a quantitative RT-PCR (qRT-PCR) assay to examine OsARF11 expression in wild-type plants. Real time qPCR analysis showed that OsARF11 transcripts accumulated in all examined tissues, including sheaths, young panicles, roots, stems and leaves (Fig. 7b). These results indicated that *OsARF11* showed a very similar expression pattern to *DS1*. We further, found that *OsARF11* expression was induced by auxin treatment (Fig. 7c), suggesting that its function might be regulated by auxin. We next examined the subcellular localization of OsARF11 protein. A construct harboring a OsARF11-GFP fusion was introduced into tobacco epidermal cells by Agrobacterium. Fluorescent signals of OsARF11-GFP fusion protein were observed exclusively in the nucleus of tobacco cells (Fig. 7d), indicating that OsARF11 is a nuclear protein, consistent with the function of OsARF11.

DS1 and OsARF11 co-regulate *D61/OsBRI1* expression

As shown in Fig. 2c, the *ds1* mutant was insensitive to BL treatment. We investigated the effect of *DS*1 on the expression of BR-related genes and the results showed that disruption of *DS1* led to significantly increased expression of *D2* and *DWARF4*, two important

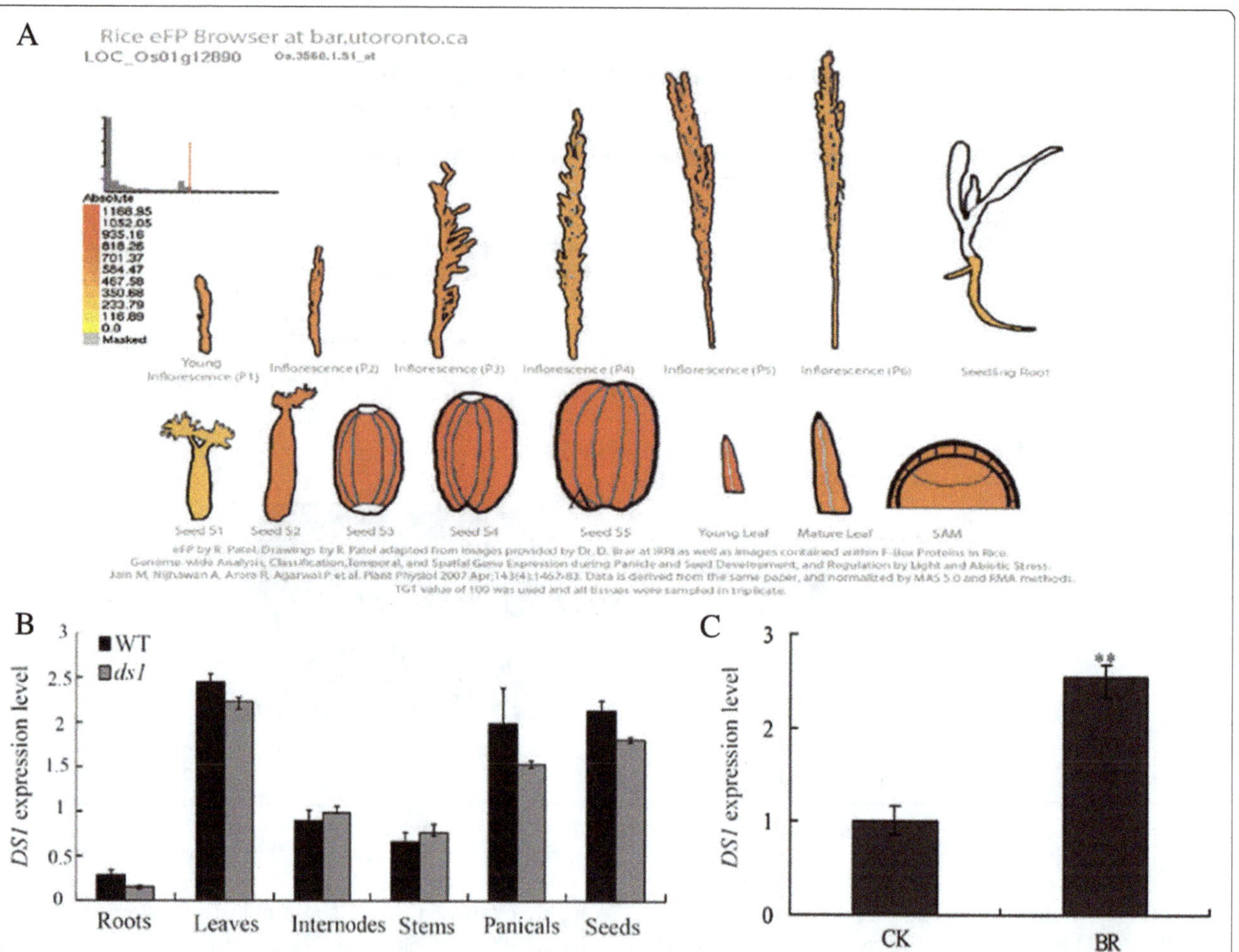

Fig. 5 Spatiotemporal expression of *DS1*. **a** Expression pattern of *DS1* based on the Rice eFP Browser. **b** Quantitative RT-PCR analysis of *DS1* expression in various rice tissues. Roots and leaves were harvested from plants at the early booting stage. Internodes and stems were collected from 5-week-old seedlings. Panicles were collected when they had reached 2 cm length. Seeds were collected at grain filling. **c** Effect of 24-epiBL on *DS1* expression. 5-day-old seedlings of the wild type (WT) were harvested after treatment with 1 μM 24-epiBL. CK represents the mock-treated control

brassinosteroid biosynthesis rate-limiting enzymes (Hong et al. 2003; Sakamoto et al. 2006). By contrast, *DS1* had large effects on expression of BR signal genes, including *D61*/*OsBRI1*, *OsBZR1*, *OsBU1* and *XTR1* (Fig. 8). This indicated that *DS1* could be involved in BR signaling.

Previous studies showed that *OsARF11* promotes plant growth and leaf angle by specific binding to an auxin response element (AuxRE) in the promoter of *OsBRI1* (Sakamoto et al. 2013), and that OsARF19 binds to an AuxRE element in the promoters of *OsBRI1* and *OsGH3–5* (Zhang et al. 2015a). We analyzed the promoters of four BR signal genes (*OsBRI1*, *OsBZR1*, *OsBU1* and *XTR1*), and found that only the promoter of *OsBRI1* contained the auxin response element. To confirm the binding of DS1 or OsARF11 to the *OsBRI1* promoter, we performed yeast one hybrid assays using in vitro-expressed DS1 and OsARF11. As shown in Fig. 9a, OsARF11 bound to the *OsBRI1* promoter but DS1 failed to bind the promoter of *OsBRI1*.

To examine the regulation of DS1 and OsARF11 on the expression of its target gene *OsBRI1*, we performed transient expression assays using ~2 kb of the *OsBRI1* promoter fused to the LUC gene as a reporter. Effector constructs for OsARF11 were expressed under control of the 35S promoter and transfected together with the reporter construct into rice protoplasts. Higher LUC activity was detected when OsARF11 protein was transfected with the reporter construct compared with the internal control (Fig. 9b). These findings strongly support the hypothesis that DS1 and OsARF11 coregulate *OsBRI1* in controlling rice plant architecture.

Discussion

Plant architecture is a major factor underlying grain yield and determination of planting density and photosynthesis efficiency in rice. The molecular mechanisms of plant architecture have been extensively studied and many critical genes have been identified. Most of these

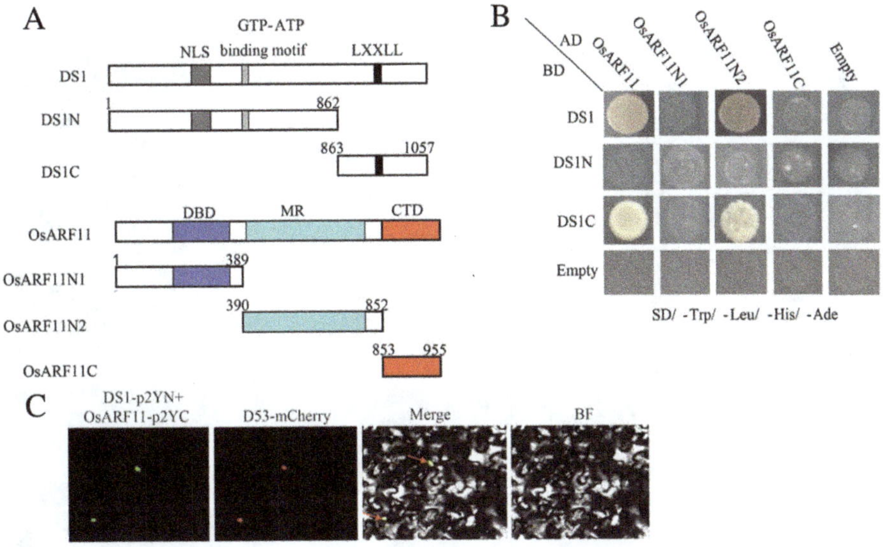

Fig. 6 DS1 physically interacts with OsARF11. **a** Sketches showing the domain structures of DS1 and OsARF11 and various deletions. **b** Yeast two-hybrid assays showing the interactions between DS1, OsARF11, and their derivatives. Transformed yeast cells were grown on SD/–Trp/–Leu/-His/–Ade. **c** BiFC assay showing DS1-YFP^{N} and OsARF11-YFP^{C} interaction to form a functional YFP in the nucleus

genes are involved in the Auxin and BR pathways. Auxin and BR signaling share many transcriptional target genes in regulating rice plant architecture (Yin et al. 2002; Nemhauser et al. 2004), suggesting that interactions between TFs involved in auxin and BR signaling might act as the points of auxin–BR crosstalk, but only a few examples have been reported. One is the response of *SAUR15* to auxin and BR. *SAUR15* expression depends on the simultaneous interaction of BES1 and ARF5 within the promoter of the *SAUR15*, gene that produces a hormone with an Up at Dawn (HUD)-type E-box and AuxRE cis elements (Walcher and Nemhauser 2012). Another example is that BZR1 and ARF6 were shown to interact and regulate a large number of common target genes (Oh et al. 2014).

In this study, we characterized mutant *ds1*, which showed dwarfing and a reduced leaf angle compared to the WT. Through map-based cloning and genetic transformation, *DS1* was identified as a new allele of *OsEMF1*. In rice, *OsEMF1* functions in influencing the level of the H3K27me3 at the *OsMADS58* locus, and repressing the expression of *OsMADS58* to control palea development (Zheng et al. 2015; Yan et al. 2015). *ds1* became less sensitive to BL treatment relative to the wild type in a lamina joint inclination test. We found that the expression level of *OsBRI1* was down-regulated 5 times in the *ds1* mutant. *OsEMF1* inhibits the expression of target genes by mediating H3K27me3, which is not consistent the down-regulation of *OsBRI1* expression in the *ds1* mutant. *DS1* may influence the expression of *OsBRI1* through histone methylation or other unknown mechanisms. We demonstrated that DS1 physically interacted with OsARF11 to positively co-regulate expression of *OsBRI1*. Indeed, *DS1* and *OsARF11* showed similar expression profiles.

Following the initial discovery of the effect of auxin-BR crosstalk on lamina joint bending, several genes have now been reported to be involved (Luo et al. 2016). For instance, *LPA1* negatively regulates lamina joint bending by inhibiting BR signaling and increasing the expression of auxin transporters (Liu et al. 2016). In a recent study, the SMOS1-SMOS2/DLT complex positively regulated expression of *OsPHI-1*, and treatment of *smos* mutants with auxin and BR still showed a synergistic effect on lamina joint bending (Hirano et al. 2017). In the current study, we showed that DS1-OsARF11 is invoved in auxin-BR crosstalk in rice. It is plausible that lamina joint bending is determined by a transcriptional network consisting of DS1-OsARF11, SMOS1-SMOS2/DLT, LPA1, and other factors regulated by auxin and BR signaling.

In high-density plantings, characteristic semi-dwarf phenotypes with erect leaves are ideal phenotypes for improving grain yield (Lu et al. 2013; Wu et al. 2008). Further, identification of other downstream genes of *DS1/OsEMF1* will be needed for fully understanding the network regulating plant growth and development. In addition, the BR signaling pathway is necessary for auxin function in controlling leaf angle, indicating that the roles of auxin and BR are interdependent. Our study uncovers a previously unknown regulatory factor in plant architecture and the coordinating network of auxin and

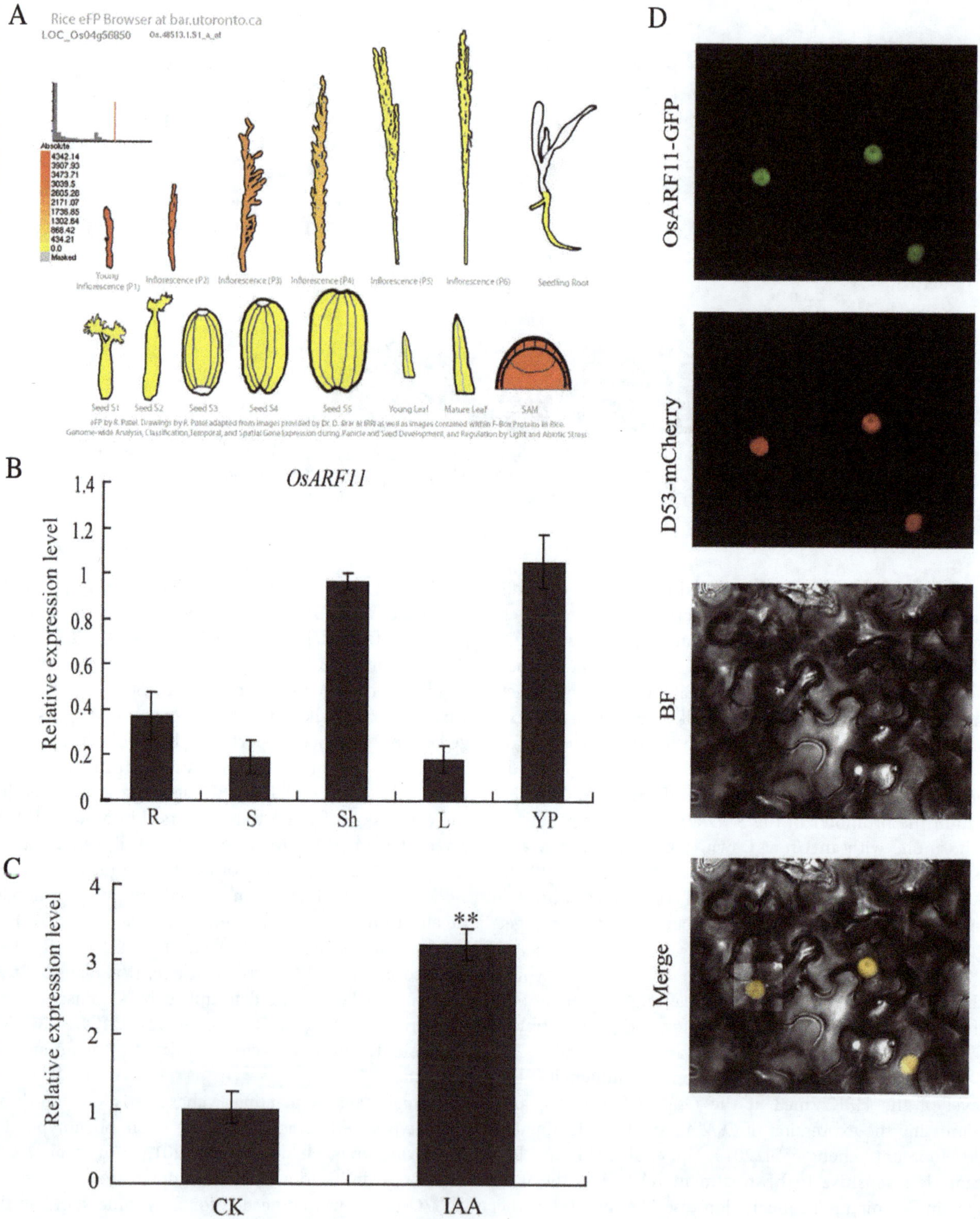

Fig. 7 Expression pattern and subcellular localization of OsARF11. **a** Expression pattern of *OsARF11* based on the Rice eFP Browser. **b** Expression pattern of *OsARF11* detected by qRT-PCR. The experiment was repeated three biological times with similar results. Error bars represent standard deviations (SD) ($n = 3$). R, root; S, stem; L, leaf; YP, young panicles; Sh, Leaf sheath. **c** *OsARF11* expression under auxin treatments. One-week-old WT seedlings were grown in solution containing 10 μM IAA for 4 h. ** indicate significant difference at $P < 0.01$. Error bars represent the SD (n = 3). **d** Subcellular localization of the OsARF11-GFP fusion protein in tobacco epidermal cells

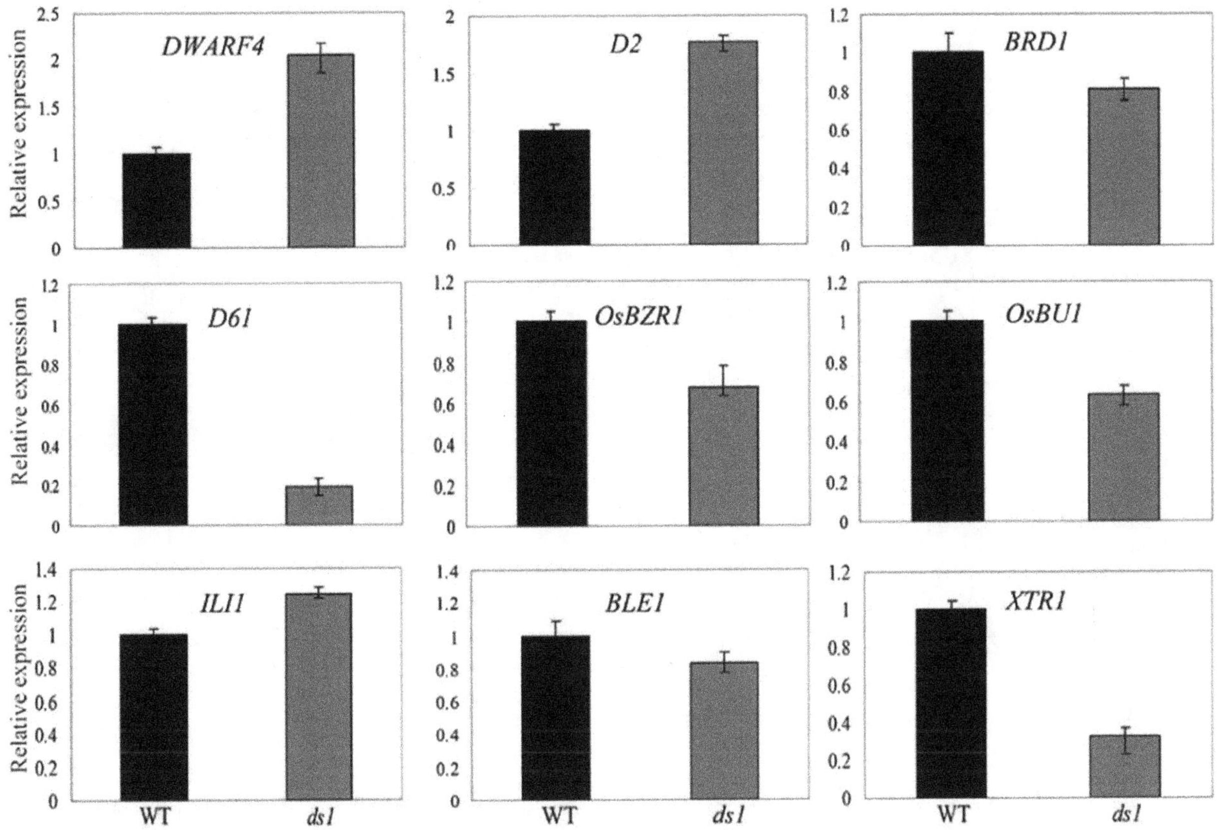

Fig. 8 Expression of BR biosynthesis and signal transduction genes. Real-time PCR analysis of BR biosynthesis and signal transduction genes in the wild type and *ds1* seedling at 15 days after germination. Data are presented as means ± SD (n = 3)

BR signaling. These findings might be useful for manipulating optimal rice architectures.

Conclusions

In this study, we found that DS1/OsEMF1 functions in BR signaling by interacting with OsARF11. The underlying mechanisms might be useful for engineering rice architecture to increase rice yield.

Methods

Plant material and growth conditions

The *ds1* mutant obtained from an irradiated population of *japonica* variety Nanjing 35 was stably inherited. The mutant was self-pollinated for several generations until shown to be heritable. Crosses were made between the *ds1* mutant and variety Nanjing 35 for genetic analysis. 93–11 (*indica*) was employed as a control in mapping of *DS1*. Pre-germinated seeds were sown in nursery beds in the field and the 1-month-old seedlings were transplanted into a paddy field in Nanjing.

Trait measurement

The grain length, width, and thickness were measured by an electronic digital display vernier caliper and fully filled grains were used for measuring the 1000-grain weight. The plant height, primary and secondary branch number, panicle length, and grain number per panicle were obtained from the measurement of the main stem. Numbers of tillers per plant were recorded at the time of harvest. Three biological repeats were measured for each sample.

Histological analysis

The second internodes and lamina joints of the flag leaves were harvested at heading date, respectively. All samples fixed in glutaric dialdehyde for 48 h. The fixed internode samples were soaked in 2% (*w*/*v*) OsO_4 for 2 h, dehydrated in a graded ethanol series (70, 80, 90, 100, and 100%), cleared in a xylene series, and then sputter coated with platinum. Second internode sections were observed by SEM (HITACHI, S-3000 N). Fixed lamina joints were embedded in paraplast. After sectioning, 10 μm thick sections were dewaxed with xylene, rehydrated, stained with 1% toluidine blue, and observed

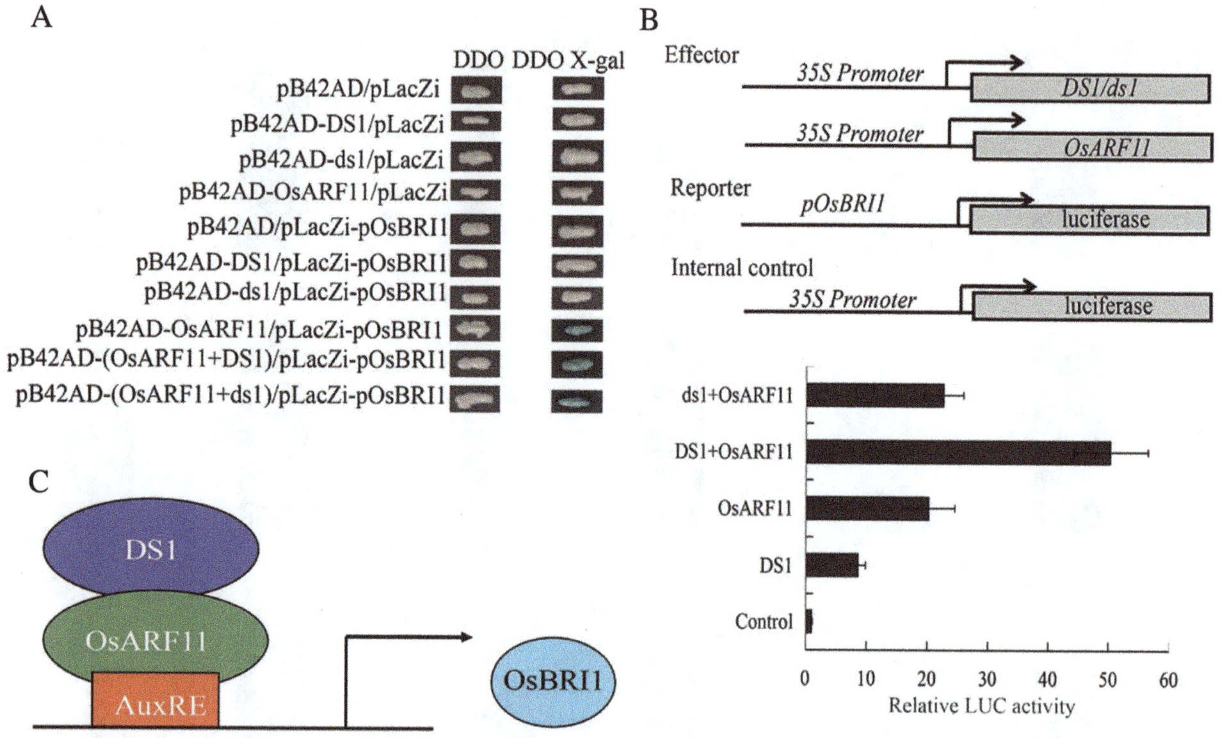

Fig. 9 DS1 interacts with OsARF11 to promote the expression of *OsBRI1*. **a** Yeast one-hybrid (Y1H) analysis of DS1, OsARF11 and *OsBRI1* promoter. The bait vector containing the *OsBRI1* promoter fragment-fused lacZ reporter gene, and the prey vector containing DS1/OsARF11-fused GAL1 activation domain were co-transformed into yeast cells (EGY48). **b** Transient expression assays of *OsBRI1* transcriptional activity modulated by DS1/ds1 and OsARF11 in rice protoplasts. Various constructs used in transient expression assays are shown in the upper panel. p*OsBRI1*:LUC was cotransformed with either the effector or empty vector (control) into rice protoplasts. Relative LUC activity indicating the level of *OsBRI1* expression activated by the effector is shown in the lower panel. Values are means ± SD of three biological replicates. **c** Model representing mechanism of action of DS1 in regulating *OsBRI1*

with a Leica DM5000B microscope. Cell lengths of each organ were measured with IMAGEJ software.

Lamina joint bending assay

Brassinolide (BL) was dissolved in ethanol. For BL treatment, sterile water was added to reach final concentrations of 0.01, 0.1, and 1 μM. Germinated seedlings of WT and *ds1* were grown in the light for 10 d at 30 °C. Lamina joint segments were floated on distilled water for 72 h containing various concentrations of brassinolide in the light. The lamina joint bending angles were then photographed and measured. Twenty plants were used for each treatment.

Genetic analysis and map-based cloning of *DS1*

Populations for genetic analysis and mapping were from crosses between *ds1* and *japonica* cv. Nanjing 35 and *indica* cv. 93–11. For mapping of *DS1*, 7680 F_2 plants were generated from a cross between *ds1* mutant and *indica* cv. 93–11, and 568 individuals with the clear *ds1* phenotypes were selected for mapping. Additional molecular markers were designed from comparative sequences of Nipponbare and 93–11. Molecular markers chosen for mapping are listed in Additional file 1: Table S2. The PCR procedure was: 95 °C for 5 min, followed by 33 cycles of 95 °C for 30 s, annealing for 30 s, 72 °C for 40 s and a final elongation step at 72 °C for 5 min. All the populations were grown in the field at Nanjing in Jiangsu province, and Lingshui County in Hainan province.

Complementation of the *ds1* mutant

To test whether the mutation in *LOC_Os01g12890* was responsible for *ds1*, an ~ 2 kb promoter and full length cDNA of *DS1* were cloned into the pCUbi1390 binary vector (primer sequences are listed in Additional file 1: Table S3). The resulting binary plasmid was then introduced into *Agrobacterium tumefaciens* strain EHA105. Calli induced from homogeneous *ds1* seeds were used for transformation. The transgenic plants were grown in a glasshouse to observe the phenotypes.

RNA extraction and qRT-PCR analysis

Total RNAs were isolated from various plant organs using Trizol reagent (Invitrogen) according to the manufacturer's instructions. First strand cDNAs were synthesized from 2 μg total RNA using a PrimeScript 1st Strand cDNA Synthesis Kit (Takara, Dalian).

For qRT-PCR, SYBR Premix *Ex* TaqTM kit (Takara) was added to the reaction system and run on an ABI Prism 7500 real-time PCR system and the *OsActin* gene was used as an internal control. The relative expression level was calculated by $2^{-\Delta\Delta Ct}$. The primers used for BR-related genes and cell wall synthesis-related genes are listed in Additional file 1: Table S3. The PCR procedure was: 95 °C for 30 s, followed by 40 cycles of 95 °C for 5 s, 60 °C for 34 s.

Yeast two-hybrid assay

For yeast two-hybrid analysis, various fragments were cloned into pGBKT7 and pGADT7 to construct BD-DS1/N/C/ and AD-OsARF11/N1/N2/C/, respectively. The reported gene assay was performed following the manufacturer's instructions (Clontech). Yeast transformation and screening procedures were performed according to the manufacturer's instructions (Clontech). The yeast cells were transformed with the rice cDNA library constructed from young panicles of the japonica cultivar "Nipponbare". Positive transformants were screened for growth on SD/–Trp/–Leu/-His/–Ade. The yeast clones were cultured and sequenced. The primers used for yeast two-hybrid analysis were listed in Additional file 1: Table S3.

BiFC assay

The full-length cDNA of DS1 was cloned into the p2YN vector to construct the DS1-YFP^N fusion protein. OsARF11 was cloned into the p2YC vector to produce an OsARF11-YFP^C fusion protein, (primer sequences are listed in Additional file 1: Table S3). BiFC analyses were performed in tobacco, as described previously (Waadt and Kudla 2008). The primers used for BiFC assay are listed in Additional file 1: Table S3.

Transient expression and yeast one-hybrid assays

To generate the p*OsBRI1*:LUC reporter construct, ~ 2 kb of the *OsBRI1* promoter was cloned into pGreenII-0800-LUC. *OsARF11* and *DS1* full length cDNA were cloned into the modified pAN580 binary vector, respectively. The construct vectors were co-transformed into rice protoplasts. Rice protoplasts were prepared, transfected, and cultured as previously described (Wang et al. 2016). The luciferase activity assay was calculated following the manufacturer's instructions (Promega) and the data presented are the averages of three biological replicates.

For yeast one-hybrid (Y1H) assays, the full length cDNA of *DS1*, *OsARF11* and promoter of *OsBRI1* were amplified using gene-specifc primers (Additional file 1: Table S3), and the amplifed products were cloned into the pB42AD vector and reporter plasmid pLacZi, respectively. Plasmids were co-transformed into yeast strain EGY48 according to the manufacturer's manual (Clontech).

Abbreviations

BiFC: Bimolecular fluorescence complementation; GFP: Green fluorescent protein; OsARF11: Auxin response factor 11; qRT-PCR: Quantitative real-time PCR; SEM: Scanning Electron Microscope; WT: Wild type; Y1H: Yeast one-hybrid

Acknowledgements

We are thankful to Dr. Chunming Wang for critical comments on the manuscript.

Funding

This research was supported by the Key Laboratory of Biology, Genetics and Breeding of Japonica Rice in Mid-lower Yangtze River, Ministry of Agriculture, P. R. China, and Jiangsu Collaborative Innovation Center for Modern Crop Production, and grants from The National Key Research and Development Program of China (2016YFD0100101–08), Jiangsu Science and Technology Development Program (BE2017368), and Agricultural Science and Technology Innovation Fund project of Jiangsu Province (CX(16)1029).

Authors' contributions

LX, YCY, JL and WJM conceived and designed the experiments. LX, YCY, MR, LJ, CPH, ZCL, ZXJ, MCL, HYS and NTL performed the experiments and analyzed the data. LSJ and TYL were responsible for material plant and field management. LX and YCY wrote the manuscript. JL and WJM revised the manuscript. All authors read and approved the manuscript.

Author information

Correspondence and requests for materials should be addressed to Ling Jiang (jiangling@njau.edu.cn) and Jianmin Wan (wanjm@njau.edu.cn).

Competing interests

The authors declare that they have no competing interests.

Author details

[1]State Key Laboratory for Crop Genetics and Germplasm Enhancement, Jiangsu Plant Gene Engineering Research Center, Nanjing Agricultural University, Nanjing 210095, China. [2]National Key Facility for Crop Gene Resources and Genetic Improvement, Institute of Crop Science, Chinese Academy of Agricultural Sciences, Beijing 100081, China.

References

Aubert D, Chen L, Moon YH, Martin D, Castle LA, Yang CH, Sung ZR (2001) EMF1, a novel protein involved in the control of shoot architecture and flowering in *Arabidopsis*. Plant Cell 13:1865–1875

Ayano M, Kani T, Kojima M, Sakakibara H, Kitaoka T, Kuroha T, Angeles-Shim RB, Kitano H, Nagai K, Ashikari M (2014) Gibberellin biosynthesis and signal transduction is essential for internode elongation in Deepwater rice. Plant Cell Environ 37:2313–2324

Calonje M, Sanchez R, Chen L, Sung ZR (2008) *EMBRYONIC FLOWER1* participates in polycomb group -mediated *AG* gene silencing in *Arabidopsis*. Plant Cell 20: 277–291

Feng ZM, Wu C, Wang C, Roh J, Zhang L, Chen J, Zhang S, Zhang H, Yang C, Hu J et al (2016) *SLG* controls grain size and leaf angle by modulating brassinosteroid homeostasis in rice. J Exp Bot 67:4241–4253

Fujisawa Y, Kato T, Ohki S, Ishikawa A, Kitano H, Sasaki T, Asahi T, Iwasaki Y (1999) Suppression of the heterotrimeric G protein causes abnormal morphology, including dwarfism, in rice. Proc Natl Acad Sci U S A 96:7575–7580

Heery DM, Kalkhoven E, Hoare S, Parker MG (1997) A signature motif in transcriptional co-activators mediates binding to nuclear receptors. Nature 387:733–736

Hirano K, Yoshida H, Aya K, Kawamura M, Hayashi M, Hobo T, Sato-Izawa K, Kitano H, Ueguchi-Tanaka M, Matsuoka M et al (2017) SMALL ORGAN SIZE 1 and SMALL ORGAN SIZE 2/ DWARF and LOW-TILLERING form a complex to integrate auxin and brassinosteroid signalling in rice. Mol Plant 10:590–604

Hong Z, Ueguchi-Tanaka M, Umemura K, Uozu S, Fujioka S, Takatsuto S, Yoshida S, Ashikari M, Kitano H, Matsuoka M (2003) A rice brassinosteroid-deficient mutant, *ebisu dwarf* (*d2*), is caused by a loss of function of a new member of cytochrome P450. Plant Cell 15:2900–2910

Ikeda M, Miura K, Aya K, Kitano H, Matsuoka M (2013) Genes offering the potential for designing yield-related traits in rice. Curr Opin Plant Biol 16:213–220

Liu JM, Park SJ, Huang J, Lee EJ, XuanYH JBI, Kumar V, Priatama RA, Raj KV, Kim SH, Min MK, Cho JH, Kim TH, Chandran AK, Jung KH, Takatsuto S, Fujioka S, Han CD (2016) Loose plant Architecture1 (LPA1) determines lamina joint bending by suppressing auxin signaling that interacts with C-22-hydroxylated and 6-deoxo brassinosteroids in rice. J Exp Bot 67:1883–1895

Lu Z, Yu H, Xiong G, Wang J, Jiao Y, Liu G, Jing Y, Meng X, Hu X, Qian Q, Fu X, Wang Y, Li J (2013) Genome-wide binding analysis of the transcription activator *ideal plant architecture 1* reveals a complex network regulating rice plant architecture. Plant Cell 25:3743–3759

Luo X, Zheng J, Huang R, Huang Y, Wang H, Jiang L, Fang X (2016) Phytohormones signaling and crosstalk regulating leaf angle in rice. Plant Cell Rep 35:2423–2433

Nakamura A, Higuchi K, Goda H, Fujiwara MT, Sawa S, Koshiba T, Shimada Y, Yoshida S (2003) Brassinolide induces *IAA5*, *IAA19*, and *DR5*, a synthetic auxin response element in *Arabidopsis*, implying a cross talk point of brassinosteroid and auxin signaling. Plant Physiol 133:1843–1853

Nemhauser JL, Mockler TC, Chory J (2004) Interdependency of brassinosteroid and auxin signaling in *Arabidopsis*. PLoS Biol 2:e258

Ning J, Zhang B, Wang N, Zhou Y, Xiong L (2011) *Increased leaf angle1*, a Raf-like MAPKKK that interacts with a nuclear protein family, regulates mechanical tissue formation in the lamina joint of rice. Plant Cell 23:4334–4347

Oh E, Zhu JY, Bai MY, Arenhart RA, Sun Y, Wang ZY (2014) Cell elongation is regulated through a central circuit of interacting transcription factors in the *Arabidopsis* hypocotyl. Elife 3:e03031

Park T, Cho H, Hwang I, Choe S (2011) Auxin stimulates *DWARF4* expression and brassinosteroid biosynthesis in Arabidopsis. Plant J 66:564–578

Sakamoto T, Morinaka Y, Inukai Y, Kitano H, Fujioka S (2013) Auxin signal transcription factor regulates expression of brassinosteroid receptor gene in rice. Plant J 73:676–688

Sakamoto T, Morinaka Y, Ohnishi T, Sunohara H, Fujioka S, Ueguchi-Tanaka M, Mizutani M, Sakata K, Takatsuto S, Yoshida S, Tanaka H, Kitano H, Matsuoka M (2006) Erect leaves caused by brassinosteroid deficiency increase biomass production and grain yield in rice. Nat Biotechnol 24:105–109

Sakamoto T, Ohnishi T, Fujioka S, Watanabe B, Mizutani M (2012) Rice CYP90D2 and CYP90D3 catalyze C-23 hydroxylation of brassinosteroids in vitro. Plant Physiol Biochem 58:220–226

Sasaki A, Ashikari M, Ueguchi-Tanaka M, Itoh H, Nishimura A, Swapan D, Ishiyama K, Saito T, Kobayashi M, Khush GS, Kitano H, Matsuoka M (2002) Green revolution: a mutant gibberellin-synthesis gene in rice. Nature 416:701–702

Shen C, Wang S, Bai Y, Wu Y, Zhang S, Chen M, Guilfoyle TJ, Wu P, Qi Y (2010) Functional analysis of the structural domain of ARF proteins in rice (*Oryza sativa* L.). J Exp Bot 61:3971–3981

Sun Y, Yang Y, Yuan Z, Muller JL, Yu C, Xu Y, Shao X, Li X, Decker EL, Reski R, Huang H (2010) Overexpression of the *Arabidopsis* gene *UPRIGHT ROSETTE* reveals a homeostatic control for indole-3-acetic acid. Plant Physiol 153:1311–1320

Torchia J, Rose DW, Inostroza J, Kamei Y, Westin S, Glass CK, Rosenfeld MG (1997) The transcriptional coactivator p/CIP binds CBP and mediates nuclear-receptor function. Nature 387:677–684

Waadt R, Kudla J (2008) In planta visualization of protein interactions using bimolecular fluorescence complementation (BiFC). CSH protoc 2008:t4995

Wada K, Marumo S, Ikekawa N, Morisaki M, Mori K (1981) Brassinolide and homobrassinolide promotion of lamina inclination of rice seedlings. Plant Cell Physiol 22:323–325

Walcher CL, Nemhauser JL (2012) Bipartite promoter element required for auxin response. Plant Physiol 158:273–282

Wang L, Xu Y, Zhang C, Ma Q, Joo SH, Kim SK, Xu Z, Chong K (2008) *OsLIC*, a novel CCCH-type zinc finger protein with transcription activation, mediates rice architecture via brassinosteroids signaling. PLoS One 3:e3521

Wang Y, Wang C, Zheng M, Lyu J, Xu Y, Li X, Niu M, Long W, Wang D, Wang H, Terzaghi W, Wang Y, Wan JM (2016) *WHITE PANICLE1*, a val-tRNA synthetase regulating chloroplast ribosome biogenesis in rice, is essential for early chloroplast development. Plant Physiol 170:2110–2123

Wu CY, Trieu A, Radhakrishnan P, Kwok SF, Harris S, Zhang K, Wang J, Wan JM, Zhai H, Takatsuto S, Matsumoto S, Fujioka S, Feldmann KA, Pennell RI (2008) Brassinosteroids regulate grain filling in rice. Plant Cell 20:2130–2145

Yamamuro C, Ihara Y, Wu X, Noguchi T, Fujioka S, Takatsuto S, Ashikari M, Kitano H, Matsuoka M (2000) Loss of function of a rice *brassinosteroid insensitive1* homolog prevents internode elongation and bending of the lamina joint. Plant Cell 12:1591–1606

Yan D, Zhang X, Zhang L, Ye S, Zeng L, Liu J, Li Q, He Z (2015) *CURVED CHIMERIC PALEA 1* encoding an *EMF1*-like protein maintains epigenetic repression of *OsMADS58* in rice Palea development. Plant J 82:12–24

Yin Y, Wang ZY, Mora-Garcia S, Li J, Yoshida S, Asami T, Chory J (2002) BES1 accumulates in the nucleus in response to brassinosteroids to regulate gene expression and promote stem elongation. Cell 109:181–191

Zhang B, Wang X, Zhao Z, Wang R, Huang X, Zhu Y, Yuan L, Wang Y, Xu X, Burlingame AL, Gao Y, Sun Y, Tang W (2016) *OsBRI1* activates BR signaling by preventing binding between the TPR and kinase domains of *OsBSK3* via phosphorylation. Plant Physiol 170:1149–1161

Zhang LY, Bai MY, Wu J, Zhu JY, Wang H, Zhang Z, Wang W, Sun Y, Zhao J, Sun X, Yang H, Xu Y, Kim SH, Fujioka S, Lin WH, Chong K, Lu T, Wang ZY (2009) Antagonistic HLH/bHLH transcription factors mediate brassinosteroid regulation of cell elongation and plant development in rice and *Arabidopsis*. Plant Cell 21:3767–3780

Zhang S, Wang S, XuY YC, Shen C, Qian Q, Geisler M, Jiang dA, Qi Y (2015a) The auxin response factor, OsARF19, controls rice leaf angles through positively regulating *OsGH3-5* and *OsBRI1*. Plant Cell Environ 38:638–654

Zhang X, Sun J, Cao X, Song X (2015b) Epigenetic mutation of *RAV6* affects leaf angle and seed size in rice. Plant Physiol 169:2118–2128

Zhao SQ, Hu J, Guo LB, Qian Q, Xue HW (2010) Rice leaf inclination2, a VIN3-like protein, regulates leaf angle through modulating cell division of collar. Cell Res 20:935–947

Zheng M, Wang Y, Wang Y, Wang C, Ren Y, Lv J, Peng C, Wu T, Liu K, Zhao S, Liu X, Guo X, Jiang L, Terzaghi W, Wan JM (2015) *DEFORMED FLORAL ORGAN1* (*DFO1*) regulates floral ORGAN identity by epigenetically repressing the expression of *OsMADS58* in rice (*Oryza sativa*). New Phytol 206:1476–1490

15

Identification and expression analysis of genes with pathogen-inducible *cis*-regulatory elements in the promoter regions in *Oryza sativa*

Weiwen Kong[1,2*], Li Ding[1], Jia Cheng[1] and Bin Wang[1]

Abstract

Background: Complex co-regulatory networks in plants may elicit responses during pathogen infections. A number of genes are activated when these responses take place. Identification of these genes would shed new light on understanding the mechanisms of rice response to pathogen infections and the elucidation of crosstalk among diverse signaling networks in rice disease resistance/susceptibility.

Results: Here we report the identification of genes with pathogen-inducible *cis*-regulatory elements (PICEs) (AS-1, G-box, GCC-box, and H-box) in the promoter regions in rice. Our results showed that a set of 882 rice genes contained these four elements in their promoter regions. Of these genes, 190 encode disease resistance/susceptibility related proteins, and 70 encode transcription factors. Analyses of the available microarray data demonstrated that 357 transcripts were differentially expressed after pathogen infections. 48 out of 53 differentially expressed transcription factors are up-regulated or down-regulated by more than 1.1-fold in response to pathogen infections. Analyses of the public mRNA-Seq data showed that 327 transcripts were differently expressed after pathogen infections. A total of 100 up-regulated genes and 37 down-regulated genes were found in common between the microarray and mRNA-Seq data.

Conclusions: We report here a set of rice genes that contain the four PICEs, i.e., AS-1, G-box, GCC-box, and H-box, in their promoter regions, of which, 53.5% were up- or down-regulated when pathogens attack. The PICEs in the gene promoters are critical for rice response to pathogen infections. They are also useful markers for identification of rice genes involved in response to pathogen infections.

Keywords: Rice, Pathogen-inducible *cis*-regulatory elements (PICEs), Disease resistance, Disease susceptibility

Background

Rice plants are exposed to a number of pathogens that may negatively affect their growth and development. To adjust to pathogen infections in their environment, rice plants have evolved complex innate immunity systems. It has been established that plants evolved pathogen-associated molecular pattern (PAMP) -triggered immunity (PTI) and effector-triggered immunity (ETI) to defend against pathogen infections (Jones and Dangl 2006; Chisholm et al. 2006). PTI is the first layer of plant defense, and it can be activated by the perception of the PAMPs through plasma membrane-localized pattern recognition receptors (PRRs) (Li et al. 2016; Macho and Zipfel 2014). Then the downstream components such as mitogen-activated protein kinase (MAPK) cascades are activated and responsible for leading to defense responses (Chisholm et al. 2006). The second layer of plant defense is ETI, which would be activated by the specific recognition of pathogen effectors through resistance (R) proteins, e.g., nucleotide-binding and leucine-rich repeat (NB-LRR) proteins, in plants. The defense response mediated by the R proteins is

* Correspondence: wwkong@yzu.edu.cn
[1]School of Horticulture and Plant Protection, Yangzhou University, Yangzhou 225009, Jiangsu, China
[2]Joint International Research Laboratory of Agriculture and Agri-Product Safety of the Ministry of Education, Yangzhou University, Yangzhou 225009, Jiangsu, China

characteristic of hypersensitive response (HR), which is usually linked to the cellular production of reactive oxygen species (ROS) (Maekawa et al. 2011). Now, it is known that multiple signaling mechanisms, such as those mediated by salicylic acid (SA), jasmonic acid (JA), and ethylene (ET), are necessary in the transduction of the signal of pathogen perception into rapid defense responses (Vlot et al. 2009; Robert-Seilaniantz et al. 2011).

Numerous genes in rice plants are rapidly or strongly activated at the transcriptional level by various pathogen attacks. Some of these genes are vital in the perception of pathogen infection (Nasir et al. 2017). Some other genes can be induced by plant hormones that include SA, JA, ET and brassinosteroid (Shigenaga and Argueso 2016; Yang et al. 2013). These plant hormones have been demonstrated as secondary signals after pathogen attack. Some genes are implicated in different primary and secondary metabolic pathways (Cho and Lee 2015). Another group of genes, named pathogenesis-related (*PR*) genes, has often been found to be induced in defense against pathogens (van Loon et al. 2006). Some genes, including transcription factor genes and *PR* genes, in other species, were observed to be strongly activated after inoculation with pathogens (Zander et al. 2014; Sohn et al. 2006; Meng et al. 2013). Similarly, previous study has shown that in incompatible interaction between rice and blast pathogen, rice genes implicated in defense were more rapidly and strongly induced at early infection stage, and this is in contrast to the genes in compatible interaction (Wang et al. 2014). It was reported that OsWRKY45 could be strongly activated after blast pathogen infection, and it could be also induced by salicylic acid and benzothiazole, leading to strongly increased resistance to *Magnisea oryzae* or *Xanthomonas oryzae* pv. *oryzae* (Shimono et al. 2007).

Many pathogen-inducible genes were found to contain specific *cis*-acting element sequences within their promoters (Yin et al. 1997; Lebel et al. 1998; Rushton and Somssich 1998). Deletion analyses of pathogen-inducible promoters have identified such sequences that are sufficient to confer responses to pathogens. Up to date, much effort has been made to identify diverse *cis*-acting elements within pathogen-inducible promoters. AS-1 (activation sequence-1, TGACG) *cis*-element was originally found in some viral and bacterial T-DNA promoters (Lam et al. 1989; Bouchez et al. 1989). The equivalent and homologous sequences are also named *ocs* or *nos* cis-regulatory elements (Ellis et al. 1993; Kim et al. 1993). This element is characterized by two TGACG motifs. G-box (with the core sequence ACGT) was first characterized in the promoters of plant light-regulated genes (Giuliano et al. 1988). Since then, as a ubiquitous *cis*-element, this *cis*-element motif was identified in the promoters in many various plant genes (Menkens et al. 1995). An H-box *cis*-element was found to be adjacent to G-box in the promoter of one soybean chalcone synthase encoding gene. H-box [CCTACC(N)$_7$CT] was first identified in the promoter of the bean *chs15* gene, which encodes chalcone synthase (Loake et al. 1992). This *cis*-element is critical for the induction of the expression of the *chs15* gene by elicitors and other stress stimuli. GCC-box (AGCCGCC) was found over-represented in the promoters of many *PRs* (Sato et al. 1996; Zhou et al. 1997; Manners et al. 1998). The *cis*-element has been shown to be responsive to ET and jasmonate, and functions after pathogen infection, followed by the induction of plant defense responses (Brown et al. 2003; Manners et al. 1998; Ohme-Takagi et al. 2000).

There are a number of *cis*-regulatory elements found in pathogen-inducible plant promoters. Some of the elements are occurring with high frequency, others with low frequency. The above four *cis*-regulatory elements have been shown present with moderate frequency in the pathogen-inducible promoters. Also, they are together involved in SA, ET, and JA signaling pathways.

Rice is one of the important food crops. Currently, the rice genome sequence is available, which provides an excellent platform to discover pathogen-inducible *cis*-regulatory elements (PICEs) within the promoters at the genome level. In this paper, we identified the AS-1, G-box, GCC-box and H-box *cis*-regulatory elements in the whole promoters in *Oryza sativa* cv. *japonica* (*O. sativa*) using in silico approaches. Analyses on the functional classification of the putative proteins encoded by the identified transcripts revealed a number of disease resistance/susceptibility proteins, transcription factors, protein kinases, and unknown/hypothetical proteins.

To test the significance of these four PICEs present in the promoter regions, we compared the transcript levels of the rice genes controlled by the promoters with these four elements after pathogen infections to those of these genes after mock inoculation. We found that the transcripts from a great number of genes controlled by the promoters with these four elements were up-regulated or down-regulated after pathogen infections, indicating that the rice genes containing these four PICEs in their promoters are responsive to pathogen infections. Our results indicated that the genes with PICEs in their promoters are important for rice response to pathogen infections.

Results

Identification of rice genes with PICEs in the promoter regions

The pathogen-inducible genes in host play an important role in disease resistance/susceptibility. To identify all the genes in *O. sativa* that contain AS-1, G-box, GCC-box and H-box *cis*-regulatory elements (Table 1)

Table 1 Pathogen-inducible *cis*-elements, sequences and interacting factors

Cis element name	Sequence	Promoter[a]	Interacting factor	Reference
AS-1	TGACG	GSTs; PRs	bZIP	(Ulmasov et al. 1994); (Yang et al. 1997); (Strompen et al. 1998); (Chen and Singh 1999)
G-box	ACGT	CHS	bZIP/bHLH/MYC	(Droge-Laser et al. 1997); (Toledo-Ortiz et al. 2003); (Boter et al. 2004)
GCC-box	AGCCGCC	PRs	AP2/ERF	(Sato et al. 1996); (Buttner and Singh 1997); (Zhou et al. 1997); (Manners et al. 1998); (Rushton et al. 2002)
H-box	CCTACC(N)$_7$CT	CHS	bZIP	(Loake et al. 1992); (Yu et al. 1993); (Droge-Laser et al. 1997)

[a]Model promoters containing the specific *cis*-elements are listed here
GSTs glutathione S-transferases, *PRs* pathogenesis-related proteins, *CHS* chalcone synthase

in their promoter regions, upstream sequences of 2 kb of 44,609 genes of *O. sativa* were downloaded from the EnsemblPlants database. The transposable elements, pseudo-genes, and noncoding RNA genes were discarded from the sequence dataset.

The above four elements were identified collectively by custom Perl scripts in upstream genomic sequences. Application of this search led to the identification of total 882 sequences with the four *cis*-regulatory elements (Fig. 1). *Cis*-regulatory elements may occur on the direct strands or on the reverse strands of the promoter sequence contexts. In this study, we identified 487 sequences with the four *cis*-regulatory elements on the direct strand of the promoters, and 407 sequences with the four *cis*-regulatory elements on the reverse strand of the promoters. We found there were 12 sequences overlap between the two groups. We defined these 882 sequences as putative pathogen-inducible promoters (PIPs) in rice. Our results showed that the number of the identified PIPs present in each rice chromosome varied, ranging from 39 to 120 (Fig. 1). Then, we identified 833 genes (encoding 882 transcripts) driven by the PIPs from the *O. sativa* genomic sequences (IRGSP-v1.0.37) (Additional file 1: Table S1).

Functional classification of identified genes driven by PIPs

The sequences and annotations of these genes were obtained from the rice genome annotation project database (RGAP, http://rice.plantbiology.msu.edu/). Also, they were annotated through the InterProScan program. There were 743 transcripts whose annotations were obtained. The identified candidate genes/transcripts were classified on the basis of their putative functions. Analysis on Gene Ontology (GO) biological process of these transcripts showed that the maximum percentage of transcripts implicated in cellular process (GO:0009987) (28.7%), followed by metabolic process (GO:0008152) (27.5%), biosynthetic process (GO:0009058) (15.9%), nucleobase, nucleoside, nucleotide and nucleic acid metabolic process (GO:0006139) (13.6%), response to stress (GO:0006950) (13.2%), and cellular protein modification process (GO:0006464) (11.8%) (Fig. 2). These data suggest that the rice response to pathogen infection may activate metabolic and biosynthetic processes, as well as stimuli response. The pathogen infection may be a stress factor that promotes rice reproductive process. Classification analysis on GO molecular function presented that the top three group transcripts were protein binding (GO:0005515) (14.1%), catalytic activity (GO:0003824) (14.0%), and binding (GO:0005488) (14.0%). Also, transcripts encoded by candidate genes driven by PIPs involved in nucleotide binding (GO:0000166), hydrolase activity (GO:0016787) (9.8%), kinase activity (GO:0016301) and transcription factor activity (GO:0003700) (8.3%) (Additional file 2: Figure S1A). According to the GO cellular component annotation, the maximum percentage of transcripts

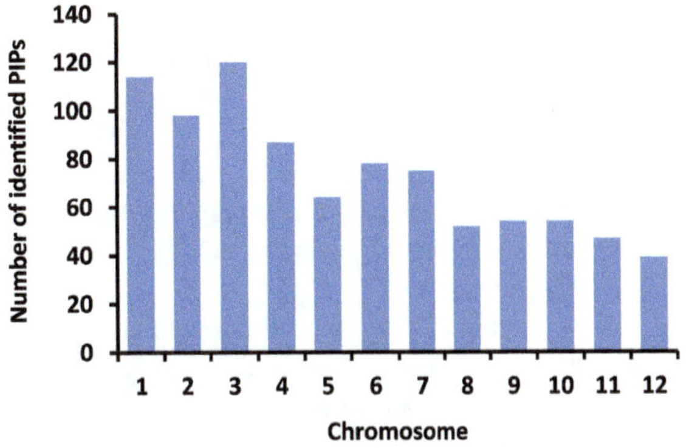

Fig. 1 Chromosomal distribution of putative rice PIPs. PIPs: Pathogen-inducible promoters

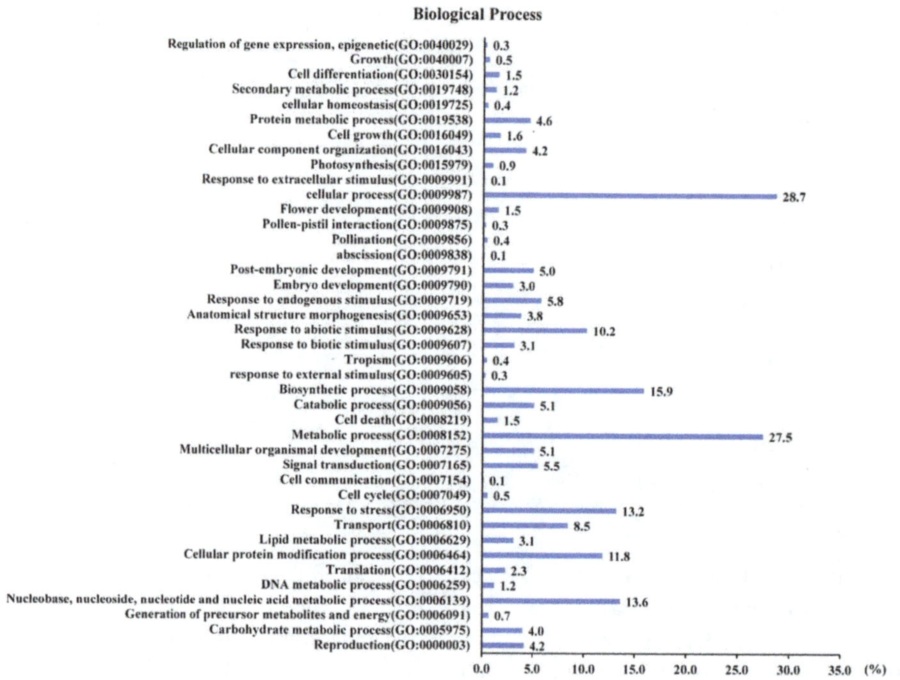

Fig. 2 Functional classification based on GO-biological process for 743 identified candidate rice genes driven by PIPs

localize to membrane (GO:0016020) (14.0%), plastid (GO:0009536) (12.1%) and plasma membrane (GO:0005886) (10.9%), followed by nucleus (GO:0005634) (10.2%) (Additional file 2: Figure S1B). Of the transcripts, 190 disease resistance/susceptibility-related (DRR/DSR) proteins and 70 transcription factors were identified through InterProScan annotation (Additional file 3: Table S2, Additional file 4: Table S3). Out of the DRR/DSR proteins, the maximum belongs to P-loop proteins, then followed by protein kinase domain proteins, serine-threonine/tyrosine-protein kinases, armadillo-type fold proteins, leucine-rich repeat domain proteins, cytochrome P450 and so on (Additional file 3: Table S2). Thus rice protein kinase, receptor and R protein encoding genes are involved in defense response to pathogen infections. It may be a rational explanation that the PIPs contribute to the activation of these genes during pathogen infections. Of the predicted transcription factors, zinc finger family proteins accounted for approximately 44% (31/70) and were the maximum group, followed by ethylene insensitive 3 or similar proteins (7%, 5/70) (Additional file 4: Table S3). Therefore, the zinc finger proteins are essential for rice response to pathogen infections, and ethylene signaling would be activated by pathogen infections.

A further functional analysis of the identified DRR/DSR proteins showed they were probably involved in many disease resistance mechanisms (Fig. 3). 16.79% (92) were involved in signaling pathways, including MAPK, SA, JA, ET, GA, ABA, etc. A group of 77 proteins (14.05%) was related in defense responses. Another group of 77 proteins (14.05%) was implicated in ETI. 12.59% (69) of the proteins were related to programmed cell death (PCD) or HR. 9.31% (51) were identified as transcription factors. 7.48% (41) were associated with degradation. Two groups of 25 proteins (4.56%) were implicated in PTI and perception of pathogen, respectively. Other proteins were involved in other than the above mechanisms, including R proteins, PRR, trichome formation, redox, and so forth (Fig. 3).

Expression analysis of genes controlled by PIPs through microarray and mRNA-Seq data

We have identified 882 rice genes with the PICEs in their promoters. To answer the question whether the PICEs lead to the transcriptional activation or repression of these genes after pathogen infections, we analyzed the microarray and mRNA-Seq data available in GEO database. We searched 38 experiments (32 from microarray data and 6 from mRNA-Seq data) querying pathogen-activated or repressed gene expression changes. Some experiments were rejected for the reason that few differentially expressed genes were obtained from these experiments. In the end, 27 experiments were employed and these experiments consisted of 21 microarray as well as six mRNA-Seq data sets, including interactions between rice and *Magnaporthe oryzae*, *Rhizoctonia solani*, *Xanthomonas oryzae* pv. *oryzae*, *Meloidogyne incognita*, and *Meloidogyne graminicola* (Additional file 5: Table S4).

In the microarray data analyses, a total of 357 differentially expressed transcripts were identified. Of these, 209 transcripts were detected to be up-regulated, and 148

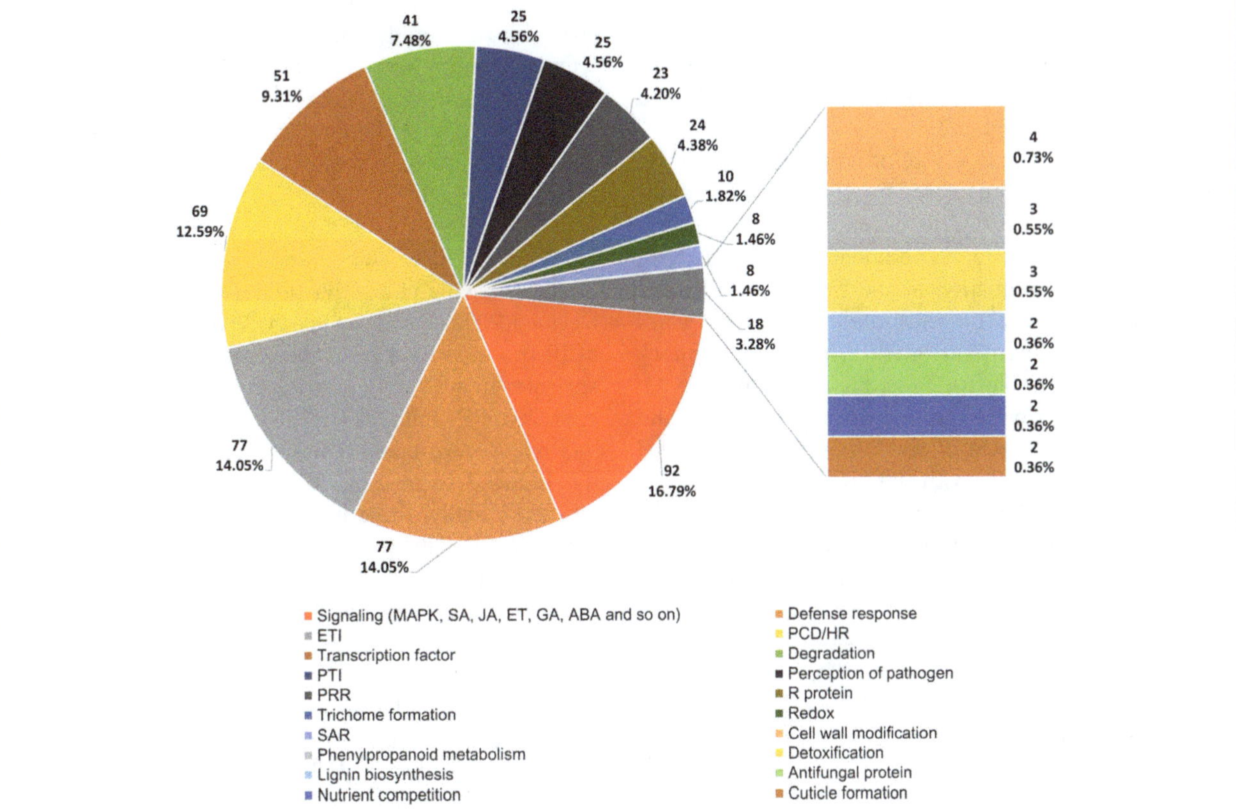

Fig. 3 Analysis on disease resistance mechanisms of the identified DRR/DSR proteins encoded by the rice genes under the control of PIPs

were down-regulated (Additional file 6: Table S5). Analyses of mRNA-Seq data detected 327 differentially expressed transcripts. Out of these, 225 transcripts were shown to be up-regulated, and 102 were down-regulated (Additional file 7: Table S6). It is notable that 212 differentially expressed transcripts were identified in common from the microarray and mRNA-Seq data (Additional file 8: Table S7). However, the expression patterns of some of the transcripts, e.g., Os01t0229200–01 and Os01t0628000 –01, are not consistent between these two experiments. This inconsistency may be caused by different conditions between the microarray and the mRNA-Seq experiments. The PICEs within the 2000 bp upstream regions of the genes encoding the 212 transcripts were highlighted (Additional file 9: Figure S2). A total of 100 up- and 37 down-regulated transcripts were discovered to be overlapped between the microarray and mRNA-Seq data (Fig. 4).

The predicted 70 transcription factors were separately checked for relative fold change in transcription during

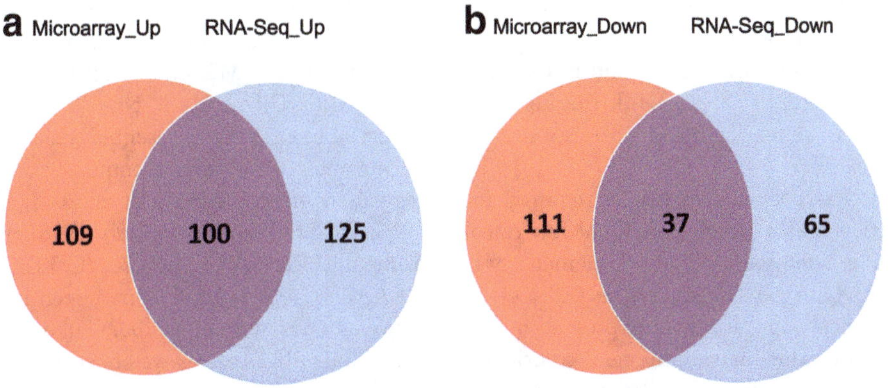

Fig. 4 Overlap of differentially expressed genes between rice microarray and RNA-Seq data. (A) Overlap of up-regulated genes between rice microarray and RNA-Seq data. (B) Overlap of down-regulated genes between rice microarray and RNA-Seq data

pathogen infections. The results showed that 53 transcription factors were differentially expressed, and a total of 90% (48/53) predicted candidate transcription factors were expressed more than 1.1 fold, by induction with pathogen infections (Table 2).

We randomly selected a pool of 50 genes without the four PICEs in their promoters from different chromosomes as random controls to examine their expression levels in these microarray experiments. In our analysis, no significant differentiation at the transcriptional level was found for most of the random control genes between the pathogen infected samples and mock samples (Additional file 10: Table S8, Additional file 11: Table S9). We performed an enrichment analysis, and found a higher incidence of the differentially expressed rice genes driven by PIPs with the PICEs in the total rice genes driven by PIPs than that of the differentially expressed rice genes in the total rice genes (Fig. 5; Additional file 12: Table S10, Additional file 13: Table S11). Thus there is a strong association between the PICEs and the differential expression of rice genes driven by PIPs after pathogen attack. Therefore, we infer that the PICEs in the promoters led to the differential expression of the genes under the control of PIPs after pathogen infection.

Discussion

In this research, we performed a promoterome-wide search for AS-1, G-box, GCC-box and H-box *cis*-elements in rice and identified 833 genes, which potentially encode 882 putative transcripts under the control of pathogen-inducible promoters with these four *cis*-elements. Of the proteins encoded by these transcripts, DRR/DSR proteins account for 21.5% (190/882), and transcription factors constitute 7.9% (70/882). More than half (472/882) of the transcripts, encoded by the genes with PICEs in their promoter regions, were attested to be pathogen-inducible by the analyses of the available microarray and mRNA-Seq data. It can be expected that more of the rice genes, with the PICEs in the promoter regions, would be demonstrated to be pathogen-inducible if more analyses are carried out on the increasing microarray and mRNA-Seq data. Therefore, the PICEs are useful markers for identification of rice genes involved in response to pathogen infections.

AS-1 *cis*-element has been shown to be required for transcriptional activation of some genes. A functional plant *ocs*-element, a homologue of AS-1 element, was identified in the soybean *Gmhsp26-A* gene (Ellis et al. 1993). AS-1 *cis*-element was also found present in the promoters of some SA-inducible glutathione S-transferase genes (*GSTs*), which are considered as defense genes and can be activated by SA, auxins, methyl jasmonate, hydrogen peroxide and pathogens (Ulmasov et al. 1994; Chen and Singh 1999; Yang et al. 1997; Xiang et al. 1996; Dudler et al. 1991). In addition, some pathogenesis-related protein encoding genes (*PRs*) were also found to contain AS-1 *cis*-elements in their promoters (Strompen et al. 1998). As has been demonstrated, the AS-1 *cis* element is an oxidative stress-responsive element and can be activated by SA with the help of oxidative species (Garretón et al. 2002). In plants, it can be bound by the TGA family of the basic/Leu zipper transcription factors (bZIPs) (Table 1) (Xiang et al. 1997; Johnson et al. 2001; Kim and Delaney 2002; Sarkar et al. 2017; Després et al. 2000). bZIP proteins have been demonstrated to play important roles in activating a number of defense genes (Alves et al. 2013; E et al. 2014).

G-boxes have been demonstrated to be implicated in the induced expression of some genes after pathogen attacks. Usually, these boxes play roles in concert with other *cis*-regulatory elements. G-box and H-box have been revealed to shape in combination the complex patterns of the bean *chs15* gene expression (Loake et al. 1992). These two boxes are also implicated in pathogen induction and other defense genes. Two proteins, named KAP-1 and KAP-2 respectively, were identified to bind to H-boxes in the promoter of the bean *chs15* gene (Yu et al. 1993). G-box and H-box could be bound by a bZIP protein, designated G/HBF-1, for activation of initial plant defense against pathogen attack (Droge-Laser et al. 1997). G-box also might be bound by the basic/helix-loop-helix (bHLH) proteins or the bHLH-leu zipper MYC proteins (Table 1) (Toledo-Ortiz et al. 2003; Boter et al. 2004).

ERFs (ethylene-responsive element binding proteins, i.e., ethylene response factors), belonging to APETALA2 (AP2) /ERF family, are ethylene-inducible DNA binding proteins that could recognize and bind the GCC-box element (Table 1) (Ohme-Takagi and Shinshi 1995; Brown et al. 2003; Ohme-Takagi et al. 2000). In addition, AP2/ERFs have been also revealed to modulate the expression levels of jasmonate-responsive genes through the interaction with the GCC-box (Brown et al. 2003).

Other than the above four *cis*-elements, GT-1 (GAAA AA; GGTTAA), MRE [A(A/C)C(A/T)A(A/C)C; ACC(A/T)A(A/C)(T/C)], and W-box (TTGACY), which were observed to be over-represented within the rice promoters in our analysis, are also common pathogen-inducible *cis*-regulatory elements (Park et al. 2004; Gurr and Rushton 2005; Rushton and Somssich 1998). Elicitor-responsive element (ERE) [AATTGACC; (C/T)TGAC(C/T); GTCA GAAAGTCAG] is another pathogen-inducible *cis*-regulatory element (Heise et al. 2002; Fukuda and Shinshi 1994). These four PICEs are as well occurring within the 2000 bp upstream sequences that match the 212 differentially expressed transcripts captured from the microarray and RNA-Seq data analyses (Additional file 9: Figure S2).

Table 2 The differentially expressed transcription factors encoded by the rice genes under the control of PIPs

InterPro annotation	Transcript-ID	Fold change[a]	adj. *P*.val/FDR[b]
Zinc finger	Os01t0123700–00	1.47	1.15E-02
	Os01t0960500–01	−1.15	2.59E-02
	Os02t0735900–01	1.4	0
	Os06t0192800–01	8.98	5.88E-03
	Os06t0687200–01	18.2	9.98E-03
	Os09t0419500–02	1.25	4.11E-02
	Os09t0468800–00	1.4	1.00E-03
	Os09t0535500–01	1.13	3.74E-02
	Os11t0572500–01	−1.8	3.96E-04
	Os03t0786400–01	1.16	1.19E-02
	Os03t0838800–00	1.26	3.36E-03
	Os10t0162300–01	−27.77	3.87E-03
	Os12t0583700–00	1.61	2.77E-02
	Os02t0727300–02	−2	0
	Os04t0686000–01	1.74	1.48E-02
	Os10t0467800–01	−1.13	3.54E-03
	Os04t0629100–01	4.28	4.67E-02
	Os04t0629100–02	4.28	4.67E-02
	Os03t0639600–01	2.5	0
	Os08t0159500–01	−6.19	2.52E-02
	Os06t0677700–00	−1.91	4.32E-04
	Os08t0491700–00	−3.36	1.40E-02
	Os09t0385700–01	15.9	0
	Os02t0724000–01	17.84	1.85E-03
	Os11t0292050–00	1.73	4.55E-02
	Os03t0619151–00	−1.77	3.46E-02
Myb/SANT	Os04t0665600–01	−1.19	2.53E-02
	Os05t0553400–01	1.99	3.24E-02
	Os03t0141100–01	−1.46	1.71E-06
	Os04t0463600–01	−1.98	2.87E-02
	Os02t0648300–01	1.57	1.97E-02
EIN3	Os03t0324300–02	1.4	0
	Os03t0324300–03	1.4	0
	Os03t0324300–04	1.4	0
	Os03t0324300–06	1.4	0
HLH	Os01t0195801–00	−1.47	4.59E-02
	Os02t0795800–00	− 1.16	3.87E-02
	Os09t0468700–00	−2.46	4.70E-02
HTH	Os03t0745000–01	−1.24	3.80E-04
	Os03t0745000–02	−1.24	3.80E-04
	Os09t0526600–03	1.5	0
AP2/ERF	Os02t0747600–01	−1.66	0
	Os05t0497200–01	1.47	1.74E-03

Table 2 The differentially expressed transcription factors encoded by the rice genes under the control of PIPs *(Continued)*

InterPro annotation	Transcript-ID	Fold change[a]	adj. *P*.val/FDR[b]
bZIP	Os02t0175100–01	1.15	2.50E-02
GRAS	Os07t0567700–03	−1.96	9.36E-03
NAM	Os03t0327100–01	1.68	1.17E-02
TFIID	Os03t0603300–01	1.6	1.00E-03
WRKY	Os05t0567200–00	1.17	4.24E-02

[a]Minus indicates down-regulation
[b]The values in blue are FDR (False Discovery Rate) measured in the RNA-Seq data analysis

It is well established on the basis of various experiments that promoters with particular *cis*-elements respond to specific triggers. The above PICEs are the target binding sites for specific transcription factors, such as bZIPs/bHLHs/MYC, AP2/ERFs, MYB (with recognition of MRE), and WRKY (interacting with W-box), which direct pathogen-inducible expression. We think that the combinatorial interactions of these PICEs in the rice promoters with their corresponding transcription factors are pivotal processes regulating spatio-temporal expressions of rice *DRR/DSR* genes during pathogen attacks. In other words, these genes respond to pathogen attacks through specific transcription factors which interact with the PICEs present in their promoters (Fig. 6).

Rice genes with the PICEs in their promoter regions are likely implicated in multiple defense mechanisms. A great number of the identified DRR/DSR proteins are protein kinase, protein kinase-like and phosphatase, which are potentially involved in signaling pathways, including mitogen-activated protein kinases. Many genes, e.g., encoding F-box-like domain proteins and ubiquitin-like domain proteins, are related to degradation. Some proteins, e.g., serine/threonine kinases, and NB-ARC domain proteins are likely involved in ETI. The leucine-rich repeat domain proteins are probably relevant to the perception of pathogen and PTI. There are some proteins functioning in defense response (e.g. cytochrome P450, thaumatin, chalcone/stilbene synthase). A lot of proteins belong to transcription factors, e.g., zinc finger proteins, heat shock transcription factors, bZIP proteins, and bHLH domain proteins. Several proteins, e.g., SKP1/BTB/POZ domain proteins, homeobox domain proteins, and ERF domain proteins, may be involved in SA, JA and ET signaling pathways, respectively. Other proteins, e.g., thioredoxin-like proteins, WAT1-related protein, and Barwin domain proteins, participate in redox reactions, cell wall modification, and HR, respectively. In our analysis, there are 17.1% (151/882) of the transcripts with unknown function. The genes encoding these transcripts contain PICEs in their promoter regions, indicating some unknown response mechanisms of rice to pathogen infections remain to be revealed. Our study sheds

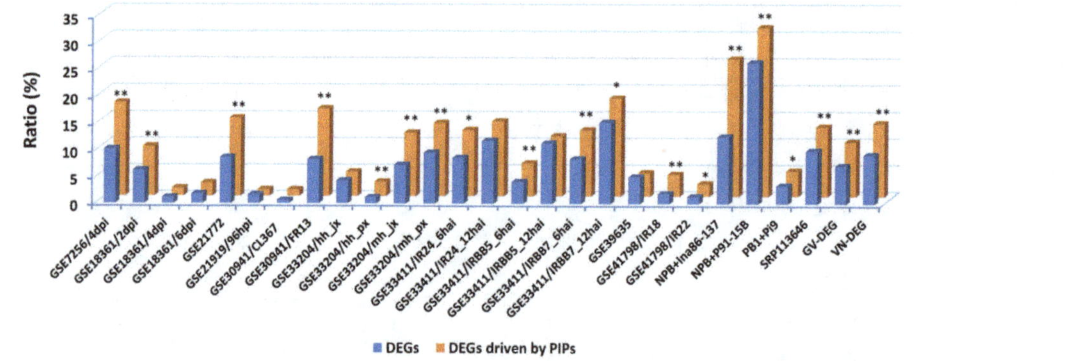

Fig. 5 The enrichment analysis of the differentially expressed rice genes driven by PIPs. *$P < 0.05$, **$P < 0.01$ by post hoc Fisher's exact test comparing each pair of data. The numbers of total reference genes were listed in Additional file/Table S10. *DEGs* differentially expressed genes

new light on rice molecular response mechanisms to pathogen infections.

The genes with the PICEs in the promoter regions in rice are likely involved in more than one regulation mechanism concerning disease resistance. All the rice genes identified here contain the PICEs in their promoter regions, they, therefore, can be potentially targeted by both bZIP and ERF transcription factors. It has been shown that bZIP and ERF transcription factors are implicated in SA, JA, and ET signaling pathways. So, the identified rice genes are likely to be activated through the hormone signaling pathways after pathogen infections. It seems that cross-coupling occurs between bZIP and ERF transcription factors and may, therefore, be of great importance in the modulation of the expression levels of the genes during the rice response to pathogen infections.

Identification and investigation of rice PIPs are valuable and useful in many aspects. Genetic engineering is an economical and convenient approach to improving rice disease resistance. Accurate regulation of exogenous gene expression is paramount to the engineering of rice plants with increased disease resistance. Many previous attempts to increase disease resistance employed constitutive over-expression of transgenes but again and again this caused poor quality or yield plants. It is now known that the poor quality or yield is attributed to the extensive cellular reprogramming connected with defence once unregulated defence reactions are triggered in uninfected cells. Therefore, compared with the constitutive expression of exogenous genes, gene expression under the control of PIPs is preferred in generating transgenic rice plants with resistance to pathogen attack, for this avoids

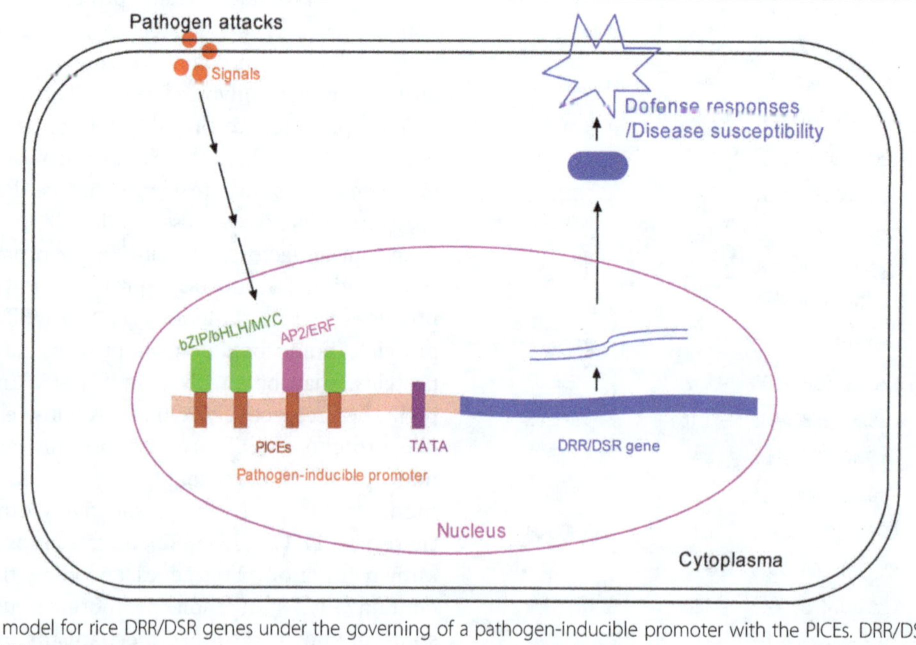

Fig. 6 A working model for rice DRR/DSR genes under the governing of a pathogen-inducible promoter with the PICEs. DRR/DSR gene: disease resistance/susceptibility-related gene

the additional cost of resistance through restricting the expression to infected cells when needed. Therefore, it is useful to search promoters inducible by pathogens for rice disease resistance engineering. Our results would be a benefit for developing rice disease resistance engineering. The defining of a set of rice genes that are putatively regulated by PIPs will help elucidate the mechanisms by which rice plants respond to pathogen infections. It provides markers useful for subsequent study and will enable the elucidation of crosstalk among diverse signaling networks in rice disease resistance/susceptibility.

Conclusions

A set of rice genes under the control of PIPs that contain the PICEs, AS-1, G-box, GCC-box, and H-box, were identified. More than half of the genes were activated or repressed after pathogen infection. The identified rice genes may be implicated in diverse disease resistance mechanisms. The PICEs in the promoters are important for rice response to pathogen infections. They are also valuable tags for identification of rice genes in response to pathogen infections.

Methods

Retrieval of upstream sequences of *O. sativa* genes

For identification of genes with PICEs in the promoter regions, upstream sequences of 2 kb of 44,609 genes of *O. sativa* were downloaded from the Rice Annotation Project database (Os-Nipponbare-Reference-IRGSP-1.0) (https://rapdb.dna.affrc.go.jp/).

Identification of PIPs with PICEs and genes controlled by PIPs

The putative PIPs with AS-1, G-box, GCC-box and H-box *cis*-regulatory elements were screened out from the upstream sequences of 2 kb of 44,609 rice genes by custom Perl scripts. On the basis of identification of the putative PIPs from the rice upstream sequences, we further screened out the genes (or transcripts) controlled by these PIPs with PICEs from the whole rice genes.

Functional annotation for genes controlled by PIPs

To infer the molecular functions of these transcripts, which genes were controlled by PIPs, we performed homology searches against the protein databases (Panther-12.0 (Mi et al. 2017), PfamA-31.0, PRINTS-42.0, ProDom-2006.1, SuperFamily-17.5, and TIGRFAM15.0) using the InterProScan ver. 5.25 program (Jones et al. 2014; Finn et al. 2017). To assign further functional information to these transcripts, we conducted a search using the Rice Annotation Project database, and then gene ontology terms were obtained and assigned to each identified transcript.

Expression analysis of genes controlled by PIPs

The rice microarray data on pathogen inoculation vs mock were extracted from GEO (Gene Expression Omnibus) database of NCBI (https://www.ncbi.nlm.nih.gov/geo/). GEO2R, an interactive GEO-based web tool was used for the identification of significantly regulated genes across pathogen vs mock (GEO accession numbers: GSE7256, GSE18361, GSE21772, GSE21919, GSE30941, GSE33204, GSE33411, GSE39635, and GSE41798). We entered these GEO accession numbers and then defined groups as inoculation and mock with given number of replicates discovered in GEO. The false discovery rate method (Benjamini and Hochberg 1995) was applied to correct for multiple testing.

mRNA-Seq data (GSE81906: GSM2177582, GSM 2177583, GSM2177584, GSM2177585, GSM2177586, and GSM2177587) on pathogen inoculation vs mock in SRA format were downloaded from the GEO Database and processed as follows using a combination of publicly available tools and custom Perl scripts. Quality was assessed using FastQC. Sequences were aligned onto the latest *O. sativa* genome assembly (released 37) using TopHat version 2.1.1(Langmead et al. 2009). Reads that map to a unique location in the *O. sativa* genome were adopted for differential expression analysis. The number of single mapping reads that locate each annotated gene in the *O. sativa* gene model annotations release-37 was counted by HTSeq-Count (version 0.9.1). Then, these data were provided as inputs to DESeq (R package, version 1.26.0) for statistical analysis of differential gene expression (Anders and Huber 2010).

Three studies presented complete information on differential gene expression. Therefore, some differentially expressed genes were obtained from these published data (Bagnaresi et al. 2012; Kawahara et al. 2012; Zhang et al. 2017).

Additional files

Additional file 1: Table S1. Chromosomal distribution of the rice genes controlled by putative PIPs.

Additional file 2 Figure S1. Annotation of the identified rice transcripts driven by the PIPs. (A) Annotation of Go molecular function of the identified rice transcripts driven by the PIPs. (B) Annotation of Go cellular component of the identified rice transcripts driven by the PIPs.

Additional file 3: Table S2. InterPro annotation for 190 putative disease resistance/susceptibility-related protein-encoding transcripts (119 groups) encoded by the rice genes under the control of PIPs.

Additional file 4: Table S3. The putative transcription factors encoded by the rice genes under the control of PIPs.

Additional file 5: Table S4. Pathogen-infected rice microarray and RNA-Seq data used in this study.

Additional file 6: Table S5. Differentially expressed transcripts driven by PIPs with the PICEs identified from the rice microarray data analysis.

Additional file 7: Table S6. Differentially expressed transcripts driven by PIPs with the PICEs identified from the rice RNA-Seq data analysis.

Additional file 8: Table S7. Differentially expressed transcripts in common from the rice microarray and RNA-Seq data analyses.

Additional file 9: Figure S2. The upstream sequences and putative pathogen-inducible *cis*-elements (PICEs) in rice. This document-like figure contains 212 upstream sequences (2000 bp) that correspond to the 212 differentially expressed transcripts obtained from the microarray and mRNA-Seq data analyses. The PICEs (AS-1, G-box, GCC-box and H-box) are highlighted in yellow. W-box, GT1, ERE (elicitor-responsive element) and MRE *cis*-elements are highlighted in green, cyan, magenta and red, respectively.

Additional file 10: Table S8. Differentially expressed genes without the PICEs in the promoters identified from the rice microarray data analysis.

Additional file 11: Table S9. Differentially expressed genes without the PICEs in the promoters identified from the rice RNA-Seq data analysis.

Additional file 12: Table S10. The proportion of the differentially expressed genes in the total rice genes.

Additional file 13: Table S11. The enrichment analysis of the differentially expressed rice genes driven by PIPs through Fisher's exact test comparing each pair of the proportions of the differentially expressed genes in the total genes from Additional file 8: Table S10.

Abbreviations
AS-1: activation sequence-1; bZIPs: the basic/Leu zipper transcription factors; DRR: disease resistance-related; DSR: disease susceptibility-related; ERFs: ethylene-responsive element binding proteins; ET: ethylene; ETI: effector-triggered immunity; GEO: Gene Expression Omnibus; GSTs: glutathione S-transferases; HR: hypersensitive response; JA: jasmonic acid; MAPK: mitogen-activated protein kinase; NB-LRR: nucleotide-binding and leucine-rich repeat; PAMP: pathogen-associated molecular pattern; PICEs: pathogen-inducible *cis* regulatory elements; PR: pathogenesis-related; PRRs: pattern recognition receptors; PTI: PAMP-triggered immunity; R protein: resistance protein; ROS: reactive oxygen species; SA: salicylic acid

Acknowledgements
This work was supported by a grant from the National Natural Science Foundation of China (30300207).

Authors' contributions
WK conceived and supervised the design of this study. LD, JC and BW analyzed the data. WK, LD and JC discussed the data. WK wrote the manuscript. All authors read and approved the final version of the manuscript.

Competing interests
The authors declare that they have no competing interests.

References

Alves MS, Dadalto SP, Goncalves AB, De Souza GB, Barros VA, Fietto LG (2013) Plant bZIP transcription factors responsive to pathogens: a review. Int J Mol Sci 14(4):7815–7828. https://doi.org/10.3390/ijms14047815

Anders S, Huber W (2010) Differential expression analysis for sequence count data. Genome Biol 11(10):R106. https://doi.org/10.1186/gb-2010-11-10-r106

Bagnaresi P, Biselli C, Orru L, Urso S, Crispino L, Abbruscato P, Piffanelli P, Lupotto E, Cattivelli L, Vale G (2012) Comparative transcriptome profiling of the early response to *Magnaporthe oryzae* in durable resistant vs susceptible rice (*Oryza sativa* L.) genotypes. PLoS One 7(12):e51609. https://doi.org/10.1371/journal.pone.0051609

Benjamini Y, Hochberg Y (1995) Controlling the false discovery rate: a practical and powerful approach to multiple testing. J R Stat Soc Ser B Methodol 57(1):289–300

Boter M, Ruiz-Rivero O, Abdeen A, Prat S (2004) Conserved MYC transcription factors play a key role in jasmonate signaling both in tomato and *Arabidopsis*. Genes Dev 18(13):1577–1591. https://doi.org/10.1101/gad.297704

Bouchez D, Tokuhisa JG, Llewellyn DJ, Dennis ES, Ellis JG (1989) The *ocs*-element is a component of the promoters of several T-DNA and plant viral genes. EMBO J 8(13):4197–4204

Brown RL, Kazan K, McGrath KC, Maclean DJ, Manners JM (2003) A role for the GCC-box in jasmonate-mediated activation of the PDF1.2 gene of *Arabidopsis*. Plant Physiol 132(2):1020–1032. https://doi.org/10.1104/pp.102.017814

Buttner M, Singh KB (1997) *Arabidopsis thaliana* ethylene-responsive element binding protein (AtEBP), an ethylene-inducible, GCC box DNA-binding protein interacts with an *ocs* element binding protein. Proc Natl Acad Sci U S A 94(11):5961–5966

Chen W, Singh KB (1999) The auxin, hydrogen peroxide and salicylic acid induced expression of the *Arabidopsis* GST6 promoter is mediated in part by an *ocs* element. Plant J 19(6):667–677. https://doi.org/10.1046/j.1365-313x.1999.00560.x

Chisholm ST, Coaker G, Day B, Staskawicz BJ (2006) Host-microbe interactions: shaping the evolution of the plant immune response. Cell 124(4):803–814. https://doi.org/10.1016/j.cell.2006.02.008

Cho MH, Lee SW (2015) Phenolic phytoalexins in rice: biological functions and biosynthesis. Int J Mol Sci 16(12):29120–29133. https://doi.org/10.3390/ijms161226152

Després C, DeLong C, Glaze S, Liu E, Fobert PR (2000) The *Arabidopsis* NPR1/NIM1 protein enhances the DNA binding activity of a subgroup of the TGA family of bZIP transcription factors. Plant Cell 12(2):279–290. https://doi.org/10.1105/tpc.12.2.279

Droge-Laser W, Kaiser A, Lindsay WP, Halkier BA, Loake GJ, Doerner P, Dixon RA, Lamb C (1997) Rapid stimulation of a soybean protein-serine kinase that phosphorylates a novel bZIP DNA-binding protein, G/HBF-1, during the induction of early transcription-dependent defenses. EMBO J 16(4):726–738. https://doi.org/10.1093/emboj/16.4.726

Dudler R, Hertig C, Rebmann G, Bull J, Mauch F (1991) A pathogen-induced wheat gene encodes a protein homologous to glutathione-S-transferases. Mol Plant-Microbe Interact 4(1):14–18

E ZG, Zhang YP, Zhou JH, Wang L (2014) Mini review roles of the bZIP gene family in rice. Genet Mol Res 13(2):3025–3036. https://doi.org/10.4238/2014.April.16.11

Ellis JG, Tokuhisa JG, Llewellyn DJ, Bouchez D, Singh K, Dennis ES, Peacock WJ (1993) Does the *ocs*-element occur as a functional component of the promoters of plant genes? Plant J 4(3):433–443. https://doi.org/10.1046/j.1365-313X.1993.04030433.x

Finn RD, Attwood TK, Babbitt PC, Bateman A, Bork P, Bridge AJ, Chang H-Y, Dosztányi Z, El-Gebali S, Fraser M, Gough J, Haft D, Holliday GL, Huang H, Huang X, Letunic I, Lopez R, Lu S, Marchler-Bauer A, Mi H, Mistry J, Natale DA, Necci M, Nuka G, Orengo CA, Park Y, Pesseat S, Piovesan D, Potter SC, Rawlings ND, Redaschi N, Richardson L, Rivoire C, Sangrador-Vegas A, Sigrist C, Sillitoe I, Smithers B, Squizzato S, Sutton G, Thanki N, Thomas PD, Tosatto Silvio CE, Wu CH, Xenarios I, Yeh L-S, Young S-Y, Mitchell AL (2017) InterPro in 2017—beyond protein family and domain annotations. Nucleic Acids Res 45(D1):D190–D199. https://doi.org/10.1093/nar/gkw1107

Fukuda Y, Shinshi H (1994) Characterization of a novel *cis*-acting element that is responsive to a fungal elicitor in the promoter of a tobacco class I chitinase gene. Plant Mol Biol 24(3):485–493

Garretón V, Carpinelli J, Jordana X, Holuigue L (2002) The *as-1* promoter element is an oxidative stress-responsive element and salicylic acid activates it via oxidative species. Plant Physiol 130(3):1516–1526. https://doi.org/10.1104/pp.009886

Giuliano G, Pichersky E, Malik VS, Timko MP, Scolnik PA, Cashmore AR (1988) An evolutionarily conserved protein binding sequence upstream of a plant light-regulated gene. Proc Natl Acad Sci U S A 85(19):7089–7093

Gurr SJ, Rushton PJ (2005) Engineering plants with increased disease resistance: how are we going to express it? Trends Biotechnol 6(23):283–290

Heise A, Lippok B, Kirsch C, Hahlbrock K (2002) Two immediate-early pathogen-responsive members of the AtCMPG gene family in Arabidopsis thaliana and the W-box-containing elicitor-response element of AtCMPG1. Proc Natl Acad Sci U S A 99(13):9049–9054. https://doi.org/10.1073/pnas.132277699

Johnson C, Boden E, Desai M, Pascuzzi P, Arias J (2001) In vivo target promoter-binding activities of a xenobiotic stress-activated TGA factor. Plant J 28(2):237–243

Jones JD, Dangl JL (2006) The plant immune system. Nature 444(7117):323–329. https://doi.org/10.1038/nature05286

Jones P, Binns D, Chang H-Y, Fraser M, Li W, McAnulla C, McWilliam H, Maslen J, Mitchell A, Nuka G, Pesseat S, Quinn AF, Sangrador-Vegas A, Scheremetjew M, Yong S-Y, Lopez R, Hunter S (2014) InterProScan 5: genome-scale protein function classification. Bioinformatics 30(9):1236–1240. https://doi.org/10.1093/bioinformatics/btu031

Kawahara Y, Oono Y, Kanamori H, Matsumoto T, Itoh T, Minami E (2012) Simultaneous RNA-seq analysis of a mixed transcriptome of rice and blast fungus interaction. PLoS One 7(11):e49423. https://doi.org/10.1371/journal.pone.0049423

Kim HS, Delaney TP (2002) Over-expression of TGA5, which encodes a bZIP transcription factor that interacts with NIM1/NPR1, confers SAR-independent resistance in *Arabidopsis thaliana* to *Peronospora parasitica*. Plant J 32(2):151–163

Kim S, Kim Y, An G (1993) Identification of methyl jasmonate and salicylic acid response elements from the nopaline synthase (nos) promoter. Plant Physiol 103(1):97–103

Lam E, Benfey PN, Gilmartin PM, Fang RX, Chua NH (1989) Site-specific mutations alter in vitro factor binding and change promoter expression pattern in transgenic plants. Proc Natl Acad Sci U S A 86(20):7890–7894

Langmead B, Trapnell C, Pop M, Salzberg SL (2009) Ultrafast and memory-efficient alignment of short DNA sequences to the human genome. Genome Biol 10(3):R25. https://doi.org/10.1186/gb-2009-10-3-r25

Lebel E, Heifetz P, Thorne L, Uknes S, Ryals J, Ward E (1998) Functional analysis of regulatory sequences controlling *PR-1* gene expression in *Arabidopsis*. Plant J 16(2):223–233. https://doi.org/10.1046/j.1365-313x.1998.00288.x

Li B, Meng X, Shan L, He P (2016) Transcriptional regulation of pattern-triggered immunity in plants. Cell Host Microbe 19(5):641–650. https://doi.org/10.1016/j.chom.2016.04.011

Loake GJ, Faktor O, Lamb CJ, Dixon RA (1992) Combination of H-box [CCTA CC(N)7CT] and G-box (CACGTG) cis elements is necessary for feed-forward stimulation of a chalcone synthase promoter by the phenylpropanoid-pathway intermediate *p*-coumaric acid. Proc Natl Acad Sci U S A 89(19):9230–9234. https://doi.org/10.1073/pnas.89.19.9230

Macho Alberto P, Zipfel C (2014) Plant PRRs and the activation of innate immune signaling. Mol Cell 54(2):263–272. https://doi.org/10.1016/j.molcel.2014.03.028

Maekawa T, Kufer TA, Schulze-Lefert P (2011) NLR functions in plant and animal immune systems: so far and yet so close. Nat Immunol 12(9):818–826. https://doi.org/10.1038/ni.2083

Manners JM, Penninckx IA, Vermaere K, Kazan K, Brown RL, Morgan A, Maclean DJ, Curtis MD, Cammue BP, Broekaert WF (1998) The promoter of the plant defensin gene *PDF1.2* from *Arabidopsis* is systemically activated by fungal pathogens and responds to methyl jasmonate but not to salicylic acid. Plant Mol Biol 38(6):1071–1080

Meng X, Xu J, He Y, Yang KY, Mordorski B, Liu Y, Zhang S (2013) Phosphorylation of an ERF transcription factor by *Arabidopsis* MPK3/MPK6 regulates plant defense gene induction and fungal resistance. Plant Cell 25(3):1126–1142. https://doi.org/10.1105/tpc.112.109074

Menkens AE, Schindler U, Cashmore AR (1995) The G-box: a ubiquitous regulatory DNA element in plants bound by the GBF family of bZIP proteins. Trends Biochem Sci 20(12):506–510. https://doi.org/10.1016/S0968-0004(00)89118-5

Mi H, Huang X, Muruganujan A, Tang H, Mills C, Kang D, Thomas PD (2017) PANTHER version 11: expanded annotation data from gene ontology and Reactome pathways, and data analysis tool enhancements. Nucleic Acids Res 45(D1):D183–D189. https://doi.org/10.1093/nar/gkw1138

Nasir F, Tian L, Chang C, Li X, Gao Y, Tran LP, Tian C (2017) Current understanding of pattern-triggered immunity and hormone-mediated defense in rice (*Oryza sativa*) in response to *Magnaporthe oryzae* infection. Semin Cell Dev Biol. https://doi.org/10.1016/j.semcdb.2017.10.020

Ohme-Takagi M, Shinshi H (1995) Ethylene-inducible DNA binding proteins that interact with an ethylene-responsive element. Plant Cell 7(2):173–182

Ohme-Takagi M, Suzuki K, Shinshi H (2000) Regulation of ethylene-induced transcription of defense genes. Plant Cell Physiol 41(11):1187–1192

Park HC, Kim ML, Kang YH, Jeon JM, Yoo JH, Kim MC, Park CY, Jeong JC, Moon BC, Lee JH, Yoon HW, Lee SH, Chung WS, Lim CO, Lee SY, Hong JC, Cho MJ (2004) Pathogen- and NaCl-induced expression of the SCaM-4 promoter is mediated in part by a GT-1 box that interacts with a GT-1-like transcription factor. Plant Physiol 135(4):2150–2161. https://doi.org/10.1104/pp.104.041442

Robert-Seilaniantz A, Grant M, Jones JD (2011) Hormone crosstalk in plant disease and defense: more than just jasmonate-salicylate antagonism. Annu Rev Phytopathol 49:317–343. https://doi.org/10.1146/annurev-phyto-073009-114447

Rushton PJ, Reinstadler A, Lipka V, Lippok B, Somssich IE (2002) Synthetic plant promoters containing defined regulatory elements provide novel insights into pathogen- and wound-induced signaling. Plant Cell 14(4):749–762

Rushton PJ, Somssich IE (1998) Transcriptional control of plant genes responsive to pathogens. Curr Opin Plant Biol 1(4):311–315

Sarkar S, Das A, Khandagale P, Maiti IB, Chattopadhyay S, Dey N (2017) Interaction of *Arabidopsis* TGA3 and WRKY53 transcription factors on *Cestrum* yellow leaf curling virus (CmYLCV) promoter mediates salicylic acid-dependent gene expression *in planta*. Planta. https://doi.org/10.1007/s00425-017-2769-6

Sato F, Kitajima S, Koyama T, Yamada Y (1996) Ethylene-induced gene expression of osmotin-like protein, a neutral isoform of tobacco PR-5, is mediated by the AGCCGCC cis-sequence. Plant Cell Physiol 37(3):249–255

Shigenaga AM, Argueso CT (2016) No hormone to rule them all: interactions of plant hormones during the responses of plants to pathogens. Semin Cell Dev Biol 56:174–189. https://doi.org/10.1016/j.semcdb.2016.06.005

Shimono M, Sugano S, Nakayama A, Jiang CJ, Ono K, Toki S, Takatsuji H (2007) Rice WRKY45 plays a crucial role in benzothiadiazole-inducible blast resistance. Plant Cell 19(6):2064–2076. https://doi.org/10.1105/tpc.106.046250

Sohn KH, Lee SC, Jung HW, Hong JK, Hwang BK (2006) Expression and functional roles of the pepper pathogen-induced transcription factor RAV1 in bacterial disease resistance, and drought and salt stress tolerance. Plant Mol Biol 61(6):897–915. https://doi.org/10.1007/s11103-006-0057-0

Strompen G, Gruner R, Pfitzner UM (1998) An *as*-1-like motif controls the level of expression of the gene for the pathogenesis-related protein 1a from tobacco. Plant Mol Biol 37(5):871–883

Toledo-Ortiz G, Huq E, Quail PH (2003) The *Arabidopsis* basic/helix-loop-helix transcription factor family. Plant Cell 15(8):1749–1770

Ulmasov T, Hagen G, Guilfoyle TJ (1994) The *ocs* element in the soybean GH2/4 promoter is activated by both active and inactive auxin and salicylic acid analogues. Plant Mol Biol 26(4):1055–1064

van Loon LC, Rep M, Pieterse CM (2006) Significance of inducible defense-related proteins in infected plants. Annu Rev Phytopathol 44:135–162. https://doi.org/10.1146/annurev.phyto.44.070505.143425

Vlot AC, Dempsey DA, Klessig DF (2009) Salicylic acid, a multifaceted hormone to combat disease. Annu Rev Phytopathol 47:177–206. https://doi.org/10.1146/annurev.phyto.050908.135202

Wang Y, Kwon SJ, Wu J, Choi J, Lee YH, Agrawal GK, Tamogami S, Rakwal R, Park SR, Kim BG, Jung KH, Kang KY, Kim SG, Kim ST (2014) Transcriptome analysis of early responsive genes in rice during *Magnaporthe oryzae* infection. Plant Pathol J 30(4):343–354. https://doi.org/10.5423/PPJ.OA.06.2014.0055

Xiang C, Miao Z, Lam E (1996) Coordinated activation of *as*-1-type elements and a tobacco glutathione S-transferase gene by auxins, salicylic acid, methyl-jasmonate and hydrogen peroxide. Plant Mol Biol 32(3):415–426

Xiang C, Miao Z, Lam E (1997) DNA-binding properties, genomic organization and expression pattern of TGA6, a new member of the TGA family of bZIP transcription factors in *Arabidopsis thaliana*. Plant Mol Biol 34(3):403–415

Yang DL, Yang Y, He Z (2013) Roles of plant hormones and their interplay in rice immunity. Mol Plant 6(3):675–685. https://doi.org/10.1093/mp/sst056

Yang Y, Shah J, Klessig DF (1997) Signal perception and transduction in plant defense responses. Genes Dev 11(13):1621–1639

Yin S, Mei L, Newman J, Back K, Chappell J (1997) Regulation of sesquiterpene cyclase gene expression. Characterization of an elicitor- and pathogen-inducible promoter. Plant Physiol 115(2):437–451

Yu LM, Lamb CJ, Dixon RA (1993) Purification and biochemical characterization of proteins which bind to the H-box *cis*-elementimplicated in transcriptional activation of plant defense genes. Plant J 3(6):805–816

Zander M, Thurow C, Gatz C (2014) TGA transcription factors activate the salicylic acid-suppressible branch of the ethylene-induced defense program by regulating ORA59 expression. Plant Physiol 165(4):1671–1683. https://doi.org/10.1104/pp.114.243360

Zhang J, Chen L, Fu C, Wang L, Liu H, Cheng Y, Li S, Deng Q, Wang S, Zhu J, Liang Y, Li P, Zheng A (2017) Comparative transcriptome analyses of gene expression changes triggered by *Rhizoctonia solani* AG1 IA infection in resistant and susceptible rice varieties. Front Plant Sci 8:1422. https://doi.org/10.3389/fpls.2017.01422

Zhou J, Tang X, Martin GB (1997) The Pto kinase conferring resistance to tomato bacterial speck disease interacts with proteins that bind a *cis*-element of pathogenesis-related genes. EMBO J 16(11):3207–3218. https://doi.org/10.1093/emboj/16.11.3207

16

Heterologous Expression and Functional Analysis of Rice GLUTAMATE RECEPTOR-LIKE Family Indicates its Role in Glutamate Triggered Calcium Flux in Rice Roots

Jun Ni[1*], Zhiming Yu[1], Guankui Du[2], Yanyan Zhang[1], Jemma L. Taylor[3], Chenjia Shen[1], Jing Xu[1], Xunyan Liu[1], Yifeng Wang[4] and Yunrong Wu[5]

Abstract

Background: Tremendous progress has been made in understanding the functions of the GLUTAMATE RECEPTOR-LIKE (GLR) family in *Arabidopsis*. Still, the functions of OsGLRs in rice, especially the ion channel activities, are largely unknown.

Results: Using the aequorin-based luminescence imaging system, we screened the specificity of amino acids involved in the induction of Ca^{2+} flux in rice roots. Of all the amino acids tested, glutamate (Glu) was the only one to trigger Ca^{2+} flux significantly in rice roots. Detailed analysis showed a dose response of Ca^{2+} increase to different concentrations of Glu. In addition, the Ca^{2+} spike response to Glu was rapid, within 20 s after the application. A desensitization assay and pharmacological tests showed that the Glu-triggered Ca^{2+} flux is mediated by OsGLRs. Whole genome analysis identified 13 *OsGLR* genes in rice, and these genes have various expression patterns in different tissues. Subcellular localization studies showed that all the OsGLRs examined are likely localized to the plasma membrane. Bacteria growth assays showed that at least OsGLR2.1 and OsGLR3.2 have the potential to mediate ion uptake in bacteria. Further analysis using Fura-2-based Ca^{2+} imaging revealed a Glu-triggered Ca^{2+} increase in OsGLR2.1-expressing human embryonic kidney (HEK) cells.

Conclusions: Our work provides a molecular basis for investigating mechanisms of Glu-triggered Ca^{2+} flux in rice.

Keywords: Aequorin, Glu, GLR, Calcium, HEK cells, Rice

Background

Glutamate (Glu), as a fast excitatory neurotransmitter, has an important role in the animal nervous system (Watkins and Jane 2006). Glu activates the ionotropic glutamate receptors (iGluRs) to mediate the Na^+, K^+ and Ca^{2+} entry into cells, catalyzing the signaling across the neurosynaptic gaps (Dingledine et al. 1999). Since the discovery of *GLUTAMATE RECEPTOR-LIKE* (*GLR*) genes in plants (Lacombe et al. 2001; Lam et al. 1998), which are closely related to the mammalian *iGluRs*, more and more evidence has indicated the existence of similar signaling pathways analogous to those found in animals (Forde 2014).

Arabidopsis has a total of 20 *GLR* genes, and phylogenetic analysis has shown that these genes can be subdivided into three clades (Chiu et al. 1999; Chiu et al. 2002; Lacombe et al. 2001). The basic domain structures of plant GLRs and animal iGluRs are conserved and phylogenetic analysis suggested that the separation of GLRs can be traced back to the last universal common ancestor of plants, metazoans and bacteria (Price et al. 2012). Recently, molecular genetics and electrophysiological approaches have made exciting advances in our understanding of various GLR functions in plants. AtGLR1.1 integrates and regulates the different aspects of carbon, nitrogen and water balance that are required for normal plant growth and development (Kang et al.

* Correspondence: nijun@hznu.edu.cn
[1]College of Life and Environmental Sciences, Hangzhou Normal University, Hangzhou 310018, China
Full list of author information is available at the end of the article

2004; Kang and Turano 2003). AtGLR1.2 mediates a plant signaling mechanism between male gametophytes and pistil tissue that is important for the pollen tube growth (Michard et al. 2011). AtGLR1.4 accounts for methionine (Met)-induced membrane depolarization in *Arabidopsis* leaves (Tapken et al. 2013). AtGLR3.3 is involved in plant defense signaling and the control of root gravitropism (Manzoor et al. 2013; Miller et al. 2010). AtGLR3.5 modulates cytosolic Ca^{2+} level to counteract effect of abscisic acid in seed germination (Kong et al. 2015). AtGLRs also mediate mechanical wound signaling and elicitor/pathogen-mediated defense signaling in *Arabidopsis* (Manzoor et al. 2013; Mousavi et al. 2013). Moreover, AtGLR3.2 and AtGLR3.4 were reported to form heteromeric channels to affect lateral root development via Ca^{2+} signaling in the phloem (Vincill et al. 2013).

In contrast with Glu or glycine (Gly) activated iGluRs, *Arabidopsis* GLRs have similar amino acid-gated ion channel activities, but with a broader agonist profile (Forde 2014). Six amino acids are considered to be the agonists of AtGLR3.3 dependent Ca^{2+} influx in *Arabidopsis*, and a mutation of *AtGLR3.3* gene inhibits the rise of Ca^{2+} triggered by these amino acids (Qi et al. 2006). Further research demonstrates that these six effective amino acids are not equivalent agonists, but grouped into hierarchical classes based on their ability to desensitize the response mechanism (Stephens et al. 2008). Human embryonic kidney (HEK) cells expressing AtGLR3.4 showed wide agonist profile with asparagine (Asn) and serine (Ser) as strong agonists and Gly less so (Vincill et al. 2012). Interestingly, AtGLR1.4 expressing in *Xenopus* oocytes, functioned as a cation channel that responded to an even broader range of amino acids with Met being the most effective and most potent agonist (Tapken et al. 2013). An aequorin-based luminescence recording system showed that strong Ca^{2+} responses were induced by all the amino acids tested, with the highest Ca^{2+} amplitude for cysteine (Cys) and lowest one for Asn (Zhu et al. 2013).

Compared with the great progresses of GLRs in dicotyledonous *Arabidopsis*, less information has been obtained from the monocot, rice. The only gene characterized was *OsGLR3.1*, which positively regulates cell proliferation in the root apical meristem and is required for survival of meristematic cells (Li et al. 2006). However, the ion channel activities of OsGLRs are still completely unknown.

In our previous research, we established an aequorin-based luminescence imaging system in rice, which was able to reflect the concentration of cytosolic free Ca^{2+} ($[Ca^{2+}]_i$) level in rice roots (Zhang et al. 2015). Here, using this system, we present the Glu-specific triggered Ca^{2+} response in rice roots. Further analysis using Fura-2-based Ca^{2+} imaging revealed a Glu triggered Ca^{2+} increase in OsGLR2.1-expressing HEK cells. Our results indicated an OsGLR mediated Ca^{2+} changes in response to Glu in rice roots.

Results

Glu Specifically Triggered Ca^{2+} Influx in Rice Roots

To investigate the amino acid-triggered Ca^{2+} influx in rice, we employed an aequorin-based luminescence imaging system, which is able to reflect the $[Ca^{2+}]_i$ level in rice roots (Zhang et al. 2015). After the reconstitution of aequorin by spraying seedlings with coelenterazine, the aequorin luminescence of rice roots was recorded using a photo-counting camera by treating plants with nine different amino acids respectively. As shown in Fig. 1, Glu induced a very strong $[Ca^{2+}]_i$ increase, while the other eight amino acids had little effect on the $[Ca^{2+}]_i$ induction compared with the control treatment. This showed that Glu is able to trigger Ca^{2+} influx specifically in rice roots.

The Characteristics of $[Ca^{2+}]_i$ Response to Glu in Rice Roots

In order to investigate the $[Ca^{2+}]_i$ increase in response to Glu in rice roots, we examined the aequorin-based luminescence under various concentrations of Glu treatments. Overall, the more Glu that was applied, the stronger the luminescence signals would be (Fig. 2a). Detailed analysis showed that the concentration-dependent $[Ca^{2+}]_i$ increase was in agreement with Michaelis-Menten function with the Km value of 9.60 mM (Fig. 2b). To investigate the time courses of $[Ca^{2+}]_i$ responses induced by different concentrations of Glu in rice roots, the average luminescence intensity of continuous images with exposure time of 20 s, were analyzed and a comparison of basic parameters (amplitudes, durations and phases) of $[Ca^{2+}]_i$ responses were made. As shown in Fig. 2c, a strong luminescence signal was detected in the first image, which collected the luminescence signal within the first 20 s. However, almost no luminescence signal was detected after the first 20 s. This showed that Glu rapidly induced a sharp spike of $[Ca^{2+}]_i$ within 20 s, which quickly declined to the basal level after the spike. The amplitudes of luminescence signals varied according to different concentrations of Glu, while both the durations and phases were the same. These results were similar to the NaCl triggered $[Ca^{2+}]_i$ increase in rice and osmotic triggered $[Ca^{2+}]_i$ increase in *Arabidopsis* (Yuan et al. 2014; Zhang et al. 2015), indicating the existence of sensory channels to mediate Glu-triggered $[Ca^{2+}]_i$ increase in rice roots.

Glu-Triggered $[Ca^{2+}]_i$ Increase is Mediated by OsGLRs

It is believed that GLRs are key components of a Ca^{2+} influx mechanism in response to amino acids in plants

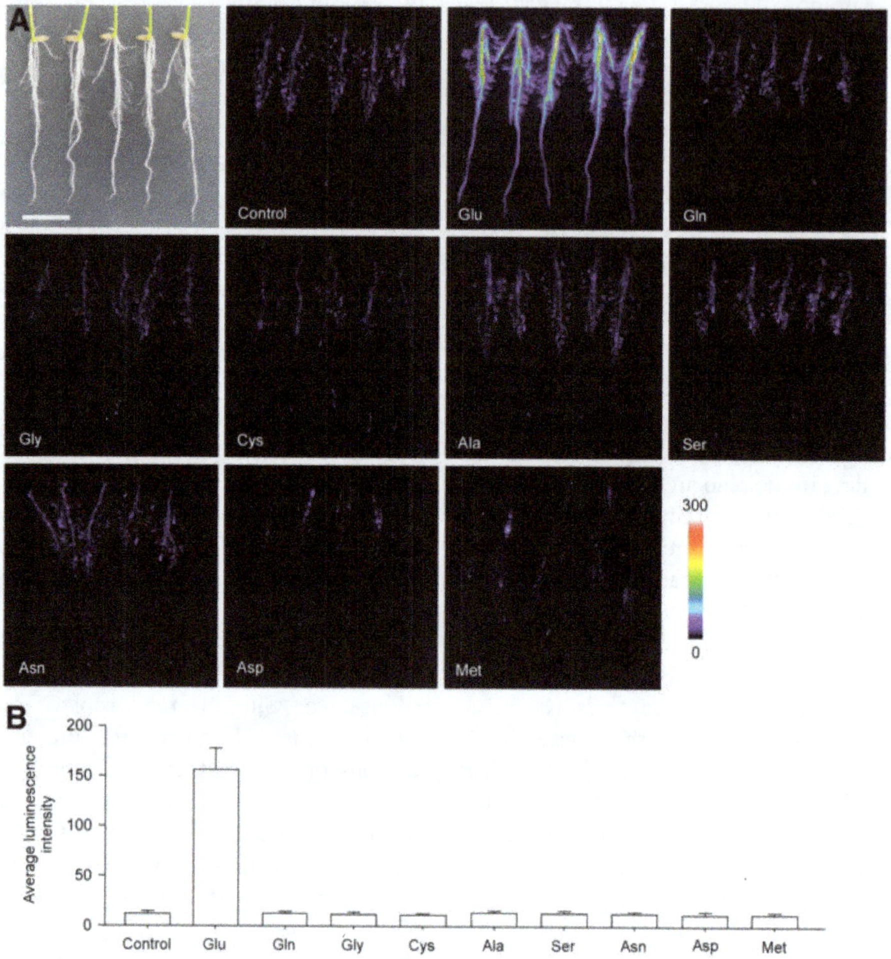

Fig. 1 Glutamate specifically triggers Ca^{2+} influx in rice roots. **a** Pseudocolor images of aequorin luminescence in roots treated with different amino acids. The relationship between luminescence intensity and the pseudocolor images are scaled by a pseudocolor bar and the numbers next to the pseudocolor bar are maximum and minimum values of luminescence intensity. The image on the left of the top is the treated roots view in white light. Bar = 2 cm. Water was used instead of amino acid in the control treatment. **b** The bar chart of luminescence signal intensity of every treatment. The results were obtained from at least three independent experiments (mean ± sd; $n = 10$)

(Dennison and Spalding 2000). Thus, we examined whether the Glu-triggered $[Ca^{2+}]_i$ increase in rice roots is also mediated by OsGLRs. Desensitization is an important mode of ligand-gated Ca^{2+} channels at synapses in the animal nervous system (Jones and Westbrook 1996). Research in *Arabidopsis* showed that the phenomenon of desensitization also exists in the Glu activated, GLR mediated Ca^{2+} entry (Meyerhoff et al. 2005; Qi et al. 2006). To determine whether the Glu- triggered $[Ca^{2+}]_i$ increase is the manifestation of the same mechanism (e.g. the activation of OsGLRs), we examined the $[Ca^{2+}]_i$ increases in response to two sequential Glu treatments. The first application of Glu triggered an obvious $[Ca^{2+}]_i$ increase, while after one hour's recovery, a second application of Glu with the same concentration, caused about 70 % reduction in the $[Ca^{2+}]_i$ increase compared with the first one (Additional file 1). To determine the dependence of *de novo* protein synthesis in the recovery of desensitization, we pretreated rice roots with cycloheximide (CHX), a well-known translation inhibitor. As a result, we observed a more serious inhibition (about 90 % reduction) in the $[Ca^{2+}]_i$ increase in response to a second application of Glu (Additional file 1). These results were very similar to that of AtGLR3.4-expressing mesophyll cells (Meyerhoff et al. 2005), indicating the possibility that OsGLRs mediate the Glu-triggered $[Ca^{2+}]_i$ increase in rice roots.

To confirm the participation of OsGLRs in the Glu-triggered $[Ca^{2+}]_i$ increase, we conducted pharmacological tests in the aequorin-based luminescence imaging system. La^{3+} and Gd^{3+} are agonists of Ca^{2+}, and they have been used as Ca^{2+} channel blockers to inhibit Ca^{2+} flux (Tracy et al. 2008). In our experiment, both $LaCl_3$ and $GdCl_3$ had significant inhibitory effects in Glu-triggered $[Ca^{2+}]_i$ increase. Furthermore, all

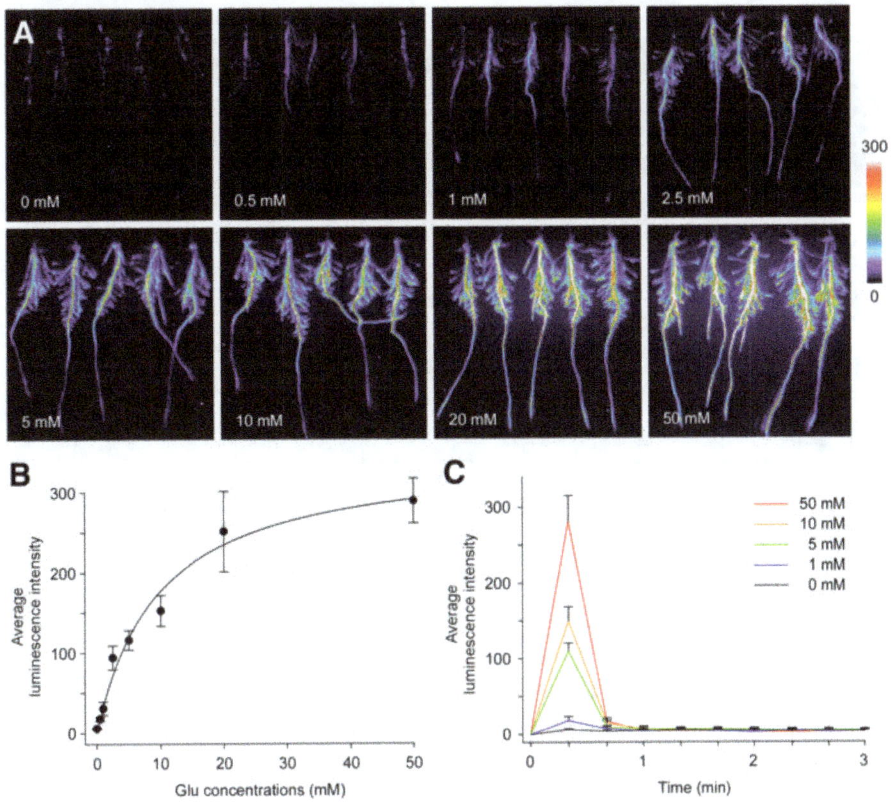

Fig. 2 The characteristics of $[Ca^{2+}]_i$ response to Glu in rice roots. **a** Pseudocolor images of aequorin luminescence in roots treated with different concentrations of Glu. The relationship between luminescence intensity and the pseudocolor images are scaled by a pseudocolor bar and the numbers next to the pseudocolor bar are maximum and minimum values of luminescence intensity. **b** Concentration dependence of $[Ca^{2+}]_i$ increase in rice roots. Data were fitted by the Michaelis-Menten function. **c** The time courses of $[Ca^{2+}]_i$ changes induced by different concentrations of Glu in rice roots. The results were obtained from at least three independent experiments (mean ± sd; $n = 10$)

the concentrations tested had similar inhibitory effects, indicating an effective inhibition function of $LaCl_3$ and $GdCl_3$ treatments (Fig. 3). 6-Cyano-7-Nitroquinoxaline-2,3-dione (CNQX) and 6,7-Dinitroquinoxaline-2,3-dione (DNQX) are GLR antagonists, and application of these drugs inhibits the channel activities of GLRs in plants (Meyerhoff et al. 2005; Michard et al. 2011; Tapken et al. 2013). In our experiment, both antagonists had significant inhibitory effects. In addition, dosage effects were observed in the inhibition of Glu-triggered $[Ca^{2+}]_i$ increase (Fig. 3). CNQX and DNQX were dissolved by Dimethyl sulfoxide (DMSO) in the stock solutions, and the highest concentration of DMSO in the working solution was 0.5 % (*v/v*). In order to investigate the effect of DMSO in the Glu-induced $[Ca^{2+}]_i$ increase, we examined the luminescence after various concentrations of DMSO pretreatments, and found that the effect 0.5 % of DMSO had no significant differences to the water control (Additional file 2). These results above showed that Glu-triggered $[Ca^{2+}]_i$ increase in rice roots is likely to be mediated by OsGLRs.

Characterization of *OsGLR* Gene Family in Rice

Since there is no comprehensive information about rice OsGLR family, searches of public genomic data were carried out. A total of 13 *OsGLR* genes that are most closely related to *Arabidospis* GLRs were identified by phytozome (http://www.phytozome.net/). Because there was no standard annotation assigned to these newly-identified genes, we named these genes according to their position in the phylogenetic tree (Fig. 4). The names of the *OsGLR* genes, the locus ID, the chromosome location, the isoelectric point (pI), the molecular weight (Mw) and predicted subcellular localizations are shown in Additional file 3. The decoded polypeptides were predicted to contain 3–6 transmembrane helices (TMHs), and most of them have the signature domains of animal iGluRs, including the 'three-plus-one' transmembrane domains (Lam et al. 1998) (Additional file 4). To detect the evolutionary relationships of GLRs, a phylogenetic tree was generated from alignments of 13 OsGLRs in rice and 20 AtGLRs in *Arabidopsis* (Fig. 4 and Additional file 5). We grouped the GLRs into three clades according to the previously reported work in

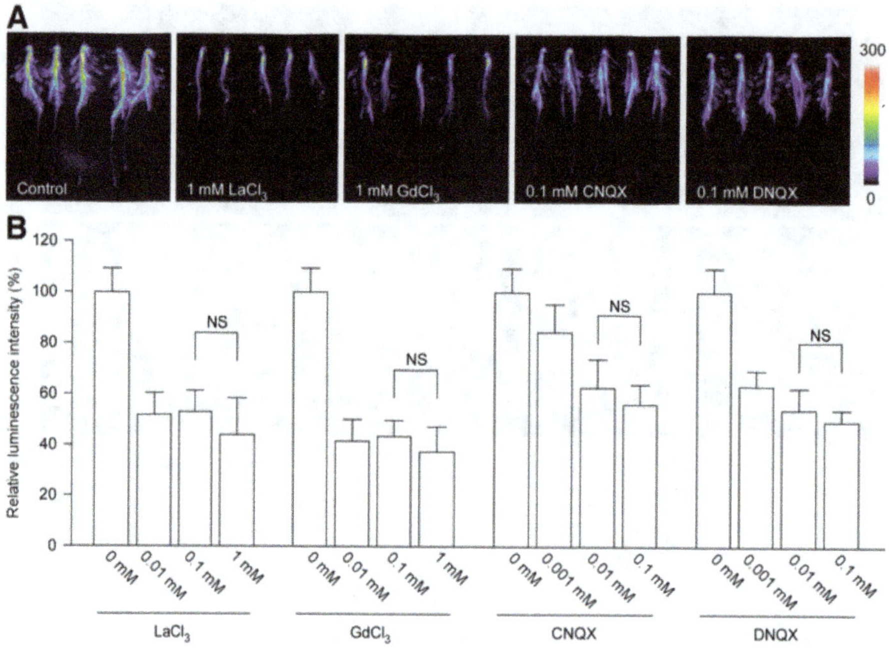

Fig. 3 The pharmacological test of Glu-induced $[Ca^{2+}]_i$ increases. **a** Treatment of Ca^{2+} agonists ($LaCl_3$ and $GdCl_3$) and GLR-specific antagonists (CNQX and DNQX) reduced Glu-induced $[Ca^{2+}]_i$ increases. The relationship between luminescence intensity and the pseudocolor images are scaled by a pseudocolor bar and the numbers next to the pseudocolor bar are maximum and minimum values of luminescence intensity. **b** The effects of different concentrations of Ca^{2+} agonists and GLR-specific antagonists in Glu-induced $[Ca^{2+}]_i$ increases. The results were obtained from at least three independent experiments (mean ± sd; $n = 10$; NS, not significant $P > 0.05$; Student's *t*-test)

Arabidopsis (Chiu et al. 2002). Detailed analysis showed that the phylogenetic relationships between *Arabidopsis* and rice are different among the three clades of GLRs. GLRs in clade I and clade II separated very early, while GLRs in clade III are evolutionary very close to each other (Fig. 4).

Tissue-Specific Expression of *OsGLR* Genes

We examined the expression patterns of identified *OsGLR* genes in the five different tissues by real-time quantitative reverse transcription polymerase chain reaction (qRT-PCR). The results revealed variations in the expression of *OsGLRs* in different tissues. In general, a number of *OsGLRs* were expressed at higher levels in all the tissues examined, compared with some members with overall lower levels. Interestingly, there were some *OsGLRs* with tissue-specific expression patterns with the highest expression levels seen in roots (Fig. 5). In clade I, high transcript levels of *OsGLR1.1* and *OsGLR1.3* were observed in all the tissues, in contrast with *OsGLR1.2* and *OsGLR1.4*, which had relatively higher expression in roots. In clade II, *OsGLR2.2* and *OsGLR2.4* had relatively low transcript levels in all the tissues, while *OsGLR2.1* and *OsGLR2.3* had very obvious tissue-specific expression patterns. *OsGLR2.1* was expressed most highly in the root, with moderate expression in shoot, stem-base and young panicle tissues, and almost no expression in the stem. *OsGLR2.3* had high expression in both root and stem-base tissues, and almost no expression in the other tissues. For clade III, all the genes except *OsGLR3.1*, had very high expression levels in all the tissues examined (Fig. 5).

We also examined the transcript level of *OsGLRs* after treatment of Glu and CNQX. We found that neither Glu nor CNQX treatment altered the transcript level of *OsGLRs* (Additional file 6). This indicated that although Glu and CNQX regulate the OsGLR mediated $[Ca^{2+}]_i$ increase in rice roots, they do not affect the transcript level of *OsGLRs*.

OsGLRs Mediate ion Uptake in Bacteria

In order to investigate the functions of OsGLRs, we amplified the coding sequences (CDS) of *OsGLR1.1*, *OsGLR1.3*, *OsGLR2.1*, *OsGLR2.3*, *OsGLR3.2*, *OsGLR3.3* and *OsGLR3.4*, which have higher expression levels according to the previous qRT-PCR results. Interestingly, we found that *E. coli* transfected with plasmids inserted with *OsGLR2.1*, *OsGLR2.3*, *OsGLR3.2* had a slower growth rate than the others (Additional file 7A). To further investigate the ion transport functions of OsGLRs in bacteria, plasmids with *OsGLRs* were transfected into the K^+ uptake-deficient *E. coli* strain LB650 (Stumpe and Bakker 1997). The drop assays on solid medium showed that all the transfected *E. coli* grew well on the selective

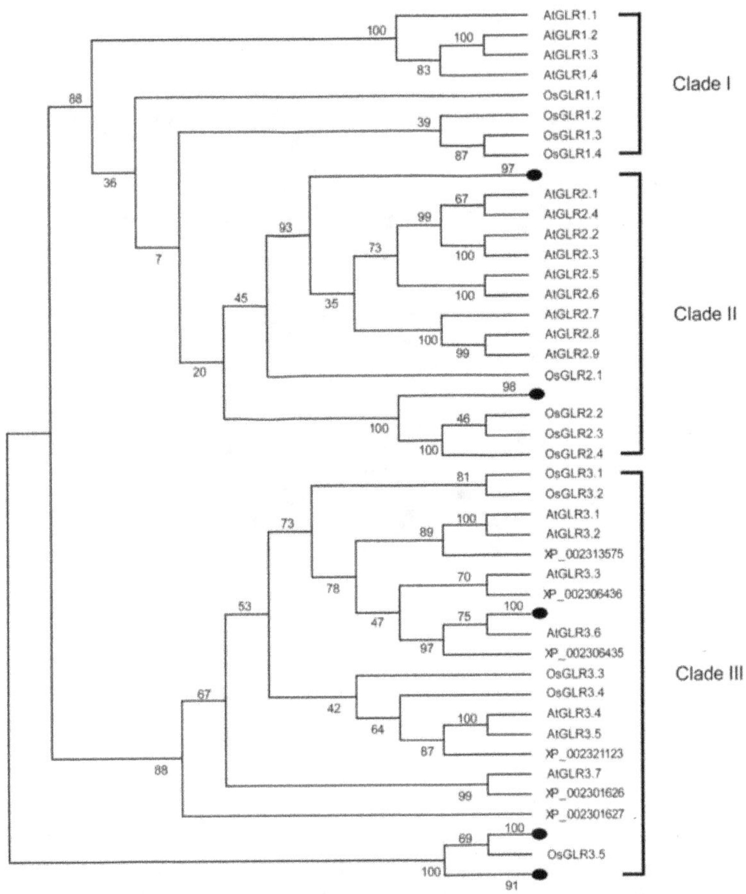

Fig. 4 Phylogenetic relationships of GLRs in *Arabidopsis*, rice and poplar. Unrooted phylogenetic tree was constructed using MEGA6 by Maximum-Likelihood method. Three different GLR clades were marked by *lines*. *Black dots* represent compressed subtree of GLRs from poplar. [GenBank: XP_002313575, GenBank: XP_002306436, GenBank: XP_002306435, GenBank: XP_002321123, GenBank: XP_002301626 and GenBank: XP_002301627] are accession numbers of GLRs from poplar. Bootstrap values from 500 replicates are indicated at each branch

medium with high K^+ level, indicating a successful transfection of these vectors. Interestingly, when grown on the medium with low K^+ level, only the *E. coli* transfected with *OsGLR2.1* and *OsGLR3.2* survived, indicating a complementation of OsGLR2.1 and OsGLR3.2 to the K^+ uptake-deficiency of LB650 (Additional file 7B).

Subcellular Localization of OsGLRs Expressed in Plant and Mammalian Cells

In order to investigate the subcellular localization of OsGLR proteins, the CDS (without stop codon) of seven *OsGLR* genes were cloned into *pGWB405* (Nakagawa et al. 2007) and *pDEST47* (Invitrogen) respectively. In all the vectors, the Green Fluorescent Protein (GFP) was fused to the C terminus of OsGLRs, and these vectors were transformed into plant cells and mammalian cells. Tobacco leaf epidermal cells expressing fusion proteins showed similar fluorescence at the cell periphery, indicating plasma membrane localizations of OsGLRs (Fig. 6a). To confirm our results, we transiently expressed OsGLR1.1-GFP in onion epidermal cells. After plasmolysis, the GFP signal was observed to localize on the plasma membrane (Fig. 6b). We also checked the GFP fluorescence after treated the onion epidermal cells with Glu and CNQX. We found that neither Glu nor CNQX treatment altered the GFP fluorescence in onion epidermal cells (Fig. 6b).

The adherent mammalian HEK293 cells expressing OsGLR-GFP fusions showed similar localization to the previously published example of AtGLR3.4-GFP fusion protein, which was detected on the plasma membrane as well as internal membranes in HEK cells (Vincill et al. 2012) (Additional file 8). In order to confirm the subcellular localization of OsGLRs in HEK cells, we treated the HEK cells with trypsin to depart the adherent HEK cells to petri dish. After the treatment, the flat irregular shaped adherent HEK cell turned to globose suspended cell, and the GFP signal was detected at the edge of the cell (Additional file 8 inserted).

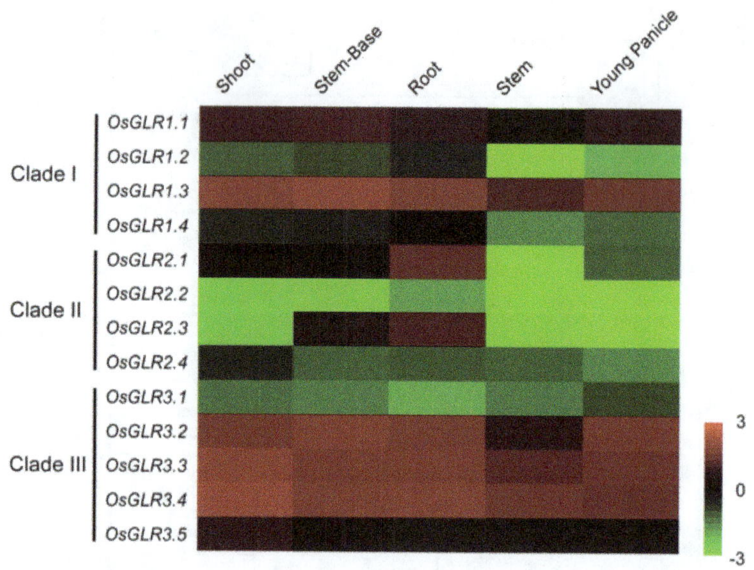

Fig. 5 Tissue-specific expression patterns of *OsGLR* genes. The expression patterns of 13 *OsGLR* genes in five different tissues were analyzed. The name of *OsGLR* genes are indicated on the left of each panel, and different tissues are indicated on the top of the panel. The different colors correspond to the log-transcription values of the gene change-fold ratio shown in the bar at the right

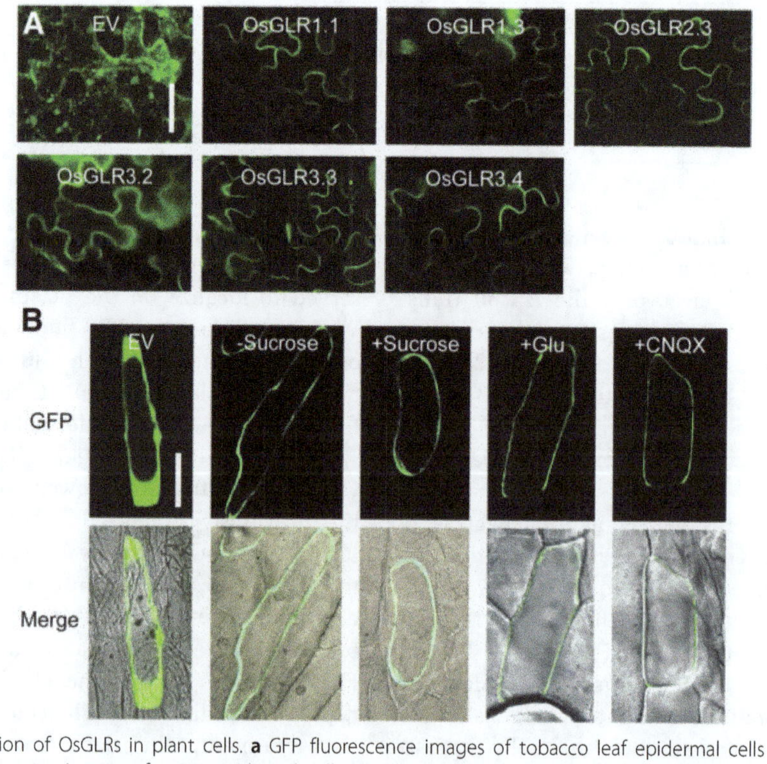

Fig. 6 Subcellular localization of OsGLRs in plant cells. **a** GFP fluorescence images of tobacco leaf epidermal cells expressing *OsGLR-GFP*. Bar = 50 μm. **b** GFP fluorescence images of onion epidermal cells expressing *OsGLR1.1-GFP*. GFP fluorescence images before (−Sucrose) and after plasmolysis (+Sucrose) are shown. Treatments of Glu (20 mM for half an hour) and CNQX (0.1 mM for half an hour) did not alter the GFP fluorescence in onion epidermal cells. Bar = 100 μm. EV, empty vector control

OsGLR2.1 Expression is Involved in Glu-Induced $[Ca^{2+}]_i$ Increase in HEK Cells

HEK cells display little or no endogenous amino acid-gated channel activity (Zhang and Huganir 1999), and are a standard functional expression system for both mammalian iGluR and plant GLR channels (Chazot et al. 1999; Vincill et al. 2012). Much important knowledge about iGluRs and GLRs has been learned by studying them in HEK cells (Dravid et al. 2008; Keinanen et al. 1990; Monyer et al. 1992; Vincill et al. 2012; Vincill et al. 2013). To examine whether any of the newly isolated OsGLRs can mediate Glu-induced $[Ca^{2+}]_i$ increase in HEK cells, we examined the Glu-gated ion channel activities of OsGLRs in HEK cells. HEK cells were transfected with plasmids carrying OsGLR cDNAs, and analyzed their activity using Fura-2-based Ca^{2+} imaging. As a result, the application of Glu triggered a $[Ca^{2+}]_i$ increase in HEK cells expressing OsGLR2.1 (Additional file 9), and pretreatment of La^{3+}, a Ca^{2+} channel blocker, inhibited the $[Ca^{2+}]_i$ increase. Furthermore, aspartate (Asp), which has similar chemical properties with Glu, was unable to induce a $[Ca^{2+}]_i$ increase in HEK cells expressing OsGLR2.1 (Fig. 7). This showed that at least OsGLR2.1 is able to mediate Glu-induced $[Ca^{2+}]_i$ increase in HEK cells. This also indicated that OsGLR2.1 may participate in the Glu-triggered Ca^{2+} influx in rice roots.

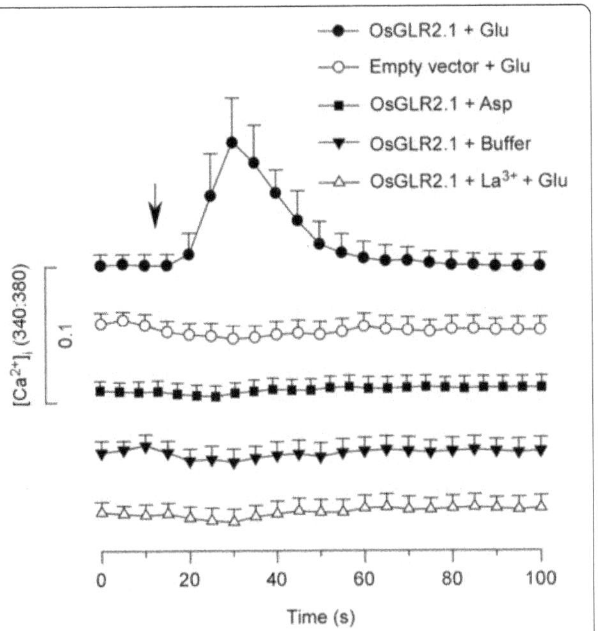

Fig. 7 OsGLR2.1 expression is involved in Glu-induced $[Ca^{2+}]_i$ increase in HEK cells. $[Ca^{2+}]_i$ increase was analyzed by Fura-2 emission ratios (F340 nm:F380 nm). *Arrow* indicates the applications of Glu, Asp or buffer control. The baselines of different treatments were the same (F340 nm:F380 nm ≈ 0.136), and the lines were separated for clearness. Application of Glu to OsGLR2.1-expressing HEK cells was able to induce the $[Ca^{2+}]_i$ increase, while other treatments had no effect to change the $[Ca^{2+}]_i$ level in HEK cells. Fluorescence data were collected every 5 s. Data for independent cells in one experiment are shown (mean ± sd; $n = 20$)

Discussion

Glu is the primary natural ligand of iGluRs in the central nervous system in animals (Traynelis et al. 2010). In this research, the effectiveness of different amino acids was tested using the rice expressing aequorin. Of all the amino acids tested, Glu was clearly the most effective agonist to evoke the $[Ca^{2+}]_i$ increase in rice roots. This is different from the previous reports that a broad range of agonist of amino acids were able to induce Ca^{2+} responses in *Arabidopsis* (Qi et al. 2006; Vincill et al. 2012; Zhu et al. 2013). We tested the six amino acids which were reported to induce Ca^{2+} responses in aequorin-based luminescence system in *Arabidopsis* (Zhu et al. 2013). None of them, except Glu, induced induce Ca^{2+} in rice. It is worth noting that our aequorin-based luminescence imaging system was only able to reflect the $[Ca^{2+}]_i$ level in rice roots in contrast with the whole seedlings in *Arabidopsis* (Zhang et al. 2015). It is possible that other amino acids instead of Glu are able to induce $[Ca^{2+}]_i$ increase in rice shoots. Alternatively, rice may have evolved different approaches to cope with amino acid signals in different environmental conditions after the divergence of monocots and dicots. It was reported that acidic pH could also cause $[Ca^{2+}]_i$ increase (Zhu et al. 2013). In our experiment, we used glutamic acid monosodium salt solution for glutamate treatments, and the pH value was very close to 7 and would not cause a pH effect in our experiment. In addition, according to our time course experiment, the $[Ca^{2+}]_i$ response was very rapid (within 20 s). This is different from the pH effect, which lasts for several minutes (Zhu et al. 2013). According to our previous results, 100 mM of NaCl had no significant effect to $[Ca^{2+}]_i$ increase in rice roots (Zhang et al. 2015). While in our experiment, the concentration of Na^+ was only 10 mM. Therefore, the $[Ca^{2+}]_i$ increase in rice roots was caused by glutamate, not by pH or salt stress effects.

We investigated the $[Ca^{2+}]_i$ changes in response to different concentrations of Glu and found that the response was very similar to previously reported $[Ca^{2+}]_i$ changes in response to Glu in *Arabidopsis* leaves, which fitted well by the Michaelis-Menten function, indicating similar mechanisms between these two species (Meyerhoff et al. 2005). It is reported that the concentrations of individual amino acids in the bulk soil solution are in the 0.01–10 μM range (Jones et al. 2005). What's more, the concentrations of amino acids in different tissues of plants are below 10 mM (Pilot et al. 2004). Considering that the Km value of fitted curve was 9.60 mM, which fall into the range of *in planta* Glu concentration, we propose that the Ca^{2+} signaling system may participate in a sensory mechanism for *in planta* Glu concentration. In the pharmacological tests of Glu-triggered $[Ca^{2+}]_i$ increase, $LaCl_3$ and $GdCl_3$ inhibited about 50 % of $[Ca^{2+}]_i$ increase, and no significant differences were observed among different concentrations of blockers, indicating a saturation of Ca^{2+} channels blocked by $LaCl_3$ and $GdCl_3$. This is different from the inhibition effect of NaCl and H_2O_2 induced $[Ca^{2+}]_i$ increases in rice roots, which reached almost 100 and 90 % respectively (Zhang et al. 2015). Moreover, research in Glu-triggered $[Ca^{2+}]_i$ increase in *Arabidopsis* leaves showed about 90 % of inhibition by $LaCl_3$ (Meyerhoff et al. 2005). Stephens et al. (2008) presented a model of GLR channels based on three types of qualitatively different, heteromeric GLR channel subtypes. We reasoned that different GLRs are able to assemble a large number of different channels, and the sensitivities of these channels to $GdCl_3/LaCl_3$ may different. The relative weak effects of Ca^{2+} channel blockers compared to that of *Arabidopsis* may also be due to our calculation method of aequorin luminescence signal. In order to obtain better repeatability, we only chose the region of primary root to calculate the level of luminescence signal. After the pretreatment with $GdCl_3$ and $LaCl_3$, most of the luminescence signal came from the region of primary root, thus our method may result to the underestimate of the effects of these blockers. We also observed that CNQX and DNQX inhibited the Glu-triggered $[Ca^{2+}]_i$ increase significantly. CNQX and DNQX were initially used to inhibit the activities of iGluR channels in animals (Armstrong and Gouaux 2000; Honore et al. 1988). Subsequent experiments showed that they were also efficient inhibitors of GLRs in plants (Dubos et al. 2003; Manzoor et al. 2013; Michard et al. 2011; Vatsa et al. 2011). It was also shown that the efficiency of these inhibitors was about 50 %, which is similar to our results (Vatsa et al. 2011).

Ca^{2+} channels in plants belong to large gene families, and these channels are regulated by different effectors (Jammes et al. 2011). For instance, mechanosensitive Ca^{2+}-permeable channels (MSCC) are controlled by mechanical stimulation (Monshausen and Gilroy 2009), cyclic nucleotide gated channels (CNGC) are controlled by cyclic nucleotides (Kaplan et al. 2007), two-pore channels (TPC) are controlled by membrane potential (Pottosin and Schonknecht 2007) and GLR is controlled by amino acids (Price et al. 2012). In our research, we found that Glu specifically triggered $[Ca^{2+}]_i$ increase in rice roots. Together with the desensitization assay and pharmacological tests, we propose that the Glu-triggered $[Ca^{2+}]_i$ increase in rice roots is mediated by OsGLR family. Compared with 20 *AtGLR* genes in *Arabidopsis* (Chiu et al. 2002; Davenport 2002), the number of *OsGLR* genes in rice is still uncertain. It was stated that there are 13 *OsGLR* genes in rice (Ward et al. 2009). However, it was also demonstrated that rice has at least 21 *OsGLR* genes (ARAMEMNON plant membrane protein database: release 7) (Forde 2014). In this research, we chose a total of 13 *OsGLR* genes that have the closest relationship to the *Arabidopsis* 20 *AtGLR* genes for further analysis. According to the phylogenetic tree, we found that *GLRs* from *Arabidopsis* and rice separated to each other in clade I and clade II, which indicated independent evolution of GLRs in these clades after the divergence of monocots and dicots.

It is reported that all 20 *AtGLR* genes are expressed in *Arabidopsis*, and there are no distinct clade-specific organ expression patterns among the three *AtGLR* clades (Chiu et al. 2002). On the basis of our qRT-PCR experiment, we found expression diversities among members of the *OsGLR* family. It is worth noting that members in clade III had relatively higher expression levels compared with the other two clades, except for *OsGLR3.1*, which had very low expression levels in all the tissues examined. It is also reported that all 20 *AtGLR* genes are expressed in roots and five of the nine clade II genes are root-specific in *Arabidopsis* (Chiu et al. 2002). Similarly, most of the *OsGLR* genes had higher expression levels in roots. Furthermore, two of the four clade II genes (*OsGLR2.1* and *OsGLR2.3*) are almost root-specific in rice. It is worth noting that all four *OsGLRs* in clade II are tandemly clustered in Chromosome 9, and the expression patterns of *OsGLR2.1* and *OsGLR2.3* are very similar to each other, with the highest level in the roots. This indicates that OsGLR2.1 and OsGLR2.3 may have similar functions in the roots.

For further analysis, we cloned seven of the *OsGLRs* from those which had the higher expression levels. So far, the strongest phenotype reported for a single *glr* mutant is in rice which is defective in the *OsGLR3.1* gene (Li et al. 2006). Interestingly, the expression level of *OsGLR3.1* is the lowest in clade III. Thus, for full understanding of the OsGLR family functions in rice, it is necessary to clone all the *OsGLR* genes for functional analysis in future, although some of them have very low expression levels.

In the bacteria growth assay, *E. coli* transfected with Entry clones inserted with *OsGLR2.1*, *OsGLR2.3* and *OsGLR3.2* had a slower growth rate than the others. The Entry clones have a background of *pDONR201*, which have no known promoters beside the insert region. It is possible that other elements (T1 and T2 sequences for instance) are responsible for the leaky expression of *OsGLR2.1*, *OsGLR2.3* and *OsGLR3.2*, which are enough to affect bacteria growth. In the plant cell subcellular localization analysis of OsGLR2.1, we failed to shuttle *OsGLR2.1* into the destination vector *pGWB405* by LR reaction. It was reported that 35S promoter can also be activated in *E. coli* (Assaad and Signer 1990). *OsGLR2.1* driven by 35S may have a more serious leaky effect, which results in the excessive ion channel activities and subsequently leads to the death of the transfected *E. coli*. We did not add extra glutamate into the medium in both bacteria growth assays. It is possible that tryptone and yeast extract in the medium contain a small amount of glutamate, which is sufficient to operate the channels formed by OsGLRs. Considering the non-selective nature of GLR channels in *Arabidopsis* (Jammes et al. 2011), our bacterial results showed that OsGLRs may also form non-selective channels at least in *E. coli*.

We expressed OsGLR-GFP fusion proteins in tobacco leaf epidermal cells and onion epidermal cells to examine the subcellular localizations of OsGLRs. As a result, all the members examined had similar subcellular localization patterns. We observed that the GFP signal was detected at the cell periphery when expressed in both the tobacco leaf epidermal cells and onion epidermal cells, which indicated that these OsGLR proteins are very likely plasma membrane localized proteins. Until now, to our knowledge, many of the AtGLRs examined are plasma membrane localized proteins (Meyerhoff et al. 2005; Tapken et al. 2013; Vincill et al. 2012; Vincill et al. 2013). However, interestingly, some AtGLRs were also reported to localize to other organelles. It was reported that AtGLR3.4 has a dual localizations to plastids and plasma membrane (Teardo et al. 2011). Splicing variant of AtGLR3.5 targets the inner mitochondrial membrane, while the other variant localizes to the chloroplasts (Teardo et al. 2015). From this point of view, although plasmolysis analysis in onion epidermal cells provided additional evidence of plasma membrane localization of OsGLRs, further experiments are still needed (Co-localization analysis by a plasma membrane marker for instance) to provide unequivocal evidence of subcellular localizations of OsGLR proteins. We also expressed OsGLR-GFP fusion proteins in mammalian HEK cells, and the GFP fluorescence was detected on the plasma membrane as well as internal membranes in HEK cells. The altered subcellular localization may probably due to the different systems between plant and animal cells. The HEK cell results also showed that the OsGLRs can be expressed in animal cells and a Ca^{2+} imaging approach in HEK cells was worth pursuing.

We screened the Ca^{2+} channel activities of these newly isolated OsGLRs in HEK cell system and found that OsGLR2.1 was able to mediate Glu-triggered $[Ca^{2+}]_i$ increase in HEK cells. Furthermore, pretreatment of a Ca^{2+} channel blocker inhibited the increase. We found that the change in Fura-2 ratio in OsGLR2.1-expresing cell is relatively small. We reasoned that this may due to the heterologous expression of OsGLR2.1 in HEK cells. Alternatively, OsGLR2.1 may need to interact with other partners to from efficient ion channels. Our results indicated that OsGLR2.1 is able to form a Ca^{2+} permeable channel in HEK cells. Considering the root expression of *OsGLR2.1*, It could be responsible for Glu-triggered $[Ca^{2+}]_i$ increase in rice roots. However, it is not necessarily that OsGLR2.1 is sufficient to assume the Glu-triggered $[Ca^{2+}]_i$ increase in rice roots. It is possible that rice roots may employ more complex GLR channels rather than those formed by OsGLR2.1. Animal iGluRs often combine as heterotetramers to form amino acid gated ion channels (Traynelis et al. 2010), and it is generally assumed that plant GLRs will assemble in a similar way (Forde 2014). Using desensitization analysis and knock out mutant, the model of different AtGLR subtypes was proposed (Stephens et al. 2008). Furthermore, the interactions between different subunits were also reported (Price et al. 2013). HEK cells transfected with *AtGLR3.4* alone showed amino acid triggered ion channel activities (Vincill et al. 2012). Further research showed that AtGLR3.4 physically interacts with AtGLR3.2 to form heteromeric ion channels to affect lateral root development in the phloem (Vincill et al. 2013). From this point of view, it is possible that OsGLR2.1 may interact with other OsGLRs to from heteromeric ion channels in rice roots.

Conclusions

We described a Glu-specific triggered $[Ca^{2+}]_i$ increase in rice roots. A desensitization assay and pharmacological tests showed that this response is mediated by OsGLR channels. We identified 13 *OsGLRs* and cloned seven of them. Further research showed that OsGLR2.1 and OsGLR3.2 have the potential to mediate ion uptake in bacteria, and HEK cells expressing OsGLR2.1 are able to permeate Ca^{2+} that are controlled by Glu. Our results indicated that OsGLR2.1 may participate in the Glu-triggered Ca^{2+} influx in rice roots. For future directions, genetic analysis (loss of function of *osglr2.1* mutant) and protein-protein interaction assays (the OsGLRs interact with OsGLR2.1) are needed to demonstrate the mechanism of Glu-triggered Ca^{2+} influx in rice roots.

Methods

Aequorin Luminescence Imaging

Seeds of aequorin-expressing rice were sown and grown on plates as described previously (Zhang et al. 2015). Reconstitution of aequorin was performed in vivo by spraying transgenic seedlings with 10 μM coelenterazine and followed by incubation at 21 °C in the dark for 12–16 h. Treatments and aequorin luminescence imaging were performed at room temperature using a ChemiPro HT system as described previously (Jiang et al. 2013). For screening of the specificity of amino acids in induction of Ca^{2+} flux in rice roots, 10 mM amino acid solutions (Glutamate, Sigma G1626; Glutamine, Sigma G3126;

Glycine, Sigma G7126; Cysteine, Sigma C1276; Alanine, Sigma A7627; Serine, Sigma S4500; Asparagine, Sigma A8381; Aspartic acid, Sigma A9256; Methionine, Sigma M9625) were used and luminescence images were acquired for 3 min. For the analysis of time courses of increase in $[Ca^{2+}]_i$, each exposure time was 20 s and the images were taken continuously for 3 min. Water was used instead of amino acid solution in the control experiment. The fitted curve was calculated by SigmaPlot 12.5. For desensitization analysis, plants were first treated with 10 mM Glu for aequorin luminescence imaging. After the imaging, plants were washed with water several times to remove the residual Glu and transferred to a new petri dish. The plants were treated with or without 0.1 mM of CHX and placed in the dark for 1 h. The second application of 10 mM Glu was carried out after 1 h. For pharmacological tests, rice roots were treated with different concentrations of $LaCl_3$, $GdCl_3$, CNQX and DNQX, respectively for 30 min before 10 mM Glu treatment was applied. CNQX and DNQX were first dissolved in DMSO (due to the limited solubility, the concentration of stock solutions were 20 mM), the stock solutions were then diluted with water to final concentrations. For calculations of average luminescence intensity, luminescence signal in the region of primary roots were manually and equivalently selected, and the average luminescence intensity was analyzed. To avoid the interference of chloroplast auto-fluorescence signal in the aequorin luminescence imaging, all the treatments were performed in complete darkness. WinView/32 and Meta Morph 7.7 were used to analyze recorded luminescence images.

Identification of *OsGLR* Gene Family in Rice

To identify potential members of the *OsGLR* gene family in rice, we performed multiple database searches. Twenty *Arabidopsis AtGLR* sequences (Chiu et al. 2002) were used as queries in BLAST searches against the rice genome sequence (http://www.phytozome.net/). The Hidden Markov Model (HMM) profiles of the GLR family (Pfam 01094: Receptor family ligand binding region; Pfam 00060: Ligand-gated ion channel; Pfam 00497: Bacterial extracellular solute-binding proteins, family 3; Pfam 13458: Periplasmic binding protein) were used to confirm the OsGLR gene family in rice.

Phylogenetic Tree Building, Transmembrane Helices and Subcellular Localization Predictions

Multiple sequence alignments were performed on the AtGLR and OsGLR proteins using ClustalW with default parameters, and the alignments were then adjusted manually. The phylogenetic tree was constructed using MEGA6.06 employing the Maximum-Likelihood (ML) method. To construct better phylogenetic tree, we added 61 GLRs form poplar (Ward et al. 2009) along with GLRs from rice and *Arabidopsis*. The CDS of *OsGLRs* were mostly obtained from a public database (http://rice.plantbiology.msu.edu/), some sequences were modified based on our sequencing data. The transmembrane domains were estimated using TMHMM2 (www.cbs.dtu.dk/services/TMHMM/). The subcellular localization predictions were estimated using PSORT (http://wolfpsort.org/).

RNA Isolation and Quantitative RT-PCR

Total RNA from different tissues including shoots, stem-base and roots of 5 day old seedlings, and the stem and young panicle of adult plants were extracted using a Plant RNeasy Mini kit (Qiagen) according to the manufacturer's instructions. First-strand cDNA was synthesized with SuperScript first-strand synthesis kit (Invitrogen) and the transcript levels of each gene were measured by qRT-PCR using Mx3000p QPCR system (Agient) with iQ SYBR Green Supermix (Bio-Rad). The program for the qRT-PCR was as follows: 5 min at 95 °C, and 45 cycles of 15 s at 95 °C, 20 s at 60 °C, 20 s at 72 °C. The heat map representation was performed according to the previous description (Shen et al. 2014). Briefly, the 1/1000 of *OsACTIN* expression level was considered as an internal standard to calculate the relative fold differences based on the comparative cycle threshold ($2^{-\Delta\Delta Ct}$) values. The logarithm of expression level change fold data were used by Treeview to visualize as a heat map. Red color represents higher expression level, black represents similar expression level and green color represents lower expression level. All the expression analysis was carried out for at least three biological repeats and the values shown in figures represents the average values of these repeats. For the expression analysis of different *OsGLRs* after treatment with Glu, rice roots were treated with 20 mM Glu for different time phrases. For the inhibitor pretreatment, rice roots were treated with 0.1 mM CNQX for half an hour. The expression analysis was carried out for three biological repeats. The primers used in this experiment were listed in Additional file 10.

Cloning and Recombinant Vector Construction

Full-length CDS (with or without stop codon) of *OsGLRs* were isolated from total RNA by RT-PCR and amplified using KOD-FX DNA polymerase (TOYOBO); the primers were listed in Additional file 10. The PCR products were cloned into the *pDONR201* entry vector using the BP Clonase™ Enzyme Mix (Invitrogen) according to the manufacturer's instructions. For the complementation of LB650 and the imaging of $[Ca^{2+}]_i$ in HEK 293 cells, the *pDONR201* entry vector containing CDS (with stop codon) of *OsGLRs* were shuttled into the Gateway destination vectors *pDEST14* and *pcDNA3.2* respectively using the LR Clonase™ Enzyme Mix (Invitrogen) according to the manufacturer's instructions. To avoid the toxic effect of *ccdB* gene to the *E.coli*, the CDS of *GFP* was used to replace the poison gene *ccdB*, and the vectors containing *GFP* were considered as empty vectors in the measurement of bacteria growth rate and complementation of LB650. For the subcellular localization analysis of OsGLRs in the plant and mammalian cells, the *pDONR201* entry vector containing CDS (without stop codon) of *OsGLRs* were shuttled into the Gateway destination vectors *pGWB405* (Nakagawa et al. 2007) and *pDEST47* (Invitrogen) by LR reactions respectively, which fuse the GFP tag to the C terminus of the translated gene product. The *pcDNA3.2* containing *GFP* was considered as an empty vector in the subcellular localization analysis in the mammalian cells. For the empty vector in the subcellular localization analysis in the plant cells, CDS of *GFP* was shuttled into the Gateway destination vector *pGWB402* (Nakagawa et al. 2007) by LR reaction, which had a cauliflower mosaic virus (CaMV) 35S promoter to drive the *GFP* gene.

Bacterial Grown Assay

For the measurement of bacteria growth rate, transformants were first grown at 37 °C in a shaking incubator with 3 ml liquid LB medium containing 10 g/L tryptone, 10 g/L NaCl and 5 g/L yeast extract. After approximately 12–16 h of growth, we adjusted all the cultures to the same density. Cell culture (1 ml) with a density of 0.6 OD_{600} was added to 100 ml LB medium and continued to grow at 37 °C and 220 rpm. The cell density was measured 12 times at 1 h intervals after the transfer. *GFP* gene was cloned into *pDONR201* to act as the empty vector comparison.

For the complementation of a K^+-uptake mutant of *E. coli* by OsGLRs, the K^+ uptake-deficient *E. coli* strain LB650 with mutations in the *TRK* H and *TRK* G genes was used for complementation studies (Stumpe and Bakker 1997). The *pDEST14* containing with *OsGLRs* (or *GFP* as empty vector) were transformed into the *E. coli* LB650 strain. The complementation assay was carried out as previously reported (Ali et al. 2006).

HEK Cell Culture and Transfection

HEK293T cells were grown and maintained in DMEM medium (GIBCO) supplemented with 10 % (*v/v*) fetal bovine serum (GIBCO) and 0.1 % (*w/v*) Penicillin-Streptomycin antibiotics (GIBCO), and placed in a 37 °C incubator with 95 % (*v/v*) air and 5 % (*v/v*) CO_2. For transfection, cells were first seeded onto poly-lysine-coated eight-well chambered coverglass (Nunc) and cultured with DMEM medium supplemented with 5 % (*v/v*) fetal bovine serum. After growth for 24 h, cells were transfected with plasmid DNA using Lipofectamine 2000 reagent (Invitrogen) as described previously (Yuan et al. 2014). The transfected HEK cells were grown for additional 24 h before subsequent experiments.

Subcellular Localization Analysis of OsGLR Proteins

For subcellular localization analysis in tobacco leaf epidermal cells, *Agrobacterium tumefaciens*–mediated transient transformation of *N. benthamiana* was performed as described (Li et al. 2008; Voinnet et al. 2003). Briefly, *A. tumefaciens* strain EHA105 carrying binary constructs was grown to OD_{600} values of 0.2 to 0.4 at 28 °C in LB medium. Agrobacterium cells were collected by centrifugation at 4000 rpm for 10 min at 25 °C, and resuspended in 10 mM MES (pH 5.7), 10 mM $MgCl_2$ and 150 mg/ml acetosyringone. Leaves of 4–5-week-old plants were infiltrated with bacterial cultures, and the fluorescence of GFP was observed under a Nikon A1 confocal microscope. For subcellular localization analysis in onion epidermal cells, plasmid-coated gold microparticles ($\Phi = 1.0$ μm) were accelerated by helium pressure (1100 psi) propeling the macrocarrier to onion by PDS-1000/He system (Bio-Rad, USA). The onion was then incubated at 26 °C in the dark. After 16 h cultivation, the onion epidermis was imaged by confocal laser scanning microscope (Nikon A1). The 0.3 g/ml of sucrose was used in plasmolysis, the 20 mM of Glu was used in Glu treatment (treated for half an hour) and the 0.1 mM of CNQX was used in blocker treatment (treated for half an hour).

For subcellular localization analysis of OsGLR proteins in HEK cells, cells transfected with plasmid DNA were grown in poly-lysine-coated eight-well chambered coverglass for 24 h and the fluorescence of GFP was observed by Zeiss Axio Observer A1 fluorescence microscope directly. To confirm the subcellular localization of OsGLRs in HEK cells, the HEK cells were treated with trypsin (GIBCO) and incubated for 5 min. After the departure of HEK cells, the trypsin was replaced with PBS (GIBCO) and the globose suspended HEK cells were observed for subcellular localizations.

Imaging of $[Ca^{2+}]_i$ in HEK Cells

HEK cells transfected with plasmid DNA were loaded with Ca^{2+} sensitive dye Fura-2 AM (Sigma), and a Fura-2 based Ca^{2+} imaging assay was performed using the Axiovert 200 fluorescence microscope. Emission ratiometric images (F340 nm:F380 nm) were collected using MetaFluor Fluorescence Ratio Imaging Software (Molecular Devices). For the Glu treatments, Fura-2 loaded HEK293 cells were incubated in a standard buffer (pH 7.4) containing 130 mM NaCl, 3 mM KCl,

0.6 mM $MgCl_2$, 10 mM glucose, 10 mM HEPES and 1 mM Ca^{2+} for 30 min. One hundred microliter Glu solution (30 mM) was dropped carefully into the 200 μl standard buffer (the final concentration of Glu was 10 mM), and the Fura-2 ratiometric images were collected. For Ca^{2+} channel blocker treatment, HEK cells were pretreated with $LaCl_3$ (1 mM) and incubated for 30 min before the Glu treatments. For control treatments, standard buffer solution and Asp solution were used instead of Glu. For data analysis, about 20 cells in one experiment were selected manually based on the increases in $[Ca^{2+}]_i$. All the experiments were repeated for at least three times.

Additional Files

Additional file 1: Desensitization of Glu activated Ca^{2+} channels in rice roots. (A) Pseudocolor images of aequorin luminescence in response to two sequential Glu treatments. From left to right, the first application of Glu (left), the second application of Glu after1 hour's recovery (middle) and second application of Glu pretreated with CHX (right). The relationship between luminescence intensity and the pseudocolor images are scaled by a pseudocolor bar and the numbers next to the pseudocolor bar are maximum and minimum values of luminescence intensity. (B) The bar chart of relative luminescence signal intensity of every treatment. The results were obtained from at least three independent experiments (mean ± sd; $n = 10$).

Additional file 2: The effect of DMSO in the Glu-induced $[Ca^{2+}]_i$ increase. (A) Pseudocolor images of aequorin luminescence in roots treated by different percentages of DMSO. The relationship between luminescence intensity and the pseudocolor images are scaled by a pseudocolor bar and the numbers next to the pseudocolor bar are maximum and minimum values of luminescence intensity. (B) Relative luminescence signal intensity of every treatment. The results were obtained from at least three independent experiments (mean ± sd; $n = 10$; NS, not significant $P > 0.05$; Student's *t*-test).

Additional file 3: The information of *OsGLR* genes in rice.

Additional file 4: Transmembrane helices structure predictions of OsGLR proteins. The transmembrane domains were estimated using TMHMM2: www.cbs.dtu.dk/services/TMHMM/. The red peaks showed the predicted transmembrane regions of proteins; the blue peaks showed the inside transmembrane domains; the pink peaks showed the outside transmembrane domain.

Additional file 5: The sequence alignment of 20 AtGLRs and 13 OsGLRs

Additional file 6: qRT-PCR analysis of *OsGLRs* after treatment of Glu and CNQX. For Glu treatment, rice roots were treated with 20 mM Glu for 5 min or 30 min. For blocker treatment, rice roots were treated with 0.1 mM CNQX for 30 min. In addition, rice roots were also pretreated with 0.1 mM CNQX for 30 min, then treated with 20 mM Glu for 30 min (CNQX + Glu). Data for independent experiments are shown (mean ± sd; $n = 3$).

Additional file 7: OsGLRs mediate ion uptake in bacteria. (A) Bacterial growth rate of *E. coli*, transfected with different *OsGLRs*. Expression of OsGLR2.1, OsGLR2.3 and OsGLR3.2 reduced the growth rate of *E. coli* compared with the empty vector control. (B) K^+ uptake-deficient *E. coli* mutant (strain LB650) transformed with different *OsGLRs*, was grown on solid selective medium with either high K^+ (100 mM) or low K^+ (2 mM). Expression of OsGLR2.1 and OsGLR3.2 enhance the growth of LB650 on low K^+ medium compared with the empty vector control.

Additional file 8: Subcellular localization of OsGLRs in HEK293 cells. GFP fluorescence images of HEK293 cells expressing *OsGLR-GFP*. HEK cells treated with trypsin were inserted with red frames. EV, empty vector control. Bar = 10 μm.

Additional file 9: The application of Glu triggered a $[Ca^{2+}]_i$ increase in HEK cells expressing OsGLR2.1 (movie in the ppt file).

Additional file 10: The sequences of primers used in this research.

Abbreviations

Asn: asparagine; Asp: aspartate; $[Ca^{2+}]_i$: concentration of cytosolic free Ca^{2+}; CDS: coding sequence; CHX: cycloheximide; CNQX: 6-Cyano-7-Nitroquinoxaline-2,3-dione; Cys: cysteine; DMSO: dimethyl sulfoxide; DNQX: 6,7-Dinitroquinoxaline-2,3-dione; GFP: green fluorescent protein; GLR: GLUTAMATE RECEPTOR-LIKE; Glu: glutamate; Gly: glycine; HEK: human embryonic kidney; HMM: hidden Markov model; iGluR: ionotropic glutamate receptor; LB: Luria Bertani; Met: Methionine; Mw: molecular weight; NJ: neighbor-joining; pI: isoelectric point; qRT-PCR: quantitative reverse transcription polymerase chain reaction; Ser: serine; TMH: transmembrane helices.

Competing interests

The authors declare that they have no competing interests.

Authors' contributions

JN carried out the aequorin luminescence imaging, imaging of $[Ca^{2+}]_i$ in HEK cells and draft of the manuscript. ZY carried out the subcellular localization analysis of OsGLRs in plant cells. GD carried out the subcellular localization analysis of OsGLRs in animal cells. YZ carried out the DNA cloning and bacteria experiments. JT participated in revising the manuscript. CS carried out the sequence alignment. JX and XL carried out the RT-PCR analysis. YW and YW participated in the growing of transgenic rice. All authors read and approved the final manuscript.

Acknowledgement

We thank Prof. Jianhua Zhang (Chinese University of Hong Kong, China) for critically reading of this manuscript. We thank Prof. Chuanzao Mao (Zhejiang University, China) and Dr. Yingjia Wang (Hangzhou Normal University, China) for technical assistants. We also thank Prof. Zhen-Ming Pei and Dr. Feihua Wu (Hangzhou Normal University, China) for helpful comments. This work was funded by National Natural Science Foundation of China under grant No. 31301000, Zhejiang Provincial Natural Science Foundation of China under grant No. LQ13C020005, National Natural Science Foundation of China under grant No. 31200913 and National Natural Science Foundation of China under grant No. 31400229.

Author details

[1]College of Life and Environmental Sciences, Hangzhou Normal University, Hangzhou 310018, China. [2]Department of Biochemistry, Hainan Medical College, Haikou 571199, China. [3]School of Life Sciences, Gibbet Hill Campus, University of Warwick, Coventry CV4 7AL, UK. [4]State Key Laboratory of Rice Biology, China National Rice Research Institute, Chinese Academy of Agricultural Sciences, Hangzhou 311400, China. [5]State Key Laboratory of Plant Physiology and Biochemistry, College of Life Science, Zhejiang University, Hangzhou 310058, China.

References

Ali R, Zielinski RE, Berkowitz GA (2006) Expression of plant cyclic nucleotide-gated cation channels in yeast. J Exp Bot 57:125–138

Armstrong N, Gouaux E (2000) Mechanisms for activation and antagonism of an AMPA-sensitive glutamate receptor: crystal structures of the GluR2 ligand binding core. Neuron 28:165–181

Assaad FF, Signer ER (1990) Cauliflower mosaic virus P35S promoter activity in Escherichia coli. Mol Gen Genet 223:517–520

Chazot PL, Cik M, Stephenson FA (1999) Transient expression of functional NMDA receptors in mammalian cells. Methods Mol Biol 128:33–42

Chiu J, DeSalle R, Lam HM, Meisel L, Coruzzi G (1999) Molecular evolution of glutamate receptors: a primitive signaling mechanism that existed before plants and animals diverged. Mol Biol Evol 16:826–838

Chiu JC, Brenner ED, DeSalle R, Nitabach MN, Holmes TC, Coruzzi GM (2002) Phylogenetic and expression analysis of the glutamate-receptor-like gene family in Arabidopsis thaliana. Mol Biol Evol 19:1066–1082

Davenport R (2002) Glutamate receptors in plants. Ann Bot 90:549–557
Dennison KL, Spalding EP (2000) Glutamate-gated calcium fluxes in Arabidopsis. Plant Physiol 124:1511–1514
Dingledine R, Borges K, Bowie D, Traynelis SF (1999) The glutamate receptor ion channels. Pharmacol Rev 51:7–61
Dravid SM, Prakash A, Traynelis SF (2008) Activation of recombinant NR1/NR2C NMDA receptors. J Physiol 586:4425–4439
Dubos C, Huggins D, Grant GH, Knight MR, Campbell MM (2003) A role for glycine in the gating of plant NMDA-like receptors. Plant J 35:800–810
Forde BG (2014) Glutamate signalling in roots. J Exp Bot 65:779–787
Honore T, Davies SN, Drejer J, Fletcher EJ, Jacobsen P, Lodge D, Nielsen FE (1988) Quinoxalinediones: potent competitive non-NMDA glutamate receptor antagonists. Science 241:701–703
Jammes F, Hu HC, Villiers F, Bouten R, Kwak JM (2011) Calcium-permeable channels in plant cells. FEBS J 278:4262–4276
Jiang Z, Zhu S, Ye R, Xue Y, Chen A, An L, Pei ZM (2013) Relationship between NaCl- and H_2O_2-Induced Cytosolic Ca^{2+} Increases in Response to Stress in *Arabidopsis*. PLoS ONE 8:e76130
Jones DL, Shannon D, Junvee-Fortune T, Farrar JF (2005) Plant capture of free amino acids is maximized under high soil amino acid concentrations. Soil Biol Biochem 37:179–181
Jones MV, Westbrook GL (1996) The impact of receptor desensitization on fast synaptic transmission. Trends Neurosci 19:96–101
Kang J, Mehta S, Turano FJ (2004) The putative glutamate receptor 1.1 (AtGLR1.1) in Arabidopsis thaliana regulates abscisic acid biosynthesis and signaling to control development and water loss. Plant Cell Physiol 45:1380–1389
Kang J, Turano FJ (2003) The putative glutamate receptor 1.1 (AtGLR1.1) functions as a regulator of carbon and nitrogen metabolism in Arabidopsis thaliana. Proc Natl Acad Sci U S A 100:6872–6877
Kaplan B, Sherman T, Fromm H (2007) Cyclic nucleotide-gated channels in plants. FEBS Lett 581:2237–2246
Keinanen K, Wisden W, Sommer B, Werner P, Herb A, Verdoorn TA, Sakmann B, Seeburg PH (1990) A family of AMPA-selective glutamate receptors. Science 249:556–560
Kong D, Ju C, Parihar A, Kim S, Cho D, Kwak JM (2015) Arabidopsis glutamate receptor homolog3.5 modulates cytosolic Ca^{2+} level to counteract effect of abscisic acid in seed germination. Plant Physiol 167:1630–1642
Lacombe B, Becker D, Hedrich R, DeSalle R, Hollmann M, Kwak JM, Schroeder JI, Le Novere N, Nam HG, Spalding EP, Tester M, Turano FJ, Chiu J, Coruzzi G (2001) The identity of plant glutamate receptors. Science 292:1486–1487
Lam HM, Chiu J, Hsieh MH, Meisel L, Oliveira IC, Shin M, Coruzzi G (1998) Glutamate-receptor genes in plants. Nature 396:125–126
Li J, Zhu S, Song X, Shen Y, Chen H, Yu J, Yi K, Liu Y, Karplus VJ, Wu P, Deng XW (2006) A rice glutamate receptor-like gene is critical for the division and survival of individual cells in the root apical meristem. Plant Cell 18:340–349
Li X, Chanroj S, Wu Z, Romanowsky SM, Harper JF, Sze H (2008) A distinct endosomal Ca^{2+}/Mn^{2+} pump affects root growth through the secretory process. Plant Physiol 147:1675–1689
Manzoor H, Kelloniemi J, Chiltz A, Wendehenne D, Pugin A, Poinssot B, Garcia-Brugger A (2013) Involvement of the glutamate receptor AtGLR3.3 in plant defense signaling and resistance to Hyaloperonospora arabidopsidis. Plant J 76:466–480
Meyerhoff O, Muller K, Roelfsema MR, Latz A, Lacombe B, Hedrich R, Dietrich P, Becker D (2005) AtGLR3.4, a glutamate receptor channel-like gene is sensitive to touch and cold. Planta 222:418–427
Michard E, Lima PT, Borges F, Silva AC, Portes MT, Carvalho JE, Gilliham M, Liu LH, Obermeyer G, Feijo JA (2011) Glutamate receptor-like genes form Ca^{2+} channels in pollen tubes and are regulated by pistil D-serine. Science 332:434–437
Miller ND, Durham Brooks TL, Assadi AH, Spalding EP (2010) Detection of a gravitropism phenotype in glutamate receptor-like 3.3 mutants of Arabidopsis thaliana using machine vision and computation. Genetics 186:585–593
Monshausen GB, Gilroy S (2009) Feeling green: mechanosensing in plants. Trends Cell Biol 19:228–235
Monyer H, Sprengel R, Schoepfer R, Herb A, Higuchi M, Lomeli H, Burnashev N, Sakmann B, Seeburg PH (1992) Heteromeric NMDA receptors: molecular and functional distinction of subtypes. Science 256:1217–1221
Mousavi SAR, Chauvin A, Pascaud F, Kellenberger S, Farmer EE (2013) GLUTAMATE RECEPTOR-LIKE genes mediate leaf-to-leaf wound signalling. Nature 500:422–426
Nakagawa T, Suzuki T, Murata S, Nakamura S, Hino T, Maeo K, Tabata R, Kawai T, Tanaka K, Niwa Y, Watanabe Y, Nakamura K, Kimura T, Ishiguro S (2007) Improved Gateway binary vectors: high-performance vectors for creation of fusion constructs in transgenic analysis of plants. Biosci Biotechnol Biochem 71:2095–2100
Pilot G, Stransky H, Bushey DF, Pratelli R, Ludewig U, Wingate VP, Frommer WB (2004) Overexpression of GLUTAMINE DUMPER1 leads to hypersecretion of glutamine from Hydathodes of Arabidopsis leaves. Plant Cell 16:1827–1840
Pottosin II, Schonknecht G (2007) Vacuolar calcium channels. J Exp Bot 58:1559–1569
Price MB, Jelesko J, Okumoto S (2012) Glutamate receptor homologs in plants: functions and evolutionary origins. Front Plant Sci 3:235
Price MB, Kong D, Okumoto S (2013) Inter-subunit interactions between glutamate-like receptors in Arabidopsis. Plant Signal Behav 8:e27034
Qi Z, Stephens NR, Spalding EP (2006) Calcium entry mediated by GLR3.3, an Arabidopsis glutamate receptor with a broad agonist profile. Plant Physiol 142:963–971
Shen C, Yue R, Yang Y, Zhang L, Sun T, Xu L, Tie S, Wang H (2014) Genome-wide identification and expression profiling analysis of the Aux/IAA gene family in Medicago truncatula during the early phase of Sinorhizobium meliloti infection. PLoS One 9:e107495
Stephens NR, Qi Z, Spalding EP (2008) Glutamate receptor subtypes evidenced by differences in desensitization and dependence on the GLR3.3 and GLR3.4 genes. Plant Physiol 146:529–538
Stumpe S, Bakker EP (1997) Requirement of a large K^+-uptake capacity and of extracytoplasmic protease activity for protamine resistance of Escherichia coli. Arch Microbiol 167:126–136
Tapken D, Anschutz U, Liu LH, Huelsken T, Seebohm G, Becker D, Hollmann M (2013) A plant homolog of animal glutamate receptors is an ion channel gated by multiple hydrophobic amino acids. Sci Signal 6:ra47
Teardo E, Carraretto L, De Bortoli S, Costa A, Behera S, Wagner R, Lo Schiavo F, Formentin E, Szabo I (2015) Alternative Splicing-Mediated Targeting of the Arabidopsis GLUTAMATE RECEPTOR3.5 to Mitochondria Affects Organelle Morphology. Plant Physiol 167:216–227
Teardo E, Formentin E, Segalla A, Giacometti GM, Marin O, Zanetti M, Lo Schiavo F, Zoratti M, Szabo I (2011) Dual localization of plant glutamate receptor AtGLR3.4 to plastids and plasmamembrane. Biochim Biophys Acta 1807: 359–367
Tracy FE, Gilliham M, Dodd AN, Webb AA, Tester M (2008) NaCl-induced changes in cytosolic free Ca^{2+} in Arabidopsis thaliana are heterogeneous and modified by external ionic composition. Plant Cell Environ 31:1063–1073
Traynelis SF, Wollmuth LP, McBain CJ, Menniti FS, Vance KM, Ogden KK, Hansen KB, Yuan H, Myers SJ, Dingledine R (2010) Glutamate receptor ion channels: structure, regulation, and function. Pharmacol Rev 62:405–496
Vatsa P, Chiltz A, Bourque S, Wendehenne D, Garcia-Brugger A, Pugin A (2011) Involvement of putative glutamate receptors in plant defence signaling and NO production. Biochimie 93:2095–2101
Vincill ED, Bieck AM, Spalding EP (2012) Ca^{2+} conduction by an amino acid-gated ion channel related to glutamate receptors. Plant Physiol 159:40–46
Vincill ED, Clarin AE, Molenda JN, Spalding EP (2013) Interacting glutamate receptor-like proteins in Phloem regulate lateral root initiation in Arabidopsis. Plant Cell 25:1304–1313
Voinnet O, Rivas S, Mestre P, Baulcombe D (2003) An enhanced transient expression system in plants based on suppression of gene silencing by the p19 protein of tomato bushy stunt virus. Plant J 33:949–956
Ward JM, Maser P, Schroeder JI (2009) Plant ion channels: gene families, physiology, and functional genomics analyses. Annu Rev Physiol 71:59–82
Watkins JC, Jane DE (2006) The glutamate story. Br J Pharmacol 147(Suppl 1): S100–S108
Yuan F, Yang H, Xue Y, Kong D, Ye R, Li C, Zhang J, Theprungsirikul L, Shrift T, Krichilsky B, Johnson DM, Swift GB, He Y, Siedow JN, Pei ZM (2014) OSCA1 mediates osmotic-stress-evoked Ca^{2+} increases vital for osmosensing in Arabidopsis. Nature 514:367–371
Zhang S, Huganir RL (1999) Calmodulin modification of NMDA receptors. Methods Mol Biol 128:103–111
Zhang Y, Wang Y, Taylor JL, Jiang Z, Zhang S, Mei F, Wu Y, Wu P, Ni J (2015) Aequorin-based luminescence imaging reveals differential calcium signalling responses to salt and reactive oxygen species in rice roots. J Exp Bot 66: 2535–2545
Zhu X, Feng Y, Liang G, Liu N, Zhu JK (2013) Aequorin-based luminescence imaging reveals stimulus- and tissue-specific Ca^{2+} dynamics in Arabidopsis plants. Mol Plant 6:444–455

17

Metabolic and transcriptomic changes induced in host during hypersensitive response mediated resistance in rice against the Asian rice gall midge

Ruchi Agarrwal[1], Ayyagari Phani Padmakumari[2], Jagadish S. Bentur[2,3] and Suresh Nair[1*]

Abstract

Background: An incompatible interaction between rice (*Oryza sativa*) and the Asian rice gall midge (AGM, *Orseolia oryzae* Wood-Mason), that is usually manifested through a hypersensitive response (HR), represents an intricate relationship between the resistant host and its avirulent pest. We investigated changes in the transcriptome and metabolome of the host (*indica* rice variety: RP2068-18-3-5, RP), showing HR when attacked by an avirulent gall midge biotype (GMB1), to deduce molecular and biochemical bases of such a complex interaction. Till now, such an integrated analysis of host transcriptome and metabolome has not been reported for any rice-insect interaction.

Results: Transcript and metabolic profiling data revealed more than 7000 differentially expressed genes and 80 differentially accumulated metabolites, respectively, in the resistant host. Microarray data revealed deregulation of carbon (C) and nitrogen (N) metabolism causing a C/N shift; up-regulation of tetrapyrrole synthesis and down-regulation of chlorophyll synthesis and photosynthesis. Integrated results revealed that genes involved in lipid peroxidation (LPO) were up-regulated and a marker metabolite for LPO (azelaic acid) accumulated during HR. This coincided with a greater accumulation of GABA (neurotransmitter and an insect antifeedant) at the feeding site. Validation of microarray results by semi-quantitative RT-PCR revealed temporal variation in gene expression profiles.

Conclusions: The study revealed extensive reprogramming of the transcriptome and metabolome of RP upon GMB1 infestation leading to an HR that was induced by the generation and release of reactive oxygen species i.e. singlet oxygen and resulted in LPO-mediated cell death. RP thus used HR as a means to limit nutrient supply to the feeding maggots and simultaneously accumulated GABA, strategies that could have led to maggot mortality. The integrated results of transcript and metabolic profiling, for the first time, provided insights into an HR+ type of resistance in rice against gall midge.

Keywords: Aminotransferase, GABA, Hypersensitive response, Insect-plant interaction, Rice-gall midge interaction, Metabolic profiling, Singlet oxygen, GC-MS, Microarray, Transcript profiling

Background

The integrated analyses of transcriptomic and metabolomic data obtained from two biological levels i.e. transcript level and metabolite level, respectively, is a useful way of examining the complexity of biological systems. An integrated study of gene expression with metabolomics provides a better insight into the cellular complexity at molecular level. Such an approach was witnessed in the last decade where the emergence of integrated 'omics' approach helped us gain valuable insights into the complex route from gene expression (transcriptomics) to function (metabolomics) in living systems under various kinds of stress. Recent literature is replete with examples where data derived from microarray is integrated with gas chromatography–mass spectrometry (GC-MS) based metabolomics data to study various aspects of plant biology such

* Correspondence: suresh@icgeb.res.in
[1]International Centre for Genetic Engineering and Biotechnology (ICGEB), Aruna Asaf Ali Marg, New Delhi 110067, India
Full list of author information is available at the end of the article

as the differences in root cells and border cells in *Medicago* (Watson et al. 2015), senescence in citrus fruits (Ding et al. 2015), triacylglycerol levels in red alga (Sumiya et al. 2015) and rice response to bacterial blight pathogen (Sana et al. 2010). However, there are only a few reports that describe an integrated transcriptomics and metabolomics approach to study insect-plant interactions and none describing rice-insect interactions.

Plant resistance against invading pests is a complex phenomenon revolving around the success of the plant defense response mounted against the invaders. The incompatible interaction between rice (*Oryza sativa*) and the Asian rice gall midge (AGM, *Orseolia oryzae* Wood-Mason; Diptera, Cecidomyiidae) is such an example of successful host defense that results in maggot mortality which could be accompanied by local necrotic death of host tissue (hypersensitive response, HR+) or without death of host tissue (HR-). The resistance in rice against AGM is governed by a gene-for-gene interaction involving the products of resistance gene (*R*-gene) of the host and avirulence gene (*avr*-gene) of the pest (Nair et al. 2011) and their interaction is shown to be closer to plant-pathogen interaction (Stuart et al. 2012, Rawat et al. 2013). Till now 11 *R*-genes (*Gm1* through *Gm11*), against AGM, have been characterized in the rice germplasm in India (Himabindu et al. 2010) and all of them are dominant *R*-genes except *gm3* and these *R*-genes are being utilized for developing gall midge resistant rice varieties through breeding. However, over-use of resistant rice varieties has led to emergence of new virulent biotypes of AGM (Bentur et al. 2003). Therefore, till date none of the *R*-genes is able to confer resistance against all the seven biotypes of gall midge prevalent in India and thus represents a constant co-evolving mechanism of resistance in rice and virulence in gall midge.

Out of the 11 *R*-genes (against AGM) known in rice, resistance associated with *Gm1* and *Gm8* is without invoking HR while resistance associated with the remaining 9 *R*-genes is manifested with HR (Bentur et al. 2011). Though resistance against gall midge is predominantly manifested with HR and results in maggot mortality, HR is not the prerequisite for maggot mortality (Bentur and Kalode 1996). In a pioneering study (Bentur and Kalode 1996), it was shown that an HR+ *indica* rice variety, Phalguna (harboring *Gm2*), was able to resist a secondary infestation by virulent larva but without the expression of HR following its incompatible interaction with gall midge biotype 1 (GMB1). The current study is based on another HR+ *indica* rice variety RP2068-18-3-5 (referred to as RP henceforth) that shows HR upon attack by gall midge biotype 1 (GMB1). RP harbors *gm3* which confers resistance, manifested with HR, against five of the seven GMBs found in India. Mapped to chromosome 4 of rice, *gm3* is the only recessive *R*-gene identified against gall midge thus far and efforts are underway to clone, characterize and functionally validate this gene (Sama et al. 2014). Thus, the resistance shown by RP against GMB1 is a complex trait which involves expression of HR, inability of the maggots to survive on the resistant host and association with a recessive resistance gene. The interaction between RP and GMB1 has never been studied before and therefore, evaluating the molecular and biochemical changes occurring in the host upon GMB1 attack will help us gain insights into the mechanism of resistance conferred by *gm3*.

Moreover, as resistance against AGM in rice is governed by a single gene understanding the mechanism of resistance will provide us valuable inputs for increasing the durability of each *R*-gene. Till date our understanding of rice-gall midge interaction at the molecular level was based on host transcriptomics involving compatible interaction between *indica* rice varieties and gall midge biotypes i.e. Kavya-GMB4M and TN1-GMB4 and similarly for the incompatible interaction between Kavya-GMB1 and Suraksha-GMB4 (Rawat et al. 2012a,b, 2013). This understanding was furthered by resorting to a metabolomics approach wherein metabolite biomarkers for resistance, susceptibility and infestation were identified during interactions between GMB1 and three different hosts i.e. TN1, Kavya and RP (Agarrwal et al. 2014) but did not provide us a picture of changes occurring in host metabolome at different stages of the interaction. Transcriptomics of the gall midge maggots feeding on susceptible or resistant host provided an insight into the changes occurring in the insect when interacting with its host (Sinha et al. 2012). However, thus far, there has been no integrated approach to study data obtained from the transcriptomics and metabolomics of rice-gall midge interaction. Therefore, in the current study, to understand the resistance mechanism of RP against GMB1 we carried out host transcript and metabolic profiling after GMB1 infestation to investigate the interaction at molecular and biochemical levels. We discuss the results of this study and highlight the complexity of the incompatible interaction, involving HR, and also generate a model based on our current findings and information available in the literature to provide an integrated understanding of resistance in RP against GMB1.

Results

Differential expression of genes and accumulation of metabolites in resistant host upon gall midge infestation

Gene expression study was performed on the host (*indica* rice variety RP2068-18-3-5; RP) tissue to evaluate the changes occurring in them upon gall midge infestation during an incompatible interaction with the avirulent biotype (GMB1). The un-infested tissue served as control. In the microarray experiment, the total number of probe sets detected were 57,381 out of which 43,738 probe sets filtered out and passed the quality check after data pre-processing and normalization. A total number of 7598 probes (differentially expressed genes, DEGs) showed statistically significant differential expression between infested (I) and un-infested (UI) host tissues at a p-value ≤0.05. In all, 2861 DEGs were retained in the list after applying a fold change filter (threshold = 2), out of which 1494 were up-regulated in infested tissue while 1367 were down-regulated as compared to un-infested tissues (RP-I/ RP-UI; Table 1, Additional file 1: Figure S1). Thus, the number of genes up-regulated upon infestation was marginally higher than the ones down-regulated. Gene Ontology (GO) analysis of the microarray data annotated the DEGs to 285 GO terms. However, after filtering the data at p-value ≤0.05 (corrected p-value) the number of GOs reduced to 249 (Additional file 2: Table S1). Out of these 249 GOs, only those GOs have been plotted which accounted for more than 10 % of total transcripts (Additional file 3: Figure S2). This figure signifies the dominance of biological process followed by metabolic process as GO categories to which the maximum number of transcripts were mapped i.e. more than 80 % and 60 %, respectively, of the total transcripts from the microarray data.

The metabolic profiles of the host tissue, at three distinct time points i.e. 24 hai (hours after infestation), 48 hai, and 72 hai, could distinguish RP-I from RP-UI when they were subjected to the Partial Least Squares-Discriminant Analysis (PLS-DA; Additional file 4: Figure S3). Of the total number of metabolites detected in metabolic profiles of host tissues, 45 metabolites showed higher accumulation and 41 metabolites showed lower accumulation in RP-I as compared to RP-UI at one or more time points. Thus, suggesting that a substantial metabolic reprogramming occurred in the host upon gall midge infestation. The metabolites showing differential accumulation (2-fold or more in either direction) in infested tissue as compared to un-infested tissue are discussed below (Additional file 5: Table S2).

Regulation of carbon metabolism in resistant host upon gall midge infestation

MapMan software (Thimm et al. 2004) was used to map the 2861 DEGs onto several pathways using *Oryza*-specific mapping files. Among the up-regulated pathways were tricarboxylic acid cycle (TCA), oxidative pentose phosphate pathway, fermentation, gluconeogenesis, mitochondrial electron transport and minor and major carbohydrate (CHO) metabolism (Fig. 1, Additional file 6: Table S3). The TCA cycle intermediates such as cis-aconitic acid, 2-oxo-glutaric acid and fumaric acid showed enhanced levels in host tissues upon infestation and so did the transcripts for citrate synthase, the enzyme catalyzing the very first step of this cycle while malic acid showed declined levels. The transcripts involved in Calvin cycle and photosynthesis were down-regulated while those involved in tetrapyrrole synthesis pathway (TSP) were highly up-regulated except a few transcripts, encoding enzymes involved in chlorophyll branch of TSP, which were notably down-regulated (Fig. 2, Additional file 7: Table S4). Of the total transcripts mapped to glycolysis and one-carbon metabolism, 50 % were up-regulated and 50 % were down-regulated (Fig. 1, Additional file 6: Table S3). Pyruvic acid, an important molecule in the glycolytic pathway, showed decline in its levels in infested tissues at 48 hai in accordance with the up-regulation shown by the transcripts encoding pyruvate phosphate dikinase, an enzyme utilizing pyruvate as substrate. Most of the organic acids such as glyceric acid, benzoic acid, malonic acid, adipic acid, oxo-butyric acid, lactic acid and methyl succinic acid involved in various metabolic pathways occurring in plant cells showed a decline in their levels in host tissue upon infestation. The other acids such as carbamic acid, glucuronic acid, threonic acid and hydroxy indole acetic acid showed enhanced accumulation in RP-I as compared to RP-UI. Myristic acid showed an early (at 24 hai) depletion in its levels in infested tissues but at 72 hai its levels went up as high as 17.82 times those found in un-infested tissues. Levels of azelaic acid, a known cellular metabolite in response to stress, were found to be depleted in infested tissues early during incompatible interaction while these levels increased at later stages i.e. 72 hai, the time point which coincides with HR in the host.

Differential accumulation of sugars and derivatives

Sugars are the main storage form of photosynthetically fixed carbon in plants. Levels of monosaccharides such as fructose and mannose declined in RP-I tissues as compared to RP-UI. However, at 48 hai, fructose and glucose levels went up in infested tissues as compared to un-infested tissues. The phosphorylated derivatives of hexoses (galactose and glucose) showed enhanced accumulation in infested tissue as compared to un-infested and so did the transcripts encoding for hexokinases and galactokinases. The levels of disaccharide sucrose and turanose declined in infested tissues as compared to un-infested tissues (Additional file 5: Table S2). A corresponding

Table 1 Summary of microarray results

Number of probes detected	Total number of probes filtered	Total number of DEGs ($p \le 0.05$)	Up-regulated[a] genes	Down regulated[a] genes
57381	43738	7598	1494	1367

DEGs: Differentially Expressed Genes
[a]Up- or down- regulated genes in rice tissues infested with the Asian rice gall midge biotype 1 (GMB1) as compared to un-infested tissues, at a fold change filter of 2. An *indica* rice variety RP2068-18-3-5 (RP) was used for the current study

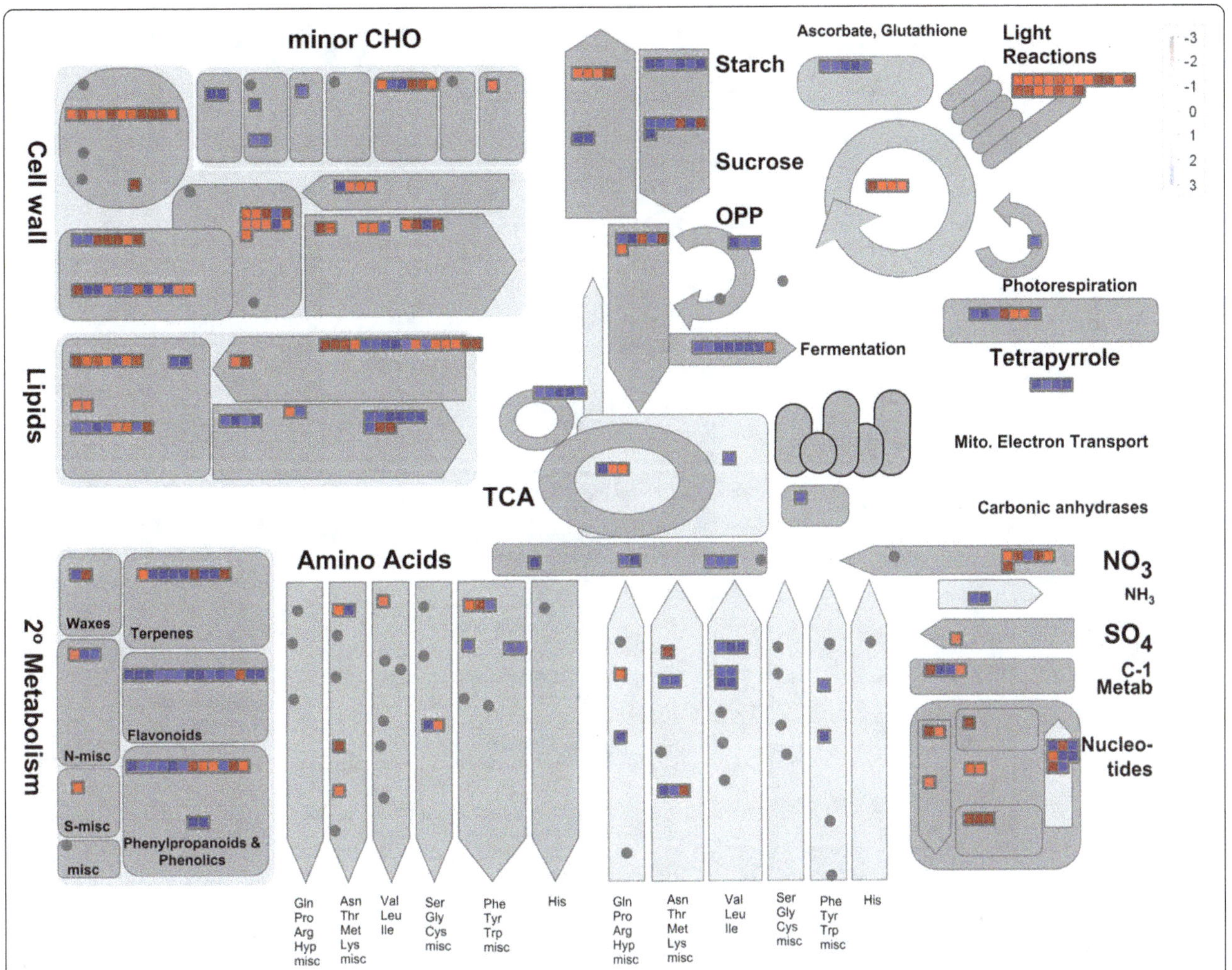

Fig. 1 MapMan-based representation of differentially expressed genes (DEGs) onto metabolic pathways. Mapping and distribution of DEGs in *indica* rice variety RP2068-18-3-5 (RP), challenged with the Asian rice gall midge biotype 1 (GMB1), onto metabolic pathway using MapMan software. Transcripts showing more than 2-fold difference between infested (RP-I) and un-infested (RP-UI) tissues of the resistant host RP have been mapped. Transcripts significantly up- and down-regulated are indicated in blue or red, respectively. Grey circles represent genes whose expression did not change more than 2-fold. For detailed information on these transcripts refer Additional file 6: Table S3

up-regulation of the transcripts encoding invertases, a class of enzyme that catalyzes the breakdown of sucrose to constituting monosaccharides, was observed. Trehalose, another disaccharide, showed elevated levels which was reflected in down-regulation of transcripts involved in trehalose metabolism. The levels of sugar alcohols such as sorbitol, arabinitol and myo-inositol decreased in host tissues upon gall midge challenge which is in line with the up-regulation of transcripts involved in minor CHO metabolism (Fig. 1, Additional file 6: Table S3).

Regulation of nitrogen metabolism in resistant host upon gall midge infestation

Microarray results revealed that the transcripts involved in polyamine and amino acid metabolism, nitrogen metabolism (especially those encoding for nitrate reductase) were up-regulated (Fig. 1, Additional file 6: Table S3). The analysis of metabolic profiles of host tissues revealed differential accumulation of amino acids and derivatives (Additional file 5: Table S2). Among the amino acids aspartic acid and the un-branched amino acids such as glycine and alanine showed depletion in their levels upon infestation while other amino acids including cysteine, isoleucine, valine, histidine and glutamine showed enhanced accumulation. Levels of leucine, lysine and threonine were elevated which probably triggered the up-regulation of transcripts involved in leucine, lysine and threonine degradation pathway. Aromatic amino acids such as phenylalanine and tyrosine (along with their precursor shikimic acid) and tryptophan, showed an increase in their levels upon infestation which was reflected in the up-regulation of the transcripts involved in the biosynthesis of these aromatic amino acids while the transcripts for shikimate kinase and chorismate synthase were down-

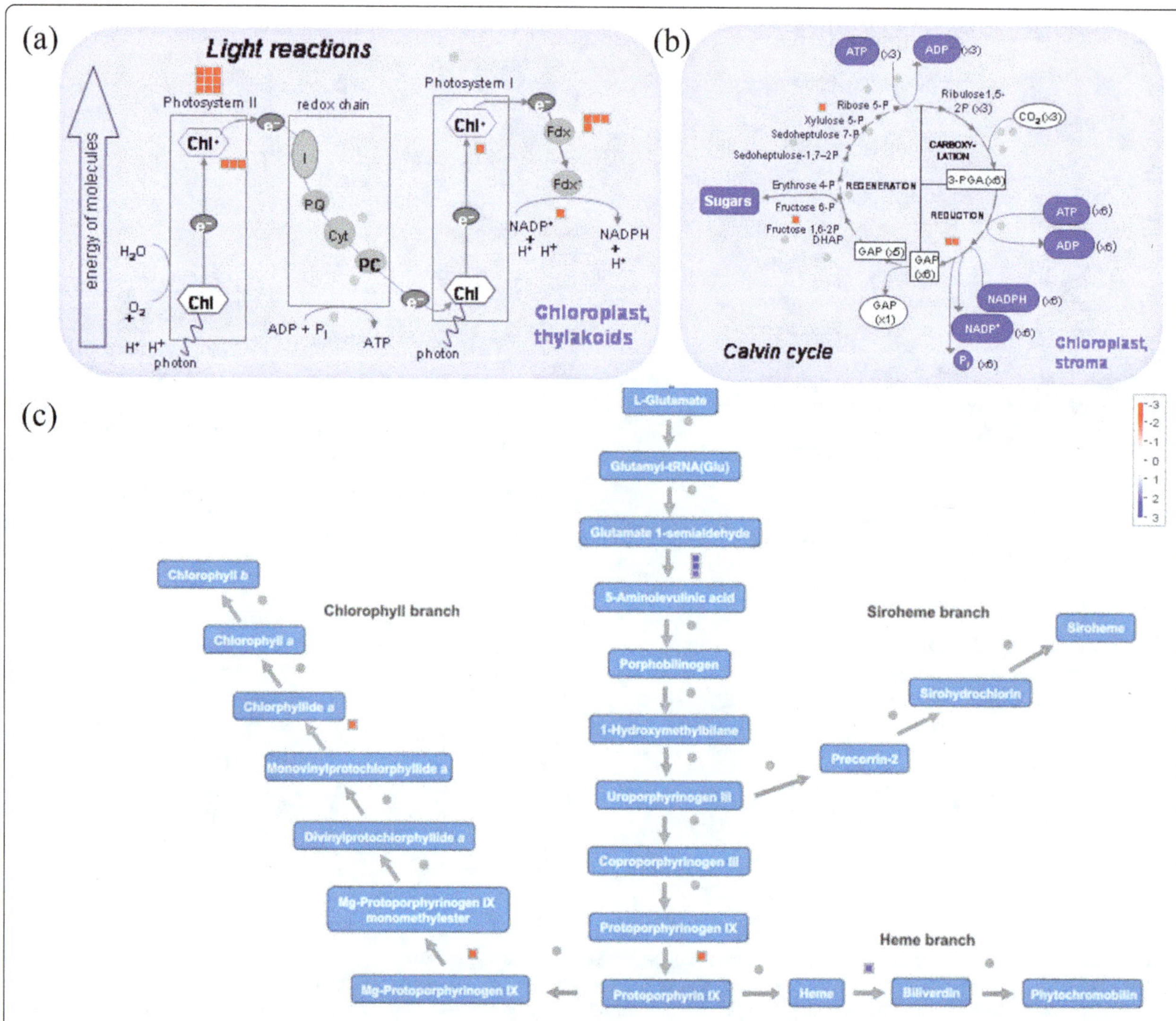

Fig. 2 Summary of differentially expressed genes (DEGs) involved in photosynthesis and tetrapyrrole synthesis pathway. Mapping and distribution of DEGs, involved in photosynthesis and tetrapyrrole synthesis pathway (TSP), in *indica* rice variety RP2068-18-3-5 (RP), challenged with the Asian rice gall midge biotype 1 (GMB1). (a) DEGs involved in light reaction of photosynthesis; (b) DEGs involved in Calvin cycle; (c) DEGs involved in TSP. Transcripts showing more than 2-fold change between infested (RP-I) and un-infested (RP-UI) tissues of the resistant host RP have been mapped. Transcripts significantly up- or down-regulated are indicated in blue or red, respectively. Grey circles represent genes whose expression did not change more than 2-fold. For detailed information on these transcripts refer Additional file 7: Table S4

regulated upon infestation. Transcripts of simple phenol metabolism were found to be up-regulated which was reflected by the levels of ferulic acid and esculin (tyrosine-derived phenolic acid and coumarin glucoside, respectively) that had depleted in infested tissues as compared to un-infested tissues. The levels of other amino acid precursor/derivative such as acetyl lysine and oxo-proline (or L-pyroglutamate) also went up in infested tissue. The other nitrogenous metabolites such as ornithine and gamma amino butyric acid (GABA) had increased accumulation whereas beta-alanine and urea levels had fallen in infested tissue (Additional file 5: Table S2). In accordance, GABA transaminase, a key enzyme in GABA-shunt was highly up-regulated in the resistant host.

Regulation of fatty acid and lipid metabolism in resistant host upon gall midge infestation

Transcripts responsible for fatty acid synthesis, fatty acid elongation, fatty acid desaturation, and phospholipid and glycolipid synthesis were also down-regulated upon infestation during incompatible interaction between RP and GMB1 (Additional file 8: Table S5). In addition, transcripts playing a role in lipid degradation and metabolism of 'exotics' such as steroids and squalene and xenobiotic degradation were found to be up-regulated (Additional file 8: Table S5). Levels of most of the fatty acids including saturated fatty acids such as arachidic acid, tetradecanoic acid, margaric acid, palmitic acid and stearic acid, and unsaturated fatty acids such as elaidic acid and oleic acid declined in RP-I when compared to RP-UI while docosanoic acid was the only fatty acid to show increased levels in infested host tissues (Additional file 5: Table S2). This was well reflected in down-regulation of transcripts involved in fatty acid synthesis and elongation. Dodecanoic acid was the only fatty acid to show a fluctuation in its levels depending upon the stages of gall midge feeding (Additional file 5: Table S2). Levels of the membrane lipid ethanolamine and its phosphorylated form were enhanced in the infested tissue (Additional file 5: Table S2).

Regulation of cellular redox state and cell development in resistant host upon gall midge infestation

The transcripts related to redox were up-regulated (Fig. 3). Among the transcripts coding for enzyme families, oxidases were mostly down-regulated while peroxidases were mostly up-regulated (Additional file 8: Table S5). A majority of transcripts involved in cell organization, cell division and cell cycle, cellular vesicle transport as well as storage proteins required during cell development were also found to be down-regulated along with the transcripts involved in cell wall precursor synthesis, cellulose synthesis and cell wall degradation as well as those encoding cell wall proteins including proline rich proteins and pectin esterases. The transcripts representing genes involved in nucleotide metabolism, DNA synthesis and repair, maintenance of chromatin assembly, RNA processing and splicing (including RNA helicases and ribonucleases) were found to be down-regulated (Additional file 8: Table S5).

Regulation of differential gene expression by transcription factors

Overall the transcription machinery was down-regulated which included several transcription factors such as E2F-DP, ORPHAN, SBP, Chromatin remodelling factors, B3, DNA methyl transferases (MT), general transcription factors, Jumonji, PHD finger, Polycomb group (PcG), silencing group, high mobility group (HMG), SET-domain containing transcription factors. Several transcripts involved in DNA synthesis and repair including histone and histone acetyl transferase, nucleosome/chromatin assembly factors were also down-regulated upon gall midge challenge in RP (Additional file 9: Table S6). Ethylene-responsive element binding protein (EREBP) family, basic helix-loop-helix (bHLH) family, WRKY domain containing family, NAC domain containing family, C3H, C2C2-CO-like, C2H2 and C2C2-Dof zinc finger family, homeobox (HB), MYB-related, MADS box, heat shock factor (HSF), bZIP, Argonaute, auxin and auxin-responsive family of transcription factors were the few up-regulated transcription factors (Fig. 3).

Semi-quantitative RT-PCR results reveal temporal variation in gene expression in host upon gall midge infestation

Since the microarray results were obtained from pooled RNA samples (see material and methods), the temporal variation in gene expression could not be observed. Therefore, the expression patterns of 14 selected genes showing greater than 10-fold up- or down-regulation were studied further at three time points *viz.*, 24, 48, and 72 hai by semi-quantitative RT-PCR (Additional file 10:

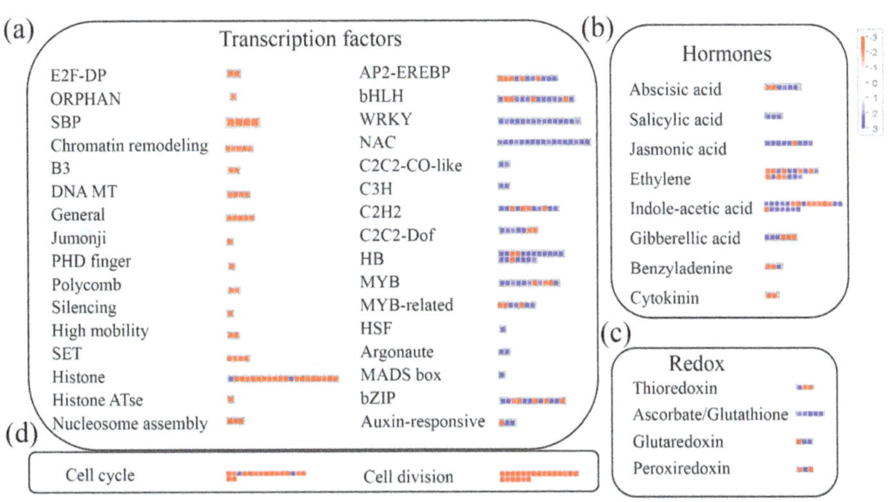

Fig. 3 Summary of differentially expressed genes (DEGs) involved in regulation of cellular functioning. Mapping and distribution of DEGs, involved in regulation of cellular functioning, in *indica* rice variety RP2068-18-3-5 (RP), challenged with the Asian rice gall midge biotype 1 (GMB1). (a) DEGs encoding transcription factors; (b) DEGs involved in hormone synthesis; (c) DEGs involved in maintaining cellular redox state; (d) DEGs involved in cellular organization. Transcripts showing more than 2-fold change between infested (RP-I) and un-infested (RP-UI) tissues of the resistant host RP have been mapped. Transcripts significantly up- or down-regulated are indicated in blue or red, respectively. For detailed information on these transcripts refer Additional file 9: Table S6

Table S7). The expression pattern of three genes out of the six most down-regulated genes showed variability at different time points (Fig. 4). Two of these (CytoP450 and prolyl endopeptidase) showed up-regulation at 48 hai while the third gene (Histone H3) showed down-regulation at early hours of infestation (24 hai) but thereafter its relative expression increased in infested tissues as compared to un-infested tissues. Out of the eight most up-regulated genes, two (LEA and an uncharacterized gene) showed down-regulation at 72 hai and 24 hai, respectively (Fig. 4). The expression of aminotransferase y4uB gene was highly up-regulated in RP at all three time points studied. However, when expression patterns were monitored in other rice varieties showing compatible or incompatible interaction (HR-) with GMB1, we did not observe a distinctive expression profile (data not shown).

Discussion

The over-representation of GO terms such as biological process and metabolic process (Additional file 3: Figure S2 and Additional file 2: Table S1) among the deregulated genes (includes both up- and down-regulated genes) in the host indicates perturbation in the biological and metabolic processes of the resistant host upon gall midge infestation. Concurrent differential accumulation of more than 80 metabolites at one or more time points and the clear distinction in metabolic profiles of infested and un-infested tissues harvested at different time points (as revealed by PLS-DA: Additional file 4: Figure S3), highlights the observed perturbations in the metabolic processes of the resistant host during its interaction with an avirulent gall midge biotype. Our results thus show that extensive reprogramming of the host transcriptome as well as metabolome occurs upon gall midge challenge. This reprogramming of both the biological and metabolic processes could be an attribute of the underlying defense response that is induced in the resistant host upon initiation of insect feeding along with the concomitant recognition of the avirulent gall midge biotype. Furthermore, overlapping results obtained from the microarray and metabolic profiling experiments reveal that host carbon and nitrogen (C/N) metabolism is differentially regulated upon gall midge infestation. The observed decline in levels of carbon-containing compounds and a corresponding increase in nitrogen-containing compounds (Additional file 5: Table S2) indicate a shift in host carbon and nitrogen metabolism (C/N shift) towards N-containing compounds upon gall midge infestation. The coordinated regulation of TCA cycle, glycolysis, minor and major CHO metabolism, amino acid metabolism and nitrogen metabolism (Fig. 1) may have contributed to this C/N shift in the host cells during the incompatible interaction between RP and GMB1. This is in contrast to earlier observations made for the Hessian fly-wheat interaction by Zhu et al. (2008) where a more dramatic C/N shift towards nitrogen in the host was observed during compatible interaction as compared to incompatible interaction with the pest. These observations indicate that C/N metabolism is not only vital to plant growth and development but has a role in plant defense as well. This is further supported by the observation that the ATL31, a C/N metabolism regulator, positively regulates the basal defense response in *Arabidopsis* (Maekawa et al. 2012) by controlling papilla formation during fungal attack (Maekawa et al. 2014). Thus, based on our current findings, we propose that C/N shift is a major reflection of the transcriptomic and metabolomic reprogramming of the host brought about by its interaction with an avirulent pest. In addition, we suggest

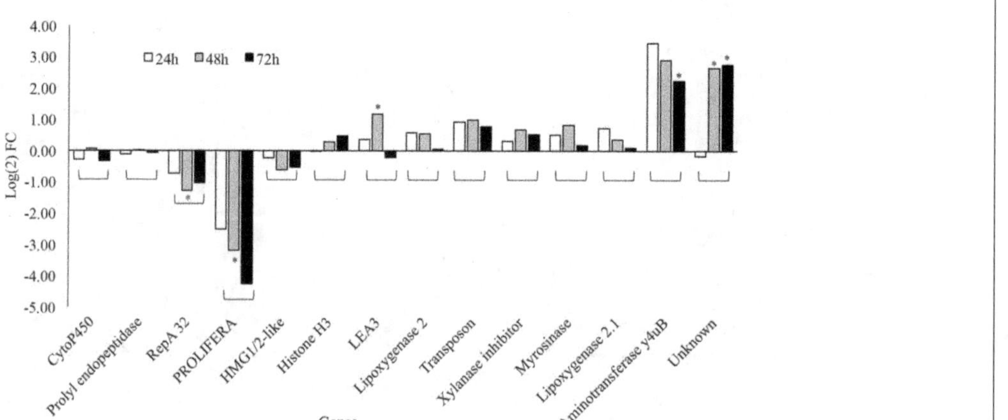

Fig. 4 Validation of microarray results by semi-quantitative RT-PCR. Graph represents the expression patterns of genes in *indica* rice variety RP2068-18-3-5 (RP), challenged with the Asian rice gall midge biotype 1 (GMB1), studied at three distinct time points of infestation i.e. 24 h, 48 h and 72 h. Y-axis depicts the log2 values of fold change (FC) in relative expression values (REV) of genes (mentioned on X-axis) between infested (RP-I) and un-infested (RP-UI) samples. Genes with significantly different (p-value ≤0.05) REVs between infested and un-infested samples have been indicated by (*) on the bars. RP: *indica* rice variety RP2068-18-3-5; h: hai

that though the observed C/N shift may be responsible for the induction of a basal defense in resistant and susceptible hosts alike, this basal defense may be further augmented by other defense responses in an incompatible interaction.

GABA shunt plays a key role in plant C/N metabolism and the growing evidences suggest that it is closely linked to TCA cycle occurring in the cell (reviewed by Fait et al. 2008, Michaeli and Fromm 2015). The three enzymatic reactions occurring in GABA shunt are catalysed by glutamate decarboxylase (GAD), GABA transaminase (GABA-T) and succinic semialdehyde dehydrogenase (SSADH). In our microarray data, the highly up-regulated gene (LOC_Os04g52440; an aminotransferase y4uB), was mapped as GABA-T and was more than 27-fold up-regulated in RP-I as compared to RP-UI (Additional file 6: Table S3), a trend that was validated by semi-quantitative RT-PCR (Fig. 4). The GABA-T mutants of *Arabidopsis* accumulate GABA at much higher levels than wild type (Palanivelu et al. 2003) and are thus resistant to bacterial pathogens (Park et al. 2010). However, in our current study, despite several fold up-regulation of GABA-T transcripts upon infestation, the GABA levels increased in infested tissues of the resistant host (Additional file 5: Table S2), thus suggesting that the observed higher accumulation of GABA in infested tissues is the unutilized substrate after GABA-T activity. Induced accumulation of GABA in pepper plants was suggested to contribute to *R*-gene mediated resistance against *Xanthomonas* and enhanced the cell death mediated defense response in the host against an avirulent pathogen (Kim et al. 2013). Additionally, elevated levels of GABA were shown to contribute towards soybean resistance against leaf roller (Ramputh and Bown 1996) as well as resistance against phytophagous insects (MacGregor et al. 2003) and nematodes (McLean et al. 2003) in transgenic tobacco. Since GABA acts as a neurotransmitter in vertebrates and invertebrates, it is probable that GABA ingested by plant-feeding insects may interfere with normal development of the larva and eventually resulting in larval mortality (Ramputh and Bown 1996). Therefore, we suggest that up-regulation of genes involved in GABA-shunt and induced accumulation of GABA in rice likely play a crucial role in *R*-gene mediated resistance against gall midge.

The resistance shown by RP (conferred by *gm3*) against GMB1 is a complex trait which involves expression of HR and the subsequent inability of the maggots to survive on the resistant host. When attacked by biotrophic pathogens, plants use HR as a strategy to limit the invading pathogen by sacrificing some of their own cells. HR is triggered by oxidative burst within the cells that releases several reactive oxygen species (ROS) such as hydrogen peroxide, superoxide radical and singlet oxygen in the cells under attack. ROS, by-products of aerobic metabolism (Halliwell 2006), are generated in peroxisomes and chloroplasts in plants cells (Foyer and Noctor 2003, Asada 2006). In mammals, ROS production is mostly NADPH-oxidase dependent (Apel and Hirt 2004) while in plants, class III peroxidases may also be responsible for ROS generation. These class III peroxidases have been shown to be effective in mounting a hydrogen peroxide-induced defense response in resistant wheat and in the non-host rice against another Cecidomyiid, the Hessian fly (Liu et al. 2010). While the major ROS scavenging genes were not affected during the Hessian fly-wheat interaction, the genes involved in redox, including the major ROS scavenging enzymes, were mostly up-regulated in the present study (Fig. 3 and Additional file 9: Table S6) and thereby, in all probability, allowing effective scavenging of H_2O_2 molecules during rice-gall midge interaction. This indicates that HR observed in RP during its interaction with GMB1 is induced by ROS other than hydrogen peroxide unlike what was observed during the Hessian fly-wheat interaction. It has been shown that the over-expression of ascorbate peroxidase (H_2O_2 scavenger) enhanced the intensity of stress response that was mediated by singlet oxygen in *Arabidopsis* (Laloi et al. 2007). Thus, our results point towards the possibility of the presence of an alternate pathway for induction of HR during rice-gall midge interaction that is likely induced by singlet oxygen.

Singlet oxygen-responsive gene networks are, however, not independent of H_2O_2 signaling (Murgia et al. 2004) but the interaction between the two reactive species has been described to be antagonistic (Laloi et al. 2007). Singlet oxygen is produced by photosystem II when energy is transferred from excited chlorophyll to oxygen (Krieger-Liszkay 2005) or from molecules acting as potent photosensitizers [protoporphyrin IX, magnesium protoporphyrinogen IX and protochlorophyllide: substrates of the enzymes involved in chlorophyll branch of tetrapyrrole synthesis pathway (TSP)] to oxygen (Meskauskiene et al. 2001, op den Camp et al. 2003). In the current study, the TSP was overall up-regulated in the resistant host upon gall midge attack but, notably, the transcripts coding for enzymes involved in chlorophyll branch of TSP were all down-regulated (Fig. 2). This indicates reduced chlorophyll synthesis that is also reflected by the observed down-regulation of transcripts involved in photosynthesis and Calvin cycle (Fig. 2). This is in line with the reported reduction in photosynthetic activity after initiation of HR in *Arabidopsis* upon *Pseudomonas syringae* infection (Berger et al. 2007). The progressive disintegration of the photosynthetic machinery during HR led to enhanced production of singlet oxygen and singlet oxygen-mediated lipid peroxidation (LPO; Zoeller et al. 2012). Photosynthesis and chlorophyll synthesis are localized in plastids and plastids are present in the dividing apical meristematic cells apart from leaf cells. Rice gall midge maggots feed inside rice plants at the apical meristem, where there is probably little photosynthesis as compared to leaves. But data from our present study clearly showed down-regulation of photosynthesis and chlorophyll branch of TSP and an up-regulation of other transcripts involved in TSP (Fig. 2) in this region of the rice plant after gall midge infestation when compared with un-infested plants. Thus, we suggest that disintegration of photosynthetic machinery leads to generation of singlet oxygen which in turn triggers HR that is observed in the host cells upon gall midge attack.

Under biotic stress, the singlet oxygen-mediated LPO in *Arabidopsis* was further enhanced by enzymatic (lipoxygenases, LOX) and non-enzymatic (free radical induced) peroxidation of membrane lipid leading to accumulation of marker metabolites such as azelaic acid (Triantaphylidès et al. 2008, Zoeller et al. 2012). Though the extent of LPO occurring in host cells was not quantified in the present investigation, azelaic acid levels in infested tissues were found to be lower than that in un-infested tissues at early stages of gall midge feeding (24 hai) but subsequently (72 hai), these levels increased in infested tissues as compared to un-infested tissues (Additional file 5: Table S2). This increase in levels of azelaic acid at later stages of gall midge infestation is indicative of LPO occurring during late stages of infestation when the host cells express HR and die. Additionally, the microarray results revealed several fold up-regulation of transcripts encoding for lipoxygenases (Additional file 8: Table S5) including LOX2 and LOX2.1. Their expression patterns were also confirmed by semi-quantitative RT-PCR (Fig. 4) which showed higher transcript abundance in infested tissues as compared to un-infested tissues at three distinct time points. In accordance with the findings of Zoeller et al. (2012), we suggest that host cell death occurring during HR in RP upon gall midge attack, is due to oxidative damage to cellular membranes triggered by singlet oxygen and which is likely further enhanced by LOX enzymatic activities. Furthermore, microarray results in the current investigation showed several fold up-regulation of transcripts involved in lipid degradation. Previously, it has been shown that mobilization of membrane lipids through lipolysis in resistant wheat plants constituted an important link in its defense against the Hessian fly (Khajuria et al. 2013). Similarly, the oxidation of lipid membranes may in turn help in mediating defense response of rice host against the avirulent gall midge maggots.

To generate a stress response, singlet oxygen signaling is coordinated with other ROS-signaling (Laloi et al. 2007) or phytohormone-signaling (Danon et al. 2005). The involvement of different phytohormones (ethylene, jasmonate, salicylic acid) in modulating the singlet oxygen-mediated cell death during stress in *Arabidopsis* mutants has also been demonstrated earlier (Danon et al. 2005, Ochsenbein et al. 2006). Similarly, a concurrent activation of phytohormone synthesis was observed in the current study which was reflected in up-regulation of genes involved in synthesis of three major phytohormones i.e. ethylene, jasmonate and salicylate (Fig. 3 and Additional file 9: Table S6). Additionally, singlet oxygen has also been shown to control the expression of nuclear encoded genes by means of several transcription factors. In the current investigation, the transcription factors such as ethylene responsive factors (EREBP), WRKY transcription factors, zinc finger proteins, NAC domain containing and MYB were up-regulated in the host after gall midge attack (Fig. 3). Several of these transcription factors have been reported to show quick and enhanced expression in plants upon release of singlet oxygen (Gadjev et al. 2006, Kim et al. 2008) and are known to be involved in plant defense response against biotic stress (van Verk et al. 2009). Thus, results of the current study indicate that the release of singlet oxygen not only triggers HR and lipid peroxidation in the host cells but also controls the host defense against gall midge *via* phytohormones signaling and/or up-regulation of several transcription factors.

The stress response induced by singlet oxygen could be due to its direct cytotoxicity (Meskauskiene et al. 2001) or the singlet oxygen triggered genetic reprogramming of the cells (Wagner et al. 2004). In the current study, many genes involved in DNA synthesis and repair, maintenance of chromatin assembly, cell cycle and cell organization showed several fold down-regulation in RP-I as compared to RP-UI (Fig. 3 and Additional file 9: Table S6) indicating the disorganization of cellular machinery and the preparations of the cells ahead of programmed cell death or HR. The reduced expression of these genes in the resistant host, upon gall midge challenge, was confirmed by semi-quantitative RT-PCR (Fig. 4) and one of them, *Prolifera*, showed statistically significant down-regulation in RP-I as compared to RP-UI. *Prolifera* is a member of mini chromosome maintenance (MCM) protein family involved in DNA replication and was found to be expressed in giant cells in *Arabidopsis* upon successful nematode infection (Huang et al. 2003). Root-knot nematodes alter cell cycle regulation and establish the feeding sites termed as giant cells to draw nutrients from host cells (Goverse et al. 2000). Similarly, genes involved in chromatin remodeling and maintenance were observed to be over-expressed at late stages of giant cell formation in plants during compatible interaction with root-knot nematodes (Portillo et al. 2013). Several genes related to cell organization and DNA synthesis were also found to be up-regulated in the rice host during compatible interactions with gall midge (Rawat et al. 2012b). However, down-regulation of the genes involved in cell organization i.e. *Prolifera*, in the gall midge resistant variety RP during HR, indicates that the avirulent maggots are unable to induce expression of host genes that are involved in establishment/development of feeding site. This may be because, in accordance to plant vigour hypothesis (Price 1991), gall midge (like other galling insects) requires a healthy host for successful establishment of feeding sites and formation of nutritive tissue (Agarrwal et al. 2014) which is not available in the resistant host undergoing cell death, at the maggot feeding site, during HR. Thus, the resistant host is able to prevent the feeding maggots from continued feeding by disrupting the source-sink nutrient channeling by killing some of its own cells at the insect-feeding site during HR.

The pre-dominance of HR in rice gall midge resistance is revealed by the fact that resistance associated with 9 of the 11 *R*-genes known (exceptions *Gm1* and *Gm8*) is manifested with HR (Bentur et al. 2011). However, the HR-induced host cell death at the gall midge feeding site is not the cause of maggot mortality (Bentur and Kalode 1996) and actual death of maggots could be a result of the inability of the maggots to continue feeding on resistant hosts. A recent study on incompatible interaction between another gall midge (*Dasineura marginemtorquens*) and willow (*Salix viminalis*) revealed that the loci for HR and the locus for resistance were not co-located (Höglund et al. 2015). The authors proposed that the resistance of the plant host against the galling insect could be explained by absence of galls i.e. lack of induced susceptibility. Loss-of-susceptibility rather than induced defense has often been used to explain the phenomena of recessive resistance (reviewed by Pavan et al. 2010). However, studies have also shown that recessive resistance is not always associated with loss-of-susceptibility but may also be associated with induced defense response (Gonzalez-Ibeas et al. 2012). Recessive resistance could also mean that the product of the dominant allele acts as a negative regulator of plant immunity and when not expressed in individuals harboring the recessive allele, it leads to the de-repression of the plant defense and consequently resulting in resistance (Pavan et al. 2010). This would probably explain the gall midge resistance phenomenon in RP conferred by *gm3* – the only recessive *R*-gene identified against gall midge thus far.

Thus, the integrated study of rice host transcriptome and metabolome during its interaction with an avirulent biotype of AGM provides important revelations with regard to the resistance mechanism of the host that is associated with a recessive *R*-gene and manifested through HR. Based on our combined analysis of results obtained from transcriptomics and metabolomics, we propose that by expressing HR the host does not kill the invader directly but does so by reducing the nutrient availability to the feeding maggots, by sacrificing some of its own cells, and then hampering the development of the insect by accumulating inhibitory neurotransmitters such as GABA. However, to prove this hypothesis experimentally, an artificial diet for feeding rice gall midge maggots needs to be established. This diet could then be supplemented with antifeedants such as GABA to study their effects on survival rates of maggots. Additionally, the present study revealed novel genes, e.g. the unknown gene (LOC_Os01g40290), that showed maximum fold change in microarray analyses results. These could be evaluated further for understanding their role in conferring resistance against gall midge in particular and also possibly other stresses in general.

Conclusions

Based on the results of our study, we conclude that extensive reprogramming of the transcriptome and metabolome occurs in the rice host (RP) upon gall midge (GMB1) infestation leading to an array of events that likely induce hypersensitive response (HR)-mediated host resistance against gall midge. The transcriptomic and metabolic reprogramming lead to a shift in C/N metabolism including induction of enzymes involved in GABA shunt – an important link between C/N metabolism and accumulation of GABA, that could be toxic to the feeding maggots, along with the concomitant release of reactive oxygen species, such as singlet oxygen, that trigger HR in the host. The cell death during HR is likely mediated by lipid peroxidation, releasing free fatty acids which are further involved in defense signaling. The host cell death, as a consequence of HR, and the breakdown of cellular machinery at the feeding site in the resistant host RP do not allow the maggots to continue feeding, leading to their starvation – a demonstration of one of the several defense strategies employed by RP to defend against GMB1 (Fig. 5).

Methods

Gall midge infestation and plant sample collection

The Asian rice gall midge (*Orseolia oryzae* Wood-Mason) biotype 1 (GMB1) was used to challenge the resistant rice variety RP2068-18-3-5 (RP henceforth) under greenhouse conditions [at the Indian Institute of Rice Research, (formerly Directorate of Rice Research) Hyderabad] for the current study. The RP seeds were sown in 4 rows, 5 cm apart, with 25 plants in each row, in plastic trays (60 x 30 x 30 cm) that were filled with puddle nutrient enriched soil. After 15 days, plants in some trays were exposed to GMB1 females ($n = 25$) and males ($n = 10$) and plants in other trays were left un-exposed but all trays were treated alike. After 48 h of insect release, all the trays were transferred to humidity chamber (>90 % relative humidity) and were maintained there to facilitate egg hatching and maggot

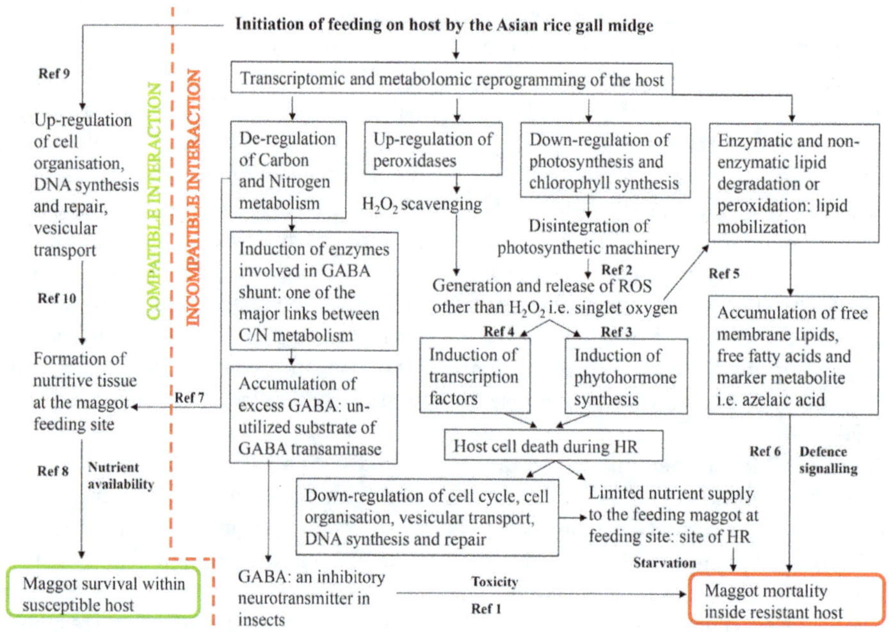

Fig. 5 Model representing flow of events occurring during rice-gall midge compatible and incompatible interactions. Representation of flow of events related to rice-gall midge compatible and incompatible interactions (separated by dotted lines). Text within green and red boxes represents outcome of compatible and incompatible interactions, respectively. The attack by the Asian rice gall midge on a resistant host induces extensive transcriptomic and metabolomic reprogramming leading to an array of events that induce hypersensitive response (HR)-mediated host resistance against gall midge. The deregulation of carbon and nitrogen metabolism induces overexpression of genes involved in GABA shunt and accumulation of GABA that could be toxic to the feeding maggots (Ref 1) along with the concomitant release of reactive oxygen species, i.e. singlet oxygen, that trigger HR in the host. The disintegration of photosynthetic machinery leads to generation of singlet oxygen (Ref 2), which further mediates defense signaling involving phytohormones (Ref 3) and transcription factors (Ref 4). Singlet oxygen also induces lipid peroxidation (Ref 5) causing lipid mobilization that plays a role in defense signaling through release of free fatty acids (Ref 6). On the other hand, C/N shift towards nitrogen, during compatible interaction (Ref 7), favors maggot growth by promoting formation of nutritive tissue (Ref 8). Up-regulation of genes involved in cell cycle and organization during compatible interaction (Ref 9) also promotes establishment of nutritive tissue at feeding sites (Ref 10). In contrast, host cellular machinery breaks down at the site of HR during incompatible interaction. Therefore, the host cell death, during HR at feeding site, limits nutrient supply to the feeding maggots and prevents formation of nutritive tissue. The flow of events in the given model is re-constructed based on information available from the current study (text within black boxes) and earlier studies on other insect-plant interactions for which appropriate references have been cited in Additional file 11 : References S1

establishment. The plants were periodically checked for maggot establishment and once it was ascertained, this time point was considered 24 hai (hours after infestation). As previously described (Rawat et al. 2010, 2012a) plant tissues (RP-I) were harvested from the feeding site (1 cm region of the meristematic tissue at the maggot feeding site) at 24, 48 and 72 hai. Maggots, both live (at 24 hai and 48 hai) and dead (at 72 hai), were carefully removed from the plant tissues. The tissues from un-infested plants (RP-UI) served as controls and were also excised at the same time points as the infested plants. This exercise was repeated for two consecutive years during *kharif* (monsoon) season (2012 and 2013).

RNA extraction and microarray for gene expression analysis
Only plant samples (RP-I and RP-UI) from 2013 *kharif* season were used for RNA extraction. The frozen plant tissues (50–100 mg) were ground to a fine powder in liquid nitrogen using a micro-pestle. RNA extraction was carried out using Plant Total RNA Miniprep Kit (mdi Membrane Technologies, USA) as per manufacturer's instructions. The RNA extracted from tissues representing different time points were pooled and an aliquot of 16 μL (200 ng/μL) was utilized for microarray experiments. Three independent biological replicates were used for transcriptome analyses on GeneChip Rice Genome Array (Affymetrix, USA). The quality of RNA (RIN values) was checked using Agilent Bioanalyzer 2100 (Agilent Technologies, USA). Labeling of RNA, target hybridization, washing-staining-scanning of arrays were carried out by M/s Innovative Life Discoveries (Gurgaon, India). Pre-processing of raw data (.CEL files) was performed using Robust Multichip Average (RMA) algorithm on GeneSpring (Gx13.1) software (Agilent Technologies, USA) and after normalization, only those probe sets having intensities between 10–100 percentile were included in the data. To identify the statistically significant differentially expressed genes (DEGs),

moderated *t*-test method (Smyth 2004) was applied and *p*-value cut off 0.05 was considered significant. Only statistically significant DEGs ($p \leq 0.05$) were subjected to fold change analysis and a filter of 2-fold change between the two conditions i.e. infested versus un-infested was applied and a volcano plot (Cui and Churchill 2003) was constructed (Additional file 1: Figure S1). MapMan software (Thimm et al. 2004) was used to visualize the distribution of DEGs into different bins and their appearance in different metabolic pathways using *Oryza* - specific mapping file. The sequence of events involved in processing of data is presented in Additional file 12: Method S1.

cDNA synthesis and validation of gene expression results by semi-quantitative RT-PCR

The RNA samples from tissues representing different time points (24, 48 and 72 hai) were used to synthesize the first strand cDNA using Verso cDNA synthesis kit (Thermo Scientific, USA). The protocol was slightly modified and is as follows: the reaction mix containing 1 μL of total RNA along with other constituents of the kit (5X cDNA synthesis buffer, Anchored Oligo dT, dNTP mix, RT enhancer) was incubated at 70 °C for 10 min followed by a quick transfer to 42 °C for another 10 min. Next, the verso enzyme mix was added and the cDNA synthesis was carried out at 42 °C for 90 min after which it was terminated by heating the reaction mix at 95 °C for 2 min. The cDNA, thus synthesized, was cleaned up using QIAquick PCR purification kit (Qiagen, Netherlands) and quantified using NanoVue (GE Healthcare Lifesciences, USA). To validate the expression pattern of highly up-regulated and down-regulated genes, semi-quantitative RT-PCR was performed with cDNA in two technical replicates for each of the three biological replicates. Gene-specific primers for the semi-quantitative RT-PCR were designed using Primer Express Software (Applied Biosystems, USA). The PCR mix (1X; 25 μL) contained 20 ng of cDNA, 0.5 μM primers (forward and reverse each), 200 μM each of the dNTPs, 1.2 Units of Taq polymerase and Taq buffer (Bangalore Genei Pvt Ltd, India). The optimum PCR conditions including cycle number and cDNA amounts were standardized for each gene separately. After the PCR, products were run on 2 % agarose gel (stained with EtBr) at 60 V for 1 h. These gels were then photographed under ultraviolet rays, using the Alpha Imager EP system (Cell Biosciences, USA). The captured images were analyzed using ImageJ software (Schneider et al. 2012). A total of 15 genes were analyzed by semi-quantitative RT-PCR including 8 up-regulated, 6 down - regulated genes and one reference gene. Rice ubiquitin (GenBank accession number NM_001049450; Forward primer- GCGTAGGCTCCTGTTCTTTGG; Reverse primer- AGGGCATCACAATCTTCACAGA) was used as a reference gene for normalization and the fold change values were calculated between the relative expression values (REVs) of infested (RP-I) and un-infested (RP-UI) tissues. A graph was constructed based on the log (2) values of the fold change using MS Excel (Microsoft, USA). The amplification products were cloned and sequenced (M/s Macrogen Inc, South Korea) to confirm their identity.

Gas chromatography–mass spectrometry (GC-MS) based metabolic profiling

The metabolic profiling was carried out in two independent GC-MS experiments performed on rice tissues, obtained from two consecutive *kharif* season crops, exposed to gall midge under greenhouse conditions. The tissues (RP-I) were exposed to gall midge biotype 1 (GMB1) and plant tissues were excised from stem region at three time points 24, 48 and 72 hai. The stem tissue excised from rice plants not exposed to GMB1 (un-infested, RP-UI) were used as control. The current study involved three independent biological replicates from each year's sample collection. The infested and un-infested tissues were treated alike and the extracted metabolites after derivatization were analyzed by gas chromatography mass spectrometry (GC-MS). Metabolite extraction and derivatization was performed using the method described earlier (Agarrwal et al. 2014). Briefly, the plant tissue was ground to a fine powder using liquid nitrogen and the metabolites were extracted in cold solvent system comprising of water–methanol-chloroform. The upper water-methanolic phase was dried and derivatized by sequentially using methoxyamine hydrochloride solution in pyridine and N-Methyl-N-(trimethylsilyl)trifluoroacetamide (MSTFA) solution. Similarly, a standard solution containing 18 commonly available compounds was also dried and derivatized (see Agarrwal et al. 2014). The derivatized samples, along with the standard solution, were then analyzed using GC-MS conditions mentioned earlier (Agarrwal et al. 2014). The data was acquired on GCMS 2010QP-PLUS (Shimadzu Corp., Japan) and processed by GCMS solution post run analysis software (Shimadzu Corp., Japan) and peak identification was carried out using NIST 05, NIST 08 and WILEY 08 libraries. The putative identification of metabolites was based on two dimensions i.e. retention index (RI) for GC and similarity index (SI) for MS.

Statistical analyses of metabolic profiles

The processed data was exported to MS Excel (Microsoft, USA) and the statistical analyses of the metabolic data was carried out using Metaboanalyst (Xia et al. 2015). The fold change in levels of accumulation of various metabolites between the infested and un-infested tissues (RP-I/RP-UI) was calculated followed by a *t*-test.

The metabolic data was subjected to multivariate statistical analysis (PLS-DA) which could differentiate the metabolic profiles of infested tissues from those of un-infested tissues at all the three distinct time points.

Accession numbers for microarray data

The microarray data reported in this article have been deposited in the Gene Expression Omnibus database (http://www.ncbi.nlm.nih.gov/geo/) with accession number GSE70526.

Additional files

Additional file 1: Figure S1. Volcano plot depicting log2-fold change (x-axis) versus –log10 p-value (y-axis) in gene expression profiles of rice upon GMB1 challenge. RP-I/RP-UI represents the comparison between infested (RP-I) and un-infested (RP-UI) rice tissues.

Additional file 2: Table S1. Overview of GO analysis.

Additional file 3: Figure S2. Bar diagram showing results of Gene Ontology (GO) analysis for the distribution of transcripts, identified in rice challenged with the Asian rice gall midge biotype 1 (GMB1), across various biological and cellular processes. Only those GOs accounting for more than 10 % transcripts have been plotted in the graph.

Additional file 4: Figure S3. Two-dimensional score plots between selected components on the basis of Partial Least Squares–Discriminant Analysis performed on metabolic profiles of infested (RP-I) and un-infested (RP-UI) samples of rice obtained at different time points after gall midge infestation a) 24 h b) 48 h c) 72 h. The X-axis represents components that have maximum variability and y-axis represents components having second highest variability. RP: *indica* rice variety RP2068-18-3-5; h: hai.

Additional file 5: Table S2. Overview of metabolic profiling results.

Additional file 6: Table S3. Overview of deregulation of transcripts, involved in metabolism (both primary and secondary metabolism), in the resistant host (RP) upon gall midge biotype 1 (GMB1) infestation. Detailed information such as fold change in relative expression of these transcripts upon gall midge infestation (Fold change RPI/RPUI), their gene locus ids, protein name and function, GO annotation, pathways, Affymetrix probe ids is provided. RP: *indica* rice variety RP2068-18-3-5, I: infested, UI: un-infested

Additional file 7: Table S4. Overview of deregulation of transcripts, involved in light reaction and Calvin cycle of photosynthesis and tetrapyrrole synthesis pathway, in the resistant host (RP) upon gall midge biotype 1 (GMB1) infestation. Detailed information such as fold change in relative expression of these transcripts upon gall midge infestation (Fold change RPI/RPUI), their gene locus ids, protein name and function, GO annotation, pathways, Affymetrix probe ids is provided. RP: *indica* rice variety RP2068-18-3-5, I: infested, UI: un-infested.

Additional file 8: Table S5. Microarray analyses of rice tissues challenged with the Asian rice gall midge biotype 1 (GMB1). Transcripts have been placed into several bins using MapMan software and information relating to fold change in relative gene expression values (fold change) between infested (RPI) and un-infested tissues (RPUI), Affymetrix gene ids, description of the transcript is provided. RP: *indica* rice variety RP2068-18-3-5.

Additional file 9: Table S6. Overview of transcripts encoding transcription factors, transcripts involved in hormone synthesis, cell organization and maintaining cellular redox state, in the resistant host (RP) upon gall midge biotype 1 (GMB1) infestation. Detailed information such as fold change in relative expression of these transcripts upon gall midge infestation (RPI/RPUI), their gene locus ids, protein name and function, GO annotation, Affymetrix probe ids is provided. RP: *indica* rice variety RP2068-18-3-5, I: infested, UI: un-infested.

Additional file 10: Table S7. Details of genes shortlisted, on the basis of microarray results, for semi-quantitative RT-PCR.

Additional file 11: References S1. List of references referred to in Fig. 5.

Additional file 12: Method S1. Work flow for microarray analyses.

Abbreviations

CHO: Carbohydrate; DEGs: Differentially expressed genes; GABA: Gamma amino butyric acid; GC-MS: Gas chromatography–mass spectrometry; GMB1: Gall midge biotype 1; GO: Gene ontology; hai: Hours after infestation; HR: Hypersensitive response; LOX: Lipoxygenases; LPO: Lipid peroxidation; PLS-DA: Partial Least Squares-Discriminant Analysis; R-gene: Resistance gene; RIN: RNA integrity number; RMA: Robust Multichip Average; ROS: Reactive oxygen species; RP: *indica* rice variety RP2068-18-3-5; RP-I: RP infested; RP-UI: RP un-infested; RP-I/RP-UI: Fold change difference between RP infested and RP un-infested; RT-PCR: Reverse transcription-polymerase chain reaction; TCA: Tricarboxylic acid cycle; TSP: Tetrapyrrole synthesis pathway.

Competing interests

The authors declare that they have no competing interests.

Authors' contributions

RA designed and performed the experiments, analyzed the data and drafted the manuscript. APP contributed reagents/materials/analysis tools and critically revised the manuscript. JSB and SN conceived of the study, participated in its design and coordination and helped to draft and critically revise the manuscript. All authors read and approved the final version of the manuscript.

Acknowledgements

We acknowledge the use of GC-MS facility at Advanced Instrumentation Research Facility, Jawaharlal Nehru University, New Delhi and Dr. Ajai Kumar for providing technical help. Support of this work by core grants from ICGEB and extramural funding (to SN) from the Department of Biotechnology and Indian Council of Agricultural Research, Government of India is gratefully acknowledged. RA thanks the University Grants Commission (UGC), Government of India, for a Senior Research Fellowship.

Author details

[1]International Centre for Genetic Engineering and Biotechnology (ICGEB), Aruna Asaf Ali Marg, New Delhi 110067, India. [2]Indian Institute of Rice Research (formerly Directorate of Rice Research), Rajendranagar, Hyderabad 500030, India. [3]Present address: AgriBiotech Foundation, Rajendranagar, Hyderabad 500030, India.

References

Agarrwal R, Bentur JS, Nair S (2014) Gas chromatography mass spectrometry based metabolic profiling reveals biomarkers involved in rice-gall midge interactions. J Integr Plant Biol 56:837–848

Apel K, Hirt H (2004) Reactive oxygen species: metabolism, oxidative stress, and signal transduction. Annu Rev Plant Biol 55:373–399

Asada K (2006) Production and scavenging of reactive oxygen species in chloroplasts and their functions. Plant Physiol 141:391–396

Bentur JS, Kalode MB (1996) Hypersensitive reaction and induced resistance in rice against Asian rice gall midge *Orseolia oryzae*. Entomol Exp Appl 78:77–81

Bentur JS, Pasalu IC, Sarma NP, Prasada Rao U, Mishra B (2003) Gall midge resistance in rice. Directorate of Rice Research, Hyderabad

Bentur JS, Padmakumari AP, Jhansi Lakshmi V, Padmavathi C, Kondala Rao Y, Amudhan S, Pasalu IC (2011) Insect resistance in rice. Technical Bulletin, 51. Directorate of Rice Research, Hyderabad, India, p 85

Berger S, Benediktyová Z, Matous K, Bonfig K, Mueller MJ, Nedbal L, Roitsch T (2007) Visualization of dynamics of plant-pathogen interaction by novel combination of chlorophyll fluorescence imaging and statistical analysis: differential effects of virulent and avirulent strains of *P. syringae* and of oxylipins on *A. thaliana*. J Exp Bot 58:797–806

Cui X, Churchill GA (2003) Statistical tests for differential expression in cDNA microarray experiments. Genome Biol 4:210. doi:10.1186/gb-2003-4-4-210

Danon A, Miersch O, Felix G, Camp RG, Apel K (2005) Concurrent activation of cell death-regulating signalling pathways by singlet oxygen in *Arabidopsis thaliana*. Plant J 41:68–80

Ding Y, Chang J, Ma Q, Chen L, Liu S, Jin S, Han J, Xu R, Zhu A, Guo J, Luo Y, Xu J, Xu Q, Zeng YL, Deng X, Cheng Y (2015) Network analysis of postharvest senescence process in citrus fruits revealed by transcriptomic and metabolomic profiling. Plant Physiol 168:357–376

Fait A, Fromm H, Walter D, Galili G, Fernie AR (2008) Highway or byway: the metabolic role of the GABA shunt in plants. Trends Plant Sci 13:14–19

Foyer CH, Noctor G (2003) Redox sensing and signalling associated with reactive oxygen in chloroplasts, peroxisomes and mitochondria. Physiol Plant 119:355–364

Gadjev I, Vanderauwera S, Gechev TS, Laloi C, Minkov IN, Shulaev V, Apel K, Inzé D, Mittler R, Van Breusegem F (2006) Transcriptomic footprints disclose specificity of reactive oxygen species signaling in *Arabidopsis*. Plant Physiol 141:436–445

Gonzalez-Ibeas D, Cañizares J, Aranda MA (2012) Microarray analysis shows that recessive resistance to Watermelon mosaic virus in melon is associated with the induction of defense response genes. Mol Plant Microbe Interact 25:107–118

Goverse A, de Almeida EJ, Verhees J, van der Krol S, Helder JH, Gheysen G (2000) Cell cycle activation by plant parasitic nematodes. Plant Mol Biol 43:747–761

Halliwell B (2006) Reactive species and antioxidants. Redox biology is a fundamental theme of aerobic life. Plant Physiol 141:312–322

Himabindu K, Suneetha K, Sama VSAK, Bentur JS (2010) A new rice gall midge resistance gene in the breeding line CR57-MR1523, mapping with flanking markers and development of NILs. Euphytica 174:179–187

Höglund S, Rönnberg-Wästljung AC, Berlin S, Larsson S (2015) Willow resistance to a galling insect is driven by a lack of induced susceptibility not an induced defense. Arthropod Plant Interact 9:447–455

Huang X, Springer PS, Kaloshian I (2003) Expression of the *Arabidopsis* MCM gene PROLIFERA during root-knot and cyst nematode infection. Phytopathology 93:35–41

Khajuria C, Wang H, Liu X, Wheeler S, Reese JC, El Bouhssini M, Whitworth RJ, Chen M-S (2013) Mobilization of lipids and fortification of cell wall and cuticle are important in host defense against Hessian fly. BMC Genomics 14:423. doi:10.1186/1471-2164-14-423

Kim C, Meskauskiene R, Apel K, Laloi C (2008) No single way to understand singlet oxygen signalling in plants. EMBO Rep 9:435–439

Kim NH, Kim BS, Hwang BK (2013) Pepper arginine decarboxylase is required for polyamine and γ-aminobutyric acid signaling in cell death and defense response. Plant Physiol 162:2067–2083

Krieger-Liszkay A (2005) Singlet oxygen production in photosynthesis. J Exp Bot 56:337–346

Laloi C, Stachowiak M, Pers-Kamczyc E, Warzych E, Murgia I, Apel K (2007) Cross-talk between singlet oxygen- and hydrogen peroxide-dependent signalling of stress responses in *Arabidopsis thaliana*. Proc Natl Acad Sci USA 104:672–677

Liu X, Williams CE, Nemacheck JA, Wang H, Subramanyam S, Zheng C, Chen M-S (2010) Reactive oxygen species are involved in plant defense against a gall midge. Plant Physiol 152:985–999

MacGregor KB, Shelp BJ, Peiris S, Bown AW (2003) Overexpression of glutamate decarboxylase in transgenic tobacco plants deters feeding by phytophagous insect larvae. J Chem Ecol 29:2177–2182

Maekawa S, Sato T, Asada Y, Yasuda S, Yoshida M, Chiba Y, Yamaguchi J (2012) The *Arabidopsis* ubiquitin ligases ATL31 and ATL6 control the defense response as well as the carbon/nitrogen response. Plant Mol Biol 79:217–227

Maekawa S, Inada N, Yasuda S, Fukao Y, Fujiwara M, Sato T, Yamaguchi J (2014) The carbon/nitrogen regulator ARABIDOPSIS TOXICOS EN LEVADURA31 controls papilla formation in response to powdery mildew fungi penetration by interacting with SYNTAXIN OF PLANTS121 in *Arabidopsis*. Plant Physiol 164:879–887

McLean MD, Yevtushenko DP, Deschene A, Van Cauwenberghe OR, Makhmoudova A, Potter JW, Bown AW, Shelp BJ (2003) Overexpression of glutamate decarboxylase in transgenic tobacco plants confers resistance to the northern root-knot nematode. Mol Breed 11:277–285

Meskauskiene R, Nater M, Goslings D, Kessler F, op den Camp R, Apel K (2001) FLU: a negative regulator of chlorophyll biosynthesis in *Arabidopsis thaliana*. Proc Natl Acad Sci USA 98:12826–12831

Michaeli S, Fromm H (2015) Closing the loop on the GABA shunt in plants: Are GABA metabolism and signaling entwined? Fron Plant Sci 6:419

Murgia I, Tarantino D, Vannini C, Bracale M, Carravieri S, Soave C (2004) *Arabidopsis thaliana* plants overexpressing thylakoidal ascorbate peroxidase show increased resistance to paraquat-induced photooxidative stress and to nitric oxide-induced cell death. Plant J 38:940–953

Nair S, Bentur JS, Sama VSAK (2011) Mapping gall midge resistance genes: towards durable resistance through gene pyramiding. In: Muralidharan K, Siddiq EA (eds) Genomics and Crop Improvement: Relevance and Reservations. ANGR Agricultural University, Institute of Biotechnology, Hyderabad, pp 256–264

Ochsenbein C, Przybyla D, Danon A, Landgraf F, Gobel C, Imboden A, Feussner I, Apel K (2006) The role of EDS1 (enhanced disease susceptibility) during singlet oxygen-mediated stress responses of *Arabidopsis*. Plant J 47:445–456

op den Camp RG, Przybyla D, Ochsenbein C, Laloi C, Kim C, Danon A, Wagner D, Hideg E, Gobel C, Feussner I, Nater M, Apel K (2003) Rapid induction of distinct stress responses after the release of singlet oxygen in *Arabidopsis*. Plant Cell Online 15:2320–2332

Palanivelu R, Brass L, Edlund AF, Preuss D (2003) Pollen tube growth and guidance is regulated by POP2, an *Arabidopsis* gene that controls GABA levels. Cell 114:47–59

Park DH, Mirabella R, Bronstein PA, Preston GM, Haring MA, Lim CK, Collmer A, Schuurink RC (2010) Mutations in γ-aminobutyric acid (GABA) transaminase genes in plants or *Pseudomonas syringae* reduce bacterial virulence. Plant J 64:318–330

Pavan S, Jacobsen E, Visser RG, Bai Y (2010) Loss of susceptibility as a novel breeding strategy for durable and broad-spectrum resistance. Mol Breed 25:1–12

Portillo M, Cabrera J, Lindsey K, Topping J, Andrés MF, Emiliozzi M, Oliveros JC, García-Casado G, Solano R, Koltai H, Resnick N, Fenoll C, Escobar C (2013) Distinct and conserved transcriptomic changes during nematode-induced giant cell development in tomato compared with *Arabidopsis*: a functional role for gene repression. New Phytol 197:1276–1290

Price PW (1991) The plant vigor hypothesis and herbivore attack. Oikos 62:244–251

Ramputh AI, Bown AW (1996) Rapid γ-aminobutyric acid synthesis and the inhibition of the growth and development of oblique banded leaf-roller larvae. Plant Physiol 111:1349–1352

Rawat N, Sinha DK, Rajendrakumar P, Srivastava P, Neeraja CN, Sundaram RM, Nair S, Bentur JS (2010) Role of pathogenesis related genes in rice gall midge interactions. Curr Sci 99:1361–1368

Rawat N, Neeraja CN, Sundaram RM, Nair S, Bentur JS (2012a) A novel mechanism of gall midge resistance in the rice variety Kavya revealed by microarray analysis. Funct Integr Genomics 12:249–264

Rawat N, Neeraja CN, Nair S, Bentur JS (2012b) Differential gene expression in gall midge susceptible rice genotypes revealed by suppressive subtraction hybridization (SSH) cDNA libraries and microarray analysis. Rice 5:8. doi:10.1186/1939-8433-5-8

Rawat N, Himabindu K, Neeraja CN, Nair S, Bentur JS (2013) Suppressive subtraction hybridization reveals that rice gall midge attack elicits plant-pathogen-like responses in rice. Plant Physiol Biochem 63:122–130

Sama VSAK, Rawat N, Sundaram RM, Himabindu K, Naik BS, Viraktamath BC, Bentur JS (2014) A putative candidate for the recessive gall midge resistance gene *gm3* in rice identified and validated. Theor Appl Genet 127:113–124

Sana TR, Fischer S, Wohlgemuth G, Katrekar A, Jung KH, Ronald PC, Fiehn O (2010) Metabolomic and transcriptomic analysis of the rice response to the bacterial blight pathogen *Xanthomonas oryzae* pv. *oryzae*. Metabolomics 6:451–465

Schneider CA, Rasband WS, Eliceiri KW (2012) NIH Image to ImageJ: 25 years of image analysis. Nature Methods 9:671–675

Sinha DK, Nagaraju J, Tomar A, Bentur JS, Nair S (2012) Pyrosequencing-based transcriptome analysis of the Asian rice gall midge reveals differential response during compatible and incompatible interaction. Int J Mol Sci 13: 13079–13103

Smyth GK (2004) Linear models and empirical bayes methods for assessing differential expression in microarray experiments. Stat Appl Genet Mol Biol 3:Article3

Stuart JJ, Chen M-S, Shukle R, Harris MO (2012) Gall midges (Hessian flies) as plant pathogens. Annu Rev Phytopath 50:339–357

Sumiya N, Kawase Y, Hayakawa J, Matsuda M, Nakamura M, Era A (2015) Expression of cyanobacterial acyl-acp reductase elevates the triacylglycerol level in the red alga *Cyanidioschyzon merolae*. Plant Cell Physiol 56:1962–1980

Thimm O, Blasing O, Gibon Y, Nagel A, Meyer S, Kruger P, Selbig J, Muller LA, Rhee SY, Stitt M (2004) MAPMAN: A user-driven tool to display genomics data sets onto diagrams of metabolic pathways and other biological processes. Plant J 37:914–939

Triantaphylidès C, Krischke M, Hoeberichts FA, Ksas B, Gresser G, Havaux M, Van Breusegem F, Mueller MJ (2008) Singlet oxygen is the major reactive oxygen species involved in photooxidative damage to plants. Plant Physiol 148:960–968

van Verk MC, Gatz C, Linthorst HJ (2009) Transcriptional regulation of plant defense responses. Adv Bot Res 51:397–438

Wagner D, Przybyla D, op den Camp R, Kim C, Landgraf F, Lee KP, Wursch M, Laloi C, Nater M, Hideg E, Apel K (2004) The genetic basis of singlet oxygen-induced stress responses of *Arabidopsis thaliana*. Science 306:1183–1185

Watson BS, Bedair M, Urbanczyk-Wochniak E, Huhman DV, Yang DS, Allen SN, Li W, Tang Y, Sumner WL (2015) Integrated metabolomics and transcriptomics reveal enhanced specialized metabolism in *Medicago truncatula* root border cells. Plant Physiol 167:1699–1716

Xia J, Sinelnikov IV, Han B, Wishart DS (2015) MetaboAnalyst 3.0—making metabolomics more meaningful. Nucleic Acids Res 43:W251–W257

Zhu L, Liu X, Liu X, Jeannotte R, Reese JC, Harris M, Stuart JJ, Chen M-S (2008) Hessian fly (*Mayetiola destructor*) attack causes a dramatic shift in carbon and nitrogen metabolism in wheat. Mol Plant Microbe Interact 21:70–78

Zoeller M, Stingl N, Krischke M, Fekete A, Waller F, Berger S, Mueller MJ (2012) Lipid profiling of the *Arabidopsis* hypersensitive response reveals specific lipid peroxidation and fragmentation processes: biogenesis of pimelic and azelaic acid. Plant Physiol 160:365–378

18

Delving deeper into technological innovations to understand differences in rice quality

Mariafe Calingacion[1,2,3*], Lu Fang[3], Lenie Quiatchon-Baeza[1,4], Roland Mumm[5], Arthur Riedel[3], Robert D Hall[2,5] and Melissa Fitzgerald[3]

Abstract

Increasing demand for better quality rice varieties, which are also more suited to growth under sub-optimal cultivation conditions, is driving innovation in rice research. Here we have used a multi-disciplinary approach, involving SNP-based genotyping together with phenotyping based on yield analysis, metabolomic analysis of grain volatiles, and sensory panel analysis to determine differences between two contrasting rice varieties, Apo and IR64. Plants were grown under standard and drought-induced conditions. Results revealed important differences between the volatile profiles of the two rice varieties and we relate these differences to those perceived by the sensory panel. Apo, which is the more drought tolerant variety, was less affected by the drought condition concerning both sensory profile and yield; IR64, which has higher quality but is drought sensitive, showed greater differences in these characteristics in response to the two growth conditions. Metabolomics analyses using GCxGC-MS, followed by multivariate statistical analyses of the data, revealed a number of discriminatory compounds between the varieties, but also effects of the difference in cultivation conditions. Results indicate the complexity of rice volatile profile, even of non-aromatic varieties, and how metabolomics can be used to help link changes in aroma profile with the sensory phenotype. Our outcomes also suggest valuable multi-disciplinary approaches which can be used to help define the aroma profile in rice, and its underlying genetic background, in order to support breeders in the generation of improved rice varieties combining high yield with high quality, and tolerance of both these traits to climate change.

Keywords: Metabolomics; Rice; Climate change; Consumer acceptance; Eating quality

Background

Asia currently faces two major challenges that are predicted to have significant impacts on food security in the region: rapid and significant population growth (United Nations 2014), and climate change (IPCC 2014). The Asian population is growing at a much faster rate than those in other regions, and is predicted to peak by 2050 (UnitedNations 2014). Rice is the staple food of Asia, and to protect Asia against food shortages in the future, it is imperative that rice improvement programs develop new varieties with much higher yield. However, finding an acceptable solution is more complex than increasing yield potential alone.

Socioeconomic patterns within the Asian population are changing rapidly. While research is helping to meet the United Nations Millennium Development goals, by lifting people out of poverty (Fanzo and Pronyk 2011), Asia is currently undergoing economic transformation (Kharas and Gertz 2010), and is predicted to become home to the largest middle class by 2030. Following from this economic growth, it is expected that the majority of Asians will have more discretionary income and greater financial capacity to make food choices based on assessment of quality and safety (Goodman and Robison 2013). Therefore, the impact of population growth in Asia, on both rice production and rice improvement programs, is that new high-yielding rice varieties must contain the quality traits that increasingly-discriminatory consumers in different Asian markets require (Calingacion et al. 2014). Combining yield and quality in rice is not trivial, because the tools for measuring quality are unable to discriminate in a manner that explains the sensory

* Correspondence: m.calingacion@uq.edu.au
[1]Grain Quality and Nutrition Centre, International Rice Research Institute, DAPO 7777, Metro Manila, Philippines
[2]Laboratory of Plant Physiology, Wageningen University and Research Centre, Wageningen, The Netherlands
Full list of author information is available at the end of the article

experiences of aroma, taste, flavour and texture (Daygon and Fitzgerald 2013; Fitzgerald et al. 2009).

Further challenging the need for more and better rice, the effects of climate change are predicted to impact most strongly in the latitudes and longitudes of relevant Asia (IPCC 2014). Climate change is predicted to lead to more drought events, higher temperatures (Peng et al. 2004), and more unpredictable weather patterns (Knutson and Tuleya 2004; Trenberth 2005). In anticipation of this dramatically changing environment, rice improvement programs now include breeding and selection for climate-ready varieties of rice (Zeigler and Barclay 2008). The breeding and domestication of rice has created variation among genotypes that enables cultivation in a number of less favourable environments, including salinity-affected coastal areas, water-scarce upland areas, and lowland monsoonal regions that are prone to inundation and flood (Mackill et al. 2012). These adaptations demonstrate that the species houses genes that enable a high degree of plasticity, and thus permit productive growth in different environments (Wassmann et al. 2009). Identifying the genetic basis of different stress tolerances, in order to introgress causal genes into elite varieties is a research priority. Recent examples of gains in managing the need for stress-tolerant rices are uncovering (i) genes for submergence tolerance for rice grown in monsoonal regions (Xu et al. 2006), and (ii) genes for drought tolerance for varieties grown in conditions of water scarcity or drought (Kumar et al. 2014).

New rice varieties with tolerance to abiotic stresses offer a clear yield advantage to farmers. However, this does not always translate into profit because, next to agronomic traits, a major driver of widespread uptake of a new variety is not farmer adoption, but consumer acceptance (Calingacion et al. 2014; Daygon and Fitzgerald 2013; Boualaphanh et al. 2011; Champagne et al. 2010; Fitzgerald et al. 2009). A clear example of this exists in the case of the two varieties Apo (also known as PSBRc 9) and IR64. Apo repeatedly gives good yield under drought conditions (Venuprasad et al. 2007); it has been released in several drought-prone areas of Asia, but has not been widely adopted, because consumers consider it to have poor eating quality (Fitzgerald et al. 2009; Champagne et al. 2010). IR64, on the other hand, is susceptible to many abiotic stresses, including drought (Venuprasad et al. 2007), but regardless, it is widely accepted and has been grown annually on over a million hectares in major rice-producing countries since its release in 1985 (Fitzgerald et al. 2009).

Apo and IR64 have been assessed in a paired comparison by a sensory evaluation panel, comprising another 11 similarly matched pairs of rice varieties, where both varieties in each pair had similar physico-chemical traits of quality, but differed in their consumer acceptance/popularity (Champagne et al. 2010). The flavour of Apo rice was found to be dominated by sewer/animal, astringent, and water-like metallic notes, whereas IR64 was predominantly characterised by sweet and corn notes (Champagne et al. 2010). The compounds that individually, or in combination, are responsible for such flavours and aromas have not yet been identified. In order to assist rice improvement programs to develop varieties that meet consumer taste and flavour requirements as well as meeting the need for increased yield, newly developed phenotyping techniques such as metabolomics can be used to identify these compounds of taste and flavour in foods (Calingacion et al. 2012; Hall et al. 2008).

Our objectives here were to (i) confirm that flavour notes previously reported by Champagne et al. (2010) for Apo and IR64 grains are reproducible, and are therefore robust characteristics of the two varieties; (ii) determine how water scarcity affects these flavour notes, because the cultivation of Apo germplasm was originally developed for water-scarce environments; and (iii) identify a panel of target compounds with low odour threshold that differ between Apo and IR64, and investigate their association with the phenotypic aromas previously reported. To assemble the most detailed picture of these two important rice varieties, and to see how key traits are influenced by water availability, we have used a highly multi-disciplinary approach involving phenotyping (metabolomics, taste panel sensory analysis, yield measurements) supported by advanced multivariate statistical analyses.

Results

Response to drought treatment

Drought significantly lowered yield for Apo and IR64 by 27% and 38%, respectively (effect of removing treatment from the model: $X^2{}_1 = 20.855$, p = <0.0001) (Figure 1A). There was a significant difference in overall yield between the two varieties (effect of removing variety from the model: $X^2{}_1 = 6.5484$, p = <0.0105). The mean yield of IR64 was lower than that of Apo, but it showed no statistically significant interaction between genotype and treatment, meaning that neither variety was significantly more affected by treatment than the other. Over the duration of the experiment, there was 3.5 mm average rainfall per day, with some rain falling during grain-filling (Figure 1B), but the artificial drought imposed on the plots by draining at panicle initiation and just before flowering, was enough to significantly affect yield (Figure 1A).

Sensory evaluation

The sensory panel reproducibly detected aromas of sweet taste, corn and sweet aromatic flavours in the grains of IR64 from irrigated plots, and water-like metallic, astringent, sour silage, sewer animal and hay-like/musty aromas in Apo grains from irrigated plots (Table 1).

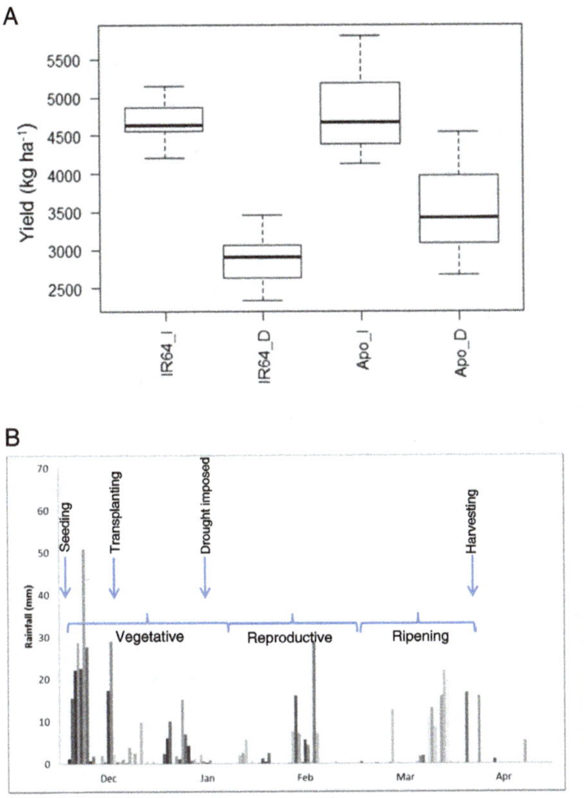

Figure 1 A. Yields (kg ha^{-1}) of Apo and IR64 rices grown under irrigated and drought conditions (Dry Season 2012), N=48. B. Rainfall (mm) at the Experimental Station of International Rice Research Institute, Philippines during the growth of the rice samples (Dry Season 2012).

When comparing the effects of drought within varieties, no difference in aroma was detected for grains of Apo (Table 1). For the two IR64 samples, the levels of corn and sweet aromatic aromas were similar in both treatments. However, higher levels of sweet taste and hay-like/musty were perceived in IR64 from the irrigated than the drought treatment. In addition, water-like metallic, astringent and sour/silage aromas were detected in IR64 grains from the drought treatment.

Analysis of headspace samples by GCxGC-TOFMS

We detected 187 volatile compounds in Apo and IR64 (Figure 2A). Principal components analysis (PCA) extracted the main axes of variation in the concentration of volatile compounds detected in Apo and IR64 from irrigated and drought treatments. PC1 explained 11% of the variation in metabolite concentrations, with IR64 forming two distinct clusters corresponding to water availability during growth (Figure 2A). Apo from both irrigation treatments clustered closely and showed considerably less variation in PC1 than IR64. Our linear mixed model analysis found no significant interactions between treatment and block effects (effect of removing block × variety interaction from the model: $X^2_3 = 5.1799$, $p = 0.1591$) in PC1, so this term was removed from the model before assessing the fixed treatment effects, but we retained the random block component to partition out variances derived from the experimental design. Critically, despite discerning no differing effects on yields between the two varieties, we found a substantive part of the variation in metabolites on PC1 was associated with variation in the differing treatment responses of the two rice varieties (effect of removing two-way interaction, variety × treatment, from the model: $X^2_1 = 22.271$, $p = <0.0001$).

PC2, encapsulating 9% of the total variation modelled, further distinguished samples according to their variety: all Apo samples loaded negatively on PC2, all IR64 samples loaded positively. Our linear mixed model found no variation between varieties among blocks (effect of removing block × variety interaction from the model: $X^2_3 = 0$, $p = 1$). Variation in scores for PC2 was found to be associated solely with differences between the two rice varieties (effect of removing variety from the model: $X^2_1 = 90.93$, $p = <2.2e\text{-}16$).

A second PCA (Figure 2B) of selected compounds with aromatic descriptors and low odour threshold compounds, showed some characteristics in common with Figure 2A.

Table 1 Comparison of flavour attributes in Apo and IR64 from irrigated and drought treatments with sensory qualities/attributes as previously reported by Champagne et al. (2010)

Flavour	Apo[a]	IR64[a]	Apo		IR64	
			Irrigated	Drought	Irrigated	Drought
Sweet taste	+	++	+	+	++	+
Corn		+			+	+
Sweet aromatic		+			+	+
Astringent	++	+	+	+		+
Water like metallic	++	+	++	++		+
Sewer/animal	++		++	++		
Sour/silage	++		+	+		+
Hay-like/musty			++	++	++	+

[a]Adapted from Champagne et al. (2010).
Eight panelists indicated the strength of the perceived aromatic characteristic of the sample using a 3 point scale.

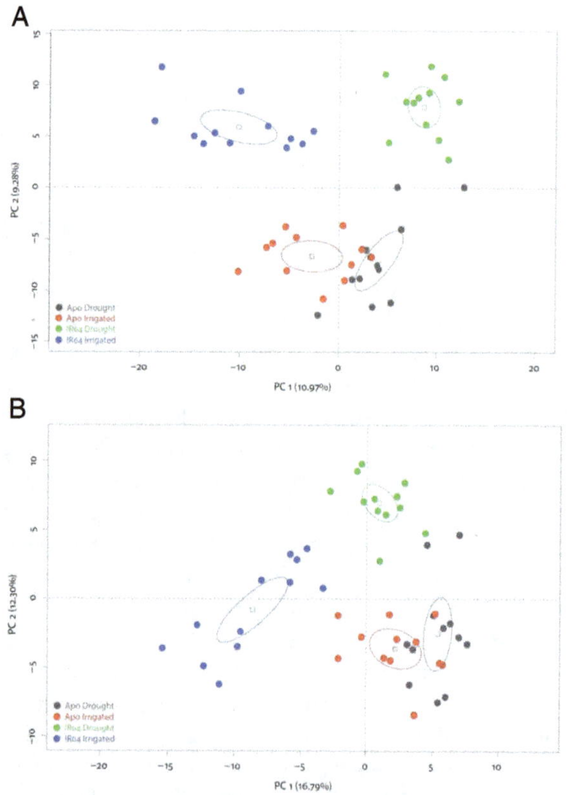

Figure 2 A. PCA of metabolites detected in 12 biological samples of each Apo and IR64 from irrigated and drought treatments, N=48. B. PCA of metabolites with known flavour and low odour threshold detected in Apo and IR64 grown in irrigated and drought treatments, N=48.

PC1, explaining 17% of the total variation in low odour threshold metabolites, described broad separation of IR64 grown in irrigated conditions, with negative scores in this dimension for all samples. IR64 grown under drought conditions showed both positive and negative scores on PC1, where both Apo treatments were closely clustered with predominantly positive scores. PC2 (12.5%) allocated positive scores to the IR64 drought-treated samples, negative scores to both Apo sample groups, and widely scattered scores for IR64 grown in irrigated conditions whose mean score values were nominally negative on PC2 (Figure 2B).

Sparse Partial Least Squares-Discriminant Analysis (sPLS-DA) identified 20 compounds for each direction through the data that discriminated between treatment groups (Table 2). The first axis discriminated samples from the IR64 irrigated treatment, the second discriminated IR64 drought samples from the two Apo treatment classes, while the third described those compounds that differed between the two treatments applied to Apo.

Discussion

As climate change exerts its influence on rice-producing countries, it is essential that science delivers new phenotyping tools to support the development and release of new varieties of rice for new and challenging environments. These new varieties must combine stress tolerance for both yield and grain quality in order to ensure consumer acceptability (Calingacion et al. 2014). The difficulty in combining both traits and then adding stress tolerance is demonstrated by the variety Apo, which has been released for drought tolerance (Espino and Cinense 2010) but its cooking and eating qualities are unacceptable, in spite having identical physico-chemical properties to IR64 and other popular varieties.

Current tools used to evaluate rice quality are not sufficiently discriminating (Calingacion et al. 2014; Daygon and Fitzgerald 2013; Fitzgerald et al. 2009). Therefore breeding programs have used a strategy to create stress-tolerant versions of already popular varieties of rice using genetic pyramiding techniques (Mackill et al. 2012). These 'upgraded' varieties, for example the *sub*1 series (Mackill et al. 2012) can be grown successfully in more challenging environments. However there is still a need to understand and control cooking and eating quality to assure these upgraded varieties continue to be accepted by consumers (Sundaram et al. 2008; Win et al. 2012; Septiningsih et al. 2009), and to make further gains in breeding targets. As genotyping technologies develop, tools enabling genetic selection become more accessible, and simultaneously, the need to push through current yield barriers becomes more pressing. Therefore bringing new science to rice quality to identify traits that consumers value is a more purposeful pathway forward in the longer term, than simply upgrading current varieties with stress tolerant genes.

In the present study, two rice varieties, the unpopular, drought-tolerant Apo and the highly popular, drought-susceptible IR64, were evaluated for yield and drought tolerance, sensory flavour and volatile metabolites in the grain, and finally genetic differences. Data obtained were used to (i) identify a panel of compounds that may be linked to the cooking and eating quality of the rices, (ii) determine the effect of drought stress on the compounds; and (iii) evaluate the potential of using a mapping population developed from these two varieties to enable future identification of genes that underlie the aromas emitted during cooking and consumption of the rices. The ultimate goal envisaged is that this work will lead to the identification of quantitative trait loci (QTLs) for key quality-related traits which will then provide new tools for breeding new, improved varieties with enhanced sensory and agronomic characteristics.

Response to drought treatment

Drought during the reproductive stage can reduce spikelet fertility, panicle exsertion and source-sink interactions which all decrease yield (Wassmann et al. 2009).

Table 2 Discriminating compounds, of known flavour and low odour threshold, in Apo and IR64 grown in irrigated and drought treatments

First axis		Second axis		Third axis	
Compound	**Loading**	**Compound**	**Loading**	**Compound**	**Loading**
Limonene	−0.391	Methylbenzoate	−0.406	1,2,3,4-Tetramethyl benzene	0.434
n-Propylbenzene	−0.372	Camphene	−0.401	Tridecane	0.425
1-Ethyl-3-methylbenzene	−0.358	Benzonitrile	−0.379	Phenethylacetate	0.366
Indole	0.338	Tetradecane	0.368	Benzophenone	−0.361
2-Heptanone	−0.297	Phenol	−0.334	3-Methylbenzaldehyde	0.262
2-Heptenal	−0.249	Isopropyldodecanoate	−0.227	Pentadecanoic acid	0.243
Hexanal	−0.233	Decanal	−0.200	Isomethyl ionone	0.225
1-Hepten-3-ol	−0.231	Benzaldehyde	−0.199	Caryophyllene	−0.200
2-Pentylfuran	−0.216	Octanal	−0.194	Naphthalene	0.199
Methylstyrene	−0.201	Heptanal	−0.193	o-Cymene	0.194
Ethyl octanoate	−0.175	1-Propene-1-thiol	0.167	Benzaldehyde	−0.161
2-Propylfuran	−0.173	Acetophenone	−0.102	1-Hexanol	0.139
α-Phellandrene	−0.168	4-Hydroxy-4-methyl-2-pentanone	−0.096	α-Farnesene	0.122
o-Cymene	−0.118	2-Pentylfuran	0.093	2-Methylheptane	0.073
β-Ocimene	−0.107	2-Methylheptane	0.091	2-Ethylhexanal	−0.045
2-Octenal	−0.071	Carvone	−0.087	p-Xylene	−0.041
2-Butylfuran	−0.027	Hexanal	0.065	4-Hydroxy-4-methyl-2-pentanone	0.036
1,3-Dimethylbenzene	−0.014	2,4-Hexadien-1-ol	−0.042	Undecane	0.028
Undecane	−0.009	Pentadecanoic acid	−0.009	Phenol	0.019
α-Pinene	−0.001	Indole	0.007	α-Phellandrene	0.006

These compounds have been extracted from each of the three axes of a sparse Partial Least Squares-Discriminant Analysis.

Yield can therefore be used as a proxy to measure the degree of drought tolerance (Kumar et al. 2008). Several studies have demonstrated the drought tolerance of Apo (Venuprasad et al. 2007; Venuprasad et al. 2009; Venuprasad et al. 2012 and susceptibility of IR64 {Venuprasad et al. 2007 #21; Kumar et al. 2008; Guan et al. 2010; Vikram et al. 2011; Ghimire et al. 2012; Palanog et al. 2014). However, in our experiment, the yield of both varieties under drought was reduced, and surprisingly there was no statistical difference between the two varieties (Figure 1A, Additional file 1: Table S1). Furthermore, drought significantly affected the aromas and flavours of the grain of IR64, but it did not have such an effect on the grain of Apo (Figures 2A and B).

Sensory analysis of the grains

Consumer acceptance of a variety of rice is driven initially by the physical appearance of the grains (Fitzgerald et al. 2009), and further cemented by, first, the cooking and second, the eating quality of rice (Cramer et al. 1993). Characteristics of cooking rice that are important to consumers are (i) the aromas generated by the processes of cooking rice, for example the sweet and floral aromas emitted when jasmine rice is cooking; (ii) cooking time; and (iii) the appearance of the cooked rice when ready for serving and eating (Bett-Garber et al. 2012). Eating quality describes the aroma, flavour, taste and texture of the rice in the mouth (Fitzgerald et al. 2009). Differences in the aroma of Apo and IR64 during and after cooking are likely to explain much of the differences in consumer acceptance of the two varieties. The aroma of IR64 grains has previously been evaluated as being high in sweet taste, and low in astringent and water-like metallic attributes, whereas Apo was described as high in sewer/animal, water-like metallic, astringent and sour/silage and low in sweet taste (Champagne et al. 2010). In that study, the rice had been grown in irrigated conditions in the same country as the rice evaluated in the present research, but 4 years earlier than this study, and in a different province. The results presented here for the samples from irrigated treatments accord well with those of the previous study. In addition, we found high levels of hay-like/musty flavours in Apo that were not previously reported by Champagne et al. (2010). Slight differences between our results with those of previous reports may derive from the harvest and storage conditions of the samples tested, but in general, the aromatic qualities of each variety are reproducible.

Sensory analysis and the volatile profile of IR64 were both affected by the drought treatment. Biochemical changes in the grain may explain much of this, either due to changes in gene expression in the grain, or changes to the regulation of source-sink transport from the plant to the grain. In IR64 grains from the drought treatment, sweetness was decreased and a water-like metallic flavour was detected which was absent in grains from the irrigated treatment. Though we cannot demonstrate the role of the underlying metabolic processes in this study, we have linked the stress caused by drought to changes in the metabolite pattern, and therefore the flavour profile of IR64. Samples of Apo from both irrigated and drought fields were indistinguishable in their characterisation by water-like metallic, sewer/animal, and hay-like/musty aromas at high levels. One previous study has investigated the effect of drainage on the sensory aroma of rice and the variety used in that study showed no difference in flavour/aroma compounds (Champagne et al. 2005). Our data suggest that a response to moisture stress in the metabolomic profile of the grain is variety dependent, as is often observed in crops e.g., fruit response to water stress that is cultivar dependent (Giné Bordonaba and Terry 2010). Understanding different varietal responses and determining the basis of stress response is important in developing climate-ready varieties.

Sensory analysis correlates with volatile metabolite profiles

Our analysis has highlighted variety x treatment interactions, with IR64 grains showing greater environmental sensitivity than Apo grains. The differences in IR64, due to differences in water availability, highlight how water stress not only affects grain yield, but also the metabolites present in the grains, and we found this to be reflected in the responses of the sensory panel. By contrast, volatile metabolites present in Apo showed less variation between the water treatments, an outcome also reflected in the uniform characterisation of Apo rice by the sensory panel. The clear differences between Apo and IR64 in their volatile profiles, which are supported by the findings of the sensory panel, reinforces that metabolite profiling is indeed capable of distinguishing between varieties (Calingacion et al. 2012), providing a tool to identify and potentially quantify important compounds of rice quality.

Volatile compounds can only be considered important in rice quality if the odour threshold, which defines the lowest concentration of a given compound that humans can perceive, is low (Leonardos et al. 1969). Of the 187 compounds detected, 65 have been putatively identified as aromatic with a low odour threshold. The PCA of this shortlist of 65 closely reflected the PCA of all 187 compounds, suggesting that much of the variation between varieties was captured with less, but more relevant data. Statistical analysis was further able to identify those flavour compounds that most discriminate between Apo and IR64. Samples of IR64 from the irrigated treatment formed a distinct cluster in the first axis, with indole as the compound most strongly correlated with that dimension (Table 2). Indole has a sweet flavour and is used at low concentrations to form synthetic sweet floral aromas (Jezussek et al. 2002), so it is a promising candidate for the sweeter taste perceived in the grains of IR64 from the irrigated treatment, and in agreement with previous reports for rice grown in France (Maraval et al. 2008).

Samples of IR64 from the drought treatment were discriminated by strong loadings of octanal and decanal, which are both described as having a low odour threshold of fruity or waxy odours (Jezussek et al. 2002; Yang et al. 2008). Hexanal, heptanal, octanal and decanal were also present, as were their primary alcohols and several alkanes, possibly representing oxidation ladders (Monsoor et al. 2004), with beneficial outcomes for the aromatic profile of the rice.

Sensory differences between irrigated and drought samples of IR64 identified an astringent aroma in the drought samples (Table 1). Phenol is present in IR64 drought and Apo grains (Table 2), and this is a compound commonly associated with astringency (Bahar and Altug 2009). Undesirable aromas such as sewer animal and sour/silage were found in Apo alone, often/typically associated with the presence of propene thiol and ocimene (O'Neill and Phillips 1992; Loper et al. 1971), both of which were detected in Apo. Hay-like/musty off-flavour aromas were most strongly detected by the sensory evaluation in drought-affected IR64 and both samples of Apo. Hexanal and heptanal could explain those aromas (Lam and Proctor 2003). The strong similarity between the sensory data of the samples and the pattern of loadings in Figure 2B indicates that delving deeper into metabolomics to explain aroma could lead to the development of a new suite of tools to select for aromatic profiles.

Our study again shows that over a hundred volatile metabolites are present in rice (Laguerre et al. 2007; Bryant and McClung 2011; Mathure et al. 2014), and while only a handful may be key to characterizing and defining rice aroma (Buttery et al. 1988) (Buttery et al. 1988; Jezussek et al. 2002; Yang et al. 2010; Calingacion et al. 2012), their relative importance needs to be better understood to enable breeders to select for this trait. Little is known about the genetic basis of aroma in rice, or the inheritance patterns of key flavour compounds, aside from 2-acetyl-1-pyrroline, which is associated with the popcorn-like aroma of jasmine and basmati rices (Bradbury et al. 2005; Kovach et al. 2009; Chen et al. 2008). Here we have taken substantial steps towards identifying additional specific compounds associated with favourable and unfavourable consumer responses to rice aromatic quality. With a better understanding of the flavour traits that influence rice acceptance, and identifying specific compounds that drive this acceptance, we gain the potential to take advantage of the genetic difference between Apo and IR64 (Additional file 2: Figures S1) for discovery of causative loci. Development of a mapping population between Apo and IR64 could facilitate the identification of QTLs associated with specific components of rice aroma. This work makes it possible to develop varieties that take the favourable aromatic/flavour characteristics of one variety and at the same time select for the drought tolerant traits of another.

Conclusions

A panel of compounds identified here that are strongly associated with the pleasant cooking and eating experience of IR64 are indole, octanal, and decanal which give sweet, fruity aromas, while Apo is identified with off-flavour compounds such as propene thiol, ocimene, heptanal and hexanal which give sewer/animal, sour/silage and hay-like/musty aromas respectively. These compounds all have low odour thresholds and associate with the aromas detected by sensory analysis in this and previous studies. Identification of the genes underlying these flavour notes, requires identifying QTLs associated with the specific compounds that appear to drive consumer responses. Such QTLs will greatly improve the efficiency and precision of market-driven breeding programs.

Methods

Plant material

Apo and IR64 were planted at the Experimental Station of the International Rice Research Institute, Philippines in the dry season of 2012. For each variety, 100 plants, at 1 plant per hill, were planted at 15 cm within and between rows in 3 blocks under irrigated and 3 blocks under drought conditions, with 4 replicated plots in each block. Inorganic fertiliser nitrogen: phosphorus: potassium (NPK) was applied to the field before transplanting at a ratio of 40:40:40 kg ha^{-1} (NPK), and the plants were top-dressed with urea 30 and 55 days after transplanting at a level of 30:0:0 kg ha^{-1}. Drought stress was artificially imposed by draining the field when plants had been assessed to have achieved maximal tillering, so that water stress coincided with the reproductive stage of the plant. Irrigated plots were maintained at a water level of approximately 5 cm until harvest, at which time they were drained. Mature grains from all 100 plants were harvested to determine total yield. The grains were dried in an oven until a moisture content of 12-14% was reached and stored at 21°C storage room until analysis. Grains were dehulled (Otake FCY2 Dehusker, Oharu, Japan), polished (test tube mill) and cryo-ground (2010SPEX SamplePrep Geno/Grinder) for 3 min at 1750 rpm. We compared yields of the four treatment classes (2 varieties × 2 treatments) in a two factor linear mixed effects model that assessed the presence of variety x treatment interactions.

Sensory evaluation of flour

Eight trained panelists participated in this study. Rice samples were prepared by placing rice flour (1 g) in screw capped vials and heating in a water bath at 80°C for 10 min. Rice was presented in this way to replicate the way the sample and its volatile compounds were later presented for headspace analysis by gas chromatography. Samples were presented to panelists unidentified and in randomised order. Panelists opened the lid of the vial carefully and smelt the aroma of the heated rices. Panelists then scored the intensity or absence of eight flavour notes on a 3-point intensity scale. The flavour notes considered were: sweet taste, corn, sweet aromatic, astringent, water-like metallic, sewer/animal, sour/silage and hay-like/musty (Champagne et al. 2010). Panelists were free to record additional aromas they detected.

Headspace analysis using two-dimensional gas chromatography time of flight-mass spectrometry (GCxGC TOF-MS)

Headspace extraction

Rice flour (1 g) was placed in a sealed vial and allowed to equilibrate overnight at room temperature. The sample was then heated at 80°C for 10 min while being agitated at 500 rpm in an agitator of autosampler of GCxGC TOF-MS. Volatile compounds in the headspace (1 mL) were collected in a syringe and 1.5 mL was injected in a GCxGC TOF-MS system consisting of an Agilent 7890 gas chromatograph (Agilent Technologies, Palo Alto, CA) and Pegasus IV TOF-MS mass spectrometer (LECO, St Joseph, MI). The primary column used was RXI-5sil (29 m × 0.25 mm id × 0.25 μm film thickness) while the secondary column was RxI-17 (1.7 m × 0.1 mm id × 0.18 μm film thickness). The secondary column was placed inside the secondary oven after the thermal modulator. The flow rate of the helium carrier gas was set to a constant flow of 1 mL min^{-1}. The primary column was set at an initial temperature of 45°C for 1 min, then ramped at 10°C min^{-1} to 250°C. The secondary column was programmed with an initial temperature of 60°C for 1 min which was then increased to 250°C at a rate of 10°C min^{-1}. The thermal modulator was set at 70°C then ramped to 260°C at a rate of 10°C min^{-1} with a 5 s modulation period. The MS mass range was 35–400 m/z with an acquisition rate of 200 spectra s^{-1}. The ion source temperature was set at 250°C and the detector voltage was 1.5 kV with electron energy of -70 eV and the temperature of the transfer line was 280°C. Headspace of a standard mixture of known compounds (benzene, toluene, ethylbenzene, p-xylene) was also injected as a quality control sample to check if peak identification is accurate and reproducible.

Data processing

The raw GCxGC data were processed using ChromaTOF software (version 4.5, LECO, St Joseph, MI). We used the automated peak finding and spectral deconvolution algorithm with baseline offset of 0.5 and signal to noise ratio of 6. Automated retention time alignment was also performed using the statistical comparison function in ChromaTOF. All samples were divided among the two water treatment classes, then data were aligned within and between the classes. Only the compounds that were present in more than 50% of the samples in a class were considered. Mass spectra were compared and matched against the NIST 2011 mass spectral library (http://www.NIST.gov).

Statistical analysis

All statistical analyses were done in R (RCoreTeam 2014), using a number of packages. First, we assessed treatment effects and interactions on yield, using a simple univariate model. For our multivariate data, to remedy the skew of several variables, they were first log transformed. To moderate the influence of large-scale variation among some compounds, and thereby allow smaller-scale variation in other metabolites to play a greater role in the overall characterisation of rice aromatic qualities, we Pareto scaled all data (van den Berg et al. 2006): for each variable, data was centred on a mean of zero and the variance scaled by dividing by the square root of the standard deviation for each variable (R package: 'MetabolAnalyze', (Gift et al. 2010)). We used Principal Components Analysis (PCA) (R package: 'FactoMineR', (Husson et al. 2013)) to extract the first two dimensions in our data that explained the greatest amount of variation. To support our description of the treatment classes by PCA, we extracted the PC scores for the 48 individual samples on these first two dimensions, and treated these condensed descriptions of the principal variation in the data as response variables in a linear model. Since our data included experimental blocks nested within growing condition (irrigated vs. drought), in all analyses we used linear mixed effects modelling to partition out the random effects from analysis of treatment effects on the metabolic profile encapsulated in the first two principal components (R package: 'lme4', (Bates et al. 2014)); as recommended for mixed models, we used Kenward-Rogers estimation of the denominator degrees of freedom (Saxton 2004). We established the significance of effects in models in the usual manner (Quinn and Keough 2002): log-likelihood ratio tests compared a full model of treatment factors, and the random experimental effects, with a nested model omitting successive effects.

In the second part of the analysis, we looked more closely at a shortlist of compounds of known low odour threshold, as these have greatest influence on rice quality and consumer preference. We created a data set comprising 65 compounds of the original 187, and again explored the treatment effects on chemical composition using PCA of those compounds that are known to influence aroma.

High dimensional data, such as is generated by GCxGC MS, is typically explored using PCA; PC loadings are one method to extract which compounds show the greatest variation, and therefore may be having greatest influence in distinguishing groups. However, high dimensional data presents a challenge in discerning which variables are more important, and the criteria for acceptance or rejection of putatively important contributions can be arbitrary and less reliable when based on the, typically small, sample sizes of these methods. Here, using a method that gives greater statistical underpinning to the choice of important variables, we used sPLS-DA (R package 'mixOmics'; (Dejean et al. 2013)), to identify those compounds likely to be most influential in discriminating between treatment effects on the two rice varieties in k-1 dimensions, i.e., three in this instance. We constrained the model to identify the 20 compounds, each with a low odour threshold, that were most influential in discriminating between the treatment groups, for each dimension in turn, extending to three dimensions, which were then extracted from the sPLS-DA model.

Competing interests

The authors declare that they have no competing interest.

Authors' contributions

MC and LQB conducted the field and genotype experiments. MF and LF carried out the metabolite profiling. MC, RM, MF and AR analysed the data. MC, AR, RH and MF wrote the paper. All authors read and approved the final manuscript.

Acknowledgments

The authors thank F Salisi, R Ilagan, L Samadio and A Madrid for their assistance in conducting the field experiment and Climate Unit, Crop and Environment Sciences Division, IRRI for the weather data. MC also thanks the Monsanto Beachell-Borlaug International Scholarship Program and University of Queensland for additional funding.

Author details

[1]Grain Quality and Nutrition Centre, International Rice Research Institute, DAPO 7777, Metro Manila, Philippines. [2]Laboratory of Plant Physiology, Wageningen University and Research Centre, Wageningen, The Netherlands. [3]School of Agriculture and Food Science, University of Queensland, St Lucia, 4072 Queensland, Australia. [4]Department of Crop Sciences, College of ACES, University of Illinois at Urbana-Champaign, Urbana, IL, USA. [5]BU Bioscience, Plant Research International, Wageningen University and Research Centre, Wageningen, The Netherlands.

References

Bahar B, Altug T (2009) Flavour characterization of sumach (*Rhus coriaria* L.) by means of GC/MS and sensory flavour profile analysis techniques. Int J Food Prop 12(2):379–387, doi:10.1080/10942910701799249

Bates D, Maechler M, Bolker B, Walker S (2014) lme4: Linear mixed-effects models using Eigen and S4. R package, http://cran.r-project.org/web/packages/lme4/index.html

Bett-Garber K, Champagne E, Thomson J, Lea J (2012) Relating raw rice colour and composition to cooked rice colour. J Sci Food Agric 92(2):283–291, doi:10.1002/jsfa.4573

Boualaphanh C, Calingacion M, Cuevas R, Jothityangkoon D, Sanitchon J, Fitzgerald M (2011) Yield and quality of traditional and improved Lao varieties of rice. Sci Asia 37(2):89–97, doi:10.2306/scienceasia1513-1874.2011.37.089

Bradbury L, Fitzgerald T, Henry R, Jin Q, Waters D (2005) The gene for fragrance in rice. Plant Biotechnol J 3(3):363–370, doi:10.1111/j.1467-7652.2005.00131.x

Bryant R, McClung A (2011) Volatile profiles of aromatic and non-aromatic rice cultivars using SPME/GC–MS. Food Chem 124(2):501–513, doi:10.1016/j.foodchem.2010.06.061

Buttery R, Turnbaugh J, Ling L (1988) Contribution of volatiles to rice aroma. J Agric Food Chem 36(5):1006–1009, doi:10.1021/jf00083a025

Calingacion M, Boualaphanh C, Daygon V, Anacleto R, Sackville Hamilton R, Biais B, Deborde C, Maucourt M, Moing A, Mumm R, de Vos R, Erban A, Kopka J, Hansen T, Laursen K, Schjoerring J, Hall R, Fitzgerald M (2012) A genomics and multi-platform metabolomics approach to identify new traits of rice quality in traditional and improved varieties. Metabolomics 8(5):771–783, doi:10.1007/s11306-011-0374-4

Calingacion M, Laborte A, Nelson A, Resurreccion A, Concepcion J, Daygon V, Mumm R, Reinke R, Dipti S, Bassinello P, Manful J, Sophany S, Lara K, Bao J, Xie L, Loaiza K, El-hissewy A, Gayin J, Sharma N, Rajeswari S, Manonmani S, Rani N, Kota S, Indrasari S, Habibi F, Hosseini M, Tavasoli F, Suzuki K, Umemoto T, Boualaphanh C, Lee H, Hung Y, Ramli A, Aung P, Ahmad R, Wattoo J, Bandonill E, Romero M, Brites C, Hafeel R, Lur H, Cheaupun K, Jongdee S, Blanco P, Bryant R, Thi Lang N, Hall RD, Fitzgerald M (2014) Diversity of global rice markets and the science required for consumer-targeted rice breeding. PLoS One 9(1):e85106, doi:10.1371/journal.pone.0085106

Champagne E, Bett-Garber K, Thompson J, Mutters R, Grimm C, McClung A (2005) Effects of drain and harvest dates on rice sensory and physicochemical properties. Cereal Chem 82(4):369–374

Champagne E, Bett-Garber K, Fitzgerald M, Grimm C, Lea J, Ohtsubo K, Jongdee S, Xie L, Bassinello P, Resurreccion A, Ahmad R, Habibi F, Reinke R (2010) Important Sensory Properties Differentiating Premium Rice Varieties. Rice 3:270–281, doi:10.1007/s12284-010-9057-4

Chen S, Yang Y, Shi W, Ji Q, He F, Zhang Z, Cheng Z, Liu X, Xu M (2008) Badh2, encoding betaine aldehyde dehydrogenase, inhibits the biosynthesis of 2-acetyl-1-pyrroline, a major component in rice fragrance. Plant Cell 20 (7):1850–1861, doi:10.1105/tpc.108.058917

Cramer G, Wailes E, Shui S (1993) Impacts of liberalizing trade in the world rice market. Am J Agric Econ 75(1):219–226, doi:10.2307/1242970

Daygon V, Fitzgerald M (2013) Application of metabolomics for providing a new generation of selection tools for crop improvement. In: Hot Topics in Metabolomics: Food and Nutrition. Future Science Book Series. Future Science Ltd, pp 6–16. doi:10.4155/ebo.13.458

Dejean S, Gonzalez I, Le Cao K, Monget P, Coquery J, Yao F, Liquet B, Rohart F (2013) mixOmics: Omics data integration project. R package

Espino A, Cinense M (2010) Water management and mulching on rice production under aerobic condition. Philippine Agric Mechanisation J 17(1):1–10

Fanzo J, Pronyk P (2011) A review of global progress toward the Millennium Development Goal 1 Hunger Target. Food Nutr Bull 32(2):144–158

Fitzgerald M, McCouch S, Hall R (2009) Not just a grain of rice: the quest for quality. Trends Plant Sci 14(3):133–139

Ghimire K, Quiatchon L, Vikram P, Swamy B, Dixit S, Ahmed H, Hernandez J, Borromeo T, Kumar A (2012) Identification and mapping of a QTL (qDTY1.1) with a consistent effect on grain yield under drought. Field Crop Res 131:88–96, doi:10.1016/j.fcr.2012.02.028

Gift N, Gormley I, Brennan L, team tRC (2010) MetabolAnalyze: probabilistic principal components analysis for metabolomic data. R package, 1.3 edn., http://www.cran.r-project.org/web/packages/MetabolAnalyze/index.html

Giné Bordonaba J, Terry L (2010) Manipulating the taste-related composition of strawberry fruits (Fragaria × ananassa) from different cultivars using deficit irrigation. Food Chem 122(4):1020–1026

Goodman D, Robison R (2013) Economic development, social status and political consciousness. The new rich in Asia. Routledge, New York

Guan Y, Serraj R, Liu S, Xu J, Ali J, Wang W, Venus E, Zhu L, Li Z (2010) Simultaneously improving yield under drought stress and non-stress conditions: A case study of rice (*Oryza sativa* L.). J Exp Bot 61(15):4145–4156

Hall R, Brouwer I, Fitzgerald M (2008) Plant metabolomics and its potential application for human nutrition. Physiol Plant 132(2):162–175, doi:10.1111/j.1399-3054.2007.00989.x

Husson F, Josse J, Le S, Mazet J (2013) FactoMineR: Multivariate exploratory data analysis and data mining with R. R package, http://cran.r-project.org/web/packages/FactoMineR/index.html

IPCC (2014) IPCC Fifth Assessment Report: Working Group II Report 'Impacts, adaptation and vulnerability'

Jezussek M, Juliano B, Schieberle P (2002) Comparison of key aroma compounds in cooked brown rice varieties based on aroma extract dilution analyses (Journal of Agricultural and Food Chemistry). J Agric Food Chem 50(5):1101–1105, doi:10.1021/jf0108720

Kharas H, Gertz G (eds) (2010) The New Global Middle Class: A Crossover from West to East. China's Emerging Middle Class: Beyond Economic Transformation. Brookings Institution Press, Washington DC, USA.

Knutson T, Tuleya R (2004) Impact of CO2-induced warming on simulated hurricane intensity and precipitation: Sensitivity to the choice of climate model and convective parameterization. J Climate 17(18):3477–3495

Kovach M, Calingacion M, Fitzgerald M, McCouch S (2009) The origin and evolution of fragrance in rice (*Oryza sativa* L.). Proc Natl Acad Sci 106(34):14444–14449, doi:10.1073/pnas.0904077106

Kumar A, Bernier J, Verulkar S, Lafitte H, Atlin G (2008) Breeding for drought tolerance: Direct selection for yield, response to selection and use of drought-tolerant donors in upland and lowland-adapted populations. Field Crop Res 107(3):221–231, doi:10.1016/j.fcr.2008.02.007

Kumar A, Dixit S, Ram T, Yadaw R, Mishra K, Mandal N (2014) Breeding high-yielding drought-tolerant rice: genetic variations and conventional and molecular approaches. Journal of Experimental Botany:1–14. doi:10.1093/jxb/eru363

Laguerre M, Mestres C, Davrieux F, Ringuet J, Boulanger R (2007) Rapid discrimination of scented rice by solid-phase microextraction, mass spectrometry, and multivariate analysis used as a mass sensor. J Agric Food Chem 55(4):1077–1083, doi:10.1021/jf0623992

Lam H, Proctor A (2003) Milled rice oxidation volatiles and odor development. J Food Sci 68(9):2676–2681, doi:10.1111/j.1365-2621.2003.tb05788.x

Leonardos G, Kendall D, Barnard N (1969) Odor threshold determinations of 53 odorant chemicals. J Air Pollut Control Assoc 19(2):91–95, doi:10.1080/00022470.1969.10466465

Loper G, Flath R, Webster J (1971) Identification of Ocimene in Alfalfa Flower Aroma by Combined GC-Mass Spectrometry1. Crop Sci 11(1):61–63, doi:10.2135/cropsci1971.0011183X001100010020x

Mackill D, Ismail A, Singh U, Labios R, Paris T, Donald L (2012) Chapter 6 - Development and rapid adoption of submergence-tolerant (Sub1) rice varieties. In: Advances in Agronomy, vol 115. Academic Press, pp 299–352. doi:10.1016/B978-0-12-394276-0.00006-8

Maraval I, Mestres C, Pernin K, Ribeyre F, Boulanger R, Guichard E, Gunata Z (2008) Odor-active compounds in cooked rice cultivars from Camargue (France) analyzed by GC – O and GC – MS (Journal of Agricultural and Food Chemistry). J Agric Food Chem 56(13):5291–5298, doi:10.1021/jf7037373

Mathure S, Jawali N, Thengane R, Nadaf A (2014) Comparative quantitative analysis of headspace volatiles and their association with BADH2 marker in non-basmati scented, basmati and non-scented rice (*Oryza sativa* L.) cultivars of India. Food Chem 142:383–391, doi:10.1016/j.foodchem.2013.07.066

Monsoor M, Proctor A, Siebenmorgen T (2004) Surface lipid and free fatty acids (FFA) content of head and broken rice produced by milling after different drying treatments. Cereal Chem 81(6):705–709

O'Neill D, Phillips V (1992) A review of the control of odour nuisance from livestock buildings: Part 3, properties of the odorous substances which have been identified in livestock wastes or in the air around them. J Agric Eng Res 53:23–50, doi:10.1016/0021-8634(92)80072-Z

Palanog A, Swamy B, Shamsudin N, Dixit S, Hernandez J, Boromeo T, Cruz P, Kumar A (2014) Grain yield QTLs with consistent-effect under reproductive-stage drought stress in rice. Field Crop Res 161:46–54

Peng S, Huang J, Sheehy J, Laza R, Visperas R, Zhong X, Centeno G, Khush G, Cassman K (2004) Rice yields decline with higher night temperature from global warming. Proc Natl Acad Sci 101(27):9971–9975, doi:10.1073/pnas.0403720101

Quinn G, Keough J (2002) Experimental design and data analysis for biologists

RCoreTeam (2014). R package, http://www.r-project.org/

Saxton A (2004) Genetic analysis of complex traits using SAS. SAS. SAS Institute, Inc., Cary, NC

Septiningsih E, Pamplona A, Sanchez D, Neeraja C, Vergara G, Heuer S, Ismail A, Mackill D (2009) Development of submergence-tolerant rice cultivars: the Sub1 locus and beyond. Ann Bot 103(2):151–160, doi:10.1093/aob/mcn206

Sundaram R, Vishnupriya M, Biradar S, Laha G, Reddy G, Rani N, Sarma N, Sonti R (2008) Marker assisted introgression of bacterial blight resistance in Samba Mahsuri, an elite indica rice variety. Euphytica 160(3):411–422, doi:10.1007/s10681-007-9564-6

Trenberth K (2005) Uncertainty in hurricanes and global warming. Science 308 (5729):1753–1754, doi:10.1126/science.1112551

United Nations (2014) World Population Prospects: The 2012 Revision vol Working Paper No. ESA/P/WP.235. United Nations, New York

van den Berg R, Hoefsloot H, Westerhuis J, Smilde A, van der Werf M (2006) Centering, scaling, and transformations: improving the biological information content of metabolomics data. BMC Genomics 7(142), doi:10.1186/1471-2164-7-142

Venuprasad R, Lafitte H, Atlin G (2007) Response to Direct Selection for Grain Yield under Drought Stress in Rice. Crop Sci 47(1):285–293, doi:10.2135/cropsci2006.03.0181

Venuprasad R, Bool M, Dalid C, Bernier J, Kumar A, Atlin G (2009) Genetic loci responding to two cycles of divergent selection for grain yield under drought stress in a rice breeding population. Euphytica 167(2):261–269, doi:10.1007/s10681-009-9898-3

Venuprasad R, Bool M, Quiatchon L, Atlin G (2012) A QTL for rice grain yield in aerobic environments with large effects in three genetic backgrounds. TAG Theor Appl Genet 124(2):323–332, doi:10.1007/s00122-011-1707-4

Vikram P, Swamy B, Dixit S, Ahmed H, Teresa Sta Cruz M, Singh A, Kumar A (2011) QDTY1.1, a major QTL for rice grain yield under reproductive-stage drought stress with a consistent effect in multiple elite genetic backgrounds. BMC Genetics 12:89.

Wassmann R, Jagadish S, Heuer S, Ismail A, Redona E, Serraj R, Singh R, Howell G, Pathak H, Sumfleth K (2009) Climate Change Affecting Rice Production: The Physiological and Agronomic Basis for Possible Adaptation Strategies. Advances in Agronomy, vol 101. Academic Press, Burlington, MA, USA.

Win K, Korinsak S, Jantaboon J, Siangliw M, Lanceras-Siangliw J, Sirithunya P, Vanavichit A, Pantuwan G, Jongdee B, Sidhiwong N, Toojinda T (2012) Breeding the Thai jasmine rice variety KDML105 for non-age-related broad-spectrum resistance to bacterial blight disease based on combined marker-assisted and phenotypic selection. Field Crop Res 137:186–194, doi:10.1016/j.fcr.2012.09.007

Xu K, Xu X, Fukao T, Canlas P, Maghirang-Rodriguez R, Heuer S, Ismail A, Bailey-Serres J, Ronald P, Mackill D (2006) Sub1A is an ethylene-response-factor-like gene that confers submergence tolerance to rice. Nature 442(7103):705–708, doi:10.1038/nature04920

Yang D, Shewfelt R, Lee K, Kays S (2008) Comparison of odor-active compounds from six distinctly different rice flavor types. J Agric Food Chem 56(8):2780–2787, doi:10.1021/jf072685t

Yang D, Lee K, Kays S (2010) Characterization and discrimination of premium-quality, waxy, and black-pigmented rice based on odor-active compounds. J Sci Food Agric 90(15):2595–2601, doi:10.1002/jsfa.4126

Zeigler R, Barclay A (2008) The relevance of rice. Rice 1(1):3–10, doi:10.1007/s12284-008-9001-z

Permissions

All chapters in this book were first published in RICE, by Springer; hereby published with permission under the Creative Commons Attribution License or equivalent. Every chapter published in this book has been scrutinized by our experts. Their significance has been extensively debated. The topics covered herein carry significant findings which will fuel the growth of the discipline. They may even be implemented as practical applications or may be referred to as a beginning point for another development.

The contributors of this book come from diverse backgrounds, making this book a truly international effort. This book will bring forth new frontiers with its revolutionizing research information and detailed analysis of the nascent developments around the world.

We would like to thank all the contributing authors for lending their expertise to make the book truly unique. They have played a crucial role in the development of this book. Without their invaluable contributions this book wouldn't have been possible. They have made vital efforts to compile up to date information on the varied aspects of this subject to make this book a valuable addition to the collection of many professionals and students.

This book was conceptualized with the vision of imparting up-to-date information and advanced data in this field. To ensure the same, a matchless editorial board was set up. Every individual on the board went through rigorous rounds of assessment to prove their worth. After which they invested a large part of their time researching and compiling the most relevant data for our readers.

The editorial board has been involved in producing this book since its inception. They have spent rigorous hours researching and exploring the diverse topics which have resulted in the successful publishing of this book. They have passed on their knowledge of decades through this book. To expedite this challenging task, the publisher supported the team at every step. A small team of assistant editors was also appointed to further simplify the editing procedure and attain best results for the readers.

Apart from the editorial board, the designing team has also invested a significant amount of their time in understanding the subject and creating the most relevant covers. They scrutinized every image to scout for the most suitable representation of the subject and create an appropriate cover for the book.

The publishing team has been an ardent support to the editorial, designing and production team. Their endless efforts to recruit the best for this project, has resulted in the accomplishment of this book. They are a veteran in the field of academics and their pool of knowledge is as vast as their experience in printing. Their expertise and guidance has proved useful at every step. Their uncompromising quality standards have made this book an exceptional effort. Their encouragement from time to time has been an inspiration for everyone.

The publisher and the editorial board hope that this book will prove to be a valuable piece of knowledge for researchers, students, practitioners and scholars across the globe.

List of Contributors

Yu-Sian Wu and Chin-Ying Yang
Department of Agronomy, National Chung Hsing University, Taichung 40227, Taiwan

Kailun Zheng, Dongzhi Lin and Yanjun Dong
Development Center of Plant Germplasm Resources, College of Life and Environmental Sciences, Shanghai Normal University, Shanghai 200234, China

Jian Zhao and Jiaying Chen
Development Center of Plant Germplasm Resources, College of Life and Environmental Sciences, Shanghai Normal University, Shanghai 200234, China
Shanghai Institute for Biological Sciences, Chinese Academy of Sciences, Shanghai 200032, China

Hua Zhou
Development Center of Plant Germplasm Resources, College of Life and Environmental Sciences, Shanghai Normal University, Shanghai 200234, China
Agricultural Faculty, Hokkaido University, Sappro 060-0817, Japan

Sheng Teng
Shanghai Institute for Biological Sciences, Chinese Academy of Sciences, Shanghai 200032, China

Jianlong Xu
Institute of Crop Sciences, Chinese Academy of Agricultural Sciences, 12 South Zhong-Guan Cun Street, Beijing 100081, China

Gustave Djedatin
Université Polytechnique d'Abomey, BP 2282 Abomey, Benin

Marie-Noelle Ndjiondjop
Africa Rice Center (AfricaRice), 01 BP2031 Cotonou, Benin

Valérie Verdier
IPME Interactions Plantes Microorganismes, Environnement, IRD - Cirad - University Montpellier, 34394 Montpellier, France

Mathias Lorieux and Alain Ghesquiere
UMR Diversité, Adaptation et Développement des plantes (DIADE), Institut de Recherche pour le Développement, 911 Avenue Agropolis BP 64501, 34394 Montpellier Cedex 5, France

Ambaliou Sanni
Université d'Abomey Calavi, 01 BP526 Cotonou, Benin

Audrey Dardou, Delphine Luquet and Tanguy Lafarge
CIRAD, UMR AGAP, F-34398 Montpellier, France

Dewi Erika Adriani
CIRAD, UMR AGAP, F-34398 Montpellier, France
Faculty of Agriculture, University of Lambung Mangkurat, Banjarbaru, Indonesia

Michael Dingkuhn
CIRAD, UMR AGAP, F-34398 Montpellier, France
Crop and Environment Science Division (CESD), International Rice Research Institute (IRRI), Los Baños, Philippines

Hélène Adam
IRD, UMR DIADE, Montpellier, France

Jamie Selby-Pham, Laura T. Moreno-Moyano and Alexander A. T. Johnson
School of BioSciences, The University of Melbourne, Parkville, Victoria, Australia

Adrian Lutz, Berin A. Boughton and Ute Roessner
School of BioSciences, The University of Melbourne, Parkville, Victoria, Australia
Metabolomics Australia, The University of Melbourne, Parkville, Victoria, Australia

Dustin G. Brown, Nora Jean Nealon and Elizabeth P. Ryan
Department of Environmental & Radiological Health Sciences, College of Veterinary Medicine and Biological Sciences, Colorado State University, 1680 Campus Delivery, Fort Collins, CO 80523, USA

Iman Zarei
Department of Environmental & Radiological Health Sciences, College of Veterinary Medicine and Biological Sciences, Colorado State University, 1680 Campus Delivery, Fort Collins, CO 80523, USA
Institute of Human Nutrition and Food, College of Human Ecology, University of the Philippines Los Baños, Los Baños 4031, Laguna, Philippines

Chang-Jin Park
Department of Plant Pathology and the Genome Center, University of California Davis, Davis, CA 95616, USA
Department of Bioresources Engineering and the Plant Engineering Research Institute, Sejong University, Seoul 05006, Republic of Korea

Rita Sharma
Department of Plant Pathology and the Genome Center, University of California Davis, Davis, CA 95616, USA
School of Computational & Integrative Sciences, Jawaharlal Nehru University, New Delhi 110067, India

Tong Wei and Pamela C. Ronald
Department of Plant Pathology and the Genome Center, University of California Davis, Davis, CA 95616, USA
Feedstocks Division, Joint BioEnergy Institute, Lawrence Berkeley National Laboratory, Berkeley, CA 94720, USA
Environmental Genomics and Systems Biology, Lawrence Berkeley National Laboratory, Berkeley, CA 94720, USA

Si-Yu Yang, Yu-Sian Wu and Chin-Ying Yang
Department of Agronomy, National Chung Hsing University, Taichung 40227, Taiwan

Chung-Tse Chen
Graduate Institute of Biotechnology, National Chung Hsing University, Taichung 40227, Taiwan

Ming-Hsin Lai and Hsing-Mu Yen
Crop Science Division, Taiwan Agricultural Research Institute, Taichung 41362, Taiwan

Sripathy K.V., Jeevan Kumar S.P., Bhojaraja Naik K., Govind Pal, Udaya Bhaskar K., Ramesh K.V. and Somasundaram G.
ICAR- Indian Institute of Seed Science, Mau, Uttar Pradesh 275103, India

Chandu Singh
ICAR- Indian Institute of Seed Science, Mau, Uttar Pradesh 275103, India
Seed Production Unit, ICAR-Indian Agricultural Research Institute, New Delhi 110012, India

Suk-Hwan Kim, Giha Song and Hee-Jong Koh
Department of Plant Science, Plant Genomics and Breeding Institute, Research Institute of Agriculture and Life Sciences, Seoul National University, Seoul 08826, Republic of Korea

Nam-Chon Paek
Department of Plant Science, Plant Genomics and Breeding Institute, Research Institute of Agriculture and Life Sciences, Seoul National University, Seoul 08826, Republic of Korea
Crop Biotechnology Institute, Institutes of Green Bio Science and Technology, Seoul National University, Pyeongchang 25354, Republic of Korea

Choon-Tak Kwon
Department of Plant Science, Plant Genomics and Breeding Institute, Research Institute of Agriculture and Life Sciences, Seoul National University, Seoul 08826, Republic of Korea
Cold Spring Harbor Laboratory, Cold Spring Harbor, NY 11724, USA

Gynheung An
Crop Biotech Institute and Graduate School of Biotechnology, Kyung Hee University, Yongin 17104, Republic of Korea

Jian Sun, Luomiao Yang, Jingguo Wang, Hualong Liu, Hongliang Zheng, Yan Jia, Hongwei Zhao and Detang Zou
College of Agriculture, Northeast Agricultural University, Harbin 150030, China

Dongwei Xie
The Institute of Industrial Crops of Heilongjiang Academy of Agricultural Sciences, Harbin 150086, China

Minghui Zhang and Mingfang Feng
College of Life Science, Northeast Agricultural University, Harbin 150030, China

Sunok Moon, Gynheung An and Ki-Hong Jung
Graduate School of Biotechnology and Crop Biotech Institute, Kyung Hee University, Yongin 446-701, South Korea

Moe Moe Oo, Sung Aeong Oh and Soon Ki Park
School of Applied Biosciences, Kyungpook National University, Daegu 702-701, South Korea

Backki Kim and Hee-Jong Koh
Department of Plant Science, Research Institute of Agriculture and Life Sciences, and Plant Genomics and Breeding Institute, Seoul National University, Seoul 151-921, South Korea

Gihwan Yi
College of Agriculture and Life Science, Daegu 702-701, South Korea

Keiki Ishiyama, Soichi Kojima, Mitsue Miyao and Toshihiko Hayakawa
Graduate School of Agricultural Science, Tohoku University, 468-1 Aoba, Aramaki-Aza, Aoba-ku, Sendai 980-8572, Japan

Miwa Ohashi
Graduate School of Agricultural Science, Tohoku University, 468-1 Aoba, Aramaki-Aza, Aoba-ku, Sendai 980-8572, Japan
Graduate School of Bioagricultural Sciences, Nagoya University, Furo-cho, Chikusa-ku, Nagoya 464-8601, Japan

Noriyuki Konishi
Graduate School of Agricultural Science, Tohoku University, 468-1 Aoba, Aramaki-Aza, Aoba-ku, Sendai 980-8572, Japan
Institute of Plant Science and Resources, Okayama University, 2-20-1 Chuo, Kurashiki 710-0046, Japan

Tomoyuki Yamaya
Graduate School of Agricultural Science, Tohoku University, 468-1 Aoba, Aramaki-Aza, Aoba-ku, Sendai 980-8572, Japan
Division for Interdisciplinary Advanced Research and Education, Tohoku University, 6-3 Aoba, Aramaki-Aza, Aoba-ku, Sendai 980-0845, Japan

Kazuhiro Sasaki
The University of Tokyo, Graduate School of Agricultural and Life Sciences, Institute of Sustainable Agro-ecosystem Services (ISAS), 1-1-1 Midori-cho, Nishitokyo, Tokyo 188-0002, Japan

Li Ding, Jia Cheng and Bin Wang
School of Horticulture and Plant Protection, Yangzhou University, Yangzhou 225009, Jiangsu, China

X. Liu, C. Y. Yang, R. Miao, C. L. Zhou, P. H. Cao, J. Lan, X. J. Zhu, C. L. Mou, Y. S. Huang, S. J. Liu, Y. L. Tian, T. L. Nguyen and L. Jiang
State Key Laboratory for Crop Genetics and Germplasm Enhancement, Jiangsu Plant Gene Engineering Research Center, Nanjing Agricultural University, Nanjing 210095, China

J. M. Wan
State Key Laboratory for Crop Genetics and Germplasm Enhancement, Jiangsu Plant Gene Engineering Research Center, Nanjing Agricultural University, Nanjing 210095, China
National Key Facility for Crop Gene Resources and Genetic Improvement, Institute of Crop Science, Chinese Academy of Agricultural Sciences, Beijing 100081, China

Weiwen Kong
School of Horticulture and Plant Protection, Yangzhou University, Yangzhou 225009, Jiangsu, China
Joint International Research Laboratory of Agriculture and Agri-Product Safety of the Ministry of Education, Yangzhou University, Yangzhou 225009, Jiangsu, China

Jun Ni, Zhiming Yu, Yanyan Zhang, Chenjia Shen, Jing Xu and Xunyan Liu
College of Life and Environmental Sciences, Hangzhou Normal University, Hangzhou 310018, China

Guankui Du
Department of Biochemistry, Hainan Medical College, Haikou 571199, China

Jemma L. Taylor
School of Life Sciences, Gibbet Hill Campus, University of Warwick, Coventry CV4 7AL, UK

Yifeng Wang
State Key Laboratory of Rice Biology, China National Rice Research Institute, Chinese Academy of Agricultural Sciences, Hangzhou 311400, China

Yunrong Wu
State Key Laboratory of Plant Physiology and Biochemistry, College of Life Science, Zhejiang University, Hangzhou 310058, China

Ruchi Agarrwal and Suresh Nair
International Centre for Genetic Engineering and Biotechnology (ICGEB), Aruna Asaf Ali Marg, New Delhi 110067, India

Ayyagari Phani Padmakumari
Indian Institute of Rice Research (formerly Directorate of Rice Research), Rajendranagar, Hyderabad 500030, India

Jagadish S. Bentur
Indian Institute of Rice Research (formerly Directorate of Rice Research), Rajendranagar, Hyderabad 500030, India
AgriBiotech Foundation, Rajendranagar, Hyderabad 500030, India

Mariafe Calingacion
Grain Quality and Nutrition Centre, International Rice Research Institute, DAPO 7777, Metro Manila, Philippines
Laboratory of Plant Physiology, Wageningen University and Research Centre, Wageningen, The Netherlands
School of Agriculture and Food Science, University of Queensland, St Lucia, 4072 Queensland, Australia

Robert D Hall
Laboratory of Plant Physiology, Wageningen University and Research Centre, Wageningen, The Netherlands
BU Bioscience, Plant Research International, Wageningen University and Research Centre, Wageningen, The Netherlands

Melissa Fitzgerald, Lu Fang and Arthur Riedel
School of Agriculture and Food Science, University of Queensland, St Lucia, 4072 Queensland, Australia

Lenie Quiatchon-Baeza
Grain Quality and Nutrition Centre, International Rice Research Institute, DAPO 7777, Metro Manila, Philippines
Department of Crop Sciences, College of ACES, University of Illinois at Urbana-Champaign, Urbana, IL, USA

Roland Mumm
BU Bioscience, Plant Research International, Wageningen University and Research Centre, Wageningen, The Netherlands

Index

O

P

R

S

T

U

X

Y

Z

www.ingramcontent.com/pod-product-compliance
Lightning Source LLC
LaVergne TN
LVHW061509220926
848485LV00017B/595